T0326318

Applied State Estimation
and Association

MIT Lincoln Laboratory Series

Perspectives on Defense Systems Analysis: The What, the Why, and the Who, but Mostly the How of Broad Defense Systems Analysis, William P. Delaney

Ultrawideband Phased Array Antenna Technology for Sensing and Communication Systems, Alan J. Fenn and Peter T. Hurst

Decision Making Under Uncertainty: Theory and Application, Mykel J. Kochenderfer

Applied State Estimation and Association, Chaw-Bing Chang and Keh-Ping Dunn

MIT Lincoln Laboratory is a federally funded research and development center that applies advanced technology to problems of national security. The books in the *MIT Lincoln Laboratory Series* cover a broad range of technology areas in which Lincoln Laboratory has made leading contributions. The books listed above and future volumes in the series renew the knowledge-sharing tradition established by the seminal *MIT Radiation Laboratory Series* published between 1947 and 1953.

Applied State Estimation and Association

Chaw-Bing Chang and Keh-Ping Dunn

The MIT Press
Cambridge, Massachusetts
London, England

© 2016 Massachusetts Institute of Technology

All rights reserved. No part of this book may be reproduced in any form by any electronic or mechanical means (including photocopying, recording, or information storage and retrieval) without permission in writing from the publisher.

This book was set in Adobe Garamond Pro and Times New Roman by Toppan Best-set Premedia Limited.

Library of Congress Cataloging-in-Publication Data

Names: Chang, Chaw-Bing, author. | Dunn, K-P. (Keh-Ping), author.
Title: Applied state estimation and association / Chaw-Bing Chang and Keh-Ping Dunn.
Description: Cambridge, MA : MIT Press, [2016] | Series: MIT Lincoln Laboratory series |
    Includes bibliographical references and index.
Identifiers: LCCN 2015038387 | ISBN 9780262034005 (hardcover : alk. paper)
ISBN 9780262548915 (paperback)
Subjects: LCSH: Systems engineering—Mathematics. | Estimation theory—Mathematics. |
    Statics—Mathematics.
Classification: LCC TA168 .C3926 2015 | DDC 620.001/1—dc23 LC record available at
    http://lccn.loc.gov/2015038387

Dedication

To
Kia-Ling Chang and Ning Dunn
Our Better Halves,
Their
Encouragement,
Patience, and
Support
Helped Make This Project a Reality

Table of Contents

Preface

This book is intended to provide readers with a rigorous introduction to the theory and application of estimation and association techniques. Skills taught in this book will prepare students for solving application problems in this technical area.

Applied estimation and association is an important area for practicing engineers in aerospace, electronics, and defense industries. A feature of this book is that it uses a unified approach in problem formulation and solution. This approach serves to help the students to build a sound theoretical foundation as well as to attain skills and tools for practical applications. Many technical subjects and examples in this book represent a collection of the most relevant and important areas in state estimation and association for practicing engineers based upon the authors' decades of experiences in this field. For this reason, this book could be used by engineering schools offering courses in this area as a textbook, as well as a reference book for students interested in engineering applications and practical solutions when taking a more theoretical course. For practicing engineers, this book can be used for self-study or as a textbook for an in-house class. It can also be used for self-study by practitioners in the area of state estimation and association.

The technical level of this book is equivalent to an advanced first- or second-year graduate course in a control or system engineering curriculum. The students are required to be familiar with the state-variable representation of systems and basic probability theory including random variables and stochastic processes. The main content of this book spans 10 chapters. Chapters 1 to 6 address the problem of estimation with a single sensor observing a single object. Chapter 7 expands from a single sensor to multiple sensors. Chapters 8 through 10 address the problem of measurement-to-track association and track-to-track correlation by expanding the problem to multiple objects. A chapter-by-chapter description is included in the Introduction with Concluding Remarks given at the end.

It is our goal that after learning the skills presented in this book, students will be able to derive solutions to problems, or to conduct further research when needed in order to solve their problems.

About the Authors

Chaw-Bing Chang received his BS degree from National Cheng Kung University, Taiwan, and MS and PhD degrees from the State University of New York at Buffalo, all in electrical engineering. He joined Lincoln Laboratory in 1974, and his initial project was on radar signal processing and trajectory estimation for ballistic missile defense (BMD). He became an Assistant Group Leader in 1984 to lead projects in air defense (AD) technology development for the US Navy. He was appointed Group Leader of the Air Defense and Sensor Technology Group in 1998, and was responsible for technology development for the Navy's airborne surveillance radar system. During this time he led a multiyear data collection and experimentation campaign supporting the Navy's Mountaintop Program. As part of the Navy AD program, he contributed to algorithm development and performance evaluation for both surface and airborne radar systems. Upon returning to the Laboratory's BMD program in 2004, he participated in advanced algorithm development and phenomenology research for radar and optical sensors and led an airborne optical sensor technology program for BMD. He has published more than 70 journal articles, conference papers, and Lincoln Laboratory reports. He is currently a senior staff member of the BMD System Integration Group.

Keh-Ping Dunn received his BS degree in control system engineering from the National Chiao Tung University in Taiwan, and MS and DSc degrees in systems science and mathematics from Washington University in St. Louis, Missouri. Before joining the Laboratory in 1976, he was with the Electronic System Laboratory of MIT in charge of a NASA project on a multiple model adaptive control system for the F-8C aircraft. At Lincoln Laboratory, he has worked in many areas of ballistic missile defense (BMD). He became the Group Leader of Systems Testing and Analysis in 1992, managing the first two campaigns of the Theater Missile Defense (TMD) Critical Measurement Program (TCMP) that conducted a series of live missile tests in the Pacific in the 1990s. He won the Missile Defense Agency's (MDAs) 2010 Technology Achievement Award for his effort on this project. He consequently managed Theater

Missile Defense (1999–2003), Advanced Concepts and Technology (2003–2008), and Missile Defense Elements (2008–2010) groups, all for MDA projects. He chaired the Panel of Tracking Parameters of the SDI Tracking Panels in the late 1980s for the Strategic Defense Initiative Organization (SDIO). He has worked on multiple target/ multiple sensor tracking, target identification, and sensor fusion and decision architecture for various BMD sensor (both optics and radar) systems at the Laboratory. He is currently a senior staff member in the BMD System Integration Group.

Acknowledgments

The authors are indebted to many of our colleagues at Lincoln Laboratory for their collaboration and technical contributions in estimation and association. It would be impossible for us to list all the names, and should we miss anyone due to our negligence, we beg for your forgiveness. Special thanks go to Dr. John Tabaczynski. He has been a mentor throughout our careers at the Laboratory and has provided an in-depth review of the manuscript with numerous suggestions, revisions, and improvements. To Dr. Steve Weiner, who has had a positive influence on our careers at the Laboratory, and has reviewed many of our papers and reports in the early days. To Dr. Michael Athans, a professor at MIT, who has taught, guided, and collaborated with us on many projects involving estimation and association. To Dr. Dan O'Connor, our long-time friend and colleague, who has provided support to this project and taught a course with us on state space modeling and estimation where much of the subject matter in this area was developed. To Drs. Mike Gruber and Robert Whiting, who provided stimulating discussions and collaborations when we first came to Lincoln Laboratory. To Dr. Robert Miller, who was our technical mentor during the early part of our careers and taught us many techniques in radar signal processing, state estimation, and system analysis. To Dr. Dieter Willner, who collaborated with us on problems of estimation and control. To Lenny Youens, an ingenious scientific programmer, whose energy and dedication helped to obtain many numerical results for our publications. To Dr. Richard Holmes, a rigorous mathematician, who played a key role in applying linear operator theory to estimation. To Drs. Lou Bellaire, Kathy Rink, and Kevin Brady, who at the beginning of their careers at the Laboratory, worked on estimation and tracking problems with us and later went on to develop a course on estimation of their own. To Dr. Steve Moore, a fine physicist and mathematician, who has developed applications of estimation techniques to tactical systems. To Ziming Qiu, who helped to implement algorithms for solving some very difficult problems. To Brian Corwin, who studied joint bias and state estimation with us in the context of space trajectories. To Dr. Hody Lambert, a stimulating colleague, who developed several new ideas on estimation and association involving multiple sensor systems. To Dr. Martin Tobias, who has contributed to algorithm studies. To Justin Goodwin, who applied many of the techniques of this book to estimation problems involving radars. To Steve Vogl, who collaborated with us on estimation problems involving passive optical sensors. To Fannie Rogal, who provided numerical examples and implementation technique using the Nassi–Shneiderman graph for implementing the multiple hypothesis tracking algorithm. To Jason Cookson, who represents the new generation of the Laboratory's expertise in estimation and association. He has developed many tools and models in estimation and association for air and space applications. Not only has Jason provided numerical results to many examples in this book, he has also provided stimulating discussions with insightful interpretations. Last but not least, we are indebted to Allison Loftin, Donna McTague, and Cheryl Nunes for their patience and dedication in preparing and reviewing the manuscript.

Introduction

This book is intended to provide the reader with a rigorous introduction to the theory and application of estimation and association techniques. Skills taught in this book will prepare the student for solving practical problems in this technical area.

Estimation and association involves the extraction of information from noisy measurements. Example applications include signal processing, tracking, navigation, and so on [1, 2, 3]. The extraction of parameter values from signals in order to estimate such attributes as time-of-arrival and sensor pointing angle is called parameter estimation. A sensor signal may have come from a moving object. Determining the kinematics of a moving object is called state estimation. Associating measurements with state estimates in a multiple object environment is a joint estimation and association problem that is known as tracking [2–5]. Example applications include sensor surveillance systems for air traffic control, guiding space vehicles toward a planet, extracting information regarding a moving object with multiple-degree-of-freedom motions, and so on.

The authors of this book, together with their colleagues, have been applying the theory and techniques of estimation and association to real-world problems for the past 40 years. They have taught classes to Lincoln Laboratory staff members who are involved in applying these skills as well as solving problems of their own. The content of this book represents their collective experience in applying estimation and association techniques. The technical level of this book is equivalent to a first year graduate course in a control or system engineering curriculum. The students are required to be familiar with the state-variable representation of systems, and basic probability theory including random variables and stochastic processes. This book can also be used for self-study by practitioners in the area of state estimation and association.

Theory and techniques developed in this book are for discrete time systems. Although all physical systems are continuous in time, the measurements are taken in discrete time and the computational system that exploits the measurements operates in discrete time. Furthermore, unique discrete time equivalence to continuous time systems can be easily derived and implemented. The use of discrete time models enables us to solve the problem without resorting to more abstract mathematics such as measure theory and Ito calculus [6]. Homework problems are included at the end of each chapter. The purpose is two-fold: (a) to develop students confidence in their derivation skills so that they are able to apply them to new problems, and (b) to build computer models so that they will have a useful set of tools for problem solving.

The theory and application of estimation has been a rich field of research for decades. The landmark papers by Kalman [4], and Kalman and Bucy [5] gave the optimal solution for state estimation of linear systems having Gaussian system and measurement noise processes. The Kalman filter (KF) algorithm using state space modeling makes it suitable for implementation with digital computers. Kalman's paper also laid the foundation for the concept of observability of a linear system, and its relationship with the Fisher information matrix and the Cramer–Rao bound (CRB) [1] for all unbiased state estimators. For this reason, it has gained enormous interest from practicing engineers. However, most of the real-world application problems are nonlinear. After Kalman's publication, considerable effort was devoted to finding the optimal filter for nonlinear systems (the counterpart of the KF for linear systems) [6]. All these studies came to the same conclusion: the solution of the optimal filter requires an infinite dimensional representation that cannot be practically constructed. Consequently, follow-on efforts focused on searching for suboptimal but practical solutions.

The approach used in this book has two features: (a) it formulates the estimation problem as an optimization problem using measurement data and a priori knowledge of the system, and (b) it develops CRB solutions for each estimation problem addressed. The first feature stresses that the solution to the estimation problem provides a best fit to the measurement data, the system model, and the a priori knowledge. It will be shown that solution algorithms for most of the estimation problems can be obtained this way. The CRB has been well known in signal processing for estimating parameters embedded in the signal [1]. It has been applied to a wide range of state estimation problems at Lincoln Laboratory [7]. In keeping with the second feature, the CRB models for parameter and state estimation are derived for the examples considered or are included as part of the homework problem assignments.

In many engineering applications, noisy measurements are obtained on some unknown variables. Variables of interest can collectively be represented as a vector. Measurements can be arranged as a measurement vector or a set of measurement vectors. In the case where the vector of interest is constant or random, it is referred to as a parameter vector. In the case where the vector of interest is time-varying and follows a set of differential equations for a continuous-time system, or difference equations for a discrete-time system, it is termed a state vector. A parameter vector is a special case of the state vector. The concept of a state vector is identical to the state vector used in the state space representation of control systems [8]. A state vector can be deterministic or random, depending on whether the system is deterministic or driven by a random process.

The estimation problem is to find a solution to the unknown vector using measurements and knowledge about the vector of interest. The measurements used in an estimator are assumed to have come from a single object or dynamic system. This assumption may not be true when multiple objects are closely spaced in sensor measurements. The problem of state association is to determine whether a measurement or a set of measurements comes from the same object.

This book has 10 chapters. Chapters 1 to 6 focus on solving the problem of estimation with a single sensor observing a single object. Chapter 7 expands consideration from a single sensor observer to multiple sensors. Chapters 8 through 10 address the problem of association by expanding the problem to multiple objects and multiple sensors. Concluding remarks and three appendices are offered at the end. They are introduced individually below.

Chapter 1: Parameter Estimation

In this text, a parameter vector can be a constant vector or a random vector with known distribution, but is never a random process. The foundation of estimation can be understood most easily by solving the problem of parameter estimation. The estimate of an unknown vector is obtained by selecting the vector that optimizes a performance criterion or a cost function given the noisy measurements. Six performance criteria are introduced in this chapter, namely, least squares, weighted least squares, maximum likelihood, maximum a posteriori probability, conditional mean, and linear least squares expressed as functions of the measurements [1, 4]. Explicit estimator solutions for linear measurements with Gaussian measurement noise are developed and the equivalence of all six estimators is discussed. It is shown that the a posteriori density function of the parameter vector conditioned on measurements contains all the information for estimating this parameter vector, regardless of whether the measurement relationship is linear or nonlinear, and the conditional mean is the minimum norm solution in the parameter space. For the linear measurement relationship, the closed form solution can be found. For nonlinear measurements, a numerical solution to the weighted least squares estimator is derived. The Cramer–Rao bounds for all cases are derived. The relationship between weighted least squares estimator, minimum variance estimator, and the conditional mean estimator is shown in the appendix.

Chapter 2: State Estimation for Linear Systems

A state vector is the solution of a first order vector differential equation for a continuous system, or difference equation for a discrete system [8]. When the initial condition is a random variable and/or when the system is driven by a random system noise process, the state vector represents a random process. For linear systems with Gaussian system and measurement noise, the a posteriori density of the state conditioned on measurements remains Gaussian, and the state estimate can therefore be completely characterized by the conditional mean and covariance. This result is known as the Kalman filter [4]. The techniques used in Chapter 1 to derive the parameter estimator are extended in this chapter to derive the KF solution for linear systems. These include the conditional mean, weighted least squares, and Bayesian recursive evolution of the a posteriori density function. The concept of smoothing is introduced, and the chapter ends with derivations for the CRB for all cases of interest.

Chapter 3: State Estimation for Nonlinear Systems

Many physical systems and measurement devices are nonlinear. As mentioned before, the conditional mean is the minimum norm estimate, and the a posteriori density function of a state conditioned on measurements contains all the information necessary for estimation. For linear systems with Gaussian noise, the a posteriori density remains Gaussian. This property is, however, no longer true for nonlinear systems even when the input and measurement noise processes are Gaussian. The recursive Bayesian relationship governing the time evolution of the a posteriori density for arbitrary nonlinear systems was published within a few years of Kalman filter [9, 10], but its exact solution for estimation remains open. For this reason, only approximated solutions for the nonlinear estimation problems have found applications. The approximated solutions include the use of the first order Taylor series expansion (the extended Kalman filter) and the addition of the second order term in the Taylor series expansion (the second order filter) [11]. Both filters are aimed at providing approximated conditional means and covariance solutions for the state estimator. Additional nonlinear estimation techniques presented in this chapter include the extension of the numerical solution for estimating a random parameter vector with nonlinear measurements of Chapter 1 to the problems of Chapter 3 to yield a single-stage iterative solution for computing state updates in extended Kalman filter (EKF). A special case for nonlinear estimation occurs when the system is deterministic. An iterative weighted least squares estimator for a deterministic nonlinear system using all the measurements is derived [12]. The numerical examples in [12] shows that the covariance of this algorithm achieves the CRB.

The EKF has gained considerable attention from practitioners due to its simplicity and direct relationship with KF: the KF algorithm becomes an EKF algorithm when the system and measurement matrices of a linear system are replaced by Jacobian matrices of the nonlinear system. The EKF does not solve all nonlinear estimation problems, nor does it provide the best answer even when it does work. Its similarity in functional form with the KF makes much of the KF analysis extendable to the EKF. For this reason, linear systems are used for discussion throughout this book. Exceptions will be noted. To conclude this chapter the CRB for nonlinear state estimation is developed.

Chapter 4: Practical Considerations in Kalman Filter Design

The previous three chapters provide the basic tools for estimation: problem definition, solution derivation, and solution algorithms. In this chapter, practical issues in filter design are discussed. Filter construction is based upon a mathematical representation of the physical process of interest. For most engineering problems, mathematical models do not exactly represent the actual physical process resulting in a less than optimal filter performance. Model differences can occur in the system equation, measurement equation, system input, measurement noise, and so on [13]. This chapter starts with a discussion about tools for estimator performance monitoring that includes the CRB, the measure of statistical behavior of the filter residual process, and the measure of filter consistency in terms of actual and computed covariance. The rest of the chapter provides detailed discussions for addressing the system model mismatch problem, measurement error uncertainty, and systems with uncertain inputs. For each subject, filter compensation methods are suggested. The issue of systems with uncertain inputs is related to tracking objects having unknown or unexpected maneuvers. A maximum likelihood estimator with its associated generalized likelihood ratio (GLR) detection algorithm is developed for systems with sudden input changes [14]. A discussion on the advantages and limitations of this approach is provided. An extension to this approach is considered by assuming that the state may be generated by one of several models (e.g., maneuvering and nonmaneuvering), and that leads into the next chapter, multiple model estimation algorithms.

Chapter 5: Multiple Model Estimation Algorithms

When the underlying true system could be one of several different models, the estimation solution is a bank of KFs with each matched to a specific model within the set of models.[1] The probability (termed the hypothesis probability) that a given filter represents the true system can be computed for each filter using the filter residual. The filter with the highest hypothesis probability is deemed to represent the truth (e.g., a target is maneuvering or not maneuvering). It can be shown that the conditional mean estimate is the weighted sum of the output of all estimators (for both state and covariance) with the hypothesis probabilities as weighting factors. This solution is optimal for linear systems when the truth stays the same as one of the models used in the filter bank [15], referred to as the constant model case. In practical problems, the true system may be changing among models. For example, a target may switch back and forth between maneuvering and nonmaneuvering at multiple instances of time. This is referred to as a switching model case. The solution to the switching model case is unbounded, that is, the true system could switch to a different model in multiple instances of time making the number of possibilities grow exponentially [16]. Approximate solutions are derived for the case when the model switching history has finite memory, for example, for a Markov process. Solutions for this problem include generalized pseudo-Bayesian (GBP) algorithm and interacting multiple model (IMM) algorithm [16, 17]. The IMM is an approximation of GBP but is simpler in implementation. In several applications, the trade-off of these two algorithms is in favor of IMM because the performance gain of GBP over IMM is small. The derivations of all these cases are included in this chapter along with a set of numerical examples.

Chapter 6: Sampling Techniques for State Estimation

In Chapter 3, several solutions to the nonlinear estimation problem were presented in the form of an approximated conditional mean and covariance computation. As stated before, the a posteriori density function of state conditioned on measurements contains all the information for estimation. In this chapter, several numerical methods for

---

1   The same concept as the bank of Doppler filters for radar signal processing.

computing the a posteriori density function are presented [18–20]. Two deterministic sampling techniques are introduced. The grid-based point-mass filter was introduced in 1969 [18], which is intuitive but computationally expensive. The second method used the unscented transformation for Gaussian noise known as the unscented Kalman filter (UKF) [19]. UKF is computationally more efficient and has gained considerable popularity since the late 1990s. Random sampling techniques by means of Monte Carlo sampling collectively known as particle filter methods [20] are introduced in this chapter. The goal is to find a numerical approximation to the a posteriori density function and hence, the conditional mean can be computed with a Monte Carlo integral. The concept of sequential Monte Carlo (SMC) sampling is based on point mass (or particle) approximation of the probability density functions involved in the estimation problem. Although the concept of this technique was first developed in the 1950s, it gained popularity in 1990s due to the availability of high-speed computers that makes its realization more feasible. Still, due to the large computational requirement, the SMC has not been used in conventional tracking/filtering problems, but is used for problems with smaller dimension and for nonconventional problems such as when the motion of the object of interest is less analytical (e.g., tracking hand motion, tracking an object moving in a maze, etc.) or the measurement equation is nonlinear and nonanalytical (e.g., hard limiting, hysteresis, etc.). Several Monte Carlo sampling techniques can be applied to these problems, such as rejection sampling, importance sampling, and Markov chain Monte Carlo (MCMC) method, among others. In this chapter, discussions are focused on techniques involving importance sampling. SMC is a current area of research and some approaches are mentioned in this chapter [21, 22].

Chapter 7: State Estimation with Multiple Sensor Systems

Problems addressed in Chapters 1 through 6 consider a single sensor observing a single object. Expansion to multiple sensors and then to multiple objects are the focus of the remainder of this book. Chapter 7 introduces the problem of estimation using multiple sensors. There are many advantages to using multiple sensors. For example, sensors distributed over a wide geographical area can provide a diversity of viewing geometry in which differences in look angle result in improved estimation accuracy. One example of a multisensor system is the air traffic control system that employs a number of radars of integrated together with a communication system. Data in a multisensor system must be fused in order to achieve the potential benefits. An example of multiple sensor architecture is to send all sensor measurements to a centralized processor for processing [23]. This is called measurement fusion. The state estimate obtained this way is considered to be global because it contains all available data on the object. Another possible architecture is to have each individual sensor process its own data to obtain its local estimate of object state. It is called a local estimate because it only uses the sensor data available locally. All local estimates are then sent to a centralized processor to obtain a joint estimate. This is termed state fusion [24]. When the transmission of local estimates does not have to be done frequently, state fusion architecture could result in a smaller communication requirement. All these algorithms are presented and discussed in this chapter. Proof that the joint estimate is less accurate than the global estimate is presented in the appendix.

Chapter 8: Estimation and Association with Uncertain Measurement Origin

Fundamental theories and algorithms for the problem of estimation were developed in Chapters 1 through 7. In all cases, an estimator was used to process a set of measurement vectors with the assumption that all measurements came from the same object. In the case where multiple individual objects are closely spaced, the assignment of measurements to a state may become ambiguous. Multiple approaches to this problem have been discussed in the literature, ranging from using a single scan to multiple scans of data, treating each track individually or jointly, making decision for pairing measurements with tracks as an assignment problem or combining all measurements probabilistically for track update, and so on. One approach to solving this problem is to exhaustively enumerate all possible solutions (including accounting for the possibility of missed and false detections) at each measurement time. This approach is known as the multiple hypothesis tracker (MHT), and is the subject of Chapter 9. The focus of Chapter 8 is to present a menu of approaches other than MHT. Mathematical representation in assigning measurements to tracks over multiple scans is developed. A practical solution starting with multiframe track initiation followed by track continuation is presented [25, 26]. An algorithm for solving the assignment problem for a single frame decision (referred to as immediate resolution) is given, and a solution for the multiple frame decision (referred to as delayed resolution) is described. Comparison of results in using single and multiple frame decisions are illustrated in a numerical example. The appendix gives a track initiation algorithm using equations in dish radar coordinates with the state vector expressed in radar centered Cartesian coordinates.

Chapter 9: Multiple Hypothesis Tracking Algorithm

In this chapter, the MHT for a multiple target tracking problem is described [27–29]. In the MHT formulation, when assigning a new measurement with tracks, the following possibilities are always considered: (a) the new measurement is the continuation of an existing track, (b) the measurement is the start of a new track, and (c) the measurement is a false alarm. Give these considerations the case that a track continues without being updated with a measurement is always one of the tracks. A track is a time sequence of measurements obtained over multiple scans. A measurement from a given scan is allowed to be used by multiple tracks. This is because a measurement may be considered as the continuation of an existing track as well as the start of a new track at the same time since the new measurement could originate from multiple objects and be unresolved due to the limited resolution of the measurement sensor. A hypothesis in MHT consists of a set of tracks that use a measurement only for one track within that hypothesis. The number of tracks can grow combinatorially in MHT. Pruning is necessary in MHT in order to limit the growth of tracks and hypotheses. A method for scoring tracks and hypotheses is developed for pruning purposes, and a numerical example for track and hypothesis scoring is discussed.

Chapter 10: Multiple Sensor Correlation and Fusion with Biased Measurements

Two fusion architectures for multiple sensor systems were introduced in Chapter 7, namely, measurement fusion and state fusion. The ability to realize benefits of a multiple sensor system is dependent on (a) the capability to handle track ambiguities, and (b) the capability to handle sensor biases. The purpose of this chapter is to present approaches to correlation and estimation for multiple sensors with biased measurements. The first half of this chapter is focused on illustrating approaches for bias estimation by means of state augmentation in the measurement fusion architecture. The results are illustrated using a space object track example [30]. In this example, the association problem is assumed solved. The approach to jointly solving the state to state correlation problem and the bias estimation problem in the state fusion architecture is the subject of the second half of the chapter [31, 32]. It is first formulated as a joint mathematical optimization problem followed by several suggested solution algorithms.

Concluding Remarks

Throughout discussions in this book, it is made evident that there are areas of unsolved problems in estimation and association. Some of these are discussed in this final chapter.

Three Appendices: Matrix Inversion Lemma (MIL), a List of Notation and Variables, and Terminologies Used in Tracking

Appendix A: MIL provides a well-known identity for matrix inversion of a specific form. It is used repeatedly in estimator derivations. It is included in this appendix with derivation for easy reference.

Appendix B: Throughout the book there are scalars, vectors, matrices, probability density functions, conditional probability density functions, statistical expectation in terms of means and covariances, hypotheses, hypothesis probabilities, indices for multiple sensors and targets, and so on. A list of symbols and notation is provided as a quick reference for readers.

Appendix C: Terminology often used in the tracking community is defined. It is included in this appendix for the purpose of cross-referencing.

References

1. H.L. Van Trees, *Detection, Estimation, and Modulation Theory*, Part 1. New York: Wiley, 1968.
2. C.B. Chang and J.A. Tabaczynski, "Application of State Estimation to Target Tracking," *IEEE Transactions on Automatic Control*, vol. AC-29, pp. 98–109, 1984.
3. Y. Bar-Shalom, X. Rong Li, and T. Kirubarajan, *Estimation with Applications to Tracking and Navigation*. New York: Wiley, 2001.
4. R.E. Kalman, "A New Approach to Linear Filtering and Prediction Problems," *Transactions of the ASME*, vol. 82, ser. D, pp. 35–45, 1960.
5. R.E. Kalman and R.S. Bucy, "New Results in Linear Filtering and Prediction Theory," *Transactions of the ASME*, vol. 83, ser. D, pp. 95–108, 1961.
6. A.H. Jazwinski, *Stochastic Processes and Filtering Theory*. New York: Academic Press, 1970.

7.   C.B. Chang, "Two Lower Bounds on the Covariance for Nonlinear Estimation," *IEEE Transactions on Automatic Control*, vol. AC-26, pp. 1294–1297, 1981.

8.   P.M. DeRusso, R.J. Roy, C.M. Close, A.A. Desrochers, *State Variables for Engineers*, 2nd ed. New York: Wiley, 1998.

9.   Y.C. Ho and R.C.K. Lee, "A Bayesian Approach to Problems in Stochastic Estimation and Control," *IEEE Transactions on Automatic Control*, vol. AC-9, pp. 333–339, 1964.

10. R.S. Bucy, "Nonlinear Filtering Theory," *IEEE Transactions on Automatic Control*, vol. AC-10, p. 198, 1965.

11. R.P. Wishner, J.A. Tabaczynski, and M. Athans, "A Comparison of Three Nonlinear Filters," *Automatica*, vol. 5, pp. 487–496, 1969.

12. C.B. Chang "Ballistic Trajectory Estimation with Angle-Only Measurements," *IEEE Transactions on Automatic Control*, vol. AC-25, pp. 474–480, 1980.

13. C.B. Chang and K.P. Dunn, "Kalman Filter Compensation for a Special Class of Systems," *IEEE Transactions on Aerospace and Electronic Systems*, vol. AES-13, pp. 700–706, 1977.

14. C.B. Chang and K.P. Dunn, "On GLR Detection and Estimation of Unexpected Inputs in Linear Discrete Systems," *IEEE Transactions on Automatic Control*, vol. AC-24. pp. 499–501, 1979.

15. D.T. Magill, "Optimal Adaptive Estimation of Sampled Stochastic Processes," *IEEE Transactions on Automatic Control*, vol. AC-10, pp. 434–439, 1965.

16. C.B. Chang and M. Athans, "State Estimation for Discrete Systems with Switching Parameters," *IEEE Transactions on Aerospace and Electronic Systems*, vol. AES-14, pp. 418–424,1978.

17. H.A.P. Blom and Y. Bar-Shalom, "The Interacting Multiple Model Algorithm for Systems with Markovian Switching Coefficients," *IEEE Transactions on Automatic Control*, vol. AC-33, pp. 780–783, 1988.

18. R.S. Bucy. "Bayes Theorem and Digital Realization for Nonlinear Filters," *Journal of Astronautical Science*, vol. 17, pp. 80–94, 1969.

19. S. Julier, J. Uhlmann, and H.F. Durrant-Whyte, "A New Method for the Nonlinear Transformation of Means and Covariances in Filters and Estimators," *IEEE Transactions on Automatic Control*, vol. AC-45, pp. 477–482, 2000.

20. S. Arulampalam, S.R. Maskell, N.J. Gordon, and T. Clapp, "A Tutorial on Particle Filters for On-line Nonlinear/Non-Gaussian Bayesian Tracking," *IEEE Transactions on Signal Processing*, vol. 50, pp. 174–188, 2002.

21. B. Ristic, S. Arulampalam, and N. Gordon, *Beyond the Kalman Filter—Particle Filters for Tracking Applications*. London: Artech House, 2004.

22. S. Challa, M.R. Morelande, D. Musicki, and R.J. Evans, *Fundamentals of Object Tracking*. Cambridge, UK: Cambridge University Press, 2011.

23. D. Willner, C.B. Chang, and K.P. Dunn, "Kalman Filter Algorithms for a Multi-Sensor System," in *Proceedings of 1976 IEEE Conference on Decision and Control,* pp. 570–574, Dec. 1976.

24. C.Y. Chong, S. Mori, W. Parker, and K.C. Chang, "Architecture and Algorithm for Track Association and Fusion," *IEEE AES Systems Magazine*, Jan. 2000.

25. C.B. Chang, K.P. Dunn, and L.C. Youens, "A Tracking Algorithm for Dense Target Environment," in *Proceedings of the 1984 American Control Conference*, pp. 613–618, June 1984.

26. C.B. Chang and L.C. Youens, "An Algorithm for Multiple Target Tracking and Data Association," MIT Lincoln Laboratory Technical Report TR-643, 13 June 1983.

27. D.B. Reid, "An Algorithm for Tracking Multiple Targets," *IEEE Transactions on Automatic Control*, vol. AC-24, pp. 843–854, 1979.

28. S.S. Blackman, "Multiple Hypothesis Tracking for Multiple Target Tracking," *IEEE Transactions on A&E Systems*, vol. 19, pp. 5–18, 2004.

29. T. Kurien, "Issues in the Design of Practical Multi-target Tracking Algorithms," in *Multitarget Multisensor Tracking: Advanced Applications*, Y. Bar-Shalom, Ed., pp. 43–83, London: Artech House, 1990.

30. B. Corwin, D. Choi, K.P. Dunn, and C.B. Chang, "Sensor to Sensor Correlation and Fusion with Biased Measurements," in *Proceedings of MSS National Symposium on Sensor and Data Fusion*, May 2005.

31. A.B. Poore, "Multidimensional Assignment Formulation of Data Association Problems Arising from Multi-Target and Multi-Sensor Tracking," *Computational Optimization and Applications,* vol. 3, pp. 27–57, 1994.

32. C.J. Humke, "Bias Removal Techniques for the Target-Object Mapping Problem," MIT Lincoln Laboratory Technical Report 1060, 9 July 2002.

1

Parameter Estimation

1.1 Introduction

Parameter estimation is defined as the estimation of an unknown vector (denoted as the parameter vector) from a set of noisy measurements. The unknown vector may be constant or may be governed by a known time-invariant probability distribution. The problem of parameter estimation is defined in Section 1.2. Section 1.3 provides estimator definitions with respect to a chosen performance index or cost function. For the case where the parameter is a random vector, a very important probability relationship, namely Bayes' rule of probability, is introduced for two random vectors. Interested readers may consult the references in this chapter for more information about probability and random variables [1–6], and for vector space, state variables, and control systems [7–9]. Measures of estimator performance in terms of bias error and the error covariance matrix are introduced. Sections 1.4 and 1.5 give the results when the measurement vector and the parameter vector are related linearly and when both the measurement noise and the a priori knowledge for the random parameter case are Gaussian. Section 1.4 addresses the problem when the parameter vector is an unknown constant. Section 1.5 presents estimators when the unknown vector is random with a known distribution. In this case, the distribution of the unknown parameter is incorporated as a priori knowledge or as an additional noisy measurement existing in the parameter vector space. The results in Section 1.5 match those of Section 1.4 when the terms representing a priori knowledge are removed. The relationships of all estimators presented in these sections are discussed in the remarks in these sections. Section 1.6 extends the results of Section 1.5 to the nonlinear measurement case. The derivation is for a weighted least squares estimator; a first order Taylor series expansion is used to approximate the nonlinear relationship. An iterative solution procedure is derived in an effort to obtain reduced estimation error. Section 1.7 derives the

Cramer–Rao bound of the unbiased estimator error covariance for the parameter estimation problem discussed in this chapter. When the measurement noise and the a priori distributions are Gaussian, error covariance of the estimators for linear cases is the same as for the Cramer–Rao bound. A numerical example is given in Section 1.8 to show the construction of an estimator and its performance against the Cramer–Rao bound. Some important mathematic derivations are included in Appendices 1.A and 1.B for the interested readers. A set of homework problems is provided at the end of the chapter in addition to a list of references. The homework problems will familiarize the reader with parameter estimation in general, and fill in some of the details in estimator derivation.

1.2   Problem Definition

Let $\mathbf{x} \in \mathbb{R}^n$ be an unknown parameter vector consisting of n components. A set of m linearly independent measurements is taken on $\mathbf{x}$ with a measurement device that is modeled by a known relationship $\mathbf{h}(.): \mathbb{R}^n \to \mathbb{R}^m$. Let the measurement vector $\mathbf{y}$ be denoted as

$$\mathbf{y} = \mathbf{h}(\mathbf{x}) + \mathbf{v} \tag{1.2-1}$$

where $\mathbf{y}, \mathbf{v} \in \mathbb{R}^m$ and the measurement noise $\mathbf{v}$ is a random vector with a known probability density function. The unknown parameter vector $\mathbf{x}$ may be constant or may be random and statistically independent of $\mathbf{v}$ with a known probability density function.

Problem for Parameter Estimation

An estimator of $\mathbf{x}$ given $\mathbf{y}$, denoted as $\hat{\mathbf{x}}$, is a function of $\mathbf{y}$. Depending on the objective of the estimation problems, there are different types of estimators. Most of the estimators are derived with the objective of optimizing certain performance criteria given the known relationship between $\mathbf{x}$ and $\mathbf{y}$ and the statistical properties of $\mathbf{v}$ and $\mathbf{x}$ by Equation (1.2-1). Different performance criteria will result in different estimators, as discussed in detail later in this chapter.

1.3   Definition of Estimators

An estimator is a function of $\mathbf{y}$ and can be derived subject to a defined performance index or cost function. Two common classes of estimators are defined below that minimize a defined performance index. One class is for constant but unknown $\mathbf{x}$. The second class is for random $\mathbf{x}$ with a known probability density function. The mathematical treatments of $\mathbf{x}$ are different, depending on whether it is a constant or a random vector; for the class of random $\mathbf{x}$, the definition of joint probability function of $\mathbf{x}$ and $\mathbf{y}$ is needed. All definitions of estimators commonly used are defined in this section. Their relationships will be discussed in the later sections when the estimators are derived.

1.3.1   Constant Parameter Estimation

When the unknown parameter is a constant vector, $\mathbf{x}$, the probability density function $p(\mathbf{y})$ can be derived from the known density function $p(\mathbf{v})$ with the relationship given by Equation (1.2-1). For each $\mathbf{y}=y$, the density function $p(\mathbf{y})$ becomes a function of $\mathbf{x}$, and the likelihood function of $\mathbf{x}$ for $\mathbf{y}=y$ and is denoted as $p(y; \mathbf{x})$.

Three commonly used estimators of $\mathbf{x}$ given $\mathbf{y}$ are discussed below: the least squares estimator (LS), the weighted least squares estimator (WLS), and the maximum likelihood estimator (ML).

Least Squares Estimator

The LS estimator of $\mathbf{x}$ given $\mathbf{y}$ is the function $\hat{\mathbf{x}}_{\mathrm{LS}} : \mathbb{R}^m \to \mathbb{R}^n$, that minimizes the norm square between $\mathbf{y}$ and $\mathbf{h}(.)$ in $\mathbb{R}^m$, or the performance index

$$J = \|\mathbf{y} - \mathbf{h}(\mathbf{x})\|^2 \qquad (1.3\text{-}1)$$

where $\|\mathbf{r}^2\| \triangleq \langle \mathbf{r}, \mathbf{r} \rangle \triangleq \mathbf{r}^T \mathbf{r}, \forall \mathbf{r} \in \mathbb{R}^m$, such that

$$\hat{\mathbf{x}}_{\mathrm{LS}} = \mathrm{ArgMin}_{\mathbf{x}} \|y - \mathbf{h}(\mathbf{x})\|^2, \forall \mathbf{y} = y.$$

Weighted Least Squares Estimator

The WLS estimator of $\mathbf{x}$ given $\mathbf{y}$ is the function $\hat{\mathbf{x}}_{\mathrm{WLS}} : \mathbb{R}^m \to \mathbb{R}^n$, which minimizes the weighted norm square between $\mathbf{y}$ and $\mathbf{h}(.)$ in $\mathbb{R}^m$. The performance index is given as

$$J = \|\mathbf{y} - \mathbf{h}(\mathbf{x})\|_{\mathbf{A}^{-1}}^2 \qquad (1.3\text{-}2)$$

where $\|\mathbf{r}\|_{\mathbf{A}^{-1}}^2 \triangleq \langle \mathbf{r}, \mathbf{A}^{-1}\mathbf{r} \rangle \triangleq \mathbf{r}^T \mathbf{A}^{-1} \mathbf{r}, \forall \mathbf{r} \in \mathbb{R}^m$ and $\mathbf{A}$ can be any positive definite $m \times m$ matrix, but is typically chosen as $\mathbf{R}$, the covariance of the noise vector $\mathbf{v}$, such that

$$\hat{\mathbf{x}}_{\mathrm{WLS}} = \mathrm{ArgMin}_{\mathbf{x}} \|y - \mathbf{h}(\mathbf{x})\|_{\mathbf{A}^{-1}}^2, \forall \mathbf{y} = y.$$

Maximum Likelihood Estimator

The ML estimator of $\mathbf{x}$ given $\mathbf{y}$ is the function $\hat{\mathbf{x}}_{\text{ML}} : \mathbb{R}^m \rightarrow \mathbb{R}^n$, which maximizes the likelihood function, $\forall \mathbf{y} = y$

$$J = p(y; \mathbf{x}) \tag{1.3-3}$$

such that

$$\hat{\mathbf{x}}_{\text{ML}} = \text{ArgMax}_{\mathbf{x}} \, p(y; \mathbf{x}), \forall \mathbf{y} = y.$$

Remarks

1. The term *estimator* is used to denote a function of the observations, $\mathbf{y}$, and *estimate* denotes the value of the estimator for a given observation, that is, $\mathbf{y} = y$. Therefore, the estimator is a random vector in $\mathbb{R}^n$, since it is a function of the random vector $\mathbf{y}$. One can construct these estimators by solving the optimization problem defined in Equations (1.3-1) through (1.3-3) for each $\mathbf{y} = y$, pointwise.

2. The geometric interpretation of WLS is the same as that for the LS estimator except that Equation (1.3-2) is defined in a different inner product space, therefore it is not expected that $\hat{\mathbf{x}}_{\text{LS}} = \hat{\mathbf{x}}_{\text{WLS}}$. The LS estimators are often used when the measurement noise is not well defined or unknown, while the WLS estimators are used for the case of zero-mean Gaussian noise vector, $\mathbf{v}$, with covariance $\mathbf{R}$, and $\mathbf{R}$ is used for $\mathbf{A}$ in Equation (1.3-2). One can show that (see Appendix 1.A and Homework Problem 2) after selecting an appropriate linear transformation, $\mathbf{L}$, in the measurement space, the transformed measurement vector has a noise vector $\mathbf{L}\mathbf{v}$ with an identity covariance matrix that is $\mathbf{L}^T \mathbf{R} \mathbf{L} = \mathbf{I}$. When minimizing with respect to the transformed measurement vector, $\mathbf{L}\mathbf{y}$, all transformed measurements are weighted equally.

1.3.2   Random Parameter Estimation

When the unknown parameter is a random vector $\mathbf{x}$, with known probability distribution, the performance index will need to be modified to reflect the known randomness of $\mathbf{x}$. Before defining the second class of estimators, it is useful to review some elements of probability theory that will be used throughout this book when both $\mathbf{x}$ and $\mathbf{y}$ are random vectors.

Conditional Probability Density Function and Bayes' Rule

Bayes' rule plays an important role when the parameter to be estimated is random. Let $\mathbf{x} \in \mathbb{R}^n$ and $\mathbf{y} \in \mathbb{R}^m$ be two random vectors each having probability density function $p(\mathbf{x})$ and $p(\mathbf{y})$, respectively, and let the joint probability density function of these two random vectors be denoted as $p(\mathbf{x,y})$. The relationship between the joint probability density function $p(\mathbf{x,y})$ and the individual density functions defines the conditional probability functions $p(\mathbf{y}|\mathbf{x})$ and $p(\mathbf{x}|\mathbf{y})$, respectively, as

$$p(\mathbf{x}, \mathbf{y}) \triangleq p(\mathbf{y}|\mathbf{x}) p(\mathbf{x})$$

and

$$p(\mathbf{x}, \mathbf{y}) \triangleq p(\mathbf{x}|\mathbf{y}) p(\mathbf{y}). \tag{1.3-4}$$

Bayes' rule of probability is derived from the above relationship as

$$p(\mathbf{x}|\mathbf{y}) = \frac{p(\mathbf{x}, \mathbf{y})}{p(\mathbf{y})} = \frac{p(\mathbf{y}|\mathbf{x})}{p(\mathbf{y})} p(\mathbf{x}). \tag{1.3-5}$$

For the problems discussed in this book, these probability density functions all have specific names:

$p(\mathbf{x})$: the a priori density of $\mathbf{x}$,

$p(\mathbf{y}|\mathbf{x})$: the conditional density of $\mathbf{y}$ given $\mathbf{x}$, a function of $\mathbf{x}$, and is called the likelihood density function of $\mathbf{x}$, and

$p(\mathbf{x}|\mathbf{y})$: the conditional density of $\mathbf{x}$ given $\mathbf{y}$, a function of $\mathbf{y}$, and is called the a posteriori density of $\mathbf{x}$ given $\mathbf{y}$.

When $\mathbf{x}$ and $\mathbf{y}$ are independent, we have

$$p(\mathbf{x,y}) = p(\mathbf{x})p(\mathbf{y}),$$

which implies $p(\mathbf{x}|\mathbf{y}) = p(\mathbf{x})$ and $p(\mathbf{y}|\mathbf{x}) = p(\mathbf{y})$.
The conditional mean of $\mathbf{x}$ given $\mathbf{y}$ is defined as the integral

$$E\{\mathbf{x}|\mathbf{y}\} = \int \mathbf{x} p(\mathbf{x}|\mathbf{y}) d\mathbf{x}. \tag{1.3-6}$$

If $\mathbf{x}$ and $\mathbf{y}$ are independent, Equations (1.3-5) and (1.3-6) yield

$$E\{\mathbf{x}|\mathbf{y}\} = E\{\mathbf{x}\}.$$

Furthermore, for any function of $\mathbf{y}$, $\mathbf{g}(\mathbf{y})$

$$E\{\mathbf{g}(\mathbf{y})|\mathbf{y}\} = \mathbf{g}\{\mathbf{y}\}.$$

Remarks

1. By the definition of the above integral, $E\{\mathbf{x}|\mathbf{y}\}$ is a function of $\mathbf{y}$ and hence it is also an estimate of $\mathbf{x}$ given $\mathbf{y}$. It is generally a nonlinear function of $\mathbf{y}$. Many important properties of this estimator and its relationship to estimators defined here will be discussed later in this chapter.

2. $E\{.|\mathbf{y}\}$ is a linear operator of any random vector defined by the integral given in Equation (1.3-6). It is important to note that $E\{\mathbf{h}(\mathbf{x})|\mathbf{y}\} \neq \mathbf{h}(E\{\mathbf{x}|\mathbf{y}\})$ in general.

Least Squares Estimator

The LS estimator of $\mathbf{x}$ given $\mathbf{y}$ is the function of $\mathbf{y}$, $\hat{\mathbf{x}}_{LS} : \mathbb{R}^m \to \mathbb{R}^n$ that minimizes the sum of norm squares between $\mathbf{y}$ and $\mathbf{h}(.)$ in $\mathbb{R}^m$, and $\bar{\mathbf{x}}_0$ and $\mathbf{x}$ in $\mathbb{R}^n$, which is the performance index

$$J = \|\mathbf{y} - \mathbf{h}(\mathbf{x})\|^2 + \|\bar{\mathbf{x}}_0 - \mathbf{x}\|^2 \tag{1.3-7}$$

where $\bar{\mathbf{x}}_0$ is the mean of $\mathbf{x}$, such that

$$\hat{\mathbf{x}}_{LS} = \text{ArgMin}_\mathbf{x} \left\{ \|\mathbf{y} - \mathbf{h}(\mathbf{x})\|^2 + \|\bar{\mathbf{x}}_0 - \mathbf{x}\|^2 \right\}, \forall \mathbf{y} = \mathbf{y}.$$

Weighted Least Squares Estimator

The WLS estimator of $\mathbf{x}$ given $\mathbf{y}$ is the function of $\mathbf{y}$, $\hat{\mathbf{x}}_{WLS} : \mathbb{R}^m \to \mathbb{R}^n$, that minimizes the sum of weighted norm squares between $\mathbf{y}$ and $\mathbf{h}(.)$ in $\mathbb{R}^m$ and $\bar{\mathbf{x}}_0$ and $\mathbf{x}$ in $\mathbb{R}^n$, which is the performance index

$$J = \|\mathbf{y} - \mathbf{h}(\mathbf{x})\|^2_{\mathbf{A}^{-1}} + \|\bar{\mathbf{x}}_0 - \mathbf{x}\|^2_{\mathbf{B}^{-1}} \tag{1.3-8}$$

where $\bar{\mathbf{x}}_0$ is the mean of $\mathbf{x}$ and $\mathbf{A}$ and $\mathbf{B}$ can be any positive definite matrices, but are typically chosen as $\mathbf{R}$, the covariance of the noise vector $\mathbf{v}$ and $\mathbf{P}_0$, the covariance of $\mathbf{x}$, respectively, such that

$$\hat{\mathbf{x}}_{WLS} = \text{ArgMin}_\mathbf{x} \left\{ \|\mathbf{y} - \mathbf{h}(\mathbf{x})\|^2_{\mathbf{A}^{-1}} + \|\bar{\mathbf{x}}_0 - \mathbf{x}\|^2_{\mathbf{B}^{-1}} \right\}, \forall \mathbf{y} = \mathbf{y}.$$

Maximum Likelihood Estimator

The ML estimator of $\mathbf{x}$ given $\mathbf{y}$ is the function $\hat{\mathbf{x}}_{\mathrm{ML}} : \mathbb{R}^m \to \mathbb{R}^n$, which maximizes the likelihood density function

$$J = p(\mathbf{y}|\mathbf{x}) \qquad (1.3\text{-}9)$$

such that

$$\hat{\mathbf{x}}_{\mathrm{ML}} = \mathrm{ArgMax}_{\mathbf{x}}\, p\left(y|\mathbf{x}\right), \forall \mathbf{y} = \mathbf{y}.$$

Maximum a Posteriori Probability Estimator

The maximum a posteriori (MAP) probability estimator of $\mathbf{x}$ given $\mathbf{y}$ is the function $\hat{\mathbf{x}}_{\mathrm{MAP}} : \mathbb{R}^m \to \mathbb{R}^n$ that maximizes the a posteriori probability

$$J = p(\mathbf{x}|\mathbf{y}) \qquad (1.3\text{-}10)$$

such that

$$\hat{\mathbf{x}}_{\mathrm{MAP}} = \mathrm{ArgMax}_{\mathbf{x}}\, p\left(\mathbf{x}|\mathbf{y}\right), \forall \mathbf{y} = \mathbf{y}.$$

Remarks

1. Each of the estimators above has its own significance. For constant parameter estimation cases, when the noise density functions are not known or only partially known, the LS estimator or the WLS estimator is most often used for solving practical problems. If the noise density functions are known explicitly, the ML estimator is often used.

2. As will be shown later, when the measurement noise vector $\mathbf{v}$ and the a priori density function of $\mathbf{x}$ are mutually independent and Gaussian with known covariance, the last three estimators, WLS, ML, and MAP, are identical. When the noise covariance $\mathbf{R}$ and $\mathbf{P}_0$ are a constant (the same constant for both $\mathbf{R}$ and $\mathbf{P}_0$) times the identity matrix, then LS, WLS, ML, and MAP are the same.

1.3.3   Properties of Estimators

Let $\hat{\mathbf{x}}$ denote an estimator of $\mathbf{x}$ given $\mathbf{y}$. Then the estimation error $\tilde{\mathbf{x}}$ is

$$\tilde{\mathbf{x}} \triangleq \mathbf{x} - \hat{\mathbf{x}}$$

and the mean and covariance of $\tilde{\mathbf{x}}$ is

$$E\{\tilde{\mathbf{x}}\} = E\{\mathbf{x}\} - E\{\hat{\mathbf{x}}\} \qquad (1.3\text{-}11)$$

$$Cov\{\tilde{\mathbf{x}}\} = E\{(\mathbf{x} - \hat{\mathbf{x}})(\mathbf{x} - \hat{\mathbf{x}})^T\}. \qquad (1.3\text{-}12)$$

The properties for each estimator will be derived explicitly and discussed later in this book. When $E\{\tilde{\mathbf{x}}\} = 0$, we refer to the estimator as unbiased, that is, $\hat{\mathbf{x}}$ is an unbiased estimator of $\mathbf{x}$ given $\mathbf{y}$. A very important measure of the estimator quality is the error covariance, $Cov\{\tilde{\mathbf{x}}\}$, which is non-negative definite by definition. The Cramer–Rao bound (CRB) is a lower bound of the error covariance of any unbiased estimator of $\mathbf{x}$. Any unbiased estimator with $Cov\{\tilde{\mathbf{x}}\}$ achieving the CRB is an efficient estimator. Van Trees [1] provides an excellent discussion and definition of the CRB with application examples. Cramer, Rao, and Fisher derived the CRB as presented in three papers [10–12]. The Cramer–Rao bound has been applied to numerous problems as a measure of estimator performance [13]. The CRB is very important for practical problems, as presented in this book. It is derived and developed later in this chapter for the parameter estimation problems, and in the following chapters for the state estimation problem.

In Section 1.3.2, the conditional mean, $E\{\mathbf{x}|\mathbf{y}\}$, is identified as an estimator of $\mathbf{x}$ given $\mathbf{y}$. Some important variants of this estimator are the conditional mean estimator (CM), and the linear least squares estimator (LLS).

Conditional Mean Estimator

The conditional mean of $\mathbf{x}$ given $\mathbf{y}$ is the function of $\mathbf{y}$ in $\mathbb{R}^n$ defined as the integral

$$\hat{\mathbf{x}}_{CM} \triangleq E\{\mathbf{x}|\mathbf{y}\} = \int \mathbf{x} p(\mathbf{x}|\mathbf{y}) d\mathbf{x}.$$

It has the following properties

1. $\hat{\mathbf{x}}_{CM} \in \mathbb{R}^n$ is the minimum variance estimator of $\mathbf{x}$ given $\mathbf{y}$ that minimizes the performance index

$$J = E\{\|\mathbf{x} - \hat{\mathbf{x}}\|^2\} = E\{(\mathbf{x} - \hat{\mathbf{x}})(\mathbf{x} - \hat{\mathbf{x}})^T\} = Cov\{\tilde{\mathbf{x}}\}, \qquad (1.3\text{-}13)$$

    such that

$$\hat{\mathbf{x}}_{CM} = \operatorname{ArgMin}_{\hat{\mathbf{x}}} E\{\|\mathbf{x} - \hat{\mathbf{x}}\|^2\}.$$

2. $\hat{\mathbf{x}}_{CM}$ is unbiased, that is,

$$E\{\mathbf{x} - \hat{\mathbf{x}}_{CM}\} = \mathbf{0}. \qquad (1.3\text{-}14)$$

3. Let $\tilde{\mathbf{x}}_{CM} = \mathbf{x} - \hat{\mathbf{x}}_{CM}$, then $\tilde{\mathbf{x}}_{CM}$ is uncorrelated with any function of $\mathbf{y}$, $\mathbf{g}: \mathbb{R}^m \to \mathbb{R}^n$, that is,

$$E\left\{\mathbf{g}(\mathbf{y})\tilde{\mathbf{x}}_{CM}^T\right\} = \mathbf{0}. \qquad (1.3\text{-}15)$$

The proof of these properties is provided in Appendix 1.B or [14].

Linear Least Squares Estimator

The LLS estimator of $\mathbf{x}$ given $\mathbf{y}$, $\hat{\mathbf{x}}_{LLS} \in \mathbb{R}^n$, is a linear function of $\mathbf{y}$, $\mathbf{L}: \mathbb{R}^m \to \mathbb{R}^n$ that minimizes the performance index

$$J = E\{\|\mathbf{x} - \mathbf{L}(\mathbf{y})\|^2\} \qquad (1.3\text{-}16)$$

such that

$$\hat{\mathbf{x}}_{LLS} = \text{ArgMin}_L E\left\{\|\mathbf{x} - \mathbf{L}(\mathbf{y})\|^2\right\}.$$

Some important properties of the LLS estimator are explored here.

Properties of the LLS Estimator

The LLS estimator, $\hat{\mathbf{x}}_{LLS}$, depends only on the first and second moments of the random vectors $\mathbf{x}$ and $\mathbf{y}$ and not on their entire probability density functions, such that

$$\hat{\mathbf{x}}_{LLS} = \bar{\mathbf{x}} + \mathbf{P}_{xy}\mathbf{P}_{yy}^{-1}(\mathbf{y} - \bar{\mathbf{y}}) \qquad (1.3\text{-}17)$$

and

$$Cov\left\{\tilde{\mathbf{x}}_{LLS}\right\} = \mathbf{P}_{xx} - \mathbf{P}_{xy}\mathbf{P}_{yy}^{-1}\mathbf{P}_{yx} \qquad (1.3\text{-}18)$$

where

$$\bar{\mathbf{x}} = E\{\mathbf{x}\}, \bar{\mathbf{y}} = E\{\mathbf{y}\},$$

$$\tilde{\mathbf{x}}_{LLS} = \mathbf{x} - \hat{\mathbf{x}}_{LLS},$$

and

$$\mathbf{P}_{ab} = Cov\{\mathbf{a},\mathbf{b}\}.$$

1. $\hat{\mathbf{x}}_{LLS}$ is linear.
2. $\hat{\mathbf{x}}_{LLS}$ is unbiased.
3. $\hat{\mathbf{x}}_{LLS} = \hat{\mathbf{x}}_{CM}$ when $\mathbf{x}$ and $\mathbf{y}$ are jointly Gaussian.

The proof of these properties for $\hat{\mathbf{x}}_{LLS}$ can be found in Appendix 1.B or [14].

Remarks

1. It is important to note that $\hat{\mathbf{x}}_{CM}$ is the minimum variance (least squares) estimator for the performance index given in Equation (1.3-13) in $\mathbb{R}^n$ for all $\mathbf{x}$ and $\mathbf{y}$, which is not the same as the minimum norm (least squares) estimator $\hat{\mathbf{x}}_{LS}$ defined pointwise in Section 1.3.2 with the performance index given in Equation (1.3-2) in $\mathbb{R}^m$. In practice, the conditional density function, $p(\mathbf{x}|\mathbf{y})$, is not generally computable, and neither is the integration in Equation (1.3-6). Approximation of both $p(\mathbf{x}|\mathbf{y})$ and $\hat{\mathbf{x}}_{CM}$ is discussed in later chapters.

2. Property 3 of $\hat{\mathbf{x}}_{CM}$ is a very important result because it provides the orthogonal relationship in the mean between the estimation error $\tilde{\mathbf{x}}_{CM}$ and any function of $\mathbf{y}$ in $\mathbb{R}^n$. We will need to use this property to relate $\hat{\mathbf{x}}_{CM}$ with the $\hat{\mathbf{x}}_{LS}$ in a later section of this chapter.

3. In the random parameter case, $p(\mathbf{x}|\mathbf{y})$ contains all the statistical information about $\mathbf{x}$ given $\mathbf{y}$ and $E\{\mathbf{x}|\mathbf{y}\}$ is the minimum variance (least squares) solution of Equation (1.3-13) for all $\mathbf{x}$ and $\mathbf{y}$. But in general, MAP and CM are not equal as shown in Figure 1.1. In cases where $p(\mathbf{x}|\mathbf{y})$ cannot be solved or computed exactly, many approximate solutions have been suggested by researchers for $p(\mathbf{x}|\mathbf{y})$ and $\hat{\mathbf{x}}_{CM}$, and some of these are discussed later in this book.

4. Before explicit equations for the estimators are derived, it is useful to summarize the definition and the associated performance indices in Table 1.1. Note that the estimator performance indices differ depending on whether $\mathbf{x}$ is an unknown constant or a random vector with known distribution.

Table 1.1   Summary of Estimator Definition

Performance Index		
Estimator Type	**x Unknown**	**x Random**
LS	Min: $J = \|\mathbf{y} - \mathbf{h}(\mathbf{x})\|^2$	Min : $J = \|\mathbf{y} - \mathbf{h}(\mathbf{x})\|^2 + \|\bar{\mathbf{x}}_0 - \mathbf{x}\|^2$
WLS	Min: $J = \|\mathbf{y} - \mathbf{h}(\mathbf{x})\|_{\mathbf{R}^{-1}}^2$	Min: $J = \|\mathbf{y} - \mathbf{h}(\mathbf{x})\|_{\mathbf{R}^{-1}}^2 + \|\bar{\mathbf{x}}_0 - \mathbf{x}\|_{\mathbf{P}_0^{-1}}^2$
ML	Max: $J = p(\mathbf{y};\mathbf{x})$	Max: $J = p(\mathbf{y}|\mathbf{x})$
MAP	Does not apply	Max: $J = p(\mathbf{x}|\mathbf{y})$
CM	Does not apply	Min: $J = E\{\|\mathbf{x} - \hat{\mathbf{x}}\|^2\}$
LLS	Does not apply	Min: $J = E\{\|\mathbf{x} - \hat{\mathbf{x}}\|^2\} \forall$ linear estimator

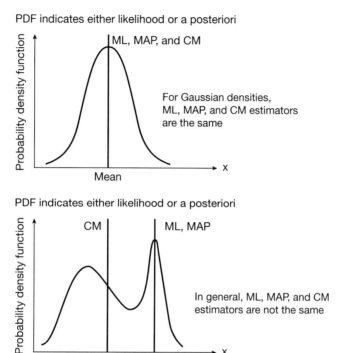

Figure 1.1   Estimator illustration—Gaussian and non-Gaussian densities.

1.3.4   Measure of Estimator Quality: Estimation Errors

The mean and covariance of the estimation error, $\tilde{\mathbf{x}}$, of an estimator $\hat{\mathbf{x}}$, given in Equations (1.3-11) and (1.3-12), are theoretical expressions. How do we know the estimators used in practice have the properties of being unbiased and efficient? One needs to carry out experiments that prove the estimators have those properties with some level of confidence. Several questions help us flesh this out. (a) How can an experiment be designed that will generate samples and represent the statistical properties of the random vectors in question? (b) If the samples of the random vectors are generated, how can the statistical properties of the estimators be extracted (e.g., the mean and covariance of $\hat{\mathbf{x}}$)? (c) Finally, after all the statistics from the experiment are collected, how can the data be related to a probability of confidence?

To address the first question, one must be able to generate a set of independent and identically distributed (i.i.d.) samples of a random variable, x, with a given cumulative distribution function (CDF), $F(s)$, such that

$$F(s) = \text{Prob}\{x \leq s\}.$$

One can draw a sequence of i.i.d. random samples $\{r_i, i = 1,2, \dots ,M\}$ from a uniform number generator that has a uniform distribution on $[0,1]$[1] and let

$$x_i = F^{-1}(r_i). \tag{1.3-19}$$

One can show that the sequence of samples $\{x_i, i = 1,2, \dots ,M\}$, is a sequence of i.i.d. random samples with CDF, $F(s)$. In Homework Problems 1–3, the reader is asked to generate random variables via a different method and with different statistical properties. It is straightforward to extend this procedure to construct a random vector $\mathbf{x}$ where each component of the vector follows the joint CDF $F(s)$ (see Homework Problem 2).

Given i.i.d. samples of $\mathbf{x}$, $\{\mathbf{x}_i: i = 1,2, \dots, M\}$ the mean of $\mathbf{x}$, $E\{\mathbf{x}\}$ can be approximated as[2]

$$E\{\mathbf{x}\} \approx \frac{1}{M}\sum_{i=1}^{M}\mathbf{x}_i = \overline{\mathbf{x}} \tag{1.3-20}$$

where $\overline{\mathbf{x}}$ is called the sample mean of $\mathbf{x}$. Likewise the covariance of $\mathbf{x}$ can be approximated as

$$\mathbf{P} = Cov\{\mathbf{x}\} \approx \frac{1}{M}\sum_{i=1}^{M}(\mathbf{x}-\mathbf{x}_i)(\mathbf{x}-\mathbf{x}_i)^T = \overline{\mathbf{P}}, \tag{1.3-21}$$

where $\overline{\mathbf{P}}$ is called the sampled mean square error of $\mathbf{x}$. As the number of samples increases, as $M \rightarrow \infty$, one should expect that $\overline{\mathbf{x}} \rightarrow E\{\mathbf{x}\}$ and $\overline{\mathbf{P}} \rightarrow Cov\{\mathbf{x}\}$.

Finally, to address the last question, one must know that the statistical properties of the random samples of $\mathbf{x}_i$ have the same properties as $\mathbf{x}$. If $\mathbf{x}$ is Gaussian, and $\{\mathbf{x}_i: i = 1, 2, \dots, M\}$ is generated by $\mathbf{x}$, then $\overline{\mathbf{x}}$ is also Gaussian and

$$\chi^2 = \frac{1}{M}\sum_{i=1}^{M}(\mathbf{x}-\mathbf{x}_i)\mathbf{P}^{-1}(\mathbf{x}-\mathbf{x}_i)^T$$

---

[1] There are very well tested uniform random number generator algorithms available for this purpose, for example, RAND in MATLAB.

[2] The interested reader can get more insight on this topic from any textbook on probability and statistics [e.g., 4, 5].

follows a χ^2 distribution with nM degrees of freedom where n is the dimension of the parameter vector $\mathbf{x}$. There is a well known χ^2 statistic test [1–5, 15] that if the hypothesis that the sequence of $\{\mathbf{x}_i: i = 1,2, \ldots , M\}$ is generated by $\mathbf{x}$ is true, it should have its χ^2 bounded by (r_1,r_2) with a probability bounded by $1 - \alpha$ as follows

$$\text{Prob}\{r_1 \leq \chi^2 \leq r_2\} \leq 1- \alpha$$

where r_1 and r_2 are determined by the degrees of freedom of the χ^2 and the value of α. This is a very useful tool to test the estimator performance. The concept of using the errors obtained in this way is introduced here, and will be discussed in depth in Section 4.2.

1.4   Estimator Derivation: Linear and Gaussian, Constant Parameter

Consider the case where the state and measurement vectors $\mathbf{x}$ and $\mathbf{y}$ are related linearly, that is, $\mathbf{h}(\mathbf{x}) = \mathbf{H}\mathbf{x}$ and

$$\mathbf{y}=\mathbf{H}\mathbf{x}+\mathbf{v},$$

$\mathbf{x} \in \mathbb{R}^n$, $\mathbf{y}$, $\mathbf{v} \in \mathbb{R}^m$, and $\mathbf{H}$ is an $m \times n$ matrix.[3] In this section, $\mathbf{x}$ is treated as an unknown constant vector. Furthermore, the measurement noise vector $\mathbf{v}$ is Gaussian with zero mean and covariance $\mathbf{R}$

$$\mathbf{v}: \sim N(\mathbf{0}, \mathbf{R}).$$

Consider the trivial case where the dimension of $\mathbf{x}$ and $\mathbf{y}$ are the same, that is, $m=n$, and if $\mathbf{H}$ is nonsingular, then solving for $\mathbf{x}$ is the same as solving for a set of linear equations, that is,

$$\hat{\mathbf{x}} = \mathbf{H}^{-1}\mathbf{y}.$$

It is left to the reader to show that $\hat{\mathbf{x}}$ is an unbiased estimate of $\mathbf{x}$ and the covariance of $\hat{\mathbf{x}}$ is $\mathbf{H}^{-1}\mathbf{R}\mathbf{H}^{-T}$.[4] This simple case will not be considered any further.

1.4.1   Least Squares Estimator

The performance index of the LS estimator is

$$J= \|\mathbf{y} - \mathbf{h}(\mathbf{x})\|^2 = (\mathbf{y} - \mathbf{H}\mathbf{x})^T(\mathbf{y} - \mathbf{H}\mathbf{x}). \tag{1.4-1}$$

---

3   Throughout the entire book, $\mathbf{H}$ is assumed to have rank of m, that is, all m equations are linearly independent.
4    $\mathbf{H}^{-T}$ denotes the transpose of $\mathbf{H}^{-1}$.

The objective is to find the **x** that minimizes J. The necessary condition for such an **x** is that it satisfies

$$\frac{\partial J}{\partial \mathbf{x}} = -2\mathbf{H}^T (\mathbf{y} - \mathbf{Hx}) = \mathbf{0}$$

or

$$\mathbf{H}^T\mathbf{Hx} = \mathbf{H}^T\mathbf{y}. \qquad (1.4\text{-}2)$$

We now consider two cases, when $m \geq n$ and when $m < n$.

1. When $m \geq n$, there are more equations than unknowns; this is referred to as an overdetermined case, if $\mathbf{H}^T\mathbf{H}$ is invertible (this may not be true in general[5]), an exact solution of $\hat{\mathbf{x}}$ is given by

$$\hat{\mathbf{x}}_{\mathrm{LS}} = \left[\mathbf{H}^T\mathbf{H}\right]^{-1}\mathbf{H}^T\mathbf{y} \qquad (1.4\text{-}3)$$

where $\hat{\mathbf{x}}_{\mathrm{LS}}$ is the least squares estimate of **x** given **y**. We will leave the proof to the reader (see Homework Problem 4) that $\hat{\mathbf{x}}_{\mathrm{LS}}$ is unbiased with covariance

$$\mathbf{P}_{\mathrm{LS}} = [\mathbf{H}^T\mathbf{H}]^{-1}[\mathbf{H}^T\mathbf{RH}][\mathbf{H}^T\mathbf{H}]^{-1} \qquad (1.4\text{-}4)$$

2. When $m < n$, the inversion of $\mathbf{H}^T\mathbf{H}$ does not exist, and there are fewer equations than unknowns; this is referred to as an underdetermined case. An infinite number of solutions can satisfy the necessary condition. The following expression is one solution that satisfies the necessary condition given in Equation (1.4-2)

$$\hat{\mathbf{x}}_{\mathrm{PI}} = \mathbf{H}^T\left[\mathbf{HH}^T\right]^{-1}\mathbf{y}, \qquad (1.4\text{-}5)$$

where $\hat{\mathbf{x}}_{\mathrm{PI}}$ is defined as the pseudo-inverse estimator of **x** given **y**. This can be shown by simple substitution of the above expression into the equation for the necessary condition Equation (1.4-2). A geometric interpretation of the result is shown in the following remark. Unfortunately, the pseudo-inverse estimator $\hat{\mathbf{x}}_{\mathrm{PI}}$ of **x** is biased. The reader is asked to show in Homework Problem 5 that the bias is

$$E\{\tilde{\mathbf{x}}_{\mathrm{PI}}\} = E\{\mathbf{x} - \hat{\mathbf{x}}_{\mathrm{PI}}\} = \left[\mathbf{I} - \mathbf{H}^T\left[\mathbf{HH}^T\right]^{-1}\mathbf{H}\right]\mathbf{x}, \qquad (1.4\text{-}6)$$

and the covariance is

$$\mathbf{P}_{\mathrm{PI}} = [\mathbf{I} - \mathbf{H}^T[\mathbf{HH}^T]^{-1}\mathbf{H}]\mathbf{xx}^T[\mathbf{I} - \mathbf{H}^T[\mathbf{HH}^T]^{-1}\mathbf{H}]^T + \mathbf{H}^T[\mathbf{HH}^T]^{-1}\mathbf{R}[\mathbf{HH}^T]^{-T}\mathbf{H}. \quad (1.4\text{-}7)$$

---

5   For $m \geq n$, $\mathbf{H}^T\mathbf{H}$ is invertible whenever the components of **y** are linearly independent.

Note that both bias and covariance are dependent on the true but unknown value of the state $\mathbf{x}$.

Remarks

Here we explore a geometric interpretation of the LLS estimator.

1. As shown in Figure 1.2, the unknown parameter, $\mathbf{x}$ in $\mathbb{R}^n$ mapped into $\mathbb{R}^m$ by a linear function $\mathbf{Hx}$ into a range space of $\mathbf{H}$ is defined as $\mathcal{R}(\mathbf{H}) \triangleq \{\mathbf{r} : \mathbf{r} = \mathbf{Hx}, \forall \mathbf{x} \in \mathbb{R}^n\}$, which is a subspace of $\mathbb{R}^m$ depicted as a straight line (in blue) in the right-hand side of Figure 1.2 as in a 2D representation. The orthogonal subspace of $\mathcal{R}(\mathbf{H})$ in $\mathbb{R}^m$ is defined as

$$\mathcal{R}^\perp(\mathbf{H}) \triangleq \{\mathbf{r}^\perp : \langle \mathbf{r}^\perp, \mathbf{r} \rangle = 0, \forall \mathbf{r} \in \mathcal{R}(\mathbf{H}), \forall \mathbf{r}^\perp \in \mathbb{R}^m\},$$

which is depicted as a perpendicular line (in blue) in Figure 1.2. The noise $\mathbf{v}$ is a random vector in $\mathbb{R}^m$ and the measurement vector $\mathbf{y}$ is depicted as a vector sum of $\mathbf{v}$ and $\mathbf{Hx}$.

2. On the other side, the $\mathbb{R}^n$ space can also be decomposed into two subspaces related to $\mathbf{H}$. They are depicted in two straight lines (in blue) in the left-hand side of Figure 1.2 as a 2D representation. First, the null space of $\mathbf{H}$ in $\mathbb{R}^n$ is defined as $\mathcal{N}(\mathbf{H}) \triangleq \{\mathbf{x}^0 : \mathbf{Hx}^0 = \mathbf{0}, \forall \mathbf{x}^0 \in \mathbb{R}^n\}$. The orthogonal subspace of $\mathcal{N}(\mathbf{H})$ in $\mathbb{R}^n$ is defined as

$$\mathcal{N}^\perp(\mathbf{H}) \triangleq \{\mathbf{x}^\perp : \langle \mathbf{x}^\perp, \mathbf{x}^0 \rangle = 0, \forall \mathbf{x}^0 \in \mathcal{N}(\mathbf{H}), \forall \mathbf{x}^\perp \in \mathbb{R}^n\}$$

3. Based on the property of Euclidean spaces $\mathbb{R}^n$ and $\mathbb{R}^m$ [7], one can show that $\mathbb{R}^n = \mathcal{N}(\mathbf{H}) \oplus \mathcal{N}^\perp(\mathbf{H})$ and $\mathbb{R}^m = \mathcal{R}(\mathbf{H}) \oplus \mathcal{R}^\perp(\mathbf{H})$, where $X = M \oplus N$ denotes that for every $\mathbf{x} \in X$ there is a unique representation of the form $\mathbf{x} = \mathbf{m} + \mathbf{n}$ where $\mathbf{m} \in M$ and $\mathbf{n} \in N$. Therefore, any $\mathbf{x}^0 \in \mathcal{N}(\mathbf{H})$ will map to $\{\mathbf{0}\}$ in $\mathbb{R}^m$, and any vector $\mathbf{x}^*$ in the linear manifold $\mathbf{x} + \mathcal{N}(\mathbf{H}) \triangleq \{\mathbf{x}^* : \mathbf{x}^* = \mathbf{x} + \mathbf{x}^0, \forall \mathbf{x}^0 \in \mathcal{N}(\mathbf{H})\}$ [which is depicted as a parallel line (in blue) to $\mathcal{N}(\mathbf{H})$ in Figure 1.2] will satisfy Equation (1.4-2). Therefore, the solution for $\hat{\mathbf{x}}_{\mathrm{LS}}$ (in green) is not unique; all vectors in the linear manifold will be a solution satisfying Equation (1.4-2). Although there are infinite solutions, there is a unique solution in the subspace, $\mathcal{N}^\perp(\mathbf{H})$ that has minimum norm, which will be defined in the following.

4. The adjoint operator of $\mathbf{H}$ which maps $\mathbb{R}^m$ back to $\mathbb{R}^n$ in the Euclidean norm spaces is $\mathbf{H}^T$ and the term $\mathbf{H}^T[\mathbf{HH}^T]^{-1}$ is also known as a pseudo-inverse operator of $\mathbf{H}$, such that $\mathcal{N}^\perp(\mathbf{H}) = \mathcal{R}\left(\mathbf{H}^T\left[\mathbf{HH}^T\right]^{-1}\right)$ [7].

Therefore, $\hat{\mathbf{x}}_{\mathrm{PI}}$ (in red) is unique and is orthogonal to the null space of $\mathbf{H}$, $\mathcal{N}(\mathbf{H})$, and it is the minimum norm solution for all $\mathbf{x} \in \mathbb{R}^n$ and it further satisfies the necessary condition Equation (1.4-2). It is important to note that a necessary condition that $\hat{\mathbf{x}}_{\mathrm{LS}}$ is unique when $\mathcal{N}(\mathbf{H}) = \{\mathbf{0}\}$ or equivalently that $\mathbf{H}$ has a full rank, is related to the observability condition of the estimation problem that will be discussed extensively later in this book.

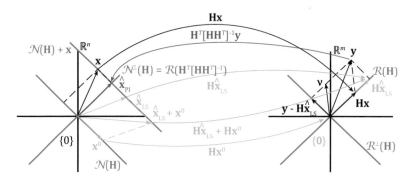

Figure 1.2   Illustration of linear operators and pseudo-inverses.

1.4.2   Weighted Least Squares Estimator

For weighted least squares estimation, the performance index shown in Equation (1.4-1) is modified to

$$J = \|\mathbf{y} - \mathbf{h}(\mathbf{x})\|_{\mathbf{R}^{-1}}^{2} = (\mathbf{y} - \mathbf{Hx})^{T}\,\mathbf{R}^{-1}(\mathbf{y} - \mathbf{Hx}). \tag{1.4-8}$$

Likewise, the necessary condition for $\mathbf{x}$ minimizing Equation (1.4-8) is

$$\mathbf{H}^{T}\mathbf{R}^{-1}\mathbf{Hx} = \mathbf{H}^{T}\mathbf{R}^{-1}\mathbf{y}. \tag{1.4-9}$$

   Similar to the LS estimator, two cases can be considered, when $m \geq n$ and when $m < n$.

1.   When $m \geq n$ and the inverse of $\mathbf{H}^{T}\mathbf{R}^{-1}\mathbf{H}$ exists, an exact solution of $\hat{\mathbf{x}}$ can be computed as

$$\hat{\mathbf{x}}_{\mathrm{WLS}} = \left[\mathbf{H}^{T}\mathbf{R}^{-1}\mathbf{H}\right]^{-1}\mathbf{H}^{T}\mathbf{R}^{-1}\mathbf{y} \tag{1.4-10}$$

where the subscript WLS denotes $\hat{\mathbf{x}}_{\mathrm{WLS}}$ as a weighted least squares estimate of $\mathbf{x}$ given $\mathbf{y}$. It can be shown (see Homework Problem 6) that $\hat{\mathbf{x}}_{\mathrm{WLS}}$ is an unbiased estimator of $\mathbf{x}$ given $\mathbf{y}$ with covariance

$$\mathbf{P}_{\mathrm{WLS}} = [\mathbf{H}^{T}\mathbf{R}^{-1}\mathbf{H}]^{-1}. \tag{1.4-11}$$

2.   When $m < n$, the inverse of $\mathbf{H}^{T}\mathbf{R}^{-1}\mathbf{H}$ does not exist, and an infinite number of solutions can satisfy the necessary condition. Therefore, there will be no unique solution for the problem posted for the performance index Equation (1.4-8). It can be shown that after selecting an appropriate linear transformation, $\mathbf{L}$, in the

measurement space (see Remark 2 of Section 1.3.1), such that $\mathbf{LL}^T = \mathbf{R}$, and the following can be derived by Equation (1.4-5) for the transformed measurement, $\mathbf{Ly}$,

$$\hat{\mathbf{x}}_{\text{WPI}} = \mathbf{H}^T \left[\mathbf{HH}^T \right]^{-1} \mathbf{y}. \qquad (1.4\text{-}12)$$

The above expression, which is identical to the non-WLS case given in Equation (1.4-5), is a solution satisfying the necessary condition, Equation (1.4-9). The subscript WPI denotes $\hat{\mathbf{x}}_{\text{WPI}}$ as a weighted pseudo-inverse estimator of $\mathbf{x}$ given $\mathbf{y}$. The same remark for Equation (1.4-5) applies here, $\hat{\mathbf{x}}_{\text{WPI}}$ is biased and the bias and error covariance are dependent on the true value of $\mathbf{x}$ as shown in Equations (1.4-6) and (1.4-7).

Remarks

1. The LS and WLS estimators for the linear measurement case form the foundation for deriving the approximate solutions for a more general nonlinear case, as will be shown later. This class of solution for the overdetermined case, although simple as they appear, is widely used in solving practical engineering problems.

2. The LS estimator is used when either (a) the measurement noise has a constant variance, that is, the measurement covariance matrix $\mathbf{R} = \sigma^2 \mathbf{I}$; or (b) the measurement noise statistics are simply not known, or $\mathbf{R}$ is not available. Satisfying condition (a) means that LS and WLS are the same. For condition (b) possible solutions become highly speculative and the right choice of estimators is always problem dependent.

3. The invertibility of the matrices $\mathbf{H}^T\mathbf{H}$ and $\mathbf{H}^T\mathbf{R}^{-1}\mathbf{H}$ for parameter estimation is the same as the observability for state estimation, and is shown in the next chapter.

   Although pseudo-inverse is a mathematically unique solution to the underdetermined problem, the results are nevertheless unsatisfying because the solution to the problems in Equations (1.4-1) and (1.4-8) is not unique as illustrated in Figure 1.2. For most of the applications of interest to the readers of this book, more independent measurements can be taken so that the solution becomes overdetermined (a necessary condition), and thus a unique solution may be found.

The derivation of multiple measurements is given in the remainder of this section. The underdetermined case is sometimes unavoidable for signal processing problems in which sufficient independent measurements simply cannot be obtained. For example, as occurs in situations of inverse convolution with a finite number of samples for system identification or input estimation for linear systems [16]. For the remainder of this book, this class of estimator will no longer be discussed.

Weighted Least Squares Estimator with Multiple Measurement Vectors

In the more general case, multiple measurements are taken on $\mathbf{x}$, that is,

$$\mathbf{y}_i = \mathbf{H}_i \mathbf{x} + \mathbf{v}_i \tag{1.4-13}$$

for $i = 1, \ldots, I$, where $\mathbf{x} \in \mathbb{R}^n$, $\mathbf{y}_i, \mathbf{v}_i \in \mathbb{R}^{m_i}$, $\mathbf{H}_i$ is an $m_i \times n$ matrix, $m_s = \sum_1^I m_i$, and the covariance of $\mathbf{v}_i$ is $\mathbf{R}_i$. The WLS estimator of $\mathbf{x}$ can be obtained in several ways. Let us first consider a direct application of the WLS estimator of Equation (1.4-9) by concatenation, that is, extending the $\mathbf{y}$ and $\mathbf{v}$ vectors and the corresponding measurement matrix.

$$\mathbf{y} = \left[\mathbf{y}_1^T, \mathbf{y}_2^T, \ldots, \mathbf{y}_I^T \right]^T$$

$$\mathbf{v} = \left[\mathbf{v}_1^T, \mathbf{v}_2^T, \ldots, \mathbf{v}_I^T \right]^T$$

$$\mathbf{H} = \left[\mathbf{H}_1^T, \mathbf{H}_2^T, \ldots, \mathbf{H}_I^T \right]^T$$

Then

$$\mathbf{y} = \mathbf{H}\mathbf{x} + \mathbf{v} \tag{1.4-14}$$

where $\mathbf{x} \in \mathbb{R}^n$ and $\mathbf{y}, \mathbf{v} \in \mathbb{R}^{m_s}$, $\mathbf{H}$ is an $m_s \times n$ matrix where $m_s = \sum_1^I m_i$, and the covariance of $\mathbf{v}$ is represented by a block diagonal matrix with submatrices $\mathbf{R}_i$, and where the $\mathbf{0}$'s represent matrices with zero entries and with appropriate dimensions,

$$\mathbf{R} = \begin{bmatrix} \mathbf{R}_1 & \mathbf{0} & \cdot & \cdot & \mathbf{0} \\ \mathbf{0} & \mathbf{R}_2 & \cdot & \cdot & \mathbf{0} \\ \cdot & \cdot & \cdot & \cdot & \cdot \\ \cdot & \cdot & \cdot & \cdot & \cdot \\ \mathbf{0} & \mathbf{0} & \cdot & \cdot & \mathbf{R}_I \end{bmatrix}$$

Then the WLS estimate of $\mathbf{x}$ is

$$\hat{\mathbf{x}}_{\mathrm{WLS}} = \left[\mathbf{H}^T \mathbf{R}^{-1} \mathbf{H} \right]^{-1} \mathbf{H}^T \mathbf{R}^{-1} \mathbf{y} \tag{1.4-15}$$

with covariance

$$\mathbf{P}_{\mathrm{WLS}} = [\mathbf{H}^T \mathbf{R}^{-1} \mathbf{H}]^{-1}. \tag{1.4-16}$$

Similar to the discussion for the WLS, the above result is valid only when $\mathbf{H}^T\mathbf{R}^{-1}\mathbf{H}$ is invertible or $m_s > n$, which is a necessary condition that can be satisfied in most applications.

An alternative approach is to process a sub-block of the measurement vectors at a time when the associated sub-block of $\mathbf{H}^T\mathbf{R}^{-1}\mathbf{H}$ is invertible. The smallest data block is a single measurement vector, $\mathbf{y}_i$. This leads to two possible cases, first, when $\mathbf{H}_i^T\mathbf{R}_i^{-1}\mathbf{H}_i$ is invertible, and second, when $\mathbf{H}_i^T\mathbf{R}_i^{-1}\mathbf{H}_i$ is not invertible or when $\mathbf{y}_i$ has smaller dimension than x.

1. When $\mathbf{H}_i^T\mathbf{R}_i^{-1}\mathbf{H}_i$ is invertible for all i, then one can derive an alternative formulation (a parallel processing architecture) for computing $\hat{\mathbf{x}}_{\text{WLS}}$ and $\mathbf{P}_{\text{WLS}}$. Let $\hat{\mathbf{x}}_{\text{WLS}}^i$ denote the WLS estimate of $\mathbf{x}$ using $\mathbf{y}_i$ with covariance $\mathbf{P}_{\text{WLS}}^i$. Applying Equations (1.4-10) and (1.4-11) one obtains

$$\hat{\mathbf{x}}_{\text{WLS}}^i = \left[\mathbf{H}_i^T\mathbf{R}_i^{-1}\mathbf{H}_i\right]^{-1}\mathbf{H}_i^T\mathbf{R}_i^{-1}\mathbf{y}_i \qquad (1.4\text{-}17)$$

$$\mathbf{P}_{\text{WLS}}^i = \left[\mathbf{H}_i^T\mathbf{R}_i^{-1}\mathbf{H}_i\right]^{-1}. \qquad (1.4\text{-}18)$$

One can show that the final estimate and covariance, $\hat{\mathbf{x}}_{\text{WLS}}$ and $\mathbf{P}_{\text{WLS}}$, can be computed with

$$\hat{\mathbf{x}}_{\text{WLS}} = \left[\sum_{i=1}^{I}\mathbf{P}_{\text{WLS}}^{i\ -1}\right]^{-1}\left(\sum_{i=1}^{I}\mathbf{P}_{\text{WLS}}^{i\ -1}\hat{\mathbf{x}}_{\text{WLS}}^i\right) \qquad (1.4\text{-}19)$$

$$\mathbf{P}_{\text{WLS}} = \left[\sum_{i=1}^{I}\mathbf{P}_{\text{WLS}}^{i\ -1}\right]^{-1}. \qquad (1.4\text{-}20)$$

2. When $\mathbf{H}_i^T\mathbf{R}_i^{-1}\mathbf{H}_i$ is not invertible, or $\mathbf{y}_i$ has smaller dimension than $\mathbf{x}$. In this case, a sub-block of data may contain multiple $\mathbf{y}_i$'s and the associated sub-block matrix of the form $\mathbf{H}^T\mathbf{R}^{-1}\mathbf{H}$ is invertible. WLS estimates of $\mathbf{x}$ can be obtained by first applying the WLS algorithm to process these sub-blocks of data; the final estimate is then obtained using Equations (1.4-19) and (1.4-20). This is a very important case because in most applications, each individual measurement vector, $\mathbf{y}_i$, is smaller in dimension than $\mathbf{x}$. Vector and matrix concatenation provides a way to obtain estimates based on sub-blocks of data first, and the final estimate is then obtained by processing the estimates obtained using the sub-blocks, then applying Equations (1.4-19) and (1.4-20). The sizes of these sub-blocks are dependent on the applications. We will later apply this idea to data compression for multiple sensor tracking problems in Chapter 7.

In Homework Problem 7, you are asked to derive Equations (1.4-19) and (1.4-20) from Equations (1.4-17) and (1.4-18).

1.4.3   Maximum Likelihood Estimator

The ML estimate of $\mathbf{x}$ given $\mathbf{y}$ is the $\mathbf{x}$ that gives the maximum of the likelihood density function,

$$J = p(\mathbf{y}|\mathbf{x}). \tag{1.4-21}$$

With Gaussian assumption, the likelihood function is[6]

$$p(\mathbf{y}|\mathbf{x}) = \frac{1}{(2\pi)^{m/2}|\mathbf{R}|^{1/2}} \exp\left\{-\frac{1}{2}(\mathbf{y} - \mathbf{Hx})^T \mathbf{R}^{-1}(\mathbf{y} - \mathbf{Hx})\right\}.$$

Taking the natural log of J, and defining $J*$ as

$$J* = -\ln(p(\mathbf{y}|\mathbf{x}))$$

yields

$$J* = \frac{1}{2}(\mathbf{y} - \mathbf{Hx})^T \mathbf{R}^{-1}(\mathbf{y} - \mathbf{Hx}) + \frac{m}{2}\ln(2\pi) + \frac{1}{2}\ln(|\mathbf{R}|).$$

Notice that maximizing J is the same as minimizing $J*$. Since the last two terms of $J*$ do not depend on $\mathbf{x}$, the maximum likelihood estimate of $\mathbf{x}$ is the same as the WLS estimate of $\mathbf{x}$ given $\mathbf{y}$, that is,

$$\hat{\mathbf{x}}_{ML} = \hat{\mathbf{x}}_{WLS} = \left[\mathbf{H}^T\mathbf{R}^{-1}\mathbf{H}\right]^{-1}\mathbf{H}^T\mathbf{R}^{-1}\mathbf{y} \tag{1.4-22}$$

with the same covariance,

$$\mathbf{P}_{ML} = \mathbf{P}_{WLS} = [\mathbf{H}^T\mathbf{R}^{-1}\mathbf{H}]^{-1} \tag{1.4-23}$$

where subscript ML indicates maximum likelihood estimate.

Remarks

The WLS and ML estimators are the same because the measurement noise is Gaussian. When this is not true, the two estimators differ. The LS or WLS estimators do not assume a distribution function and are less model dependent, thus they are generally regarded as more robust in solving practical problems.

---

6 |R| denotes the determinant of $\mathbf{R}$.

Table 1.2   Summary of Estimation Algorithms: Linear and Gaussian, with Constant Parameter

---

Estimation Algorithms: Linear and Gaussian, with Constant Parameter
Problem Definition
Estimate an unknown constant vector $\mathbf{x}$ with

$$\mathbf{y} = \mathbf{Hx} + \mathbf{v}$$

where $\mathbf{x} \in \mathbb{R}^n$, $\mathbf{y}$, $\mathbf{v} \in \mathbb{R}^m$, $\mathbf{v} \sim N(\mathbf{0}, \mathbf{R})$

Least Squares Estimator

$$\hat{\mathbf{x}}_{LS} = \left[\mathbf{H}^T\mathbf{H}\right]^{-1}\mathbf{H}^T\mathbf{y}$$

$$\mathbf{P}_{LS} = [\mathbf{H}^T\mathbf{H}]^{-1}[\mathbf{H}^T\mathbf{R}\mathbf{H}][\mathbf{H}^T\mathbf{H}]^{-1}$$

Weighted Least Squares and Maximum Likelihood Estimator

$$\hat{\mathbf{x}}_{ML} = \hat{\mathbf{x}}_{WLS} = \left[\mathbf{H}^T\mathbf{R}^{-1}\mathbf{H}\right]^{-1}\mathbf{H}^T\mathbf{R}^{-1}\mathbf{y}$$

$$\mathbf{P}_{ML} = \mathbf{P}_{WLS} = [\mathbf{H}^T\mathbf{R}^{-1}\mathbf{H}]^{-1}$$

Table 1.2 provides a summary of the basic estimation algorithms derived in this section. The extension to the multiple measurement vector case is not included.

1.5   Estimator Derivation: Linear and Gaussian, Random Parameter

Now consider the case that in addition to noisy measurements on a random parameter $\mathbf{x}$, there also exists a priori knowledge about it. Let $\mathbf{x}$ be a random vector with a Gaussian distribution with mean $\bar{\mathbf{x}}_0$ and covariance $\mathbf{P}_0$,

$$\mathbf{x} \sim N(\bar{\mathbf{x}}_0, \mathbf{P}_0),$$

and further let $\mathbf{x}$ be independent of the noise vector $\mathbf{v}$. With the addition of prior knowledge on $\mathbf{x}$, this problem will never be underdetermined because $\mathbf{P}_0$ is positive definite and the number of equations will always be larger than the number of unknowns. The estimate of $\mathbf{x}$ will be derived initially by incorporating the prior condition of $\mathbf{x}$ into the performance index. Later it will be shown that this is the same as treating the prior condition as an initial estimate of $\mathbf{x}$ with new measurements made on $\mathbf{x}$ in the form of $\mathbf{y} = \mathbf{Hx} + \mathbf{v}$ as new information about $\mathbf{x}$. Let us now consider all estimators individually as before. An alternative derivation of estimators using the definition from Equation (1.3-13) is given in Appendix 1.B.

1.5.1   Least Squares Estimator

The performance index of LS estimator with the addition of a priori knowledge on $\mathbf{x}$ can be written as

$$J = (\mathbf{y} - \mathbf{Hx})^T (\mathbf{y} - \mathbf{Hx}) + (\mathbf{x} - \overline{\mathbf{x}}_0)^T (\mathbf{x} - \overline{\mathbf{x}}_0). \tag{1.5-1}$$

The objective is to find the $\mathbf{x}$, which minimizes J. Taking the derivative of J with respect to $\mathbf{x}$ and setting it to zero yields the necessary condition for finding $\mathbf{x}$, that is,

$$\frac{\partial J}{\partial \mathbf{x}} = -\mathbf{H}^T (\mathbf{y} - \mathbf{Hx}) - (\mathbf{x} - \overline{\mathbf{x}}_0) = \mathbf{0}$$

or

$$\left[\mathbf{H}^T \mathbf{H} + \mathbf{I}\right]\mathbf{x} = \mathbf{H}^T \mathbf{y} + \overline{\mathbf{x}}_0.$$

Note that in this case, $[\mathbf{H}^T\mathbf{H} + \mathbf{I}]$ is nonsingular because of the addition of the identity matrix $\mathbf{I}$. We therefore have the LS estimate of $\mathbf{x}$ as

$$\hat{\mathbf{x}}_{LS} = \left[\mathbf{H}^T \mathbf{H} + \mathbf{I}\right]^{-1}\left(\mathbf{H}^T\mathbf{y} + \overline{\mathbf{x}}_0\right) \tag{1.5-2}$$

where the subscript LS denotes that $\hat{\mathbf{x}}_{LS}$ is the least squares estimate. The above expression takes on a very intuitive form: the estimate is obtained as an average of two pieces of information, namely, $\mathbf{y}$ and $\overline{\mathbf{x}}_0$. The average is carried out in the $\mathbf{x}$ space, thus the transformation of $\mathbf{y}$ using $\mathbf{H}^T$. The matrix $[\mathbf{H}^T\mathbf{H} + \mathbf{I}]^{-1}$ is a normalization factor, the same as dividing the sum of all measurements by the total number of samples as in averaging. Another form of the LS estimate uses the new information $\mathbf{y}$ as a correction to the prior information $\overline{\mathbf{x}}_0$. This takes the general form of

$$\hat{\mathbf{x}} = \overline{\mathbf{x}}_0 + \mathbf{K}\left(\mathbf{y} - \mathbf{H}\overline{\mathbf{x}}_0\right)$$

where the matrix $\mathbf{K}$ is generally referred to as a filter gain and the term $(\mathbf{y} - \mathbf{H}\overline{\mathbf{x}}_0)$ is referred to as the measurement residual, that is, the new information provided by the measurement. The concept of residual is important for filtering processing in general and is discussed in detail as part of the state estimation process covered in the next chapter. It can be shown (see Homework Problem 8) that

$$\hat{\mathbf{x}}_{LS} = \overline{\mathbf{x}}_0 + \left[\mathbf{H}^T \mathbf{H} + \mathbf{I}\right]^{-1} \mathbf{H}^T (\mathbf{y} - \mathbf{H}\overline{\mathbf{x}}_0). \tag{1.5-3}$$

The filter gain for the LS estimator, $\mathbf{K}_{LS}$ is

$$\mathbf{K}_{LS} = [\mathbf{H}^T\mathbf{H} + \mathbf{I}]^{-1}\mathbf{H}^T.$$

Furthermore, the LS estimator $\hat{\mathbf{x}}_{LS}$ is unbiased with covariance

$$\mathbf{P}_{LS} = [\mathbf{I} - \mathbf{K}_{LS}\mathbf{H}]\mathbf{P}_0[\mathbf{I} - \mathbf{K}_{LS}\mathbf{H}]^T + \mathbf{K}_{LS}\mathbf{R}\mathbf{K}_{LS}^T. \tag{1.5-4}$$

1.5.2   Weighted Least Squares Estimator

For WLS estimation, the performance index, Equation (1.4-8), is modified to be

$$J = (\mathbf{y} - \mathbf{H}\mathbf{x})^T \mathbf{R}^{-1}(\mathbf{y} - \mathbf{H}\mathbf{x}) + (\mathbf{x} - \overline{\mathbf{x}}_0)^T \mathbf{P}_0^{-1}(\mathbf{x} - \overline{\mathbf{x}}_0). \tag{1.5-5}$$

Taking the derivative of J with respect to $\mathbf{x}$ and setting it to zero yields the modified necessary condition

$$\left[\mathbf{H}^T\mathbf{R}^{-1}\mathbf{H} + \mathbf{P}_0^{-1}\right]\mathbf{x} = \mathbf{H}^T\mathbf{R}^{-1}\mathbf{y} + \mathbf{P}_0^{-1}\overline{\mathbf{x}}_0.$$

The WLS estimate of $\mathbf{x}$ is then

$$\hat{\mathbf{x}}_{WLS} = \left[\mathbf{H}^T\mathbf{R}^{-1}\mathbf{H} + \mathbf{P}_0^{-1}\right]^{-1}\left(\mathbf{H}^T\mathbf{R}^{-1}\mathbf{y} + \mathbf{P}_0^{-1}\overline{\mathbf{x}}_0\right). \tag{1.5-6}$$

Similar observations to the LS estimator can be made here. The above expression makes intuitive sense: the estimate is obtained as a weighted sum of two pieces of information, namely, $\mathbf{y}$ and $\mathbf{x}$. The weighting matrix is their corresponding covariance matrix with proper coordinate transformation (see Remark 2 in Section 1.3.1, there exist nonsingular matrices $\mathbf{L}$ and $\mathbf{G}$ such that $\mathbf{R} = \mathbf{L}\mathbf{L}^T$ and $\mathbf{P}_0 = \mathbf{G}\mathbf{G}^T$). The matrix $\left[\mathbf{H}^T\mathbf{R}^{-1}\mathbf{H} + \mathbf{P}_0^{-1}\right]$ is the normalization factor of the weighted sum. Similar to the LS case, the alternative form of the WLS estimator, uses the new information $\mathbf{y}$ as a correction to the prior information $\overline{\mathbf{x}}_0$, that is,

$$\hat{\mathbf{x}} = \overline{\mathbf{x}}_0 + \mathbf{K}(\mathbf{y} - \mathbf{H}\overline{\mathbf{x}}_0), \tag{1.5-7}$$

where $\mathbf{K}$ is usually referred to as the filter gain. It can be shown (Homework Problem 9) that the WLS estimate is

$$\hat{\mathbf{x}}_{WLS} = \overline{\mathbf{x}}_0 + \mathbf{P}_0\mathbf{H}^T\left[\mathbf{H}\mathbf{P}_0\mathbf{H}^T + \mathbf{R}\right]^{-1}(\mathbf{y} - \mathbf{H}\overline{\mathbf{x}}_0) \tag{1.5-8}$$

and that $\hat{\mathbf{x}}_{WLS}$ is unbiased with covariance

$$\mathbf{P}_{WLS} = \mathbf{P}_0 - \mathbf{P}_0\mathbf{H}^T\left[\mathbf{H}\mathbf{P}_0\mathbf{H}^T + \mathbf{R}\right]^{-1}\mathbf{H}\mathbf{P}_0 = \left[\mathbf{H}^T\mathbf{R}^{-1}\mathbf{H} + \mathbf{P}_0^{-1}\right]^{-1}. \tag{1.5-9}$$

The last equality can be obtained by using the Matrix Inversion Lemma (see Appendix A). The equation of choice for covariance computation is usually dependent on the size of the matrix to be inverted.

Remarks

1. In this section, it was shown that prior knowledge on $\mathbf{x}$ ($\overline{\mathbf{x}}_0$ and $\mathbf{P}_0$) is used in the LS and WLS estimators. The new estimate (either LS or WLS) is a function of prior knowledge and the measurement, and can be expressed as a sum of the prior knowledge of $\mathbf{x}$, $\overline{\mathbf{x}}_0$, and a weighted correction, $\mathbf{y} - \mathbf{H}\overline{\mathbf{x}}_0$, incorporating the new measurement $\mathbf{y}$.

2. The formulation of LS and WLS estimators using the correction term (residual and filter gain) is fundamental for state estimation when $\mathbf{x}$ is a random process. It will be shown in Chapter 2 that the discrete Kalman filter is the WLS estimator of a random process $\mathbf{x}$ takes exactly this form. All the associated derivations for filter and covariance equations can be readily extended to the state estimation when $\mathbf{x}$ is a random process.

3. Similar to the case when $\mathbf{x}$ is an unknown vector, the LS estimator is used when either (a) both the measurement noise covariance and the prior covariance are the same known constants; or (b) they are unknown. Satisfying condition (a) means that LS and WLS are the same. Condition (b) presents a very unsatisfactory condition because it says that one simply does not know the value of the prior knowledge relative to the new measurement. Should this be the practical problem that one attempts to solve, the choice of possible solutions becomes highly speculative, and is problem dependent.

4. The invertibility of $\mathbf{H}^T\mathbf{H}+\mathbf{I}$ and $\mathbf{H}^T\mathbf{R}^{-1}\mathbf{H}+\mathbf{P}_0^{-1}$ is not an issue here because the sum of positive semidefinite matrices, $\mathbf{H}^T\mathbf{H}$, and $\mathbf{H}^T\mathbf{R}^{-1}\mathbf{H}$ with positive-definite matrices, $\mathbf{I}$ and $\mathbf{P}_0$, respectively, is always positive definite. In later sections, it will be shown that $\mathbf{H}^T\mathbf{R}^{-1}\mathbf{H}$ and $\mathbf{H}^T\mathbf{R}^{-1}\mathbf{H}+\mathbf{P}_0^{-1}$ are the Fisher information matrices associated with the problems of $\mathbf{x}$, without and with prior knowledge, respectively, and their inverses are the Cramer–Rao bounds for estimation error covariance.

Recursive Weighted Least Squares Estimator with Multiple Measurement Vectors

Similar to the formulation in Section 1.5.2, consider the case

$$\mathbf{y}_i = \mathbf{H}_i\mathbf{x} + \mathbf{v}_i$$

for $i = 1...I$, where $\mathbf{x} \in \mathbb{R}^n$, $\mathbf{y}_i$, $\mathbf{v}_i \in \mathbb{R}^m$, $\mathbf{H}_i$ is an $m_i \times n$ matrix, $m_s = \sum_1^I m_i$, and $\mathbf{x}$ has a known a priori distribution $\mathbf{x} \sim N(\overline{\mathbf{x}}_0, \mathbf{P}_0)$, $\mathbf{v}_i: \sim N(\mathbf{0}, \mathbf{R}_i)$ and $\mathbf{x}$ and $\mathbf{v}_i$ are mutually independent. Let $\hat{\mathbf{x}}_{\text{WLS}}^1$ denote the WLS estimate of $\mathbf{x}$ using $\mathbf{y}_1$ with covariance $\mathbf{P}_{\text{WLS}}^1$. Applying the WLS estimator formulation in Equations (1.5-8) and (1.5-9) one obtains the results

$$\hat{\mathbf{x}}_{\text{WLS}}^1 = \overline{\mathbf{x}}_0 + \mathbf{P}_0\mathbf{H}_1^T \left[\mathbf{H}_1\mathbf{P}_0\mathbf{H}_1^T + \mathbf{R}_1\right]^{-1} (\mathbf{y}_1 - \mathbf{H}_1\overline{\mathbf{x}}_0) \tag{1.5-10}$$

and

$$\mathbf{P}_{\mathrm{WLS}}^1 = \mathbf{P}_0 - \mathbf{P}_0 \mathbf{H}_1^T \left[\mathbf{H}_1 \mathbf{P}_0 \mathbf{H}_1^T + \mathbf{R}_1 \right]^{-1}. \qquad (1.5\text{-}11)$$

The above expression can now be extended to a recursive algorithm,

$$\hat{\mathbf{x}}_{\mathrm{WLS}}^i = \hat{\mathbf{x}}_{\mathrm{WLS}}^{i-1} + \mathbf{P}_{\mathrm{WLS}}^{i-1} \mathbf{H}_i^T \left[\mathbf{H}_i \mathbf{P}_{\mathrm{WLS}}^{i-1} \mathbf{H}_i^T + \mathbf{R}_i \right]^{-1} \left(\mathbf{y}_i - \mathbf{H}_i \hat{\mathbf{x}}_{\mathrm{WLS}}^{i-1} \right) \qquad (1.5\text{-}12)$$

and

$$\mathbf{P}_{\mathrm{WLS}}^i = \mathbf{P}_{\mathrm{WLS}}^{i-1} - \mathbf{P}_{\mathrm{WLS}}^{i-1} \mathbf{H}_i^T \left[\mathbf{H}_i \mathbf{P}_{\mathrm{WLS}}^{i-1} \mathbf{H}_i^T + \mathbf{R}_i \right]^{-1} \mathbf{H}_i \mathbf{P}_{\mathrm{WLS}}^{i-1} \qquad (1.5\text{-}13)$$

for $i = 1 \dots I$ with initial conditions $\hat{\mathbf{x}}_{\mathrm{WLS}}^0 = \bar{\mathbf{x}}_0$, $\mathbf{P}_{\mathrm{WLS}}^1 = \mathbf{P}_0$ and final estimates $\hat{\mathbf{x}}_{\mathrm{WLS}} = \hat{\mathbf{x}}_{\mathrm{WLS}}^I$ and $\mathbf{P}_{\mathrm{WLS}} = \mathbf{P}_{\mathrm{WLS}}^I$.

Going back to Section 1.4.2, consider the case where the initial WLS estimate is obtained with the smallest sub-block of data, and then the subsequent data vectors can be processed using the recursive update algorithm derived in this section (see Homework Problem 10).

1.5.3   Maximum a Posteriori Probability Estimator

The maximum a posteriori probability estimate is the $\mathbf{x}$ that maximizes the performance index

$$J = p(\mathbf{x}|\mathbf{y}).$$

Applying Bayes' rule, one obtains,

$$p(\mathbf{x}|\mathbf{y}) = \frac{p(\mathbf{y}|\mathbf{x})}{p(\mathbf{y})} p(\mathbf{x}).$$

Taking the natural log of J, and defining J^* as

$$J^* = -\ln(p(\mathbf{x}|\mathbf{y})) = -\ln(p(\mathbf{y}|\mathbf{x})) - \ln(p(\mathbf{x})) + \ln(p(\mathbf{y})).$$

With Gaussian assumption, the following distribution functions are true,

$$p(\mathbf{y}|\mathbf{x}) = \frac{1}{(2\pi)^{m/2} |\mathbf{R}|^{1/2}} \exp\left\{ -\frac{1}{2} (\mathbf{y} - \mathbf{H}\mathbf{x})^T \mathbf{R}^{-1} (\mathbf{y} - \mathbf{H}\mathbf{x}) \right\}, \qquad (1.5\text{-}14)$$

$$p(\mathbf{x}) = \frac{1}{(2\pi)^{n/2} |\mathbf{P}_0|^{1/2}} \exp\left\{ -\frac{1}{2} (\mathbf{x} - \bar{\mathbf{x}}_0)^T \mathbf{P}_0^{-1} (\mathbf{x} - \bar{\mathbf{x}}_0) \right\}, \qquad (1.5\text{-}15)$$

$$p(\mathbf{y}) = \frac{1}{(2\pi)^{m/2} \left| \mathbf{H}\mathbf{P}_0\mathbf{H}^T + \mathbf{R} \right|^{1/2}} \exp\left\{ -\frac{1}{2}(\mathbf{y} - \mathbf{H}\bar{\mathbf{x}}_0)^T \left[\mathbf{H}\mathbf{P}_0\mathbf{H}^T + \mathbf{R} \right]^{-1} (\mathbf{y} - \mathbf{H}\bar{\mathbf{x}}_0) \right\}. \quad (1.5\text{-}16)$$

Substituting Equations (1.5-14) through (1.5-16) into J^* results in

$$J^* = \frac{1}{2}(\mathbf{y} - \mathbf{H}\mathbf{x})^T \mathbf{R}^{-1}(\mathbf{y} - \mathbf{H}\mathbf{x}) + \frac{1}{2}(\mathbf{x} - \bar{\mathbf{x}}_0)^T \mathbf{P}_0^{-1}(\mathbf{x} - \bar{\mathbf{x}}_0) + \text{constants}$$

where *constants* denote all terms that are independent of $\mathbf{x}$. Maximizing J is the same as minimizing J^*. Since the last two terms of J^* do not depend on $\mathbf{x}$, the MAP estimate is identical to the WLS estimate, that is,

$$\hat{\mathbf{x}}_{\text{MAP}} = \hat{\mathbf{x}}_{\text{WLS}} = \left[\mathbf{H}^T\mathbf{R}^{-1}\mathbf{H} + \mathbf{P}_0^{-1} \right]^{-1} \left(\mathbf{H}^T\mathbf{R}^{-1}\mathbf{y} + \mathbf{P}_0^{-1}\bar{\mathbf{x}}_0 \right)$$

$$= \bar{\mathbf{x}}_0 + \mathbf{P}_0\mathbf{H}^T \left[\mathbf{H}\mathbf{P}_0\mathbf{H}^T + \mathbf{R} \right]^{-1} (\mathbf{y} - \mathbf{H}\bar{\mathbf{x}}_0) \quad (1.5\text{-}17)$$

with the same covariance,

$$\mathbf{P}_{\text{MAP}} = \mathbf{P}_{\text{WLS}} = \mathbf{P}_0 - \mathbf{P}_0\mathbf{H}^T \left[\mathbf{H}\mathbf{P}_0\mathbf{H}^T + \mathbf{R} \right]^{-1} \mathbf{H}\mathbf{P}_0 = \left[\mathbf{H}^T\mathbf{R}^{-1}\mathbf{H} + \mathbf{P}_0^{-1} \right]^{-1} \quad (1.5\text{-}18)$$

where subscript MAP indicates maximum likelihood estimate.

Remarks

1. The WLS and MAP estimators are the same because of both $\mathbf{x}$ and the measurement noise $\mathbf{v}$ are mutually independent Gaussian distributed random vectors and the observation function is linear. When the observation function is not linear, the WLS and MAP estimators are not equivalent, in general.

2. When the densities are not Gaussian, the mean and covariance of the estimator cannot be characterized easily because it requires the a posteriori density function $p(\mathbf{x}|\mathbf{y})$ to perform the integrals. Various numerical techniques and algorithms have been presented in the literature and we will discuss some of them later in this book to obtain an approximated a posteriori density function $p(\mathbf{x}|\mathbf{y})$ and compute the mean and covariance. Monte Carlo simulation approaches were introduced in the 1950s [17] to compute these integrals involving this conditional density function, which is a function of $\mathbf{y}$, a random vector. Not until the late 1990s did these techniques gain considerable popularity due to the availability of faster computers and the development of particle filters for state estimation [18]. The Monte Carlo simulation techniques (i.e., the particle filter) are discussed in Chapter 6 after the Bayesian approach for state estimation has been developed.

1.5.4   Conditional Mean Estimator

The CM estimator is the statistical mean of $\mathbf{x}$ with the conditional density of $\mathbf{x}$ given $\mathbf{y}$, that is,

$$\hat{\mathbf{x}}_{CM} = E\{\mathbf{x}|\mathbf{y}\} = \int \mathbf{x} p(\mathbf{x}|\mathbf{y}) d\mathbf{x}.$$

Following the properties of $\hat{\mathbf{x}}_{CM}$ in Section 1.3.3, it is the minimum variance (least squares) estimator in $\mathbb{R}^n$ that minimizes Equation (1.3-13) and is unbiased. Let the function $\mathbf{g}(\mathbf{y}) \in \mathbb{R}^n$ be a linear estimator of $\mathbf{x}$ given $\mathbf{y}$, such that

$$\mathbf{g}(\mathbf{y}) = \mathbf{A}\mathbf{y} + \mathbf{b},$$

where $\mathbf{A}$ is a constant $n \times m$ matrix and $\mathbf{b}$ is a constant vector in $\mathbb{R}^n$. Following Property 3 of the LLS estimator, $\hat{\mathbf{x}}_{LLS}$ (see Section 1.3.3), concludes $\hat{\mathbf{x}}_{CM} = \hat{\mathbf{x}}_{LLS}$ when $\mathbf{x}$ and $\mathbf{y}$ are jointly Gaussian. Furthermore, with the optimum solution for $\mathbf{A}$ and $\mathbf{b}$ this yields (see Appendix 1.B)

$$\hat{\mathbf{x}}_{CM} = \hat{\mathbf{x}}_{LLS} = \bar{\mathbf{x}} + \mathbf{P}_{xy}\mathbf{P}_{yy}^{-1}(\mathbf{y} - \bar{\mathbf{y}}) \tag{1.5-19}$$

and

$$\mathbf{P}_{CM} = \mathbf{P}_{LLS} = Cov\{\tilde{\mathbf{x}}_{LLS}\} = \mathbf{P}_{xx} - \mathbf{P}_{xy}\mathbf{P}_{yy}^{-1}\mathbf{P}_{yx} \tag{1.5-20}$$

where

$$\bar{\mathbf{x}} = E\{\mathbf{x}\} = \bar{\mathbf{x}}_0,$$

$$\bar{\mathbf{y}} = E\{\mathbf{y}\} = \mathbf{H}\bar{\mathbf{x}}_0,$$

$$\mathbf{P}_{yy} = Cov\{\mathbf{y}, \mathbf{y}\} = \mathbf{H}\mathbf{P}_0\mathbf{H}^T + \mathbf{R}, \text{ and}$$

$$\mathbf{P}_{xy} = Cov\{\mathbf{x}, \mathbf{y}\} = \mathbf{P}_0\mathbf{H}^T.$$

Then, substituting the above into Equations (1.5-19) and (1.5-20) yields

$$\hat{\mathbf{x}}_{CM} = \hat{\mathbf{x}}_{LLS} = \bar{\mathbf{x}}_0 + \mathbf{P}_0\mathbf{H}^T\left[\mathbf{H}\mathbf{P}_0\mathbf{H}^T + \mathbf{R}\right]^{-1}(\mathbf{y} - \mathbf{H}\bar{\mathbf{x}}_0)$$
$$= \left[\mathbf{H}^T\mathbf{R}^{-1}\mathbf{H} + \mathbf{P}_0^{-1}\right]^{-1}\left(\mathbf{H}^T\mathbf{R}^{-1}\mathbf{y} + \mathbf{P}_0^{-1}\bar{\mathbf{x}}_0\right) \tag{1.5-21}$$

and

$$\mathbf{P}_{CM} = \mathbf{P}_{LLS} = \mathbf{P}_0 - \mathbf{P}_0\mathbf{H}^T\left[\mathbf{H}\mathbf{P}_0\mathbf{H}^T + \mathbf{R}\right]^{-1}\mathbf{H}\mathbf{P}_0 = \left[\mathbf{H}^T\mathbf{R}^{-1}\mathbf{H} + \mathbf{P}_0^{-1}\right]^{-1}. \tag{1.5-22}$$

Therefore, comparing Equations (1.5-17) and (1.5-18) with Equations (1.5-21) and (1.5-22), we have

$$\hat{\mathbf{x}}_{CM} = \hat{\mathbf{x}}_{LLS} = \hat{\mathbf{x}}_{WLS} = \hat{\mathbf{x}}_{MAP}.$$

Remarks

1. When the linear and Gaussian assumption is not true, the solutions are not the same. The approximate solution for the estimators with nonlinear and/or non-Gaussian problems can be derived using the methodology introduced in this chapter. The next section introduces an iterative solution for the WLS estimator with nonlinear measurements. Sometimes numerical approximation to a solution may be applied and the idea of Monte Carlo simulation, for example, particle filter [17, 18] is one such technique.

2. When there is no a priori knowledge about the random vector, $\mathbf{x}$, that is the same as setting $\mathbf{P}_0^{-1} = 0$ (no information), and one obtains the same equations as in the previous section.

Table 1.3 provides a summary of basic estimation algorithms for estimating a Gaussian parameter vector with linear measurements. The extension to the multiple measurement vector case shown in Section 1.5.2 is not included.

1.6   Nonlinear Measurement with Jointly Gaussian Distributed Noise and Random Parameter

Now consider the case in which the random vector $\mathbf{x}$ and measurement vector $\mathbf{y}$ are related through a known but nonlinear relationship denoted as

$$\mathbf{y} = \mathbf{h}(\mathbf{x}) + \mathbf{v}.$$

In this case, only the WLS estimator will be considered. Generalization to the other estimators will not be covered here. Interested readers can derive their own solutions with the techniques illustrated. Consider the performance index to be minimized as

$$J = (\mathbf{y} - \mathbf{h}(\mathbf{x}))^T \mathbf{R}^{-1} (\mathbf{y} - \mathbf{h}(\mathbf{x})) + (\mathbf{x} - \overline{\mathbf{x}}_0)^T \mathbf{P}_0^{-1} (\mathbf{x} - \overline{\mathbf{x}}_0). \tag{1.6-1}$$

Taking the derivative of J with respect to $\mathbf{x}$ and setting it to zero is the necessary condition for the estimate to satisfy

Table 1.3   Summary of Parameter Estimation: Linear and Gaussian, with Random Parameter

---

Estimation Algorithms: Linear and Gaussian, with Random Parameter
Problem Definition
Estimate a random vector $\mathbf{x}$ with observations

$$\mathbf{y} = \mathbf{H}\mathbf{x} + \mathbf{v}$$

where $\mathbf{x} \in \mathbb{R}^n$, $\mathbf{y}, \mathbf{v} \in \mathbb{R}^m$, $\mathbf{v} \colon \sim N(\mathbf{0}, \mathbf{R})$, $\mathbf{x} \colon \sim N(\overline{\mathbf{x}}_0, \mathbf{P}_0)$ and $\mathbf{x}$ and $\mathbf{v}$ are mutually independent.
Least Squares Estimator

$$\hat{\mathbf{x}}_{\mathrm{LS}} = \overline{\mathbf{x}}_0 + \left(\mathbf{H}^T\mathbf{H} + \mathbf{I}\right)^{-1}\mathbf{H}^T\left(\mathbf{y} - \mathbf{H}\overline{\mathbf{x}}_0\right)$$

$$\mathbf{P}_{\mathrm{LS}} = (\mathbf{I} - \mathbf{K}_{\mathrm{LS}}\mathbf{H})\mathbf{P}_0(\mathbf{I} - \mathbf{K}_{\mathrm{LS}}\mathbf{H})^T + \mathbf{K}_{\mathrm{LS}}\mathbf{R}\mathbf{K}_{\mathrm{LS}}^T$$

$$\mathbf{K}_{\mathrm{LS}} = [\mathbf{H}^T\mathbf{H} + \mathbf{I}]^{-1}\mathbf{H}^T$$

Weighted Least Squares, Maximum a Posteriori Probability Estimator, and the Conditional Mean Estimator

$$\hat{\mathbf{x}} = \hat{\mathbf{x}}_{\mathrm{CM}} = \hat{\mathbf{x}}_{\mathrm{MAP}} = \hat{\mathbf{x}}_{\mathrm{WLS}} = \left(\mathbf{H}^T\mathbf{R}^{-1}\mathbf{H} + \mathbf{P}_0^{-1}\right)^{-1}\left(\mathbf{H}^T\mathbf{R}^{-1}\mathbf{y} + \mathbf{P}_0^{-1}\overline{\mathbf{x}}_0\right)$$

$$\mathbf{P} = \mathbf{P}_{\mathrm{CM}} = \mathbf{P}_{\mathrm{MAP}} = \mathbf{P}_{\mathrm{WLS}} = \left(\mathbf{H}^T\mathbf{R}^{-1}\mathbf{H} + \mathbf{P}_0^{-1}\right)^{-1}$$

WLS, MAP, and CM Estimator Alternative Form

$$\hat{\mathbf{x}} = \overline{\mathbf{x}}_0 + \mathbf{P}_0\mathbf{H}^T\left(\mathbf{H}\mathbf{P}_0\mathbf{H}^T + \mathbf{R}\right)^{-1}\left(\mathbf{y} - \mathbf{H}\overline{\mathbf{x}}_0\right)$$

$$\mathbf{P} = \mathbf{P}_0 - \mathbf{P}_0\mathbf{H}^T(\mathbf{H}\mathbf{P}_0\mathbf{H}^T + \mathbf{R})^{-1}\mathbf{H}\mathbf{P}_0$$

$$-\frac{\partial J}{\partial \mathbf{x}} = 2\left\{\left[\frac{\partial \mathbf{h}(\mathbf{x})}{\partial \mathbf{x}}\right]_{\mathbf{x}}^T \mathbf{R}^{-1}\left(\mathbf{y} - \mathbf{h}(\mathbf{x})\right) - \mathbf{P}_0^{-1}\left(\mathbf{x} - \overline{\mathbf{x}}_0\right)\right\} = \mathbf{0}. \qquad (1.6\text{-}2)$$

The matrix $\left[\frac{\partial \mathbf{h}(\mathbf{x})}{\partial \mathbf{x}}\right]_{\mathbf{x}}$ is known as the Jacobian of $\mathbf{h}(\mathbf{x})$ evaluated at $\mathbf{x}$, and is computed as

$$\left[\frac{\partial \mathbf{h}(\mathbf{x})}{\partial \mathbf{x}}\right]_{\mathbf{x}} = \begin{bmatrix} \dfrac{\partial h_1}{\partial x_1} & \cdots & \dfrac{\partial h_1}{\partial x_n} \\ \vdots & \ddots & \vdots \\ \dfrac{\partial h_m}{\partial x_1} & \cdots & \dfrac{\partial h_m}{\partial x_n} \end{bmatrix}_x .$$

Note that we have used the following convention to define column vectors $\mathbf{x}$ and $\mathbf{h}$,

$$\mathbf{x} = (x_1, x_2, \ldots x_n)^T,$$

$$\mathbf{h} = (h_1, h_2, \ldots h_m)^T.$$

The notation is further simplified by using $\mathbf{H_x}$ to denote the Jacobian of $\mathbf{h}$, which is the partial derivative of $\mathbf{h}$ with respect to $\mathbf{x}$

$$\mathbf{H_x} = \left[\frac{\partial \mathbf{h}(\mathbf{x})}{\partial \mathbf{x}}\right]_{\mathbf{x}}, \tag{1.6-3}$$

where the subscript $\mathbf{x}$ denotes that the matrix is calculated with the value of $\mathbf{x}$. Clearly, one cannot explicitly solve for $\mathbf{x}$ unless the function $\mathbf{h}(\mathbf{x})$ is better specified. Typically, an approximate solution is obtained by applying a first order Taylor series expansion for $\mathbf{h}(\mathbf{x})$ evaluated about the prior knowledge of $\mathbf{x}$, $\bar{\mathbf{x}}_0$,

$$\mathbf{h}(\mathbf{x}) \approx \mathbf{h}(\bar{\mathbf{x}}_0) + \left[\frac{\partial \mathbf{h}(\mathbf{x})}{\partial \mathbf{x}}\right]_{\bar{\mathbf{x}}_0} (\mathbf{x} - \bar{\mathbf{x}}_0). \tag{1.6-4}$$

Substituting the above expression into the necessary condition from Equation (1.6-2) yields

$$\mathbf{H}_{\bar{\mathbf{x}}_0}^T \left[\mathbf{R}^{-1}\left(\mathbf{y} - \mathbf{h}(\bar{\mathbf{x}}_0) + \mathbf{H}_{\bar{\mathbf{x}}_0}(\mathbf{x} - \bar{\mathbf{x}}_0)\right)\right] - \mathbf{P}_0^{-1}(\mathbf{x} - \bar{\mathbf{x}}_0) \approx \mathbf{0}. \tag{1.6-5}$$

After collecting terms one obtains

$$\left[\mathbf{H}_{\bar{\mathbf{x}}_0}^T \mathbf{R}^{-1} \mathbf{H}_{\bar{\mathbf{x}}_0} + \mathbf{P}_0^{-1}\right]\mathbf{x} \approx \left[\mathbf{H}_{\bar{\mathbf{x}}_0}^T \mathbf{R}^{-1} \mathbf{H}_{\bar{\mathbf{x}}_0} + \mathbf{P}_0^{-1}\right]\bar{\mathbf{x}}_0 + \mathbf{H}_{\bar{\mathbf{x}}_0}^T \mathbf{R}^{-1}\left(\mathbf{y} - \mathbf{h}(\bar{\mathbf{x}}_0)\right).$$

Therefore, $\hat{\mathbf{x}}$ is an approximation of the solution for Equation (1.6-2) that minimizes Equation (1.6-1)

$$\hat{\mathbf{x}} = \bar{\mathbf{x}}_0 + \left[\mathbf{H}_{\bar{\mathbf{x}}_0}^T \mathbf{R}^{-1} \mathbf{H}_{\bar{\mathbf{x}}_0} + \mathbf{P}_0^{-1}\right]^{-1} \mathbf{H}_{\bar{\mathbf{x}}_0}^T \mathbf{R}^{-1}\left(\mathbf{y} - \mathbf{h}(\bar{\mathbf{x}}_0)\right). \tag{1.6-6}$$

Applying the matrix inversion lemma, one obtains the equivalent filter equation

$$\hat{\mathbf{x}} = \bar{\mathbf{x}}_0 + \mathbf{P}_0 \mathbf{H}_{\bar{\mathbf{x}}_0}^T \left[\mathbf{H}_{\bar{\mathbf{x}}_0} \mathbf{P}_0 \mathbf{H}_{\bar{\mathbf{x}}_0}^T + \mathbf{R}\right]^{-1}\left(\mathbf{y} - \mathbf{h}(\bar{\mathbf{x}}_0)\right). \tag{1.6-7}$$

Since $\hat{\mathbf{x}}$ obtained above is an approximate solution using a first order Taylor series expansion, a more refined or more accurate estimate may be obtained with further iteration. Rewriting the Taylor series expansion about an initial guess $\hat{\mathbf{x}}^0$ (which could be denoted as $\bar{\mathbf{x}}_0$ but we choose a different notation to distinguish it from the a priori knowledge), then

$$\mathbf{h}(\mathbf{x}) \approx \mathbf{h}(\hat{\mathbf{x}}^0) + \left[\frac{\partial \mathbf{h}(\mathbf{x})}{\partial \mathbf{x}}\right]_{\hat{\mathbf{x}}^0} (\mathbf{x} - \hat{\mathbf{x}}^0). \tag{1.6-8}$$

Substituting Equation (1.6-8) in Equation (1.6-2) yields

$$\left[\mathbf{H}_{\hat{\mathbf{x}}^0}^T \mathbf{R}^{-1} \mathbf{H}_{\hat{\mathbf{x}}^0} + \mathbf{P}_0^{-1}\right]\mathbf{x} \approx \mathbf{H}_{\hat{\mathbf{x}}^0}^T \mathbf{R}^{-1} \mathbf{H}_{\hat{\mathbf{x}}^0} \hat{\mathbf{x}}^0 + \mathbf{P}_0^{-1} \overline{\mathbf{x}}_0 + \mathbf{H}_{\hat{\mathbf{x}}^0}^T \mathbf{R}^{-1}\left(\mathbf{y} - \mathbf{h}\left(\hat{\mathbf{x}}^0\right)\right).$$

The solution to the above equation is

$$\hat{\mathbf{x}}^1 = \left[\mathbf{H}_{\hat{\mathbf{x}}^0}^T \mathbf{R}^{-1} \mathbf{H}_{\hat{\mathbf{x}}^0} + \mathbf{P}_0^{-1}\right]^{-1} \left(\mathbf{H}_{\hat{\mathbf{x}}^0}^T \mathbf{R}^{-1} \mathbf{H}_{\hat{\mathbf{x}}^0} \hat{\mathbf{x}}^0 + \mathbf{P}_0^{-1} \overline{\mathbf{x}}_0\right)$$

$$+ \left[\mathbf{H}_{\hat{\mathbf{x}}^0}^T \mathbf{R}^{-1} \mathbf{H}_{\hat{\mathbf{x}}^0} + \mathbf{P}_0^{-1}\right]^{-1} \mathbf{H}_{\hat{\mathbf{x}}^0}^T \mathbf{R}^{-1}\left(\mathbf{y} - \mathbf{h}\left(\hat{\mathbf{x}}^0\right)\right)$$

and the iterative solution is

$$\hat{\mathbf{x}}^{i+1} = \left[\mathbf{H}_{\hat{\mathbf{x}}^i}^T \mathbf{R}^{-1} \mathbf{H}_{\hat{\mathbf{x}}^i} + \mathbf{P}_0^{-1}\right]^{-1} \left(\mathbf{H}_{\hat{\mathbf{x}}^i}^T \mathbf{R}^{-1} \mathbf{H}_{\hat{\mathbf{x}}^i} \hat{\mathbf{x}}^i + \mathbf{P}_0^{-1} \overline{\mathbf{x}}_0\right)$$

$$+ \left[\mathbf{H}_{\hat{\mathbf{x}}^i}^T \mathbf{R}^{-1} \mathbf{H}_{\hat{\mathbf{x}}^i} + \mathbf{P}_0^{-1}\right]^{-1} \mathbf{H}_{\hat{\mathbf{x}}^i}^T \mathbf{R}^{-1}\left(\mathbf{y} - \mathbf{h}\left(\hat{\mathbf{x}}^i\right)\right). \tag{1.6-9}$$

Or in a more conventional form

$$\hat{\mathbf{x}}^{i+1} = \hat{\mathbf{x}}^i + \left[\mathbf{H}_{\hat{\mathbf{x}}^i}^T \mathbf{R}^{-1} \mathbf{H}_{\hat{\mathbf{x}}^i} + \mathbf{P}_0^{-1}\right]^{-1} \left(\mathbf{H}_{\hat{\mathbf{x}}^i}^T \mathbf{R}^{-1}\left(\mathbf{y} - \mathbf{h}\left(\hat{\mathbf{x}}^i\right)\right) + \mathbf{P}_0^{-1}\left(\overline{\mathbf{x}}_0 - \hat{\mathbf{x}}^i\right)\right). \tag{1.6-10}$$

The iteration can be stopped when the difference between $\hat{\mathbf{x}}^{i+1}$ and $\hat{\mathbf{x}}^i$ becomes very small. Denoting the final solution as $\hat{\mathbf{x}}$, then the covariance of the $\hat{\mathbf{x}}$ is (approximately)

$$\mathbf{P} \approx \left[\mathbf{H}_{\hat{\mathbf{x}}}^T \mathbf{R}^{-1} \mathbf{H}_{\hat{\mathbf{x}}} + \mathbf{P}_0^{-1}\right]^{-1}. \tag{1.6-11}$$

The covariance of the estimate is evaluated using the converged state estimate, $\hat{\mathbf{x}}$.

Remarks

1. Although Equations (1.6-6) and (1.6-7) are for parameter estimation, in Chapter 3 they will be shown to be the same as the extended Kalman filter for nonlinear state estimation, and Equation (1.6-10) is a multi-iterative solution to the same problem.

2. The covariance equation of $\hat{\mathbf{x}}$, Equation (1.6-11), is identical to the CRB in the functional form for the constant parameter case, and the equivalence is shown in the next section.

A summary of the solution for this nonlinear estimation problem is shown in Table 1.4.

Table 1.4   Summary of a Weighted Iterative Algorithm When the Measurement Equation Is Nonlinear

---

A Weighted Iterative Least Squares Algorithm for Nonlinear Measurements
Problem Definition
Find the LS estimate of a random vector $\mathbf{x}$ with nonlinear observations

$$\mathbf{y} = \mathbf{h}(\mathbf{x}) + \mathbf{v}$$

where $\mathbf{x} \in \mathbb{R}^n$, $\mathbf{y}, \mathbf{v} \in \mathbb{R}^m$, $\mathbf{v}: \sim N(\mathbf{0}, \mathbf{R})$, $\mathbf{x} \sim N(\bar{\mathbf{x}}_0, \mathbf{P}_0)$ and $\mathbf{x}$ and $\mathbf{v}$ are mutually independent.
A Weighted Iterative Least Squares Estimator

$$\hat{\mathbf{x}}^{i+1} = \left[\mathbf{H}_{\hat{\mathbf{x}}^i}^T \mathbf{R}^{-1} \mathbf{H}_{\hat{\mathbf{x}}^i} + \mathbf{P}_0^{-1}\right]^{-1} \left(\mathbf{H}_{\hat{\mathbf{x}}^i}^T \mathbf{R}^{-1} \mathbf{H}_{\hat{\mathbf{x}}^i} \hat{\mathbf{x}}^i + \mathbf{P}_0^{-1} \bar{\mathbf{x}}_0\right)$$
$$+ \left[\mathbf{H}_{\hat{\mathbf{x}}^i}^T \mathbf{R}^{-1} \mathbf{H}_{\hat{\mathbf{x}}^i} + \mathbf{P}_0^{-1}\right]^{-1} \mathbf{H}_{\hat{\mathbf{x}}^i}^T \mathbf{R}^{-1} \left(\mathbf{y} - \mathbf{h}(\hat{\mathbf{x}}^i)\right)$$

where $\mathbf{H}_{\hat{\mathbf{x}}^i} = \left[\frac{\partial \mathbf{h}(\mathbf{x})}{\partial \mathbf{x}}\right]_{\hat{\mathbf{x}}^i}$ and $\hat{\mathbf{x}} = \hat{\mathbf{x}}^{i+1}$ when $\left\|\hat{\mathbf{x}}^{i+1} - \hat{\mathbf{x}}^i\right\|$ is small.
Alternative Form

$$\hat{\mathbf{x}}^{i+1} = \hat{\mathbf{x}}^i + \left[\mathbf{H}_{\hat{\mathbf{x}}^i}^T \mathbf{R}^{-1} \mathbf{H}_{\hat{\mathbf{x}}^i} + \mathbf{P}_0^{-1}\right]^{-1} \left(\mathbf{H}_{\hat{\mathbf{x}}^i}^T \mathbf{R}^{-1} \left(\mathbf{y} - \mathbf{h}(\hat{\mathbf{x}}^i)\right) + \mathbf{P}_0^{-1} \left(\bar{\mathbf{x}}_0 - \hat{\mathbf{x}}^i\right)\right)$$

Covariance

$$\mathbf{P} = \left[\mathbf{H}_{\hat{\mathbf{x}}^i}^T \mathbf{R}^{-1} \mathbf{H}_{\hat{\mathbf{x}}^i} + \mathbf{P}_0^{-1}\right]^{-1}$$

1.7   Cramer–Rao Bound

The CRB is the lower bound on the error covariance for any unbiased estimator used to estimate $\mathbf{x}$ based on observations of $\mathbf{y}$ given the relationship,

$$\mathbf{y} = \mathbf{h}(\mathbf{x}) + \mathbf{v} \tag{1.7-1}$$

where $\mathbf{y}, \mathbf{v} \in \mathbb{R}^m$ and $\mathbf{x} \in \mathbb{R}^n$ [1, 9–13].

Two cases arise: when $\mathbf{x}$ is an unknown constant vector, and when $\mathbf{x}$ is a random vector with known prior distribution.

1. $\mathbf{x}$ is an unknown constant vector, the CRB is computed with log likelihood function

$$Cov\{\hat{\mathbf{x}}\} \geq \left[E\left\{\left[\frac{\partial \ln p(\mathbf{y}|\mathbf{x})}{\partial \mathbf{x}}\right]\left[\frac{\partial \ln p(\mathbf{y}|\mathbf{x})}{\partial \mathbf{x}}\right]^T\right\}\right]^{-1}. \tag{1.7-2}$$

2. $\mathbf{x}$ is a random vector with known prior distribution, the CRB is computed with log joint density (or equivalently an a posteriori) function

$$Cov\{\hat{\mathbf{x}}\} \geq \left[E\left\{\left[\frac{\partial \ln p(\mathbf{y}, \mathbf{x})}{\partial \mathbf{x}}\right]\left[\frac{\partial \ln p(\mathbf{y}, \mathbf{x})}{\partial \mathbf{x}}\right]^T\right\}\right]^{-1}. \tag{1.7-3}$$

The above definition applies to any density function of $\mathbf{v}$ and $\mathbf{x}$ and any unbiased estimator. When the covariance of an unbiased estimator is the same as the CRB, this estimator is termed efficient. If the covariance of an unbiased estimator is approaching the CRB (e.g., with decreasing measurement errors and/or increasing number of measurements, etc.), then this estimator is termed asymptotically efficient.

We first derive the CRB for the case with constant parameter vector $\mathbf{x}$ and nonlinear measurements having additive Gaussian noise $\mathbf{v}$. The partial derivative of $\ln p(\mathbf{y}|\mathbf{x})$ with respect to $\mathbf{x}$ is

$$\frac{\partial \ln p(\mathbf{y}|\mathbf{x})}{\partial \mathbf{x}} = -\left[\frac{\partial \mathbf{h}(\mathbf{x})}{\partial \mathbf{x}}\right]^{T} \mathbf{R}^{-1}(\mathbf{y} - \mathbf{h}(\mathbf{x})). \tag{1.7-4}$$

Next, multiply $\frac{\partial \ln p(\mathbf{y}|\mathbf{x})}{\partial \mathbf{x}}$ by its transpose with the shorthand notation $\mathbf{H_x}$ from Equation (1.6-3) to obtain

$$\left[\frac{\partial \ln p(\mathbf{y}|\mathbf{x})}{\partial \mathbf{x}}\right]\left[\frac{\partial \ln p(\mathbf{y}|\mathbf{x})}{\partial \mathbf{x}}\right]^{T} = \left(\mathbf{H_x}^{T}\mathbf{R}^{-1}(\mathbf{y} - \mathbf{h}(\mathbf{x}))\right)\left(\mathbf{H_x}^{T}\mathbf{R}^{-1}(\mathbf{y} - \mathbf{h}(\mathbf{x}))\right)^{T}. \tag{1.7-5}$$

Since $\mathbf{y} - \mathbf{h}(\mathbf{x})$ is the measurement noise, the expected value of Equation (1.7-5) is

$$E\left\{\left[\frac{\partial \ln p(\mathbf{y}|\mathbf{x})}{\partial \mathbf{x}}\right]\left[\frac{\partial \ln p(\mathbf{y}|\mathbf{x})}{\partial \mathbf{x}}\right]^{T}\right\} = \mathbf{H_x}^{T}\mathbf{R}^{-1}\mathbf{H_x}.$$

The above expression is known as the Fisher information matrix [10]. Let it be denoted as $\mathcal{F}$,

$$\mathcal{F} = \mathbf{H_x}^{T}\mathbf{R}^{-1}\mathbf{H_x}.$$

Thus the inverse of the Fisher information matrix yields the Cramer–Rao bound,

$$Cov\{\hat{\mathbf{x}}\} \geq \mathcal{F}^{-1} = \left[\mathbf{H_x}^{T}\mathbf{R}^{-1}\mathbf{H_x}\right]^{-1}. \tag{1.7-6}$$

The derivation of the CRB for the case with random parameter vector $\mathbf{x}$ and nonlinear measurements is very similar to Equation (1.7-6), except that the partial derivative is with respect to the random vector $\mathbf{x}$ and the density function is $p(\mathbf{y},\mathbf{x})$. We will first obtain, for a given $\mathbf{x}$,

$$\left[\frac{\partial \ln p(\mathbf{y}, \mathbf{x})}{\partial \mathbf{x}}\right]_{\mathbf{x}} = -\left[\frac{\partial \mathbf{h}(\mathbf{x})}{\partial \mathbf{x}}\right]_{\mathbf{x}}^{T} \mathbf{R}^{-1}(\mathbf{y} - \mathbf{h}(\mathbf{x})) + \mathbf{P}_0^{-1}(\mathbf{x} - \bar{\mathbf{x}}_0). \tag{1.7-7}$$

Since $\mathbf{y} - \mathbf{h}(\mathbf{x})$ is the measurement noise and is independent of the initial uncertainty, $\mathbf{x} - \bar{\mathbf{x}}_0$, the expected value of Equation (1.7-3) is thus

$$E\left\{\left[\frac{\partial\ln p(\mathbf{y},\mathbf{x})}{\partial\mathbf{x}}\right]_{\mathbf{x}}\left[\frac{\partial\ln p(\mathbf{y},\mathbf{x})}{\partial\mathbf{x}}\right]_{\mathbf{x}}^{T}\right\}=E\left\{\mathbf{H}_{\mathbf{x}}^{T}\mathbf{R}^{-1}\mathbf{H}_{\mathbf{x}}+\mathbf{P}_{0}^{-1}\right\}.$$

and the Cramer–Rao bound,

$$Cov\{\hat{\mathbf{x}}\}\geq\mathcal{F}^{-1}=\left[E\left\{\mathbf{H}_{\mathbf{x}}^{T}\mathbf{R}^{-1}\mathbf{H}_{\mathbf{x}}+\mathbf{P}_{0}^{-1}\right\}\right]^{-1}. \tag{1.7-8}$$

This equation is identical in functional form with Equation (1.6-11), the covariance for a nonlinear estimator. The difference is that Equation (1.6-11) is evaluated with the estimate of the random vector $\mathbf{x}$, $\hat{\mathbf{x}}$, while the Cramer–Rao bound, Equation (1.7-8) is the expectation over the random vector $\mathbf{x}$.

At this point, the development is completed for the Cramer–Rao bound on the covariance of any unbiased estimator of the random parameter vector $\mathbf{x}$ given a known nonlinear measurement relation where both $\mathbf{v}$ and $\mathbf{x}$ are Gaussian and mutually independent. One must go back to the general definition shown in Equation (1.7-3) to derive a solution for any given density functions.

We now consider three special cases.

1. The first case is when the measurement equation is linear, that is, $\mathbf{y} = \mathbf{Hx} + \mathbf{v}$. It is straightforward to show that $\mathbf{H_x}$ is $\mathbf{H}$. This is a very important result because it shows that the WLS, MAP, and CM estimators derived earlier are all efficient estimators and attain the smallest possible estimation error covariance.

2. It was shown in the previous section that the iterative WLS algorithm has a covariance equation that is identical to the expression for CRB.

3. When there is no a priori knowledge, that is, when initial information is nonexistent, $\mathbf{P}_0^{-1} = \mathbf{0}$ (no information), the CRB above reduces to the simpler form of the covariance equation for a constant unknown parameter vector developed earlier.

The CRB for parameter estimation developed here is extended to the state estimation in the next chapter.

1.8   Numerical Example

Let us consider a classical numerical example of estimating a sinusoidal signal in noise. This problem will appear several times in this book as more tools for estimation are introduced. Sampled noisy measurements are taken from a sinusoidal signal

$$y_k = a\sin(\omega t_k + \varphi) + v_k,$$

where y_k is the noisy measurement taken at time step t_k, and v_k is the white measurement noise sequence at t_k. There are three unknown parameters, (a, ω, φ), that are to be estimated given $y_{1:N} = \{y_1, y_2, y_3, \dots, y_N\}$. Since the parameters (ω, φ) are related to the signal and the measurement nonlinearly, the nonlinear WLS estimator is used. The unknown parameter vector $\mathbf{x}$ is

$$\mathbf{x} = [a, \omega, \varphi]^T.$$

The measurement vector $\mathbf{y} = [y_1, y_2, y_3, \dots, y_N]^T$ and noise vector $\mathbf{v} = [v_1, v_2, v_3, \dots, v_N]^T$ are related as

$$\mathbf{y} = \mathbf{h}(\mathbf{x}) + \mathbf{v},$$

where the elements of $\mathbf{h}(\mathbf{x})$ are the sampled sine wave, that is,

$$\mathbf{h}(\mathbf{x}) = [a\sin(\omega t_1 + \varphi), \, a\sin(\omega t_2 + \varphi), \, \dots, \, a\sin(\omega t_N + \varphi)]^T. \tag{1.8-1}$$

Remarks

1. For simplicity, let us assume we have uniformly spaced sample data, that is, $t_i = t_1 + (i - 1)\Delta$, for $i = 1, \dots N$, and Δ is the sampling period. It is obvious that we have the same value of $\mathbf{h}(\mathbf{x})$ in Equation (1.8-1) for all $\varphi_j = 2j\pi + \varphi$ for $j = 0, \pm 1, \dots, \pm \infty$, as well as for all $\omega_n = 2n\pi/\Delta + \omega$ for $n = 0, \pm 1, \dots, \pm \infty$. Therefore, we have to limit the parameter space for φ and ω for the possibility that they have unique estimate values. It is clear that the frequency estimate can be limited to being positive, that is, $\omega \geq 0$. It is also correct to assume the phase angle to be less than 2π, $0 \leq \varphi < 2\pi$, as it would otherwise be ambiguous.

2. The ambiguity in frequency estimate, for example, $\omega_n = 2n\pi/\Delta + \omega$, can be avoided by choosing appropriate sampling rate, $1/\Delta$, such that the unknown frequency is bounded by $0 \leq \omega \leq \pi/\Delta$. This is the well-known Nyquist rate in sampling theory [19].

The iterative weighted least squares algorithm for nonlinear measurements introduced above in Equations (1.6-9) through (1.6-11) is applied to this problem and the results are presented in the next few figures. Following the remarks above, the search for a solution is limited to within the parameter region $0 \leq \varphi < 2\pi$ and $0 \leq \omega \leq \pi/\Delta$. No a priori knowledge is assumed (i.e., $\mathbf{P}_0^{-1} = \mathbf{0}$). The true parameter values for this example are $a = 1$, $\omega = 2\pi$, $\varphi = \pi/6$. The noise covariance, $\sigma_v^2 = 0.0225$, which is equivalent to a voltage signal-to-noise ratio (SNR) of $a/\sigma_v = 6.667$.

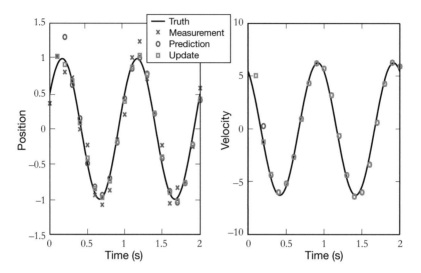

Figure 1.3   Comparison of estimated and true sinusoidal waves with amplitude and phase angle estimated, single Monte Carlo run.

We will first look at the case when ω is known, that is, the parameters to be estimated are $\mathbf{x} = [a, \varphi]^T$. The result of one Monte Carlo run for a 3-second data window with a 0.1-second sampling rate is shown in Figure 1.3. For every time step t_k, where all measurements $y_{1:k} = \{y_1, y_2, y_3, \ldots, y_k\}$ are used for estimation, the resulting estimates were used to compute the sinusoidal signal at t_k (referred to as the updated position) and its derivative (a cosine, referred to as the updated velocity) denoted by green squares. The initial guess $\hat{\mathbf{x}}^0$ of each time step can be chosen arbitrarily. For this example, the algorithm autonomously selects the converged estimate of the previous time step as the initial guess for the next time step to speed up the convergence. This strategy worked for this case, but in general, a random initial guess is advised to avoid converging to a local minimum. Before running the algorithm with one additional measurement, the resulting estimates were used to compute the sinusoidal signal for the next time step, t_{k+1} (referred to as the predictions of position and velocity in Figure 1.3, denoted by blue circles). Following the estimator presented in Equations (1.6-10) and (1.6-11), the parameter estimate at a given time t_k was computed with all past measurements (thus batch processing) from the initial time up to time t_k. The algorithm normally converged after three iterations with a few exceptions that ran for five iterations with $\varepsilon = 10^{-6}$. The estimator required at least two consecutive measurements

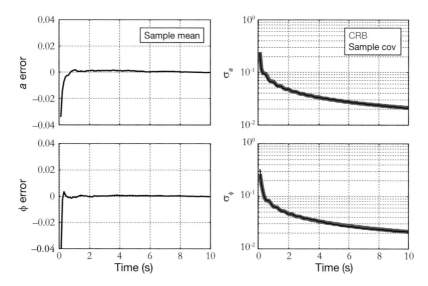

Figure 1.4 Sample mean and RMS error compared to square root of the CRB with estimated amplitude and phase angle.

($k = 2$) because there are two unknown parameters. For this reason, the first estimate is obtained for $t=0.1$ s. In Figure 1.3, the measured (red crosses), predicted (blue circles), and updated (green squares) sine wave values at each time step are shown. The predicted and updated sine wave values converge to the truth after eight time steps and further measurements do not contribute to the estimator outputs on this scale.

Figure 1.3 shows the estimator behavior in the measurement space with one Monte Carlo run. The next question to address is how the estimation errors behave statistically. Figure 1.4 shows the sample mean from Equation (1.3-20) and the root mean square (RMS) error from Equation (1.3-21) as defined in Section 1.3.3 with 1000 Monte Carlo runs. Comparing the sample mean to the true value (mean error) of the parameters in Figure 1.4, the estimator is nearly unbiased after a few time steps. The RMS error of each parameter (blue curve) is plotted in Figure 1.4 against the square root of each associated CRB (red curve) of Equation (1.7-6). It can be seen that the RMS error is nearly undistinguishable from the square root of the CRB after only a few steps.

When we consider the case with three unknown parameters, that is, $\mathbf{x} = [a, \omega, \varphi]^{T}$, the problem becomes highly ambiguous. The iterative LS algorithm presented here can easily get trapped in a local minimum or wandering around on a plateau surface

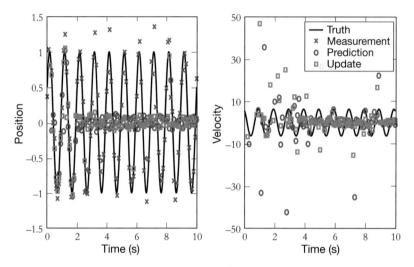

Figure 1.5   Comparison of estimated and true sinusoidal waves with amplitude, phase angle, and frequency estimated, single Monte Carlo run.

and could not converge to the true parameters. The result of one Monte Carlo run, similar to the two-parameter case, is shown in Figure 1.5. In this case the parameter estimation error seems to decrease substantially as the sample window increases. But after 2 to 3 seconds, the estimates no longer follow measurements and estimation errors become large. This is because a longer observation window washes out the sinusoidal signal and "zero" answer fits well on an ambiguous peak (or valley) of the performance index. This becomes more clear when the sample mean errors and the RMS error (black curves) of each parameter are plotted against the CRB (red curves), respectively, as shown in Figure 1.6. Note that the estimator is biased and the RMS error of each parameter is higher than the corresponding CRB. The reader is encouraged to explore this problem further by plotting the performance index surface as a function of the unknown parameters to exhibit the ambiguous nature of the index function. This example will be explored further when additional estimator options have been introduced.

Appendix 1.A   Simulating Correlated Random Vectors with a Given Covariance Matrix

The problem of generating random samples corresponding to a known probability density function often arises in practical problems. Homework Problem 1 challenges

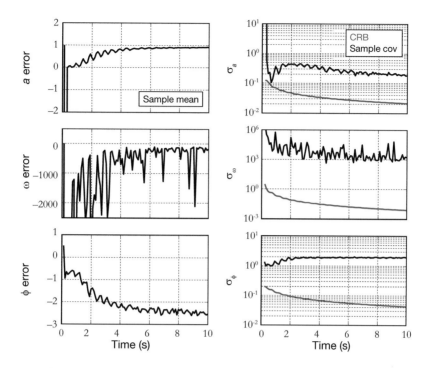

Figure 1.6 Sample mean and RMS error compared to square root of the CRB with estimated amplitude, phase angle, and frequency.

the reader to learn this skill. In Homework Problem 2, the reader is asked to generate a random vector corresponding to a given covariance matrix. This appendix provides some basic matrix fundamentals for Homework Problem 2. More discussions in this area can be found in many control system textbooks [e.g., 7, 8].

Basics

This involves the concept of eigenvalue, eigenvector, orthogonal transformation, and similarity transformation. Let $\mathbf{P}$ and $\mathbf{L}$ be two positive definite matrices with dimension $n \times n$ with the relationship

$$\mathbf{L}^{-1}\mathbf{P}\mathbf{L} = \boldsymbol{\Lambda}. \text{ (similarity transformation)} \tag{1.A-1}$$

where $\boldsymbol{\Lambda}$ is a diagonal matrix with diagonal elements equal to λ_i. We can show that λ_i are eigenvalues of $\mathbf{P}$ and the column vectors of $\mathbf{L}$ are the corresponding normalized eigenvectors.

The above equation is the same as

$$\mathbf{PL} = \mathbf{L\Lambda}.$$

Let the columns of $\mathbf{L}$ be denoted as $\boldsymbol{\ell}_i$, that is,

$$\mathbf{L} = [\boldsymbol{\ell}_1, \boldsymbol{\ell}_2, \ldots, \boldsymbol{\ell}_n]$$

Therefore,

$$\mathbf{L\Lambda} = [\lambda_1\boldsymbol{\ell}_1, \lambda_2\boldsymbol{\ell}_2, \ldots, \lambda_n\boldsymbol{\ell}_n].$$

Then,

$$\mathbf{P}\boldsymbol{\ell}_i = \lambda_i\boldsymbol{\ell}_i$$

for all i, or

$$[\mathbf{P} - \lambda_i\mathbf{I}]\boldsymbol{\ell}_i = 0$$

Then λ_i is the solution of

$$|\mathbf{P} - \lambda\mathbf{I}| = 0 \tag{1.A-2}$$

which are the eigenvalues of $\mathbf{P}$. Since $\mathbf{L}$ is the matrix that diagonalizes $\mathbf{P}$, each $\boldsymbol{\ell}_i$ is a normalized eigenvector with corresponding eigenvalue λ_i. The column vectors $\boldsymbol{\ell}_i$ for all i form an orthonormal basis,

$$<\boldsymbol{\ell}_i, \boldsymbol{\ell}_j> = 1 \ \forall i = j$$

$$= 0 \ \forall i \neq j.$$

Therefore

$$\mathbf{L}^T = \mathbf{L}^{-1} \text{ (orthonormal transformation).}$$

We can rewrite Equation (1.A-1) as

$$\mathbf{L}^T\mathbf{PL} = \mathbf{\Lambda}. \tag{1.A-3}$$

Numerical Experiment

Let $\mathbf{x}$ be an n-dimensional random vector with distribution $\mathbf{x}: \sim N(\mathbf{0}, \mathbf{P_x})$ and let $\mathbf{L}$ be the matrix of normalized eigenvectors of $\mathbf{Px}$ and $\mathbf{u} = \mathbf{L}^T\mathbf{x}$. Then

$$\mathbf{uu}^T = \mathbf{L}^T\mathbf{xx}^T\mathbf{L}$$

and

$$\mathbf{P_u} = \mathbf{L}^T \mathbf{P_x} \mathbf{L}$$

where $\mathbf{P_u}$ is a diagonal matrix with elements λ_i. The elements of $\mathbf{u} = (u_1, u_2, \ldots, u_n)^T$ have distribution $u_i{:} \sim N(0, \lambda_i)$ and $\{u_i\}$ are mutually independent.

Given $\mathbf{P_x}$, use MATLAB routines to find its eigenvalues and the corresponding normalized eigenvectors, which yield $\mathbf{L}$ and $\mathbf{P_u}$. Draw random samples from a Gaussian random number generator with $N(0, \lambda_i)$ for $i = 1, \ldots, n$ to form elements of a simulated random vector of $\mathbf{u}$. This is repeated for M times. Let the jth time be denoted as $\mathbf{u}_j$, then the corresponding $\mathbf{x}_j$ vector is $\mathbf{x}_j = \mathbf{L}\mathbf{u}_j$. One can show numerically and analytically that the sample covariance matrix of $\mathbf{x}$, $\tilde{\mathbf{P}}_\mathbf{x}$ is approaching $\mathbf{P_x}$ when M is large, that is,

$$\tilde{\mathbf{P}}_\mathbf{x} = \frac{1}{M} \sum_{j=1}^{M} \mathbf{x}_j \mathbf{x}_j^T \to \mathbf{P_x}.$$

Appendix 1.B More Properties of Least Squares Estimators

Alternative Definition of LS Estimator: Consider two jointly distributed random vectors $\mathbf{x}$ and $\mathbf{y}$ with joint probability density function $p(\mathbf{x}, \mathbf{y})$. The LS estimator of $\mathbf{x}$ given $\mathbf{y}$, $\hat{\mathbf{x}}$, is the function $\mathbf{g} : \mathbb{R}^m \to \mathbb{R}^n$ and

$$\hat{\mathbf{x}} = \text{ArgMin}_\mathbf{g} E\left\{ \|\mathbf{x} - \mathbf{g}(\mathbf{y})\|^2 \right\}. \tag{1.B-1}$$

Proposition 1: The LS estimator, $\hat{\mathbf{x}}_{CM}$ of $\mathbf{x}$ in terms of $\mathbf{y}$, satisfies Equation (1.B-1) and is the conditional mean

$$\hat{\mathbf{x}}_{CM} = E\left\{ \mathbf{x}|\mathbf{y} \right\} \tag{1.B-2}$$

of $\mathbf{x}$ given $\mathbf{y}$, and the corresponding minimum mean square error is the covariance $E\{\|\mathbf{x} - E\{\mathbf{x}|\mathbf{y}\}\|^2\}$.

Proof: For every $\mathbf{y} = y^7$ with $p(\mathbf{y} = y) > 0$, one can define a random vector in $\mathbb{R}^n$ by

$$\hat{\mathbf{x}}_{CM}(y) = E\left\{ \mathbf{x}|\mathbf{y} = y \right\} = \int x \frac{p(\mathbf{y} = y|\mathbf{x} = x)}{p(\mathbf{y} = y)} p(\mathbf{x} = x) dx. \tag{1.B-3}$$

---

7   In this book, lower case letters are used for random variables and lower case *italic* letters for their realizations; and boldface letters for vectors and non-boldface letters for scalars.

And for any vector $\mathbf{z} \in \mathbb{R}^n$, we have

$$
\begin{aligned}
E\{\|\mathbf{x} - \mathbf{z}\|^2 | \mathbf{y} = y\} &= E\{\mathbf{x}^T\mathbf{x} - 2\mathbf{z}^T\mathbf{x} + \mathbf{z}^T\mathbf{z} | \mathbf{y} = y\} \\
&= E\{\mathbf{x}^T\mathbf{x} | \mathbf{y} = y\} - 2\mathbf{z}^T E\{\mathbf{x} | \mathbf{y} = y\} + \mathbf{z}^T\mathbf{z} \\
&= E\{\mathbf{x}^T\mathbf{x} | \mathbf{y} = y\} - \|E\{\mathbf{x} | \mathbf{y} = y\}\|^2 \\
&\quad + E\{\|\mathbf{z} - E\{\mathbf{x} | \mathbf{y} = y\}\|^2 | \mathbf{y} = y\}
\end{aligned}
\tag{1.B-4}
$$

The only term in Equation (1.B-4) that depends on $\mathbf{z}$ is the last term, and $E\{\|\mathbf{x} - \mathbf{z}\|^2 | \mathbf{y} = y\}$ is uniquely minimized by setting

$$
\mathbf{z} = \hat{\mathbf{x}}_{\mathrm{CM}}(y) = E\{\mathbf{x} | \mathbf{y} = y\}.
\tag{1.B-5}
$$

The minimum value of Equation (1.B-4) is the sum of the first two terms, which is

$$
E\left\{\|\mathbf{x} - \hat{\mathbf{x}}_{\mathrm{CM}}(\mathbf{y})\|^2 | \mathbf{y} = y\right\} = E\left\{\mathbf{x}^T\mathbf{x} | \mathbf{y} = y\right\} - \|\hat{\mathbf{x}}_{\mathrm{CM}}(y)\|^2.
\tag{1.B-6}
$$

If $\mathbf{g}$ is any mapping $\mathbb{R}^m$ into $\mathbb{R}^n$, then it is clear from the way we constructed $\hat{\mathbf{x}}_{\mathrm{CM}}(\mathbf{y})$ for every $\mathbf{y} = y$ with $p(\mathbf{y} = y) > 0$, we have

$$
E\left\{\|\mathbf{x} - \hat{\mathbf{x}}_{\mathrm{CM}}(\mathbf{y})\|^2 | \mathbf{y} = y\right\} \leq E\left\{\|\mathbf{x} - \mathbf{g}(y)\|^2 | \mathbf{y} = y\right\}.
$$

The expectation of the right-hand side of the above is with respect to $\mathbf{y}$,

$$
E\{E\{\|\mathbf{x} - \mathbf{g}(\mathbf{y})\|^2 | \mathbf{y} = y\}\} = E\{\|\mathbf{x} - \mathbf{g}(\mathbf{y})\|^2\}
$$

which yields

$$
E\left\{\|\mathbf{x} - \hat{\mathbf{x}}_{\mathrm{CM}}(\mathbf{y})\|^2\right\} \leq E\left\{\|\mathbf{x} - \mathbf{g}(\mathbf{y})\|^2\right\}
$$

for all $\mathbf{y} = y$ with $p(\mathbf{y} = y) > 0$. There may be cases which for some function $\mathbf{g}$ and value y

$$
E\left\{\|\mathbf{x} - \hat{\mathbf{x}}_{\mathrm{CM}}(\mathbf{y})\|^2 | \mathbf{y} = y\right\} \geq E\left\{\mathbf{x} - \mathbf{g}(\mathbf{y})^2 | \mathbf{y} = y\right\}
$$

for $p(\mathbf{y} = y) = 0$. But $\hat{\mathbf{x}}_{\mathrm{CM}}$ is still the best "on the average" in the sense of Equation (1.B-1). Likewise, taking the expectation on both sides of Equation (1.B-6), we have the minimum of Equation (1.B-1)

$$
\mathrm{Min}_{\mathbf{g}}\, E\{\|\mathbf{x} - \mathbf{g}(\mathbf{y})\|^2\} = E\{\|\mathbf{x} - E\{\mathbf{x} | \mathbf{y}\}\|^2\}
\qquad \text{Q.E.D.}
$$

Proposition 2 (Properties of the LS Estimator): The LS estimator $\hat{\mathbf{x}}_{\mathrm{CM}} = E\{\mathbf{x} | \mathbf{y}\}$ has the following properties.

1. It is linear, that is, for any deterministic matrix $\mathbf{A}$ and deterministic vector $\mathbf{b}$ with appropriate dimensions

$$
\left(\widehat{\mathbf{A}\mathbf{x} + \mathbf{b}}\right)_{\mathrm{CM}} = E\{\mathbf{A}\mathbf{x} + \mathbf{b} | \mathbf{y}\} = \mathbf{A}E\{\mathbf{x} | \mathbf{y}\} + \mathbf{b} = \mathbf{A}\hat{\mathbf{x}} + \mathbf{b}
$$

and if $\mathbf{x}$ and $\mathbf{z}$ are random vectors with the same dimension

$$\left(\widehat{\mathbf{x}+\mathbf{z}}\right)_{CM} = E\{\mathbf{x}+\mathbf{z}|\mathbf{y}\} = E\{\mathbf{x}|\mathbf{y}\} + E\{\mathbf{z}|\mathbf{y}\} = \hat{\mathbf{x}}_{CM} + \hat{\mathbf{z}}_{CM}$$

2. It is unbiased, that is,

$$E\{\mathbf{x}-\hat{\mathbf{x}}_{CM}\} = E\{\mathbf{x}\} - E\{E\{\mathbf{x}|\mathbf{y}\}\} = E\{\mathbf{x}\} - E\{\mathbf{x}\} = \mathbf{0}.$$

In fact we have a stronger statement

$$E\{\mathbf{x}-\hat{\mathbf{x}}_{CM}|\mathbf{y}\} = E\{\mathbf{x}|\mathbf{y}\} - E\{E\{\mathbf{x}|\mathbf{y}\}|\mathbf{y}\} = \hat{\mathbf{x}}_{CM} - \hat{\mathbf{x}}_{CM} = \mathbf{0}.$$

3. Let $\tilde{\mathbf{x}}_{CM} = \mathbf{x} - \hat{\mathbf{x}}_{CM}$, then $\tilde{\mathbf{x}}_{CM}$ is uncorrelated with any function $\mathbf{g}(.)$ of $\mathbf{y}$, that is,

$$E\{\mathbf{g}(\mathbf{y})\,\tilde{\mathbf{x}}_{CM}{}^T\} = \mathbf{0}$$

and, in fact,

$$E\{\mathbf{g}(\mathbf{y})\tilde{\mathbf{x}}_{CM}{}^T|\mathbf{y}\} = \mathbf{0}.$$

Proof: Properties 1 and 2 are straightforward with the linear property of expectation. For every $\mathbf{y}=y$ with $p(\mathbf{y}=y) > 0$, we have

$$E\{\mathbf{g}(\mathbf{y})\tilde{\mathbf{x}}_{CM}{}^T|\mathbf{y}\} = E\{\mathbf{g}(\mathbf{y})(\mathbf{x}-\hat{\mathbf{x}}_{CM}(\mathbf{y}))^T|\mathbf{y}=y\} = \mathbf{g}(y)E\{(\mathbf{x}-\hat{\mathbf{x}}_{CM}(\mathbf{y}))^T|\mathbf{y}=y\}$$

$$= \mathbf{g}(y)(\hat{\mathbf{x}}_{CM}(y) - \hat{\mathbf{x}}_{CM}(y))^T = \mathbf{0}. \qquad \text{Q.E.D.}$$

Remarks

The CM operator of $\mathbf{x}$, $E\{\mathbf{x}|\mathbf{y}\}$ is a function of $\mathbf{y}$ which maps $\mathbf{y}$ into the space of $\mathbf{x}$. We have proved that the operator is a linear operator on $\mathbf{x}$, but itself is not necessarily a linear function of $\mathbf{y}$. In the following section, we will introduce estimators that are linear functions of $\mathbf{y}$.

Definition of LLS Estimator: Consider two jointly distributed random vectors $\mathbf{x}$ and $\mathbf{y}$ with joint probability density function $p(\mathbf{x}, \mathbf{y})$. The LLS estimator of $\mathbf{x}$ given $\mathbf{y}$, $\hat{\mathbf{x}}_{LLS}$ is the linear function $\mathbf{L}(.): \mathbb{R}^m \to \mathbb{R}^n$ and

$$\text{Min}_L\ E\{\|\mathbf{x} - \mathbf{L}(\mathbf{y})\|^2\}. \tag{1.B-7}$$

Proposition 3 (Properties of the LLS Estimator): The LLS estimator, $\hat{\mathbf{x}}_{\text{LLS}}$, depends only on the first moments $(\bar{\mathbf{x}}, \bar{\mathbf{y}})$ and second moments $(\mathbf{P}_{\mathbf{xx}}, \mathbf{P}_{\mathbf{xy}}, \mathbf{P}_{\mathbf{yy}})$ of the random vectors $\mathbf{x}$ and $\mathbf{y}$ and not on their entire probability density functions, such that

$$\hat{\mathbf{x}}_{\text{LLS}} = \bar{\mathbf{x}} + \mathbf{P}_{\mathbf{xy}}\mathbf{P}_{\mathbf{yy}}^{-1}(\mathbf{y} - \bar{\mathbf{y}}) \tag{1.B-8}$$

and

$$Cov\{\tilde{\mathbf{x}}_{\text{LLS}}, \tilde{\mathbf{x}}_{\text{LLS}}\} = \mathbf{P}_{\mathbf{xx}} - \mathbf{P}_{\mathbf{xy}}\mathbf{P}_{\mathbf{yy}}^{-1}\mathbf{P}_{\mathbf{yx}}, \tag{1.B-9}$$

where $\tilde{\mathbf{x}}_{\text{LLS}} = \mathbf{x} - \tilde{\mathbf{x}}_{\text{LLS}}$ and

$$\mathbf{P}_{\mathbf{ab}} = Cov\{\mathbf{a}, \mathbf{b}\}$$

Then we can conclude that

1. $\hat{\mathbf{x}}_{\text{LLS}}$ is linear,

2. $\hat{\mathbf{x}}_{\text{LLS}}$ is unbiased,

3. $\hat{\mathbf{x}}_{\text{LLS}} = \hat{\mathbf{x}}_{\text{CM}}$ when $\mathbf{x}$ and $\mathbf{y}$ are jointly Gaussian.

Proof: Let us assume $\mathbf{x}$ and $\mathbf{y}$ have zero mean and define any linear function of $\mathbf{y}$, $\mathbf{L}(\mathbf{y})$, as

$$\mathbf{L}(\mathbf{y}) = \mathbf{A}\mathbf{y} + \mathbf{b}$$

where $\mathbf{A}$ is an $n \times m$ matrix and $\mathbf{b}$ is an n-vector. The LLS estimator, $\hat{\mathbf{x}}_{\text{LLS}}$, is a linear function of $\mathbf{y}$

$$\hat{\mathbf{x}}_{\text{LLS}} = \mathbf{A}\mathbf{y} + \mathbf{b}$$

which minimizes Equation (1.B-7), or equivalently

$$\text{Min}_{\mathbf{A},\mathbf{b}} E\{\|\mathbf{x} - (\mathbf{A}\mathbf{y} + \mathbf{b})\|^2\} \tag{1.B-10}$$

Using the trace identity $\text{tr}[\mathbf{AB}] = \text{tr}[\mathbf{BA}]$ [20] and the linearity of expectation and its interchangeability with the trace operation, we obtain

$$\begin{aligned}
E\{\|\mathbf{x} - (\mathbf{A}\mathbf{y} + \mathbf{b})\|^2\} &= E\{(\mathbf{x} - (\mathbf{A}\mathbf{y} + \mathbf{b}))^T(\mathbf{x} - (\mathbf{A}\mathbf{y} + \mathbf{b}))\} \\
&= \text{tr}E\{(\mathbf{x} - (\mathbf{A}\mathbf{y} + \mathbf{b}))(\mathbf{x} - (\mathbf{A}\mathbf{y} + \mathbf{b}))^T\} \\
&= \text{tr}[E\{\mathbf{x}\mathbf{x}^T\} - AE\{\mathbf{y}\mathbf{x}^T\} - E\{\mathbf{x}\mathbf{y}^T\}\mathbf{A}^T + AE\{\mathbf{y}\mathbf{y}^T\}\mathbf{A}^T + \mathbf{b}\mathbf{b}^T] \\
&= \text{tr}[\mathbf{P}_{\mathbf{xx}} - A\mathbf{P}_{\mathbf{yx}} - \mathbf{P}_{\mathbf{xy}}\mathbf{A}^T + A\mathbf{P}_{\mathbf{yy}}\mathbf{A}^T + \mathbf{b}\mathbf{b}^T]
\end{aligned} \tag{1.B-11}$$

Taking partial derivatives of Equation (1.B-11) with respect to $\mathbf{A}$ and $\mathbf{b}$ [20], we obtain

$$\frac{\partial}{\partial \mathbf{b}} E\left\{\|\mathbf{x} - (\mathbf{A}\mathbf{y} + \mathbf{b})\|^2\right\} = \mathbf{0} = 2\mathbf{b}.$$

$$\frac{\partial}{\partial \mathbf{A}} E\left\{\|\mathbf{x} - (\mathbf{A}\mathbf{y} + \mathbf{b})\|^2\right\} = \mathbf{0} = \mathbf{A}\mathbf{P}_{yy} + \mathbf{A}\mathbf{P}_{yy}^T - \mathbf{P}_{yx} - \mathbf{P}_{xy}^T$$

which yields

$$\hat{\mathbf{x}}_{\text{LLS}} = \mathbf{P}_{xy}\mathbf{P}_{yy}^{-1}\mathbf{y}$$

and the corresponding minimum value of the variance of estimation error $\tilde{\mathbf{x}}_{\text{LLS}} = \mathbf{x} - \hat{\mathbf{x}}_{\text{LLS}}$ is,

$$E\left\{\|\tilde{\mathbf{x}}_{\text{LLS}}\|^2\right\} = \text{tr}\left[\mathbf{P}_{xx} - \mathbf{P}_{xy}\mathbf{P}_{yy}^{-1}\mathbf{P}_{yx}\right].$$

An equivalent presentation is that

$$Cov\{\tilde{\mathbf{x}}_{\text{LLS}}, \tilde{\mathbf{x}}_{\text{LLS}}\} = \mathbf{P}_{xx} - \mathbf{P}_{xy}\mathbf{P}_{yy}^{-1}\mathbf{P}_{yx}.$$

In the case where $\mathbf{x}$ and $\mathbf{y}$ have nonzero mean, $\bar{\mathbf{x}}$ and $\bar{\mathbf{y}}$, respectively, we have

$$\left(\widehat{\mathbf{x} - \bar{\mathbf{x}}}\right)_{\text{LLS}} = \mathbf{P}_{xy}\mathbf{P}_{yy}^{-1}(\mathbf{y} - \bar{\mathbf{y}})$$

or, equivalently

$$\hat{\mathbf{x}}_{\text{LLS}} = \bar{\mathbf{x}} + \mathbf{P}_{xy}\mathbf{P}_{yy}^{-1}(\mathbf{y} - \bar{\mathbf{y}}).$$

The three properties are easy to prove [14].

Q.E.D.

Homework Problems

Random Variable and Distributions

1. (a) Let x be a scalar-valued random variable uniformly distributed between [–0.5, 0.5]. Use the uniform random number generator in MATLAB (or a language of your choice) to generate hundreds of i.i.d. samples of x and plot the distribution.

   (b) For the i.i.d. x's generated above, compute $y = \frac{1}{n}\Sigma_1^n x_i$, and plot the distribution of y. Show the convergence of y to a Gaussian distribution.

(c) Let $u = \frac{1}{\lambda}e^{-\lambda z}$, numerically show the density of u when z is $U[-0.5, 0.5]$. Use the Gaussian random number generator in MATLAB (or a language of your choice) to generate z with a Gaussian distribution $N(0, \sigma^2)$ then repeat part one of this problem.

2. Let $\mathbf{x}$ be a random vector in $\mathbb{R}^n$ and with Gaussian density $N(\mathbf{0}, \mathbf{P})$. Pick a dimension for n (e.g., 3) and a positive definite matrix, $\mathbf{P}$, satisfying the property of a covariance matrix. Using the Gaussian random number generator, generate many $\mathbf{x}$ values and empirically show that the distribution of $\mathbf{x}$ is close to $N(\mathbf{0}, \mathbf{P})$. (Hint: Find the transformation that will diagonalize $\mathbf{P}$. Generate random vectors with independent components and variance commensurate with the eigenvalues of $\mathbf{P}$. Use the inverse transformation to bring those random vectors to the $\mathbf{x}$ space.)

Estimators

3. (a) Let $y_i = a + v_i$ where a is an unknown constant and y_i are the repeated measurements of a with independent measurement noise v_i. Derive $\hat{a}$ for LS, WLS, and ML estimators when (a) $v_i = N(0, \sigma^2)$ where σ is a constant for all i, (b) $v_i = N(0, \sigma_i^2)$, and (c) $v_i = U[-0.5, 0.5]$.

(b) Repeat the above for a to be a random variable with (a) $a = N(0, \sigma_a^2)$ and (c) $a = U[b_1, b_2]$.

(c) Derive a recursive estimator for a as i increases for both cases above.

4. Show that the LS estimator of Equation (1.4-3) is unbiased with covariance as in Equation (1.4-4). (Hint: Apply definitions of bias and covariance, $E\{\tilde{\mathbf{x}}_{LS}\} = E\{\mathbf{x}\} - E\{\hat{\mathbf{x}}_{LS}\}$ and $\mathbf{P}_{LS} = E\{(\mathbf{x} - \hat{\mathbf{x}}_{LS})(\mathbf{x} - \hat{\mathbf{x}}_{LS})^T\}$.)

5. Derive the mean and covariance of the pseudo-inverse estimate shown as in Equations (1.4-5) and (1.4-7).

6. Derive the covariance for the WLS estimator, Equation (1.4-10).

7. Derive the mean and covariance for WLS estimator with multiple measurement vectors, Equations (1.4-19) and (1.4-20). (Hint: Think of the estimate obtained with the initial measurement vector as a priori knowledge for the new measurements.)

8. Derive the LS estimator with a priori knowledge, Equations (1.5-3) and (1.5-4).

9. Derive the WLS estimator with a priori knowledge, Equations (1.5-6) through (1.5-9).

10. Similar to Equations (1.4-19) and (1.4-20), derive the recursive WLS with multiple measurement vectors of Equations (1.5-12) and (1.5-13).

11. Derive the CM/LLS estimator, Equation (1.5-21) and Equation (1.5-22).

12. Derive estimator equations when measurement noise is uniformly distributed.

13. In a radar system, a completely known baseband waveform denoted as $s(t)$ is transmitted, the returned signal, $y(t)$, from a point target is a replica of $s(t)$ with an unknown time delay and a complex-valued amplitude modulation (due partly to target reflectivity, atmosphere attenuation, and the modulation and demodulation process) and is denoted as $y(t) = as(t - \tau)$. A set of noisy and discrete time measurements of $y(t)$ is taken that is denoted as $z(t_k) = y(t_k) + \xi_k$, for $k=1, \ldots, N$ where ξ_k is an uncorrelated nonstationary zero mean Gaussian noise sequence $N(0, \sigma_k^2)$.

   a. Derive an estimator for τ. (Hint: the log likelihood function is

   $$\ln\left(p(z|\tau, a)\right) = -\sum_{k=1}^{N} \frac{1}{2\sigma_k^2} |z(t_k) - as(t_k - \tau)|^2 - \text{const.}$$

   where a can be treated as a nuisance parameter or estimated together with τ.)

   b. Derive CRB for σ_τ^2. Similarly, a can be treated as a nuisance parameter or estimated jointly.[8]

14. For two closely-spaced point targets (or scatterers), the returned signal is $y(t) = a_1 s(t - \tau_1) + a_2 s(t - \tau_2)$ where a_1 and a_2 are complex valued with the phase of a_1 being uniformly random and the phase of a_2 relative to a_1 is $\Delta\theta = \frac{4\pi}{\lambda}(\tau_2 - \tau_1)$.

   a. Derive an estimator for τ_1 and τ_2. One can treat the phase angles of both a_1 and a_2 as uniformly random, they are thus nuisance parameters. The fact that the phase difference reflects the relative time delay, it contains information. A more useful way is to estimate time-difference-of-arrival using the phase difference.

   b. Derive CRB for $\sigma_{\tau_1}^2$ and $\sigma_{\tau_2}^2$. Similarly, one can treat the phases as nuisance parameters or as parameters to be estimated. Explain why the CRB by treating the phases as nuisance parameters is lower than the CRB by jointly estimating phase as an additional unknown parameter.

---

8   For treatment of nuisance parameter see [1, 8].

15. Consider the case that a set of polynomials of variable t, denotes time, represented by $\{f_0(t), f_1(t), \ldots, f_k(t)\}$. Assuming that this set of polynomials can adequately represent a function $y(t)$ in the form of a weighted linear sum, $\mathbf{y}(t) = \sum_{k=0}^{K} a_k f_k(t)$. A set of noisy and discrete time measurements of $y(t)$ is taken that is denoted as $z(t_n) = y(t_n) + \xi_n$, for $n = 1, \ldots, N$ where ξ_n is an uncorrelated nonstationary zero mean noise sequence with variance σ_n^2. The objective is to find a set of a_k's that best represent $y(t)$.

a. Derive the general expression for the WLS estimator for a_k's with covariance.

b. Derive the estimator and covariance equations with the special case where

$$f_0(t) = 1, f_1(t) = t, \ldots, f_K(t) = t^K/K!.$$

c. Further simplify the result for $K=2$ and $\sigma_n^2 = \sigma^2$, a constant. Notice in this case, if one puts the center of the observation window at $t=0$, the final expressions can be greatly simplified.

d. Derive results for $K=1$ and $\sigma_n^2 = \sigma^2$.

References

1. H.L. Van Trees, *Detection, Estimation, and Modulation Theory*, Part 1. New York: Wiley, 1968.
2. R. Deutsch, *Estimation Theory*. Englewood Cliffs, NJ: Prentice Hall, 1965.
3. H. Sorenson, *Parameter Estimation*. New York: Marcel Dekker, 1980.
4. J. Gubner, *Probability and Random Processes for Electrical and Computer Engineers*. Cambridge, UK: Cambridge University Press, 2006.
5. A. Papoulis and S.U. Pillai, *Probability, Random Variables, and Stochastic Processes*, 4th ed. New York: McGraw-Hill, 2002.
6. W.B. Davenport and W.L. Root, *Random Signals and Noise*. New York: McGraw-Hill, 1958.
7. D.G. Luenberger, *Optimization by Vector Space Methods*. New York: Wiley, 1969.
8. P.M. DeRusso, R.J. Roy, C.M. Close, and A.A. Desrochers, *State Variables for Engineers*, 2nd ed. New York: Wiley, 1998.
9. W.L. Brogan, *Modern Control Theory*. Englewood Cliffs, NJ: Prentice Hall, 1990.
10. H. Cramer, *Mathematical Methods of Statistics*. Princeton, NJ: Princeton University Press, 1946.

11. R. Rao, "Information and the Accuracy Attainable in the Estimation of Statistical Parameters," *Bulletin of the Calcutta Mathematical Society*, no. 37, pp. 81–89, 1945.

12. R.A. Fisher, "On the Mathematical Foundations of Theoretical Statistics," *Philosophical Transactions of the Royal Society of London*, ser. A, vol. 222, pp. 309–368, 1922.

13. R.W. Miller and C.B. Chang, "A Modified Cramer–Rao Bound and its Applications," *IEEE Transactions on Information Theory*, vol. IT-24, pp. 398–400, May 1978.

14. B. Rhodes, "A Tutorial Introduction to Estimation and Filtering," *IEEE Transactions on Automatic Control*, vol. AC-16, pp. 688–706, Dec. 1971.

15. Y. Bar-Shalom, X. Rong Li, and T. Kirubarajan, *Estimation with Applications to Tracking and Navigation*. New York: Wiley, 2001.

16. C.B. Chang and R.B. Homes, "Non-parametric Identification of Continuous Systems with Discrete Measurements," in *Proceedings of the 23rd Conference on Decision and Control*, pp. 944–950, 1984.

17. J. M. Hammersley and K. W. Morton, "Poor Man's Monte Carlo," *Journal of the Royal Statistical Society B*, vol. 16, pp. 23–38, 1954.

18. S. Arulampalam, S. R. Maskell, N. J. Gordon, and T. Clapp, "A Tutorial on Particle Filters for On-line Nonlinear/Non-Gaussian Bayesian Tracking," *IEEE Transactions on Signal Processing*, vol. 50, pp. 174–188, 2002.

19. A.V. Oppenheim and R. W. Schafer, *Digital Signal Processing*. Englewood Cliffs, N.J.: Prentice-Hall, 1978.

20. M. Athans, "The Matrix Minimum Principle," *Information and Control*, vol. 11, pp. 592–606, 1968.

2

State Estimation for Linear Systems

2.1 Introduction

State estimation is defined as estimating state vector of a dynamic system at a given time from a time series of noisy measurements. The state vector is defined the same way as the state space representation of control systems [e.g., 1–5]. The state trajectory is often driven by a noise process (to be referred to as system noise or process noise), thus evolving as a stochastic process. A list of definitions and terminology about stochastic processes is summarized in Appendix 2.A for easy reference. Readers wanting to have a more in depth understanding of this subject should consult textbooks in this area [e.g., 6]. The problem for state estimation when there is no process noise, that is, when the state vector evolves deterministically, will also be discussed in this chapter.

This chapter focuses on problems with linear state dynamics and measurement equations. This subject is widely covered in the literature [6–13]. The notation for state dynamic and measurement equations are defined first in Section 2.2. For the purpose of simplicity (to avoid complicated mathematics) and because most of the application problems use sampled data, systems with discrete time representation will be used. A quick development of relationships between continuous time and discrete time systems is given in Section 2.2. The estimator definition in terms of filtering, smoothing, and prediction is given in Section 2.3. An important concept for estimation is observability. The development of observability criterion is given. The solution to a general estimation problem is in the evolution of the a posteriori probability density function. This is captured as the Bayesian approach for state estimation in Section 2.4. Filter equations can be derived using this approach similar to Bayes' rule of probability introduced in Chapter 1. The solution for a linear estimation problem when the initial state, system noise, and measurement noise are Gaussian is the well-known Kalman filter (KF) [1]. Section 2.5 provides a general discussion on the KF, and Sections 2.6 and 2.7 present two different approaches in deriving the KF. Section

2.8 reviews some of the fundamental discussions on estimator properties and interpretations in Kalman's original paper [1]. Section 2.9 presents algorithms for smoothing, and Section 2.10 presents the Cramer–Rao bound for state estimation. Section 2.11 presents the application of KF for the sinusoidal signal parameter estimation problem given in Section 1.8 for the case when the frequency of the sine wave is known. When the frequency is known, the sinusoidal noise problem can be formulated as a linear state estimation problem. Homework problems and a list of references conclude the chapter.

2.2   State and Measurement Equations

A prerequisite to studying this book is state space representation of control systems. The purpose of this section is to establish some basic relations and notations. Therefore, a quick derivation of discrete time systems from the continuous time counterpart is given. Readers seeking additional resources on this subject can consult textbooks in this area [e.g., 2, 3, 6].

Many underlying physical processes (the system state equation) are continuous time functions. A continuous time evolution of the state vector is represented by the linear stochastic differential equation[1]

$$\dot{\mathbf{x}}_t = \mathbf{A}_t \mathbf{x}_t + \mathbf{B}_t \boldsymbol{\xi}_t, \tag{2.2-1}$$

where $\mathbf{x}_t \in \mathbb{R}^n$, $\mathbf{A}_t$ is the system matrix, $\mathbf{B}_t$ is the distribution matrix for the system noise $\boldsymbol{\xi}_t$, both matrices are assumed to be deterministic, and $\boldsymbol{\xi}_t$ is a noise process, which is usually referred to as system noise because it is a driving function of the system. In practical problems, it is used to represent random disturbances that may exist in a system, or to represent unknown or unmodeled system dynamics. Mathematically, it will be formally defined as a zero mean white Gaussian process such that

$$E\{\boldsymbol{\xi}_t\} = \mathbf{0}, \text{ and}$$

$$E\left\{\boldsymbol{\xi}_t \boldsymbol{\xi}_s^T\right\} = \Sigma_t \delta(t-s) \ \forall t, s$$

where $\delta(t)$ is a Dirac delta function.

---

1   For more rigorous derivation related to the continuous time stochastic process, one should use the Wiener integral representation for Equation (2.2-1) and Ito calculus for the derivations of results shown in this section [6, Chapter 4].

The measurements on $\mathbf{x}_t$ are taken at discrete times denoted as

$$\mathbf{y}_k = \mathbf{H}_k \mathbf{x}_k + \mathbf{v}_k, \tag{2.2-2}$$

where $\mathbf{y}_k, \mathbf{v}_k \in \mathbb{R}^m$, $\mathbf{v}_k$ is independent of $\boldsymbol{\xi}_t$ $\forall t, k$ and a white Gaussian noise process with zero mean and covariance $\mathbf{R}_k$, that is,

$$\mathbf{v}_k : \sim N(\mathbf{0}, \mathbf{R}_k).$$

The discrete time equivalent of the continuous time[2] system, Equation (2.2-1), is defined via the state transition matrix $\boldsymbol{\Phi}_{k,k-1}$, that is,

$$\mathbf{x}_k = \boldsymbol{\Phi}_{k,k-1} \mathbf{x}_{k-1} + \boldsymbol{\mu}_{k-1}. \tag{2.2-3}$$

The general form of the state transition matrix $\boldsymbol{\Phi}_{t,t_0}$ represents the time evolution of the state vector from $\mathbf{x}_{t_0}$ to $\mathbf{x}_t$, with external driving force $\boldsymbol{\xi}_t = \mathbf{0}$ in Equation (2.2-1), that is,

$$\mathbf{x}_t = \boldsymbol{\Phi}_{t,t_0} \mathbf{x}_{t_0}. \tag{2.2-4}$$

This is the homogeneous solution of the linear differential equation, Equation (2.2-1), with $\boldsymbol{\xi}_t = \mathbf{0}$ [2]. Taking the derivative with respect to t in Equation (2.2-4) yields

$$\dot{\boldsymbol{\Phi}}_{t,t_0} \mathbf{x}_{t_0} = \dot{\mathbf{x}}_t = \mathbf{A}_t \mathbf{x}_t = \mathbf{A}_t \boldsymbol{\Phi}_{t,t_0} \mathbf{x}_{t_0}.$$

And thus the matrix differential equation for $\boldsymbol{\Phi}_{t,t_0}$ becomes

$$\dot{\boldsymbol{\Phi}}_{t,t_0} = \mathbf{A}_t \boldsymbol{\Phi}_{t,t_0}. \tag{2.2-5}$$

In the special case when $\mathbf{A}_t$ is a constant matrix, $\mathbf{A}$, one can show that the state transition matrix is simply a matrix exponential of $\mathbf{A}(t - t_0)$

$$\boldsymbol{\Phi}_{t,t_0} = e^{\mathbf{A}(t-t_0)}. \tag{2.2-6}$$

The approximate solution in terms of the Taylor series expansion is often used to compute $\boldsymbol{\Phi}_{t,t_0}$, that is,

$$\boldsymbol{\Phi}_{t,t_0} = \mathbf{I} + \mathbf{A}(t - t_0) + \text{HOT}$$

---

2   For simplicity, in the remainder of this book "discrete" will be used as "discrete time" and "continuous" as "continuous time" when referring to a system or a process.

where HOT stands for higher order terms of the series expansion. Even for a time-varying $\mathbf{A}_t$, the same approximation can be used when the time interval $t - t_0$ is small.

Several important properties of Φ_{t_1,t_2} are stated here without derivation. Interested readers can explore the derivation in textbooks on state variables and control systems [e.g., 2, 3]

$$1. \quad \Phi_{t_1,t_2}\Phi_{t_2,t_3} = \Phi_{t_1,t_3},$$
$$2. \quad \Phi_{t_1,t_2}^{-1} = \Phi_{t_2,t_1},$$
$$3. \quad \Phi_{t,t_1}\Phi_{t_1,t} = \Phi_{t,t} = \mathbf{I}. \tag{2.2-7}$$

The relationship between Equations (2.2-1) and (2.2-3) with discrete process noise μ_k is shown next. Treating ξ_t as a deterministic function and using the variation of constant formula and Equation (2.2-5), one can show the following equation is the inhomogeneous solution of Equation (2.2-1) [2, 3]

$$\mathbf{x}_t = \Phi_{t,t_0}\mathbf{x}_{t_0} + \int_{t_0}^t \Phi_{t,\tau}\mathbf{B}_\tau\xi_\tau d\tau. \tag{2.2-8}$$

Proof: Taking derivatives with respect to t on both sides of Equation (2.2-8)

$$\dot{\mathbf{x}}_t = \dot{\Phi}_{t,t_0}\mathbf{x}_{t_0} + \int_{t_0}^t \dot{\Phi}_{t,\tau}\mathbf{B}_\tau\xi_\tau d\tau + \Phi_{t,t}\mathbf{B}_t\xi_t.$$

Applying Equation (2.2-5) and properties of Φ_{t,t_0} in (2.2-7) yields

$$\dot{\mathbf{x}}_t = \mathbf{A}_t\Phi_{t,t_0}\mathbf{x}_{t_0} + \int_{t_0}^t \mathbf{A}_t\Phi_{t,\tau}\mathbf{B}_\tau\xi_\tau d\tau + \mathbf{B}_t\xi_t$$

$$= \mathbf{A}_t\left[\Phi_{t,t_0}\mathbf{x}_{t_0} + \int_{t_0}^t \Phi_{t,\tau}\mathbf{B}_\tau\xi_\tau d\tau\right] + \mathbf{B}_t\xi_t = \mathbf{A}_t\mathbf{x}_t + \mathbf{B}_t\xi_t. \qquad \text{Q.E.D.}$$

Given Equation (2.2-8), the discrete system noise μ_{k-1} can be formally related to the continuous system noise ξ_t as[3]

$$\mu_{k-1} = \int_{t_{k-1}}^{t_k} \Phi_{t,\tau}\mathbf{B}_\tau\xi_\tau d\tau$$

Let $\mathbf{Q}_{k-1}$ be the covariance of μ_{k-1} then it can be shown that $\mathbf{Q}_{k-1}$ and Σ_t are related by[4]

$$\mathbf{Q}_{k-1} = \int_{t_{k-1}}^{t_k} \Phi_{t,\tau}\mathbf{B}_\tau\Sigma_\tau\mathbf{B}_\tau^T\Phi_{t,\tau}^T d\tau. \tag{2.2-9}$$

---

3   Rigorously, this integral should be a Wiener integral and ξ_t is formally the derivative of a Wiener process [6, Chapter 4].

4   The time integral expression in Equation (2.2-9) can be derived rigorously with Ito calculus [e.g., 6].

Given that ξ_t is a continuous time zero mean and Gaussian white noise, and then μ_{k-1} is a zero mean white Gaussian sequence

$$\mu_{k-1} : \sim N(\mathbf{0}, \mathbf{Q}_{k-1}).$$

The initial state vector is denoted as $\mathbf{x}_0$, where both $\mathbf{x}_0$ and $\mu_{k-1} \in \mathbb{R}^n$ are independent $\forall k$ and it is modeled by a Gaussian density function as

$$\mathbf{x}_0 : \sim N(\overline{\mathbf{x}}_0, \mathbf{P}_0).$$

Example

Let x denote the position of an object traveling along a straight line. If the object is traveling with a constant velocity then the differential equation of motion for x is $\ddot{x} = 0$, which is sometimes referred to as a constant velocity (CV) model. The state equation of motion for this object can be represented as

$$\dot{\mathbf{x}}_t = \mathbf{A}_t \mathbf{x}_t.$$

The state vector is $\mathbf{x}_t = [x_1, x_2]^T$, with $x_1 = x$, $x_2 = \dot{x}$, and the system matrix $\mathbf{A}_t$ is

$$\mathbf{A}_t = \begin{bmatrix} 0 & 1 \\ 0 & 0 \end{bmatrix}.$$

The solution of the above equation for a given initial condition $\mathbf{x}_0$ gives the trajectory of an object traveling in a straight line. However, realistic motion is rarely a straight line. The uncertainty of such motion is modeled by the system noise, or process noise, as ξ_t of Equation (2.2-1). The system equation is thus modified to

$$\dot{\mathbf{x}}_t = \mathbf{A}_t \mathbf{x}_t + \mathbf{B}_t \xi_t$$

where the input matrix $\mathbf{B}_t$ is

$$\mathbf{B}_t = \begin{bmatrix} 0 \\ 1 \end{bmatrix}.$$

This indicates that the system uncertainty is only appearing at the acceleration level as a driver for $\ddot{x}$. Note that ξ_t is a scalar. Using the meter, kilogram, and second (MKS) convention, ξ_t will have dimensions of m/s². The discrete equivalent of this continuous system is represented in using the state transition matrix. As shown in Equation (2.2-3), the discrete representation can be written as

$$\mathbf{x}_k = \mathbf{\Phi}_{k,k-1} \mathbf{x}_{k-1} + \mu_{k-1}$$

where the time between samples is a constant, T. After applying the definition of state transition matrix in Equation (2.2-6), one obtains

$$\boldsymbol{\Phi}_{k,k-1} = \boldsymbol{\Phi}_T = \begin{bmatrix} 1 & T \\ 0 & 1 \end{bmatrix} \forall k.$$

Let the variance of ξ_t be a constant and equal to σ_ξ^2. Applying Equation (2.2-9), it can be shown that the covariance of $\boldsymbol{\mu}_{k-1}$ is

$$\mathbf{Q}_{k-1} = \mathbf{Q} = \sigma_\xi^2 \begin{bmatrix} \dfrac{1}{3}T^3 & \dfrac{1}{2}T^2 \\ \dfrac{1}{2}T^2 & T \end{bmatrix} \forall k.$$

Using the same approach, one can show that for an object traveling on a straight line with constant acceleration (CA) motion, the state vector is $\mathbf{x}_t = [x_1, x_2, x_3]^T$, with $x_1 = x$, $x_2 = \dot{x}$, $x_3 = \ddot{x}$ and the system matrix $\mathbf{A}_t$ is

$$\mathbf{A}_t = \begin{bmatrix} 0 & 1 & 0 \\ 0 & 0 & 1 \\ 0 & 0 & 0 \end{bmatrix}.$$

The state transition matrix is

$$\boldsymbol{\Phi}_{k,k-1} = \boldsymbol{\Phi}_T = \begin{bmatrix} 1 & T & 0 \\ 0 & 1 & T \\ 0 & 0 & 1 \end{bmatrix},$$

and the discrete process noise covariance is

$$\mathbf{Q}_{k-1} = \mathbf{Q} = \sigma_\xi^2 \begin{bmatrix} 0 & 0 & 0 \\ 0 & \dfrac{1}{3}T^3 & \dfrac{1}{2}T^2 \\ 0 & \dfrac{1}{2}T^2 & T \end{bmatrix} \forall k.$$

Remarks

1. It is important to note that both $\mathbf{x}_k$ and $\mathbf{y}_k$ are Gaussian processes, because any linear transformation of a Gaussian process is a Gaussian process.

2. As will be shown later, the estimator for the state and measurement system shown above is optimal if the actual process is the same as modeled and the noise processes are Gaussian, zero mean, and with known covariance. Such requirements are usually not satisfied in practical applications. The process noise covariance $\mathbf{Q}_k$ is often adjusted to fine tune a filter design in order to reduce any apparent problems in the filter performance (see Chapter 4).

3. In just about all applications with state estimation, only discrete, sampled data are taken as measurements, even though the underlying process is continuous. The state vector at the discrete measurement time can be computed using the known linear continuous system equations, via the state transition matrix, which was developed above. Nevertheless, it becomes a discrete state estimation problem. The equivalence between continuous and discrete representation for a linear system is well understood. Readers who do not already have this background from their study of control systems are encouraged to refer to a senior level or first-year graduate level control systems textbook [e.g., 2, 3].

2.3   Definition of State Estimators

The problem of state estimation is restated here as computing state vectors $\mathbf{x}_k$ for all $k = 1, \ldots, K$ from a set of noisy measurements in which the state evolves with a known stochastic process. The state and measurement equations are repeated below for the purpose of clarity,

$$\mathbf{x}_k = \mathbf{\Phi}_{k,k-1}\mathbf{x}_{k-1} + \mathbf{\mu}_{k-1}, \tag{2.3-1}$$

$$\mathbf{y}_k = \mathbf{H}_k\mathbf{x}_k + \mathbf{v}_k, \tag{2.3-2}$$

where $\mathbf{x}_k, \mathbf{x}_0, \mathbf{\mu}_{k-1} \in \mathbb{R}^n$, $\mathbf{y}_k, \mathbf{v}_k \in \mathbb{R}^m$, $\mathbf{\Phi}_{k,k-1}$ and $\mathbf{H}_k$ are known matrices, and $\mathbf{x}_0, \mathbf{\mu}_{k-1}$, and $\mathbf{v}_k$ are mutually independent Gaussian random vectors: $\mathbf{x}_0 \colon \sim N(\bar{\mathbf{x}}_0, \mathbf{P}_0)$, $\mathbf{\mu}_k \colon \sim N(\mathbf{0}, \mathbf{Q}_k)$, and $\mathbf{v}_k \colon \sim N(\mathbf{0}, \mathbf{R}_k)$.

The variables are defined as

$\mathbf{y}_{1:j}$ the collection of measurements from $1, 2, \ldots, j$, that is, $\mathbf{y}_{1:j} = \{\mathbf{y}_1, \mathbf{y}_2, \ldots, \mathbf{y}_j\}$,

$\hat{\mathbf{x}}_{i|j}$ the estimate of $\mathbf{x}_i$ given measurements $\mathbf{y}_{1:j}$, then

1. $\hat{\mathbf{x}}_{i|j}$ is the filtered estimator of $\mathbf{x}_i$ when $i = j$,

2. $\hat{\mathbf{x}}_{i|j}$ is the predicted estimator of $\mathbf{x}_i$ when $i > j$,

3. $\hat{\mathbf{x}}_{i|j}$ is the smoothed estimator of $\mathbf{x}_i$ when $i < j$.

It can be further asserted that the optimal state estimator is the statistical expectation of the state vector condition on measurements, that is,

$$\hat{\mathbf{x}}_{i|j} = E\{\mathbf{x}_i | \mathbf{y}_{1:j}\}. \tag{2.3-3}$$

2.3.1   Observability

From the definition above, it is clear that the problem of state estimation is an inverse mapping problem. Similar to the problem of solving an unknown vector from a set of linear equations, there are certain conditions that must be met in order for the solution to exist and to be unique. The condition, namely observability, for which the state vector can be uniquely computed from a set of measurements is considered. The deterministic case, that is, when both $\boldsymbol{\mu}_k$ and $\mathbf{v}_k$ are zero, $\forall k$ will be studied first. Thus the problem is equivalent to solving for $\mathbf{x}_0$ which follows a known difference equation

$$\mathbf{x}_k = \boldsymbol{\Phi}_{k,k-1}\mathbf{x}_{k-1}, \tag{2.3-4}$$

given a set of measurements

$$\mathbf{y}_i = \mathbf{H}_i\mathbf{x}_i \ i = 1, \dots, k \tag{2.3-5}$$

where k is an integer time index. To solve for $\mathbf{x}_0$, all measurements in terms of $\mathbf{x}_0$ are written as

$$\mathbf{y}_1 = \mathbf{H}_1\boldsymbol{\Phi}_{1,0}\mathbf{x}_0,$$

$$\mathbf{y}_2 = \mathbf{H}_2\boldsymbol{\Phi}_{2,0}\mathbf{x}_0,$$

$$\mathbf{y}_k = \mathbf{H}_k\boldsymbol{\Phi}_{k,0}\mathbf{x}_0.$$

A vector and a matrix are defined as

$$\mathbf{y}_{1:k} = \left(\mathbf{y}_1^T, \ \mathbf{y}_2^T, \dots, \ \mathbf{y}_k^T\right)^T$$

$$\mathbf{H}_{1:k} = \left[\left[\mathbf{H}_1\boldsymbol{\Phi}_{1,0}\right]^T, \left[\mathbf{H}_2\boldsymbol{\Phi}_{2,0}\right]^T, \dots, \left[\mathbf{H}_k\boldsymbol{\Phi}_{k,0}\right]^T\right]^T.$$

This gives

$$\mathbf{y}_{1:k} = \mathbf{H}_{1:k}\mathbf{x}_0,$$

where $\mathbf{y}_{1:k}$ is a $(k*m) \times 1$ vector, $\mathbf{H}_{1:k}$ is a $(k*m) \times n$ matrix, and $\mathbf{x}_0$ is an $n \times 1$ vector. Note that $(k*m)$ is used to denote k multiplied by m. Assuming that $(k*m)$ is at least greater than or equal to n and the rank of $\mathbf{H}_{1:k}$ is n, then $\mathbf{x}_0$ can be uniquely computed as

$$\mathbf{x}_0 = \left[\mathbf{H}_{1:k}^T\mathbf{H}_{1:k}\right]^{-1}\mathbf{H}_{1:k}^T\mathbf{y}_{1:k}.$$

Thus for the linear discrete system considered, when $\text{Rank}\left\{\mathbf{H}_{1:k}^T\mathbf{H}_{1:k}\right\} = n$, the system is observable [2–5]. Writing out $\left[\mathbf{H}_{1:k}^T\mathbf{H}_{1:k}\right]$ explicitly,

$$\left[\mathbf{H}_{1:k}^T\mathbf{H}_{1:k}\right] = \sum_{i=1}^{k}\mathbf{\Phi}_{i,0}^T\mathbf{H}_i^T\mathbf{H}_i\mathbf{\Phi}_{i,0}. \tag{2.3-6}$$

The observability condition for a linear discrete system is summarized in Table 2.1.

Remarks

It will be shown in Section 3.10 that when the above expression is modified to

$$\sum_{i=1}^{k}\mathbf{\Phi}_{i,0}^T\mathbf{H}_i^T\mathbf{R}_i^{-1}\mathbf{H}_i\mathbf{\Phi}_{i,0} \tag{2.3-7}$$

for a stochastic linear system Equations (2.3-1) and (2.3-2), it becomes the inverse of the Cramer–Rao bound (CRB) for estimating $\mathbf{x}_0$ thus connecting the observability criterion to estimation accuracy. If $\left[\mathbf{H}_{1:k}^T\mathbf{H}_{1:k}\right]$ is singular for all k, then the system is not observable and there does not exist any unbiased estimator of $\mathbf{x}_0$ with finite covariance.

Table 2.1   Summary of Observability Condition

Observability Condition

Given a deterministic linear system:

$\mathbf{x}_k = \mathbf{\Phi}_{k,k-1}\mathbf{x}_{k-1},$ (2.3-4)

$\mathbf{y}_k = \mathbf{H}_k\mathbf{x}_k.$ (2.3-5)

This system is observable if and only if the following matrix is nonsingular for a finite k, $\sum_{i=1}^{k}\mathbf{\Phi}_{i,0}^T\mathbf{H}_i^T\mathbf{H}_i\mathbf{\Phi}_{i,0}.$

(2.3-6)

2.3.2   Estimation Error

The definition of estimation error for parameter estimation described in Section 1.4 can be extended to state estimation. As defined in Section 2.3, let $\hat{\mathbf{x}}_{i|j}$ denote the estimate of $\mathbf{x}_i$ based upon measurements up to time j, that is, $\mathbf{y}_{1:j}$. For the purpose of simplicity, the discussion is limited to the filtering problem, that is, for $i = j = k$. The estimation error $\tilde{\mathbf{x}}_{k|k}$ is

$$\tilde{\mathbf{x}}_{k|k} = \mathbf{x}_k - \hat{\mathbf{x}}_{k|k}.$$

For linear Gaussian case, $\mathbf{x}_k$ and $\hat{\mathbf{x}}_{k|k}$ are both Gaussian, therefore $\tilde{\mathbf{x}}_{k|k}$ is Gaussian and fully characterized by its mean and covariance. In a more general situation, probability density function of $\hat{\mathbf{x}}_{k||:k}$ will be required. The definition of mean and covariance of $\hat{\mathbf{x}}_{k|k}$ is

$$E\{\tilde{\mathbf{x}}_{k|k}\} = E\{\mathbf{x}_k\} - E\{\hat{\mathbf{x}}_{k|k}\}, \tag{2.3-8}$$

and the covariance is

$$Cov\{\tilde{\mathbf{x}}_{k|k}\} = E\{(\mathbf{x}_k - \hat{\mathbf{x}}_{k|k})(\mathbf{x}_k - \hat{\mathbf{x}}_{k|k})^T\}. \tag{2.3-9}$$

When $E\{\tilde{\mathbf{x}}_{k|k}\} = \mathbf{0}$, then $\hat{\mathbf{x}}_{k|k}$ is an unbiased estimator. Similar to the parameter estimation case, an important measure of quality for error covariance is known as the CRB, which is a lower bound of the error covariance of any unbiased estimator of $\mathbf{x}_k$. Any unbiased estimator with covariance achieving the CRB is an efficient estimator, and $\hat{\mathbf{x}}_{k|k}$ is an efficient estimate of $\mathbf{x}_k$. The Cramer–Rao bound has been applied to numerous problems as a measure of estimator performance. The CRB for parameter estimation in Section 1.7 will be extended to state estimation in the later part of this chapter.

2.4   Bayesian Approach for State Estimation

The equivalent representation of conditional probabilities that will be used frequently in this chapter is

$$p(\mathbf{x}_k|\mathbf{y}_{1:k}) = p(\mathbf{x}_k|\mathbf{y}_k,\mathbf{y}_{1:k-1}).$$

Using the above equation and applying the Bayes' rule of probability Equation (1.3-5), we obtain the following equation to relate a priori to a posteriori density functions:

$$p(\mathbf{x}_k|\mathbf{y}_{1:k}) = \frac{p(\mathbf{y}_k|\mathbf{x}_k)}{p(\mathbf{y}_k|\mathbf{y}_{1:k-1})} p(\mathbf{x}_k|\mathbf{y}_{1:k-1}), \tag{2.4-1}$$

where $p(\mathbf{y}_k|\mathbf{x}_k,\mathbf{y}_{1:k-1}) = p(\mathbf{y}_k|\mathbf{x}_k)$ and the density $p(\mathbf{x}_k|\mathbf{y}_{k-1})$ can be obtained via

$$p(\mathbf{x}_k|\mathbf{y}_{1:k-1}) = \int p(\mathbf{x}_k|\mathbf{x}_{k-1})p(\mathbf{x}_{k-1}|\mathbf{y}_{1:k-1})d\mathbf{x}_{k-1}. \tag{2.4-2}$$

Putting these two together one obtains

$$p(\mathbf{x}_k|\mathbf{y}_{1:k}) = \frac{p(\mathbf{y}_k|\mathbf{x}_k)}{p(\mathbf{y}_k|\mathbf{y}_{1:k-1})} \int p(\mathbf{x}_k|\mathbf{x}_{k-1})p(\mathbf{x}_{k-1}|\mathbf{y}_{1:k-1})d\mathbf{x}_{k-1} \tag{2.4-3}$$

Remarks

1. $p(\mathbf{x}_k|\mathbf{y}_{1:k})$ is the a posteriori density of $\mathbf{x}_k$ given all measurements from discrete time 1 through $k, \mathbf{y}_{1:k}$, and $p(\mathbf{x}_k|\mathbf{y}_{1:k-1})$ is the a priori density of $\mathbf{x}_k$ giving all the measurements from discrete time 1 through $k-1$, $\mathbf{y}_{1:k-1}$.

2. $p(\mathbf{x}_k|\mathbf{x}_{k-1})$ is the state transition probability governed by the state equation, Equation (2.3-1), of the stochastic process $\mathbf{x}_k$. Because of the independence assumption on $\mathbf{x}_0$ and μ_{k-1}, $\mathbf{x}_k$ is a Markov process.

3. $p(\mathbf{x}_{k-1}|\mathbf{y}_{1:k-1})$ is the a posteriori density of $\mathbf{x}_{k-1}$ given all measurements, $\mathbf{y}_{1:k-1}$, from discrete time 1 through $k-1$, which is used to compute the a priori density for $\mathbf{x}_k$ as shown in Equation (2.4-2).

4. Equation (2.4-3) came from the paper by Bucy [14] where he treated the continuous system with continuous measurements. The above is the version for discrete systems and was first published by Ho and Lee [15]. This relationship is sufficiently general to represent both linear and nonlinear estimation problems. It should be emphasized here that the a posteriori probability density function, $p(\mathbf{x}_k|\mathbf{y}_{1:k})$ gives the complete statistical characterization of the estimate of the state vector $\mathbf{x}_k$, $\hat{\mathbf{x}}_{k|k}$, given all the measurements up to time k.

5. For linear and Gaussian systems, all densities in Equation (2.4-3) are Gaussian. $p(\mathbf{x}_k|\mathbf{y}_{1:k})$ can be completely represented by conditional mean and covariance of $\mathbf{x}_k$ given $\mathbf{y}_{1:k}$ that will be shown in the next section and the solution is the KF. For nonlinear and non-Gaussian systems, only numerical or approximated solutions can be found. Several such approaches, such as extended KF, unscented KF, and particle filter, will be discussed in future chapters.

2.5   Kalman Filter for State Estimation

The solution to the problem posted in Section 2.3 was published in the landmark paper by Kalman [1] for a linear system as shown in Equations (2.3-1) and (2.3-2). The state estimation problem is restated below based upon the recursive nature of state vector evolution.

The state estimation problem is to find $\hat{\mathbf{x}}_{k|k}$, the optimal estimator of $\mathbf{x}_k$ given a new measurement vector $\mathbf{y}_k$, and its covariance $\mathbf{P}_{k|k}$ with given the optimal estimator of $\mathbf{x}_{k-1}$ at the previous discrete time, $\hat{\mathbf{x}}_{k-1|k-1}$, and its covariance $\mathbf{P}_{k-1|k-1}$.

The KF gives a recursive algorithm to compute $\hat{\mathbf{x}}_{k|k}$ and its covariance $\mathbf{P}_{k|k}$ from $\mathbf{y}_k$, $\hat{\mathbf{x}}_{k-1|k-1}$, and $\mathbf{P}_{k-1|k-1}$ for $k = 1,2,\dots, k$. The KF is optimal (in the sense of weighted least squares and maximum a posteriori probability) when the following conditions are met.

1. System and measurement equations are linear.
2. $\mathbf{\Phi}_{k,k-1}$ (or $\mathbf{A}_t$) and $\mathbf{H}_k$ are completely known.
3. The statistics for initial state and system and measurement white noise processes are mutually independent and Gaussian with known mean and covariance, $\mathbf{x}_0: \sim N(\bar{\mathbf{x}}_0, \mathbf{P}_0)$, $\mathbf{\mu}_k: \sim N(\mathbf{0}, \mathbf{Q}_k)$, $\mathbf{v}_k: \sim N(\mathbf{0}, \mathbf{R}_k)$.
4. The system is observable (the ability to estimate the states with finite covariance).

Readers interested in learning more on controllability and observability should consult a control system textbook [e.g., 2–4]. The observability criterion derived in Section 2.3.1 will be revisited later through CRB derivation for state estimation. It will be shown that (a) the existence of CRB for the linear system Equations (2.3-1) and (2.3-2) and the observability of the system completely correspond to each other, and (b) the CRB gives the lowest covariance that any unbiased estimator can hope to achieve.

In the following sections, two approaches are presented to derive optimal state estimators that both result in the KF equations. The simplest derivation is an extension of the weighted least squares (WLS) estimator for parameter estimation of Section 1.5.2 to multiple time steps; this will be shown in the next section. The second approach is by applying Bayesian recursion to the linear system state estimation problem, which in later chapters will be shown as a way to solve a more general class of problems. The original paper by Kalman will be revisited to introduce the concept of orthogonal projection and its connection with the conditional mean (CM) estimator.

2.6 Kalman Filter Derivation: An Extension of Weighted Least Squares Estimator for Parameter Estimation

Recall that the WLS estimator for parameter estimation (see Section 1.6.2) considers the case that there exists a priori knowledge of the unknown parameter $\mathbf{x}$ represented by its mean and covariance, $\bar{\mathbf{x}}_0$ and $\mathbf{P}_0$. The optimum estimate of $\mathbf{x}$ given $\mathbf{y}$ is obtained as a weighted average of the measurement vector $\mathbf{y}$ and $\bar{\mathbf{x}}_0$ in Equations (1.5-8) and (1.5-9). This approach is applied to state estimation with time evolution using mathematical induction.

These two intermediate terms, $\hat{\mathbf{x}}_{k|k-1}$ and $\mathbf{P}_{k|k-1}$, are defined as

$\hat{\mathbf{x}}_{k|k-1}$ as the one-step predictor of $\mathbf{x}_k$ given all the measurements $\mathbf{y}_i$, from 1 through $k-1$, $\mathbf{y}_{1:k-1}$, and

$\mathbf{P}_{k|k-1}$ as the covariance of $\hat{\mathbf{x}}_{k|k-1}$.

Assuming that the WLS estimate for $\mathbf{x}_{k-1}$ at discrete time $k-1$ is given and represented by the conditional mean and covariance, $\hat{\mathbf{x}}_{k-1|k-1}$ and $\mathbf{P}_{k-1|k-1}$, respectively. Given that $\mathbf{x}_k$ is a linear mapping of $\mathbf{x}_{k-1}$ and $\boldsymbol{\mu}_{k-1}$ via Equation (2.3-1) and $\boldsymbol{\mu}_1, \boldsymbol{\mu}_2, \ldots, \boldsymbol{\mu}_{k-1}$ are mutually independent, the WLS estimate of $\mathbf{x}_k$ given $\mathbf{y}_{1:k-1}$, $\hat{\mathbf{x}}_{k|k-1}$, is the CM, $E\{\mathbf{x}_k|\mathbf{y}_{1:k-1}\}$ (see Proposition 1 and 2 of Appendix 1.B), this yields

$$\hat{\mathbf{x}}_{k|k-1} = \boldsymbol{\Phi}_{k,k-1}\hat{\mathbf{x}}_{k-1|k-1}, \tag{2.6-1}$$

and

$$\mathbf{P}_{k|k-1} = \boldsymbol{\Phi}_{k,k-1}\mathbf{P}_{k-1|k-1}\boldsymbol{\Phi}_{k,k-1}^T + \mathbf{Q}_{k-1}. \tag{2.6-2}$$

The new measurement taken at time k, $\mathbf{y}_k$, is applied to compute $\hat{\mathbf{x}}_{k|k}$. This can be done by choosing the $\mathbf{x}_k$ that minimizes the performance index

$$J = (\mathbf{y}_k - \mathbf{H}_k\mathbf{x}_k)^T \mathbf{R}_k^{-1}(\mathbf{y}_k - \mathbf{H}_k\mathbf{x}_k) + (\mathbf{x}_k - \hat{\mathbf{x}}_{k|k-1})^T \mathbf{P}_{k|k-1}^{-1}(\mathbf{x}_k - \hat{\mathbf{x}}_{k|k-1}).$$

Note that this equation is the same as Equation (1.5-5) in parameter estimation. $\hat{\mathbf{x}}_{k|k-1}$ is treated as the a priori knowledge of $\mathbf{x}_k$: $\sim N(\hat{\mathbf{x}}_{k|k-1}, \mathbf{P}_{k|k-1})$ given $\mathbf{y}_{1:k-1}$, so the new measurement $\mathbf{y}_k$ is incorporated by means of WLS estimation. Following the derivation in Section 1.5.2 and omitting intermediate steps, we obtain

$$\hat{\mathbf{x}}_{k|k} = \left[\mathbf{H}_k^T\mathbf{R}_k^{-1}\mathbf{H}_k + \mathbf{P}_{k|k-1}^{-1}\right]^{-1}\left(\mathbf{H}_k^T\mathbf{R}_k^{-1}\mathbf{y}_k + \mathbf{P}_{k|k-1}^{-1}\hat{\mathbf{x}}_{k|k-1}\right) \tag{2.6-3}$$

$$\mathbf{P}_{k|k} = \left[\mathbf{H}_k^T\mathbf{R}_k^{-1}\mathbf{H}_k + \mathbf{P}_{k|k-1}^{-1}\right]^{-1}. \tag{2.6-4}$$

The above equations are equivalent to Equations (1.5-6) and (1.5-9). Again, following the same derivation as in Section 1.5.2, we obtain the more familiar form of KF update equations, that is, the new measurement $\mathbf{y}_k$ is incorporated as a correction term to the predicted estimate $\hat{\mathbf{x}}_{k|k-1}$ as

$$\hat{\mathbf{x}}_{k|k} = \hat{\mathbf{x}}_{k|k-1} + \mathbf{P}_{k|k-1}\mathbf{H}_k^T \left[\mathbf{H}_k \mathbf{P}_{k|k-1}\mathbf{H}_k^T + \mathbf{R}_k \right]^{-1} (\mathbf{y}_k - \mathbf{H}_k \hat{\mathbf{x}}_{k|k-1}) \tag{2.6-5}$$

$$\mathbf{P}_{k|k} = \mathbf{P}_{k|k-1} - \mathbf{P}_{k|k-1}\mathbf{H}_k^T \left[\mathbf{H}_k \mathbf{P}_{k|k-1}\mathbf{H}_k^T + \mathbf{R}_k \right]^{-1} \mathbf{H}_k \mathbf{P}_{k|k-1}. \tag{2.6-6}$$

These equations are equivalent to Equations (1.5-8) and (1.5-9).

The filter covariance equations, Equation (2.6-2) and Equation (2.6-6), together are also known as the discrete Riccati equation (as equivalent to its counterpart in the continuous KF covariance equation). The equivalence of three ways for filter gain computation also applies and is used in Equation (2.6-5) as

$$\mathbf{K}_k = \mathbf{P}_{k|k-1}\mathbf{H}_k^T \left[\mathbf{H}_k \mathbf{P}_{k|k-1}\mathbf{H}_k^T + \mathbf{R}_k \right]^{-1},$$

$$\mathbf{K}_k = \mathbf{P}_{k|k}\mathbf{H}_k^T \mathbf{R}_k^{-1}, \text{ and}$$

$$\mathbf{K}_k = \left[\mathbf{H}_k^T \mathbf{R}_k^{-1}\mathbf{H}_k + \mathbf{P}_{k|k-1}^{-1} \right]^{-1} \mathbf{H}_k^T \mathbf{R}_k^{-1}. \tag{2.6-7}$$

Using the definition of filter gain in Equation (2.6-7), we can obtain another form to compute $\mathbf{P}_{k|k}$ that will guarantee it is a positive semidefinite matrix:

$$\mathbf{P}_{k|k} = (\mathbf{I} - \mathbf{K}_k\mathbf{H}_k)\mathbf{P}_{k|k-1}(\mathbf{I} - \mathbf{K}_k\mathbf{H}_k)^T + \mathbf{K}_k\mathbf{R}_k\mathbf{K}_k^T.$$

Information Filter

The recursive algorithm in Equations (2.6-5) through (2.6-7) can be replaced by the corresponding Fisher information matrix defined in Section 1.7 for parameter estimation. Letting $\mathcal{F}_{k|k} = \mathbf{P}_{k|k}^{-1}$ and $\mathcal{F}_{k|k-1} = \mathbf{P}_{k|k-1}^{-1}$, and using Equation (2.6-4), one obtains

$$\mathcal{F}_{k|k} = \mathbf{H}_k^T \mathbf{R}_k^{-1}\mathbf{H}_k + \mathcal{F}_{k|k-1}. \tag{2.6-8}$$

Using Equation (2.6-2), one obtains the one-step predicted information

$$\mathcal{F}_{k|k-1} = \left[\mathbf{\Phi}_{k,k-1}\mathcal{F}_{k-1|k-1}^{-1}\mathbf{\Phi}_{k,k-1}^T + \mathbf{\Phi}_{k-1} \right]^{-1}.$$

Applying Extension 1 of the matrix inversion lemma from Appendix A, Equation (A.2), one obtains

$$\mathcal{F}_{k|k-1} = \mathbf{Q}_{k-1}^{-1} - \mathbf{Q}_{k-1}^{-1}\mathbf{\Phi}_{k,k-1} \left[\mathcal{F}_{k-1|k-1} + \mathbf{\Phi}_{k,k-1}^T\mathbf{Q}_{k-1}^{-1}\mathbf{\Phi}_{k,k-1} \right]^{-1} \mathbf{\Phi}_{k,k-1}^T\mathbf{Q}_{k-1}^{-1}. \tag{2.6-9}$$

The state prediction equation is unchanged while the state update equation becomes

$$\hat{\mathbf{x}}_{k|k} = \mathcal{F}_{k|k}^{-1} \left(\mathbf{H}_k^T\mathbf{R}_k^{-1}\mathbf{y}_k + \mathcal{F}_{k|k-1}\hat{\mathbf{x}}_{k|k-1} \right). \tag{2.6-10}$$

Using Equation (2.6-7), the Kalman gain can be written using the notation of the information matrix as

$$\mathbf{K}_k = \mathcal{F}_{k|k}^{-1} \mathbf{H}_k^T \mathbf{R}_k^{-1}.$$

An advantage of implementing the KF with information filter (IF) algorithms is that the filter can be initiated without any initial condition, that is, the above algorithm can be started with $\mathcal{F}_{0|0} = \mathbf{0}$. This will be shown to be very useful in the fixed interval smoother implementation for the backward filter (see Section 2.9.1).

A summary of discrete KF equations is shown in Table 2.2.

Remarks

1. The details of deriving Equations (2.6-5) through (2.6-7) are omitted here. Readers unclear about the derivation should refer to the previous chapter and the associated homework problems.

2. Equations (2.6-1), (2.6-2), (2.6-5), and (2.6-6) are the KF equations; the first two are known as state and covariance prediction equations and the second two are known as the state and covariance update equations. The equations derived in Section 1.6.2 are all equivalent to the above, where those in Section 1.6.2 are a single stage update and thus a special case of the KF. The recursive nature of the KF in terms of the prediction-correction process is illustrated in Figure 2.1.

3. The term $(\mathbf{y}_k - \mathbf{H}_k \hat{\mathbf{x}}_{k|k-1})$ in Equation (2.6-5) is the difference between the actual measurement $\mathbf{y}_k$ and the predicted measurement $\mathbf{H}_k \hat{\mathbf{x}}_{k|k-1}$, both at time k. It is known as filter residual or the innovation process, see Kailath [12, 16]. The innovation process is a white noise process with zero mean and covariance $\mathbf{H}_k \mathbf{P}_{k|k-1} \mathbf{H}_k^T + \mathbf{R}_k$ when all the assumptions for the optimality of KF realizations are met [16]. When the unknown system behavior is modeled as part of the disturbance by the process noise, a technique was suggested by Mehra [17] to adaptively adjust $\mathbf{Q}_k$ based upon the behavior of the innovation process, and will be discussed in a later chapter.

2.7   Kalman Filter Derivation: Using the Recursive Bayes' Rule

In this section, the Bayesian recursive equation, Equation (2.4-1), is applied to derive the KF. This relationship is repeated below for easy reference:

$$p(\mathbf{x}_k | \mathbf{y}_{1:k}) = \frac{p(\mathbf{y}_k | \mathbf{x}_k)}{p(\mathbf{y}_k | \mathbf{y}_{1:k-1})} p(\mathbf{x}_k | \mathbf{y}_{1:k-1}). \tag{2.7-1}$$

Table 2.2   Summary of Discrete Kalman Filter Equations

Discrete Kalman Filter
State and Measurement Equations

$$\mathbf{x}_k = \boldsymbol{\Phi}_{k,k-1}\mathbf{x}_{k-1} + \boldsymbol{\mu}_{k-1}$$

$$\mathbf{y}_k = \mathbf{H}_k\mathbf{x}_k + \mathbf{v}_k$$

where $\mathbf{x} \in \mathbb{R}^n$, $\mathbf{y} \in \mathbb{R}^m$, $\mathbf{x}_0: \sim N(\overline{\mathbf{x}}_0, \mathbf{P}_0)$, $\boldsymbol{\mu}_k: \sim N(\mathbf{0}, \mathbf{Q}_k)$, $\mathbf{v}_k: \sim N(\mathbf{0}, \mathbf{R}_k)$, are mutually independent and $\boldsymbol{\Phi}_{k,k-1}$ and $\mathbf{H}_k$ are completely specified.

Problem Statement
Obtain an estimate of $\mathbf{x}_k$, denoted as $\hat{\mathbf{x}}_{k|k}$, with all measurements $\mathbf{y}_{1:k} = \{\mathbf{y}_1, \mathbf{y}_2, \ldots, \mathbf{y}_k\}$
Obtain the covariance of $\hat{\mathbf{x}}_{k|k}$, denoted as $\mathbf{P}_{k|k}$.

Solution Process
Given: $\hat{\mathbf{x}}_{k-1|k-1}$ and $\mathbf{P}_{k-1|k-1}$, with initial conditions $\hat{\mathbf{x}}_{0|0} = \overline{\mathbf{x}}_0$ and $\mathbf{P}_{0|0} = \mathbf{P}_0$ at $k = 1$.
Predict: compute $\hat{\mathbf{x}}_{k|k-1}$ and $\mathbf{P}_{k|k-1}$ with $\hat{\mathbf{x}}_{k-1|k-1}$ and $\mathbf{P}_{k-1|k-1}$ using conditional mean formulation.
Update: compute $\hat{\mathbf{x}}_{k|k}$ and $\mathbf{P}_{k|k}$ with $\hat{\mathbf{x}}_{k|k-1}$ and $\mathbf{P}_{k|k-1}$ and new measurement vector $\mathbf{y}_k$ using the WLS formulation.

Solution Equations
Let $k = 1$ and set $\hat{\mathbf{x}}_{0|0} = \overline{\mathbf{x}}_0$ and $\mathbf{P}_{0|0} = \mathbf{P}_0$
Predict:

$$\hat{\mathbf{x}}_{k|k-1} = \boldsymbol{\Phi}_{k,k-1}\hat{\mathbf{x}}_{k-1|k-1}$$

$$\mathbf{P}_{k|k-1} = \boldsymbol{\Phi}_{k,k-1}\mathbf{P}_{k-1|k-1}\boldsymbol{\Phi}_{k,k-1}^T + \mathbf{Q}_{k-1}$$

Update:

$$\hat{\mathbf{x}}_{k|k} = \hat{\mathbf{x}}_{k|k-1} + \mathbf{P}_{k|k-1}\mathbf{H}_k^T\left[\mathbf{H}_k\mathbf{P}_{k|k-1}\mathbf{H}_k^T + \mathbf{R}_k\right]^{-1}\left(\mathbf{y}_k - \mathbf{H}_k\hat{\mathbf{x}}_{k|k-1}\right)$$

$$\mathbf{P}_{k|k} = \mathbf{P}_{k|k-1} - \mathbf{P}_{k|k-1}\mathbf{H}_k^T\left[\mathbf{H}_k\mathbf{P}_{k|k-1}\mathbf{H}_k^T + \mathbf{R}_k\right]^{-1}\mathbf{H}_k\mathbf{P}_{k|k-1}$$

Alternative Update Equation

$$\hat{\mathbf{x}}_{k|k} = \left[\mathbf{H}_k^T\mathbf{R}_k^{-1}\mathbf{H}_k + \mathbf{P}_{k|k-1}^{-1}\right]^{-1}\left(\mathbf{H}_k^T\mathbf{R}_k^{-1}\mathbf{y}_k + \mathbf{P}_{k|k-1}^{-1}\hat{\mathbf{x}}_{k|k-1}\right)$$

$$\mathbf{P}_{k|k} = \left[\mathbf{H}_k^T\mathbf{R}_k^{-1}\mathbf{H}_k + \mathbf{P}_{k|k-1}^{-1}\right]^{-1}$$

Kalman Gain Equation

$$\mathbf{K}_k = \mathbf{P}_{k|k-1}\mathbf{H}_k^T\left[\mathbf{H}_k\mathbf{P}_{k|k-1}\mathbf{H}_k^T + \mathbf{R}_k\right]^{-1} = \mathbf{P}_{k|k}\mathbf{H}_k^T\mathbf{R}_k^{-1} = \left[\mathbf{H}_k^T\mathbf{R}_k^{-1}\mathbf{H}_k + \mathbf{P}_{k|k-1}^{-1}\right]^{-1}\mathbf{H}_k^T\mathbf{R}_k^{-1}$$

Alternative Solution via Information Filter
Let $k = 1$ and set $\hat{\mathbf{x}}_{0|0} = \overline{\mathbf{x}}_0$ and $\mathcal{F}_{0|0} = \mathbf{P}_0^{-1}$
Predict:

$$\hat{\mathbf{x}}_{k|k-1} = \boldsymbol{\Phi}_{k,k-1}\hat{\mathbf{x}}_{k-1|k-1}$$

$$\mathcal{F}_{k|k-1} = \mathbf{Q}_{k-1}^{-1} - \mathbf{Q}_{k-1}^{-1}\boldsymbol{\Phi}_{k,k-1}\left[\mathcal{F}_{k-1|k-1} + \boldsymbol{\Phi}_{k,k-1}^T\mathbf{Q}_{k-1}^{-1}\boldsymbol{\Phi}_{k,k-1}\right]^{-1}\boldsymbol{\Phi}_{k,k-1}^T\mathbf{Q}_{k-1}^{-1}$$

Update:

$$\hat{\mathbf{x}}_{k|k} = \mathcal{F}_{k|k}^{-1}\left(\mathbf{H}_k^T\mathbf{R}_k^{-1}\mathbf{y}_k + \mathcal{F}_{k|k-1}\hat{\mathbf{x}}_{k|k-1}\right) = \hat{\mathbf{x}}_{k|k-1} + \mathbf{K}_k\left(\mathbf{y}_k - \mathbf{H}_k\hat{\mathbf{x}}_{k|k-1}\right)$$

$$\mathcal{F}_{k|k} = \mathbf{H}_k^T\mathbf{R}_k^{-1}\mathbf{H}_k + \mathcal{F}_{k|k-1}$$

Kalman Gain Equation

$$\mathbf{K}_k = \mathcal{F}_{k|k}^{-1}\mathbf{H}_k^T\mathbf{R}_k^{-1}$$

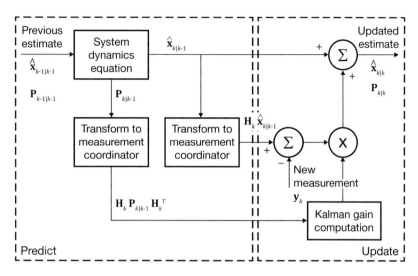

Figure 2.1   Kalman filter illustrated as a prediction-update process.

For linear Gaussian cases, the definitions and relations for the conditional mean and covariance of all the (Gaussian) probability density functions are used and then substituted into the left- and right-hand sides of Equation (2.7-1). The derivations in this section follow closely with those in Section 1.6.5. For the conditional mean and covariance of all the (Gaussian) probability density functions on the right-hand side of Equation (2.7-1),

1. $p(\mathbf{x}_k|\mathbf{y}_{1:k-1})$

$E\left\{\mathbf{x}_k|\mathbf{y}_{1:k-1}\right\} \triangleq \hat{\mathbf{x}}_{k|k-1} = \mathbf{\Phi}_{k,k-1}\hat{\mathbf{x}}_{k-1|k-1},$

$Cov\left\{\mathbf{x}_k|\mathbf{y}_{1:k-1}\right\} \triangleq \mathbf{P}_{k|k-1} = \mathbf{\Phi}_{k,k-1}\mathbf{P}_{k-1|k-1}\mathbf{\Phi}_{k,k-1}^T + \mathbf{Q}_{k-1}.$

2. $p(\mathbf{y}_k|\mathbf{x}_k)$

$E\{\mathbf{y}_k|\mathbf{x}_k\} = \mathbf{H}_k\mathbf{x}_k,$

$Cov\{\mathbf{y}_k|\mathbf{x}_k\} = \mathbf{R}_k.$

3. $p(\mathbf{y}_k|\mathbf{y}_{1:k-1})$

$E\left\{\mathbf{y}_k|\mathbf{y}_{1:k-1}\right\} = \mathbf{H}_k\hat{\mathbf{x}}_{k|k-1},$

$Cov\left\{\mathbf{y}_k|\mathbf{y}_{1:k-1}\right\} = \mathbf{H}_k\mathbf{P}_{k|k-1}\mathbf{H}_k^T + \mathbf{R}_k.$

Further, the conditional mean and covariance of the conditional density function on the left-hand side of (2.7-1), that is, $p(\mathbf{x}_k|\mathbf{y}_{1:k})$, yields

4. $p(\mathbf{x}_k|\mathbf{y}_{1:k})$

$$E\left\{\mathbf{x}_k|\mathbf{y}_{1:k}\right\} \triangleq \hat{\mathbf{x}}_{k|k},$$

$$Cov\left\{\mathbf{x}_k|\mathbf{y}_{1:k}\right\} \triangleq \mathbf{P}_{k|k}.$$

Applying the same method as used in Homework Problem 10 of Chapter 1, one obtains

$$\mathbf{P}_{k|k}^{-1} = \mathbf{P}_{k|k-1}^{-1} + \mathbf{H}_k^T \mathbf{R}_k^{-1} \mathbf{H}_k, \tag{2.7-2}$$

or

$$\mathbf{P}_{k|k} = \left[\mathbf{P}_{k|k-1}^{-1} + \mathbf{H}_k^T \mathbf{R}_k^{-1} \mathbf{H}_k\right]^{-1},$$

and

$$\hat{\mathbf{x}}_{k|k} = \mathbf{P}_{k|k}\left(\mathbf{P}_{k|k-1}^{-1}\hat{\mathbf{x}}_{k|k-1} + \mathbf{H}_k^T \mathbf{R}_k^{-1}\mathbf{y}_k\right). \tag{2.7-3}$$

Equations (2.7-2) and (2.7-3) can be rewritten into the more familiar form by simple application of the matrix inversion lemma (see Appendix A), resulting in the same measurement update algorithm as Equations (2.6-5) and (2.6-6)

$$\hat{\mathbf{x}}_{k|k} = \hat{\mathbf{x}}_{k|k-1} + \mathbf{P}_{k|k-1}\mathbf{H}_k^T\left[\mathbf{H}_k\mathbf{P}_{k|k-1}\mathbf{H}_k^T + \mathbf{R}_k\right]^{-1}\left(\mathbf{y}_k - \mathbf{H}_k\hat{\mathbf{x}}_{k|k-1}\right), \tag{2.7-4}$$

$$\mathbf{P}_{k|k} = \mathbf{P}_{k|k-1} - \mathbf{P}_{k|k-1}\mathbf{H}_k^T\left[\mathbf{H}_k\mathbf{P}_{k|k-1}\mathbf{H}_k^T + \mathbf{R}_k\right]^{-1}\mathbf{H}_k\mathbf{P}_{k|k-1}. \tag{2.7-5}$$

Thus, the KF is rederived using the Bayesian approach with the conditional mean as the state estimate (see Homework Problem 4). These results are identical to those obtained as a multistage extension of the WLS estimator.

2.8 Review of Certain Estimator Properties in the Kalman Filter Original Paper

The derivations in the original Kalman filter paper [1] (an updated derivation with noisy measurements is in [18]) will not be reproduced. However, several salient points of the paper will be summarized as they represent important concepts in estimation in linear vector spaces. The definition of orthogonal projection in a linear vector space will be introduced first.

Orthogonal Projections

Let $\{\mathbf{z}_1, \mathbf{z}_2, \ldots, \mathbf{z}_j\}$ denote a set of random vectors in $\mathbb{R}^n$ and the vector space spanned by $\{\mathbf{z}_1, \mathbf{z}_2, \ldots, \mathbf{z}_j\}$, is denoted by $\mathbb{Z}_j = \mathrm{Span}\{\mathbf{z}_1, \mathbf{z}_2, \ldots, \mathbf{z}_j\}$, furthermore, there exists a set of coefficients $\{a_1, a_2, \ldots, a_j\}$ for every $\mathbf{z} \in \mathbb{Z}_j$ such that

$$\mathbf{z} = \sum_{i=1}^{j} a_i \mathbf{z}_i.$$

A random vector $\mathbf{x}$ is orthogonal to $\mathbb{Z}_j$, if the inner product

$$\langle \mathbf{x}, \mathbf{z}_i \rangle \triangleq E\{\mathbf{x}^T \mathbf{z}_i\} = 0, \ \forall i$$

that is equivalent to the inner product of $\mathbf{x}$ with any vectors in $\mathbb{Z}_j$ is zero. Then, for any random vector, $\mathbf{x}_t$, it can be represented as $\mathbf{x}_t = \hat{\mathbf{x}} + \tilde{\mathbf{x}}$ where $\hat{\mathbf{x}}$ is the orthogonal projection of $\mathbf{x}_t$ onto $\mathbb{Z}_j$, and $\langle \tilde{\mathbf{x}}, \hat{\mathbf{x}} \rangle = 0$. Furthermore, $\tilde{\mathbf{x}}$ is orthogonal to any linear combination of $\{\mathbf{z}_1, \mathbf{z}_2, \ldots, \mathbf{z}_j\}$ as shown in Figure 2.2.

Conditional Mean Estimator

Let $L(.)$ be a scalar-valued nondecreasing and symmetric loss function (or cost function) and $L(\mathbf{0}) = 0$. Let $\mathbf{g}(\mathbf{y})$ be the estimate of $\mathbf{x}$ given $\mathbf{y}$ and $\boldsymbol{\varepsilon}$ is the estimation error, that is, $\varepsilon = \mathbf{x} - \hat{\mathbf{x}}$, then the optimal estimate, in the sense of minimizing the average loss function; $E\{L(\boldsymbol{\varepsilon})\}$ is the conditional expectation, that is,

$$\hat{\mathbf{x}} = E\{\mathbf{x} \mid \mathbf{y}\}. \tag{2.8-1}$$

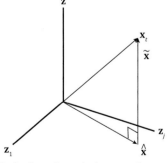

Projection of $\mathbf{x}_t$ onto $\{\mathbf{z}_1, \mathbf{z}_2, \ldots \mathbf{z}_j\}$ space

Figure 2.2   Illustration of orthogonal projection.

Typical loss functions could be square error, $\| \mathbf{\varepsilon} \|^2$, and the associated inner product as

$$\langle \mathbf{x}, \mathbf{z} \rangle \triangleq E\{\mathbf{x}^T \mathbf{z}\} \tag{2.8-2}$$

for all random vectors $\mathbf{x}, \mathbf{z} \in \mathbb{R}^n$. The proof of Equation (2.8-1) is the solution of $\mathbf{g}(\mathbf{y})$ that minimizes $E\{\| \mathbf{\varepsilon} \|^2\}$ is given in Appendix 1.B.

Remarks

Property 3 of the conditional mean estimator (see Section 1.3.3 and Proposition 2 in Appendix 1.B) can be used to show that $\mathbf{\varepsilon} = \mathbf{x} - \hat{\mathbf{x}}$ is orthogonal to all $\mathbf{g}(\mathbf{y})$ mapped into $\mathbb{R}^n$. If $\mathbf{g}(\mathbf{y})$ is limited to be linear, Figure 2.2 illustrates the relationship between $\mathbf{\varepsilon} = \tilde{\mathbf{x}}$ and $\mathbb{Z}_j$ is spanned by vectors of linear mapping of $\mathbf{y}$, or, equivalently, all linear estimators of $\mathbf{x}$ given $\mathbf{y}$. But as mentioned in Section 1.3, the CM estimator is not the same as the WLS estimator in general.

Now, applying the same orthogonal property in the measurement space, $\{\mathbf{y}_1, \mathbf{y}_2, \ldots, \mathbf{y}_k\}$ and the associated linear vector space $\mathbb{Y}_k = \mathrm{Span}\{\mathbf{y}_1, \mathbf{y}_2, \ldots, \mathbf{y}_k\}$ up to time step k. For a new measurement vector $\mathbf{y}_{k+1}$ in $\mathbb{R}^m$ at time step $k+1$, one can prove that if $\tilde{\mathbf{y}}_{k+1|k} \triangleq \mathbf{y}_{k+1} - E\{\mathbf{y}_{k+1}|\mathbb{Y}_k\}$, then

$$\tilde{\mathbf{y}}_{k+1|k} \perp \mathbb{Y}_k, \tag{2.8-3}$$

followed by the Property 3 of CM estimator in Section 1.3.3. The random vector $\tilde{\mathbf{y}}_{k+1|k}$ is the additional information conveyed by the new measurement $\mathbf{y}_{k+1}$ [1] and is called the innovation process by Kailath [16]. The conditional expectation of $E\{\mathbf{y}_{k+1}|\mathbb{Y}_k\}$ is not computable in general, for the case of Equation (2.2-2)

$$E\{\mathbf{y}_{k+1}|\mathbb{Y}_k\} = E\{\mathbf{H}_{k+1}\mathbf{x}_{k+1} + \mathbf{v}_{k+1}|\mathbb{Y}_k\} = \mathbf{H}_{k+1}E\{\mathbf{x}_{k+1}|\mathbb{Y}_k\} + E\{\mathbf{v}_{k+1}|\mathbb{Y}_k\} = \mathbf{H}_{k+1}E\{\mathbf{x}_{k+1}|\mathbb{Y}_k\}.$$

The last equality is due to the independent assumption of $\mathbf{v}_{k+1}$ in Equations (2.2-1) and (2.2-2).

Orthogonal Projection and Conditional Mean Estimator

For any random vectors $\mathbf{x}_k$ and $\mathbf{y}_{1:k}$ and $L(\mathbf{\varepsilon}) \triangleq \langle \mathbf{\varepsilon}, \mathbf{\varepsilon} \rangle$, then the orthogonal projection estimate is the conditional expectation, $\hat{\mathbf{x}}_k = E\{\mathbf{x}_k \mid \mathbf{y}_{1:k}\}$.

The above properties were true without any restrictions between $\mathbf{x}_k$ and $\mathbf{y}_{1:k}$ (see Proposition 1 of Appendix 1.B), thus a practical algorithm to compute the estimate is rather pointless. However, if $\mathbf{x}_k$ and $\mathbf{y}_k$ are related as defined by linear Equations (2.2-1) and (2.2-2) and driven by Gaussian white noise, then the KF equation can be derived, and the first step of the derivation will be shown.

Using the CM definition,

$$\hat{\mathbf{x}}_{i|j} = E\{\mathbf{x}_i|\mathbf{y}_{1:j}\} = E\{\mathbf{x}_i|\mathbf{y}_j,\mathbf{y}_{1:j-1}\} = E\{\mathbf{x}_i|\mathbf{y}_{1:j-1}\} + E\{\mathbf{x}_i|\tilde{\mathbf{y}}_j\}. \qquad (2.8\text{-}4)$$

The last equality is derived from the CM property and the orthogonal projection property in Equation (2.8-3) of the measurement space $\mathbf{y}_{1:j}$. Because $\mathbf{x}_i$ and $\mathbf{y}_{1:j-1}$ are both Gaussian random vectors, and Equations (2.2-1) and (2.2-2) are linear, yields

$$\hat{\mathbf{x}}_{i|j} = \Phi_{i,i-1}E\{\mathbf{x}_{i-1}|\mathbf{y}_{1:j-1}\} + E\{\mathbf{x}_i|\tilde{\mathbf{y}}_j\} = \Phi_{i,i-1}\hat{\mathbf{x}}_{i-1|j-1} + E\{\mathbf{x}_i|\tilde{\mathbf{y}}_j\} = \hat{\mathbf{x}}_{i|j-1} + E\{\mathbf{x}_i|\tilde{\mathbf{y}}_j\}.$$

This equation follows the familiar KF structure where $\hat{\mathbf{x}}_{i|j-1}$ is the one-step predicted estimate and the $E\{\mathbf{x}_i|\tilde{\mathbf{y}}_j\}$ is the correction term to update the estimate of $\mathbf{x}_i$ with new measurements $\mathbf{y}_j$. Details of this part of the derivation is omitted; interested readers are directed to the references [1, 16, 18]. Note that the time indices are freely chosen to be i and j without any restrictions. This says that the same structure applies generally for smoothing and prediction (see definition in Section 2.3).

2.9 Smoother

For the discussion in this chapter up to this point, the focus has been seeking the best estimate of state vector $\mathbf{x}_k$ with measurements taken from t_1 to current time t_k. It is possible to use data up to t_k, $\mathbf{y}_{1:k}$, which is beyond the time of interest t_i ($t_k > t_i$), to get improved estimates of $\mathbf{x}_i$. This is called smoothing. As defined in Section 2.3, $\hat{\mathbf{x}}_{i|k}$ is the estimate of $\mathbf{x}_i$ based upon measurements up to time t_k, $\mathbf{y}_{1:k}$, where k can be $> i$. In this section, three types of smoothing algorithms are considered, each for a different purpose, fixed interval smoothing (FIS), fixed point smoothing (FPS), and fixed lag smoothing (FLS).

Fixed Interval Smoothing
The total data interval is given as $[1, \ldots, K]$, find $\hat{\mathbf{x}}_{k|K}$ where $1 \leq k \leq K$.

Fixed Point Smoothing
Obtain the estimate at a fixed time t_k while the end of data interval K is increasing, that is, find $\hat{\mathbf{x}}_{k|K}$ as K increasing with k staying constant.

Fixed Lag Smoothing

Obtain the estimate at a fixed lag K behind the end of the data interval $k + K$, that is, find $\hat{\mathbf{x}}_{k|k+K}$ with K staying constant.

Among the three, FIS is used most often as a tool to get the best estimated target trajectory after data collection on a target is completed from a flight test, the so-called post-mission analysis to obtain the estimates that are closest to the true target state trajectory. It is also the simplest in terms of algorithm complexity. The FPS and FLS can all be implemented based upon the basic algorithm of FIS although they may not be the most efficient way. FIS is developed below with some example results for FPS and FLS algorithms stated without derivation. Since data beyond the time of interest is used, some new notation is defined.

2.9.1   Notation and Definitions

Let $[1, \ldots, K]$ be the total interval of data collection, $\mathbf{y}_{1:K}$, then the smoothed estimate of $\mathbf{x}_k$ for $(k < K)$ is given as

Smoothed Estimate of $\mathbf{x}_k$ for $(k < K)$

$$\hat{\mathbf{x}}_{k|1:K} = E\left\{\mathbf{x}_k | \mathbf{y}_{1:K}\right\} \text{ with covariance } \mathbf{P}_{k|1:K}. \tag{2.9-1}$$

Furthermore, two new sets of state estimates and covariance are defined with one that depends on past (up to now) data and the other depends on future data.

Forward Filtered Estimate of $\mathbf{x}_k$ given $\mathbf{y}_{1:k}$

$$\hat{\mathbf{x}}_{k|1:k} = E\left\{\mathbf{x}_k | \mathbf{y}_{1:k}\right\} \text{ with covariance } \mathbf{P}_{k|1:k}, \tag{2.9-2}$$

Backward Filtered Estimate of $\mathbf{x}_{k+1}$ given $\mathbf{y}_{k+1:K}$

$$\hat{\mathbf{x}}_{k+1|k+1:K} = E\left\{\mathbf{x}_{k+1} | \mathbf{y}_{k+1:K}\right\} \text{ with covariance } \mathbf{P}_{k+1|k+1:K}, \tag{2.9-3}$$

where the one-step backward predicted estimate of $\mathbf{x}_k$ is obtained as

$$\hat{\mathbf{x}}_{k|k+1:K} = \mathbf{\Phi}_{k,k+1}\hat{\mathbf{x}}_{k+1|k+1:K}, \tag{2.9-4}$$

with covariance

$$\mathbf{P}_{k|k+1:K} = \mathbf{\Phi}_{k,k+1}\mathbf{P}_{k+1|k+1:K}\mathbf{\Phi}_{k,k+1}^{T} + \mathbf{Q}_{k+1}. \tag{2.9-5}$$

The relationships of those definitions are illustrated in Figure 2.3.

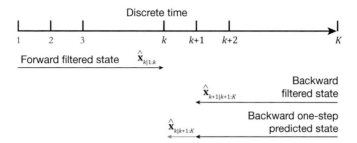

Figure 2.3   Illustration of forward and backward estimation.

2.9.2   Fixed Interval Smoother

For FIS, the smoother is first derived by combining the forward filtered and backward filtered estimates. Using the approach of WLS estimator in Section 2.6, the optimal smoothed estimate $\hat{\mathbf{x}}_{k\text{II}:K}$ is the $\mathbf{x}_k$ minimizing

$$J = \left(\mathbf{x}_k - \hat{\mathbf{x}}_{k\text{II}:k}\right)^T \mathbf{P}_{k\text{II}:k}^{-1} \left(\mathbf{x}_k - \hat{\mathbf{x}}_{k\text{II}:k}\right) + \left(\mathbf{x}_k - \hat{\mathbf{x}}_{k|k+1:K}\right)^T \mathbf{P}_{k|k+1:K}^{-1} \left(\mathbf{x}_k - \hat{\mathbf{x}}_{k|k+1:K}\right).$$

This is again a familiar quadratic form for minimization; the results are stated without derivation,

$$\hat{\mathbf{x}}_{k\text{II}:K} = \left[\mathbf{P}_{k\text{II}:k}^{-1} + \mathbf{P}_{k|k+1:K}^{-1}\right]^{-1} \left(\mathbf{P}_{k\text{II}:k}^{-1} \hat{\mathbf{x}}_{k\text{II}:k} + \mathbf{P}_{k|k+1:K}^{-1} \hat{\mathbf{x}}_{k|k+1:K}\right), \tag{2.9-6}$$

and

$$\mathbf{P}_{k\text{II}:K} = \left[\mathbf{P}_{k\text{II}:k}^{-1} + \mathbf{P}_{k|k+1:K}^{-1}\right]^{-1}. \tag{2.9-7}$$

The above algorithm can be written using the IF algorithms as

$$\hat{\mathbf{x}}_{k\text{II}:K} = \left[\mathcal{F}_{k\text{II}:K}\right]^{-1} \left(\mathcal{F}_{k\text{II}:k} \hat{\mathbf{x}}_{k\text{II}:k} + \mathcal{F}_{k|k+1:K} \hat{\mathbf{x}}_{k|k+1:K}\right),$$

and

$$\mathcal{F}_{k\text{II}:K} = \mathcal{F}_{k\text{II}:k} + \mathcal{F}_{k\text{II}+1:K},$$

where $\mathcal{F}_{K|K:K} = \mathbf{0}$ and $\hat{\mathbf{x}}_{K|K:K}$ can be arbitrary because it has no information but are conveniently set to zero.

Similar to the dual form of KF, one can apply the matrix inversion lemma to obtain the alternate form of the FIS algorithm

$$\hat{\mathbf{x}}_{k\text{II}:K} = \hat{\mathbf{x}}_{k\text{II}:k} + \mathbf{P}_{k\text{II}:K} \mathbf{P}_{k|k+1:K}^{-1} \left(\hat{\mathbf{x}}_{k|k+1:K} - \hat{\mathbf{x}}_{k\text{II}:k}\right), \tag{2.9-8}$$

$$\mathbf{P}_{k|1:K} = \mathbf{P}_{k|1:k} - \mathbf{P}_{k|1:k} \left[\mathbf{P}_{k|1:k} + \mathbf{P}_{k|k+1:K}\right]^{-1} \mathbf{P}_{k|1:k}. \tag{2.9-9}$$

Equations (2.9-8) and (2.9-9) are the FIS implementation used most often because of its simplicity.

The backward filtered estimate $\hat{\mathbf{x}}_{k+1|k+1:K}$ is obtained by running a KF backward in time starting at the end, K, using the same system and measurement equations. The initial condition of the backward filter is with zero information, similar to the parameter estimation without a priori knowledge. The FIS filter was first published by Rauch, Tung, and Striebel [19]. Their original algorithm is stated below without derivation.

$$\hat{\mathbf{x}}_{k|1:K} = \hat{\mathbf{x}}_{k|1:k} + \mathbf{A}_k \left(\hat{\mathbf{x}}_{k+1|1:K} - \hat{\mathbf{x}}_{k+1|1:k} \right),$$

$$\mathbf{A}_k = \mathbf{P}_{k|1:k} \mathbf{\Phi}_{k+1,k}^T \mathbf{P}_{k+1|1:k}^{-1},$$

$$\mathbf{P}_{k|1:K} = \mathbf{P}_{k|1:k} + \mathbf{A}_k \left[\mathbf{P}_{k+1|1:K} - \mathbf{P}_{k+1|1:k} \right] \mathbf{A}_k^T.$$

Note that this form of FIS is going backward in time as $\hat{\mathbf{x}}_{k+1|1:K}$ is used to obtain $\hat{\mathbf{x}}_{k|1:K}$.

2.9.3   Fixed Point Smoother

The FPS algorithm is stated below without derivation [7].

$$\hat{\mathbf{x}}_{k|1:K} = \hat{\mathbf{x}}_{k|1:K-1} + \mathbf{B}_{K-1} \left(\hat{\mathbf{x}}_{K|1:K} - \hat{\mathbf{x}}_{K|1:K-1} \right),$$

$$\mathbf{B}_{K-1} = \prod_{i=k}^{K-1} \mathbf{A}_i,$$

$$\mathbf{A}_i = \mathbf{P}_{i|1:i} \mathbf{\Phi}_{i+1,i}^T \mathbf{P}_{i+1|1:i}^{-1},$$

$$\mathbf{P}_{k|1:K} = \mathbf{P}_{k|1:K-1} + \mathbf{B}_{K-1} \left[\mathbf{P}_{k|1:k} - \mathbf{P}_{k|1:k-1} \right] \mathbf{B}_{K-1}^T.$$

2.9.4   Fixed Lag Smoother

The FLS algorithm is stated below without derivation [7].

$$\hat{\mathbf{x}}_{k|1:K+k} = \mathbf{\Phi}_{k,k-1} \hat{\mathbf{x}}_{k-1|1:K+k-1} + \mathbf{Q}_{k-1} \mathbf{\Phi}_{k-1,k}^T \mathbf{P}_{k-1|1:k-1}^{-1} \left(\hat{\mathbf{x}}_{k-1|1:K+k-1} - \hat{\mathbf{x}}_{k-1|1:k-1} \right)$$
$$+ \mathbf{B}_{K+k-1} \mathbf{K}_{K+k-1} \left(\mathbf{y}_{K+k} - \mathbf{H}_k \mathbf{\Phi}_{K+k,K+k-1} \hat{\mathbf{x}}_{K+k-1|1:K+k-1} \right)$$

$$\mathbf{B}_{K+k-1} = \prod_{i=k}^{K+k-1} \mathbf{A}_i,$$

$$\mathbf{A}_i = \mathbf{P}_{i|1:i} \mathbf{\Phi}_{i+1,i}^T \mathbf{P}_{i+1|1:i}^{-1},$$

$$\mathbf{P}_{k|1:K+k} = \mathbf{P}_{k|1:k-1} + \mathbf{B}_{K+k-1} \mathbf{K}_{K+k-1} \mathbf{H}_{K+k-1} \mathbf{P}_{K+k|1:K+k-1} \mathbf{B}_{K+k-1}^T - \mathbf{A}_{k-1}^{-1} \left[\mathbf{P}_{k-1|1:k-1} - \mathbf{P}_{k-1|1:K+k-1} \right] \mathbf{A}_{k-1}^{-T}$$

It should be emphasized that both FPS and FLS estimates can be obtained by repeatedly applying FIS, which is simple and less prone to mistakes in implementation. The above algorithms are stated for the purpose of completeness.

2.9.5   FIS for Deterministic Systems with Noisy Measurements

For deterministic systems, the process noise term $\boldsymbol{\mu}_k$ is zero for all k. There are plenty of such examples in the real world, for example, a ballistic object (like a satellite) flying in space is a deterministic system. Only linear systems are considered here. The extension to nonlinear systems will be discussed in a later chapter. For deterministic systems represented by state-space differential equations and that the solution of these equations is unique, then the determination of a state vector at any instance of time along the trajectory determines the entire trajectory. For the sake of generality, the state is chosen at a particular time t_k given the total time interval of data collection $[1, \ldots, K]$. This is the same as finding the $\mathbf{x}_k$ that minimizes

$$J = \sum_{i=1}^{K} (\mathbf{y}_i - \mathbf{H}_i \mathbf{x}_i)^T \mathbf{R}_i^{-1} (\mathbf{y}_i - \mathbf{H}_i \mathbf{x}_i) + (\mathbf{x}_0 - \hat{\mathbf{x}}_0)^T \mathbf{P}_0^{-1} (\mathbf{x}_0 - \hat{\mathbf{x}}_0).$$

Taking the derivative of J with respect to $\mathbf{x}_k$ and setting it to zero gives the necessary condition for a solution

$$\frac{\partial J}{\partial \mathbf{x}_k} = -\sum_{i=1}^{K} \left[\mathbf{H}_i \left[\frac{\partial \mathbf{x}_i}{\partial \mathbf{x}_k} \right] \right]^T \mathbf{R}_i^{-1} (\mathbf{y}_i - \mathbf{H}_i \mathbf{x}_i) + \left[\frac{\partial \mathbf{x}_0}{\partial \mathbf{x}_k} \right]^T \mathbf{P}_0^{-1} (\mathbf{x}_0 - \hat{\mathbf{x}}_0) = \mathbf{0}. \quad (2.9\text{-}10)$$

We will illustrate how $\frac{\partial \mathbf{x}_i}{\partial \mathbf{x}_k}$ is computed. The answer will also lead to the computation of $\frac{\partial \mathbf{x}_0}{\partial \mathbf{x}_k}$. The computation of $\frac{\partial \mathbf{x}_k}{\partial \mathbf{x}_i}$ is shown as follows

$$\frac{\partial \mathbf{x}_k}{\partial \mathbf{x}_i} = \boldsymbol{\Phi}_{k,k-1} \frac{\partial \mathbf{x}_{k-1}}{\partial \mathbf{x}_i} = \boldsymbol{\Phi}_{k,k-1} \boldsymbol{\Phi}_{k-1,k-2} \frac{\partial \mathbf{x}_{k-2}}{\partial \mathbf{x}_i} = \ldots = \boldsymbol{\Phi}_{k,k-1} \boldsymbol{\Phi}_{k-1,k-2} \ldots \boldsymbol{\Phi}_{i+1,i} = \boldsymbol{\Phi}_{k,i}$$

Recall that $\boldsymbol{\Phi}_{j,j-1}$ is the state transition matrix of $\mathbf{x}_{j-1}$ to $\mathbf{x}_j$ defined in Equations (2.2-3) and (2.2-5). With the property of $\boldsymbol{\Phi}_{k,i}$ shown in Section 2.2, time t_i could be greater or less than t_k. With this definition, $\dfrac{\partial \mathbf{x}_i}{\partial \mathbf{x}_k}$ can be written as

$$\frac{\partial \mathbf{x}_i}{\partial \mathbf{x}_k} = \boldsymbol{\Phi}_{i,i+1} \ldots \boldsymbol{\Phi}_{k-2,k-1} \boldsymbol{\Phi}_{k-1,k} = \boldsymbol{\Phi}_{i,k}. \quad (2.9\text{-}11)$$

Substituting Equation (2.9-11) into Equation (2.9-10) yields

$$-\sum_{i=1}^{K} \boldsymbol{\Phi}_{i,k}^T \mathbf{H}_i^T \mathbf{R}_i^{-1} (\mathbf{y}_i - \mathbf{H}_i \boldsymbol{\Phi}_{i,k} \mathbf{x}_k) + \boldsymbol{\Phi}_{0,k}^T \mathbf{P}_0^{-1} (\boldsymbol{\Phi}_{0,k} \mathbf{x}_k - \hat{\mathbf{x}}_0) = \mathbf{0}. \quad (2.9\text{-}12)$$

Adopting the notation defined in Section 2.9.2, the solution to Equation (2.9-12) is denoted as $\hat{\mathbf{x}}_{k\|:K}$,

$$\hat{\mathbf{x}}_{k\text{II}:K} = \left[\sum\nolimits_{i=1}^{K} \boldsymbol{\Phi}_{i,k}^{T} \mathbf{H}^{T} \mathbf{R}_{i}^{-1} \mathbf{H} \boldsymbol{\Phi}_{i,k} + \boldsymbol{\Phi}_{0,k}^{T} \mathbf{P}_{0}^{-1} \boldsymbol{\Phi}_{0,k} \right]^{-1}$$
$$\times \left(\sum\nolimits_{i=1}^{K} \boldsymbol{\Phi}_{i,k}^{T} \mathbf{H}_{i}^{T} \mathbf{R}_{i}^{-1} \mathbf{y}_{i} + \boldsymbol{\Phi}_{0,k}^{T} \mathbf{P}_{0}^{-1} \hat{\mathbf{x}}_{0} \right) \tag{2.9-13}$$

This solution takes on a familiar form: the weighted sum of the initial condition on $\mathbf{x}_0$ and measurements $\mathbf{y}_i$ for $i \in [1, 2, \ldots, K]$. Note that all vectors (measurements and initial state) are transitioned to time t_k for the solution on $\mathbf{x}_k$. The covariance of $\hat{\mathbf{x}}_{k\text{II}:K}$ is

$$\mathbf{P}_{k\text{II}:K} = \left[\sum\nolimits_{i=1}^{K} \boldsymbol{\Phi}_{i,k}^{T} \mathbf{H}^{T} \mathbf{R}_{i}^{-1} \mathbf{H} \boldsymbol{\Phi}_{i,k} + \boldsymbol{\Phi}_{0,k}^{T} \mathbf{P}_{0}^{-1} \boldsymbol{\Phi}_{0,k} \right]^{-1}. \tag{2.9-14}$$

One can derive the FIS equation for the deterministic system by setting the process noise covariance $\mathbf{Q}$ in the FIS equation for stochastic systems to zero. This is left as a homework problem (Homework Problem 6).

This algorithm has often been applied for a short segment of data to produce a state estimate, sometimes referred to as a tracklet. The advantage of doing data compression is so that the data transmission rate in a multisensor system can be reduced. This will be discussed in Chapter 7 on the multisensor estimation algorithm. The fixed interval smoother algorithms for stochastic and deterministic systems are summarized in Table 2.3.

Remarks

As illustrated above, FIS can be implemented using two KFs, one running forward in time and the other running backward. The estimate at a given time is the weighted sum of the forward filtered estimate and the backward one-step predicted estimate. Because FIS uses data of the past and the future, the estimate obtained by FIS uses all the available information in time history, therefore achieving the best accuracy. It is referred to as the best estimated trajectory (BET) in some applications. BET is produced after data collection exercises for use as "ground truth." In the case when the system is nonlinear, two extended KFs (EKF)[5] can be used in place of KF. The improved accuracy of FIS over filtering will be shown in Section 4.8 for the problem of a sinusoidal signal in noise using EKF. Interested readers looking for a comparison of filter and smoother results for a highly nonlinear system can consult the references [20, 21].

---

5   EKF is the subject of next chapter, Nonlinear Estimation.

Table 2.3 Summary of Fixed Interval Smoother Algorithms

Fixed Interval Smoother Algorithms
Definition

$$\hat{\mathbf{x}}_{k|1:K} = E\{\mathbf{x}_k | \mathbf{y}_{1:K}\} \text{ with covariance } \mathbf{P}_{k|1:K}$$

for the total data interval of $[1, \ldots, K]$, where $1 \le k \le K$.

For Stochastic System, Using Forward and Backward Filter Implementation

$$\hat{\mathbf{x}}_{k|1:K} = \left[\mathbf{P}_{k|1:k}^{-1} + \mathbf{P}_{k|k+1:K}^{-1}\right]^{-1}\left(\mathbf{P}_{k|1:k}^{-1}\hat{\mathbf{x}}_{k|1:k} + \mathbf{P}_{k|k+1:K}^{-1}\hat{\mathbf{x}}_{k|k+1:K}\right)$$

$$\mathbf{P}_{k|1:K} = \left[\mathbf{P}_{k|1:k}^{-1} + \mathbf{P}_{k|k+1:K}^{-1}\right]^{-1}.$$

We emphasize that the backward filter is to be initiated without any prior information.

Alternative Form

$$\hat{\mathbf{x}}_{k|1:K} = \hat{\mathbf{x}}_{k|1:k} + \mathbf{P}_{k|1:K}\mathbf{P}_{k|k+1:K}^{-1}\left(\hat{\mathbf{x}}_{k|k+1:K} - \hat{\mathbf{x}}_{k|1:k}\right)$$

$$\mathbf{P}_{k|1:K} = \mathbf{P}_{k|1:k} - \mathbf{P}_{k|1:k}[\mathbf{P}_{k|1:k} + \mathbf{P}_{k|k+1:K}]^{-1}\mathbf{P}_{k|1:k}$$

For Deterministic Systems, Use Batch Filter Concept

$$\hat{\mathbf{x}}_{k|1:K} = \left[\sum_{i=1}^{k}\boldsymbol{\Phi}_{i,k}^{T}\mathbf{H}_{i}^{T}\mathbf{R}_{i}^{-1}\mathbf{H}_{i}\boldsymbol{\Phi}_{i,k} + \boldsymbol{\Phi}_{0,k}^{T}\mathbf{P}_{0}^{-1}\boldsymbol{\Phi}_{0,k}\right]^{-1} \times \left(\sum_{i=1}^{k}\boldsymbol{\Phi}_{i,k}^{T}\mathbf{H}_{i}^{T}\mathbf{R}_{i}^{-1}\mathbf{y}_{i} + \boldsymbol{\Phi}_{0,k}^{T}\mathbf{P}_{0}^{-1}\hat{\mathbf{x}}_{0}\right)$$

$$\mathbf{P}_{k|1:K} = \left[\sum_{i=1}^{k}\boldsymbol{\Phi}_{i,k}^{T}\mathbf{H}_{i}^{T}\mathbf{R}_{i}^{-1}\mathbf{H}_{i}\boldsymbol{\Phi}_{i,k} + \boldsymbol{\Phi}_{0,k}^{T}\mathbf{P}_{0}^{-1}\boldsymbol{\Phi}_{0,k}\right]^{-1}$$

2.9.6   Application of FIS for Kalman Filter Initial Condition Computation

To restate the discrete linear system equations from the KF development,

$$\mathbf{x}_k = \boldsymbol{\Phi}_{k,k-1}\mathbf{x}_{k-1} + \boldsymbol{\mu}_{k-1},$$

$$\mathbf{y}_k = \mathbf{H}_k\mathbf{x}_k + \mathbf{v}_k,$$

where $\mathbf{x} \in \mathbb{R}^n$, $\mathbf{y} \in \mathbb{R}^m$, $\boldsymbol{\Phi}_{k,k-1}$, and $\mathbf{H}_k$ are known matrices, and $\mathbf{x}_0$, $\boldsymbol{\mu}_{k-1}$, and $\mathbf{v}_k$ are mutually independent and $\mathbf{x}_0 : \sim N(\bar{\mathbf{x}}_0, \mathbf{P}_0)$, $\boldsymbol{\mu}_k : \sim N(\mathbf{0}, \mathbf{Q}_k)$, and $\mathbf{v}_k : \sim N(\mathbf{0}, \mathbf{R}_k)$.

In most practical problems, there is either no prior knowledge of the initial state, or else $\bar{\mathbf{x}}_0$ and $\mathbf{P}_0$ do not exist. One will need to compute them using the measurements $\mathbf{y}_k$, $k = 1 \ldots K$. Consider the algorithm of Section 2.6.5, the FIS for a deterministic system. It described a WLS estimator for noise free system. The deterministic system assumption is valid when the data window used for the WLS is short. Applying Equations (2.9-13) and (2.9-14) by setting $\mathbf{P}_0^{-1} = \mathbf{0}$ (i.e., no a priori information) one obtains

$$\hat{\mathbf{x}}_{k|1:K} = \left[\sum_{i=1}^{K}\boldsymbol{\Phi}_{i,k}^{T}\mathbf{H}_{i}^{T}\mathbf{R}_{i}^{-1}\mathbf{H}\boldsymbol{\Phi}_{i,k}\right]^{-1}\left(\sum_{i=1}^{K}\boldsymbol{\Phi}_{i,k}^{T}\mathbf{H}_{i}^{T}\mathbf{R}_{i}^{-1}\mathbf{y}_{i}\right) \qquad (2.9\text{-}15)$$

with covariance

$$\mathbf{P}_{k\text{II}:K} = \left[\sum_{i=1}^{K} \mathbf{\Phi}_{i,k}^{T} \mathbf{H}_{i}^{T} \mathbf{R}_{i}^{-1} \ \mathbf{H} \mathbf{\Phi}_{i,k} \right]^{-1}. \tag{2.9-16}$$

The data window of length K should be taken to be as short as possible in order to make the assumption of the deterministic system close to being valid. For instance, for $\mathbf{x} \in \mathbb{R}^{n}$ and $\mathbf{y} \in \mathbb{R}^{m}$, K can be int$(n/m) + 1$, where int(n/m) denotes the integer of the fraction n/m, which is the condition for which an observable system will result in nonsingular information matrix $\left[\sum_{i=1}^{K} \mathbf{\Phi}_{i,k}^{T} \mathbf{H}_{i}^{T} \mathbf{R}_{i}^{-1} \ \mathbf{H} \mathbf{\Phi}_{i,k} \right]$ with minimum K. The resulting $\hat{\mathbf{x}}_{k\text{II}:K}$ and $\mathbf{P}_{k|1:K}$ thus form the initial condition for KF at time K.

The fixed interval smoother with deterministic system introduced here is for a linear system. As will be shown in the next chapter, the extension to the extended KF is by simply replacing the linear system and measurement matrices with Jacobian matrices (the first derivatives) of the nonlinear system and measurement equations. The use of FIS for filter initiation can thus be extended to nonlinear systems. If one wants to obtain a more accurate initial estimate, an iterative solution for nonlinear systems introduced in Chapter 3 on nonlinear estimation can be used.

A specific example of filter initial condition computation for radar measurements with a polynomial trajectory model will be shown in the Appendix in Chapter 8.

2.10   The Cramer–Rao Bound for State Estimation

The CRB for parameter estimation in Section 1.7 can be extended to state estimation. Recall the definition of CRB for estimating a random parameter vector $\mathbf{x}$ as

$$Cov\{\hat{\mathbf{x}}\} \geq \left[E \left\{ \left[\frac{\partial \ln p(\mathbf{y}, \mathbf{x})}{\partial \mathbf{x}} \right] \left[\frac{\partial \ln p(\mathbf{y}, \mathbf{x})}{\partial \mathbf{x}} \right]^{T} \right\} \right]^{-1}.$$

For state estimation, $\mathbf{x}$ evolves with time and $\mathbf{y}$ represents multiple measurements of $\mathbf{x}$ in time. The state and measurement equations for linear systems were defined in Equations (2.2-1) and (2.2-2), respectively. Because of the process noise term $\mathbf{\mu}_{k-1}$ in Equation (2.2-1), estimating $\mathbf{x}_{k}$ is no longer a static problem as in parameter estimation: the estimate of $\mathbf{x}_{k}$ at time t_{k} depends on the estimate of $\mathbf{x}$ at all other times. The straightforward extension of the definition of CRB for state estimation is

$$Cov\{\hat{\mathbf{x}}_{1:k\text{II}:k}\} \geq \text{CRB}(\mathbf{x}_{1:k}|\mathbf{y}_{1:k}) = \left[E \left\{ \left[\frac{\partial \ln p(\mathbf{y}_{1:k}; \mathbf{x}_{1:k})}{\partial \mathbf{x}_{1:k}} \right] \left[\frac{\partial \ln p(\mathbf{y}_{1:k}; \mathbf{x}_{1:k})}{\partial \mathbf{x}_{1:k}} \right]^{T} \right\} \right]^{-1}, \tag{2.10-1}$$

where the Fisher information matrix for estimating $\mathbf{x}_{1:k}$ is a $nk \times nk$ matrix as shown within the inside of the inversion bracket above

$$\mathcal{F}(\hat{\mathbf{x}}_{1:k|1:k}) \triangleq \left[E \left\{ \left[\frac{\partial \ln p(\mathbf{x}_{1:k};\mathbf{y}_{1:k})}{\partial \mathbf{x}_{1:k}} \right] \left[\frac{\partial \ln p(\mathbf{x}_{1:k};\mathbf{y}_{1:k})}{\partial \mathbf{x}_{1:k}} \right]^{T} \right\} \right].$$

Note that the notations $\mathbf{y}_{1:k}$ and $\mathbf{x}_{1:k}$ are used to represent the series of measurement and state vectors, respectively, from 1 to k. With the matrix that is growing with k, it is impractical to compute its inverse if one were to compute CRB using Equation (2.10-1) directly. A special case is considered first for when the system is deterministic; in this case the state estimation problem becomes a parameter estimation problem, thus the growing dimensions problem does not exist. The CRB for the problem of estimating the state of a linear stochastic system will be shown as a special case for nonlinear stochastic system, which will be fully explored in Section 3.10 of the next chapter.

We emphasize that because KF is a minimum variance estimator and its covariance equation is exact, a covariance bound is not meaningful. The purpose of discussing CRB for linear system is to show the fact that the KF covariance equation is indeed the same as the CRB equation.

2.10.1   For Deterministic Systems

For deterministic systems, the process noise term, $\boldsymbol{\mu}_k$ is zero for all k. Similar to the reason in Section 2.8.5 for the FIS for deterministic systems, determining the CRB for state estimate of any instance of time will determine the CRB for the entire trajectory. Letting k be the current time, we determine the CRB for the covariance of estimating $\mathbf{x}_k$, which by definition is

$$Cov\{\hat{\mathbf{x}}_{k|k}\} \geq \left[E_{\mathbf{v}_{1:k};\mathbf{x}_0} \left\{ \left[\frac{\partial \ln p(\mathbf{y}_{1:k},\mathbf{x}_0;\mathbf{x}_k)}{\partial \mathbf{x}_k} \right] \left[\frac{\partial \ln p(\mathbf{y}_{1:k},\mathbf{x}_0;\mathbf{x}_k)}{\partial \mathbf{x}_k} \right]^{T} \right\} \right]^{-1}. \quad (2.10\text{-}2)$$

Note that the subscript r of the operator $E_r\{.\}$ denotes the random variable that the statistical expectation is taken with respect to. Assuming mutual independence and Gaussian on the measurement noise and on the initial condition $\mathbf{x}_0$, the log of the density function after ignoring the additive and multiplicative constants is

$$J = \sum_{i=1}^{k} (\mathbf{y}_i - \mathbf{H}_i \mathbf{x}_i)^{T} \mathbf{R}_i^{-1} (\mathbf{y}_i - \mathbf{H}_i \mathbf{x}_i) + (\mathbf{x}_0 - \bar{\mathbf{x}}_0)^{T} \mathbf{P}_0^{-1} (\mathbf{x}_0 - \bar{\mathbf{x}}_0). \quad (2.10\text{-}3)$$

Taking derivative of J with respect to $\mathbf{x}_k$ yields

$$\frac{\partial J}{\partial \mathbf{x}_k} = \sum_{i=1}^{k} \left[\mathbf{H}_i \left[\frac{\partial \mathbf{x}_i}{\partial \mathbf{x}_k} \right] \right]^T \mathbf{R}_i^{-1} (\mathbf{y}_i - \mathbf{H}_i \mathbf{x}_i) + \left[\frac{\partial \mathbf{x}_0}{\partial \mathbf{x}_k} \right]^T \mathbf{P}_0^{-1} (\mathbf{x}_0 - \overline{\mathbf{x}}_0).$$

Multiplying $\dfrac{\partial J}{\partial \mathbf{x}_k}$ by its transpose and taking the expectation with respect to $\mathbf{v}_{1:k}$ and $\mathbf{x}_0$ yields

$$E_{\mathbf{v}_{1:k}:\mathbf{x}_0} \left\{ \left(\frac{\partial J}{\partial \mathbf{x}_k} \right) \left(\frac{\partial J}{\partial \mathbf{x}_k} \right)^T \right\} = \sum_{i=1}^{k} \left[\mathbf{H}_i \left[\frac{\partial \mathbf{x}_i}{\partial \mathbf{x}_k} \right] \right]^T \mathbf{R}_i^{-1} \left[\mathbf{H}_i \left[\frac{\partial \mathbf{x}_i}{\partial \mathbf{x}_k} \right] \right] + \left[\frac{\partial \mathbf{x}_0}{\partial \mathbf{x}_k} \right]^T \mathbf{P}_0^{-1} \left[\frac{\partial \mathbf{x}_0}{\partial \mathbf{x}_k} \right].$$

The above expression is the Fisher information matrix on estimating $\mathbf{x}_k$. Let it be denoted by $\mathcal{F}_{\hat{\mathbf{x}}_{k|k}}$. Then the CRB on estimating $\mathbf{x}_k$ is

$$Cov\{\hat{\mathbf{x}}_{k|k}\} \geq CRB(\mathbf{x}_k | \mathbf{y}_{1:k}) = [\mathcal{F}_{\hat{\mathbf{x}}_{k|k}}]^{-1}$$

$$= \left[\sum_{i=1}^{k} \left[\mathbf{H}_i \left[\frac{\partial \mathbf{x}_i}{\partial \mathbf{x}_k} \right] \right]^T \mathbf{R}_i^{-1} \left[\mathbf{H}_i \left[\frac{\partial \mathbf{x}_i}{\partial \mathbf{x}_k} \right] \right] + \left[\frac{\partial \mathbf{x}_0}{\partial \mathbf{x}_k} \right]^T \mathbf{P}_0^{-1} \left[\frac{\partial \mathbf{x}_0}{\partial \mathbf{x}_k} \right] \right]^{-1} \qquad (2.10\text{-}4)$$

Equation (2.10-4) is the fundamental result. Two special cases are considered: (a) the relationship of CRB to system observability criterion; and (b) a recursive algorithm to compute Fisher information matrix, or its inverse, the CRB.

Relationship to Observability Criterion

Re-deriving Equation (2.10-4) for the case that the state vector to be estimated is the initial state, $\mathbf{x}_0$, results in

$$Cov\{\hat{\mathbf{x}}_{0|k}\} \geq CRB(\mathbf{x}_0 | \mathbf{y}_{1:k}) = [\mathcal{F}_{\hat{\mathbf{x}}_{0|k}}]^{-1}$$

$$= \left[\sum_{i=1}^{k} \left[\mathbf{H}_i \left[\frac{\partial \mathbf{x}_i}{\partial \mathbf{x}_0} \right] \right]^T \mathbf{R}_i^{-1} \left[\mathbf{H}_i \left[\frac{\partial \mathbf{x}_i}{\partial \mathbf{x}_0} \right] \right] + \mathbf{P}_0^{-1} \right]^{-1} \qquad (2.10\text{-}5)$$

where $\left[\frac{\partial \mathbf{x}_0}{\partial \mathbf{x}_0} \right] = \mathbf{I}$. The computation of $\frac{\partial \mathbf{x}_i}{\partial \mathbf{x}_0}$ and that of $\frac{\partial \mathbf{x}_k}{\partial \mathbf{x}_i}$ are the same as defined in Section 2.9.5 and will not be repeated here.

The significance of Equation (2.10-5) is as follows, when there is no a priori knowledge on $\mathbf{x}_0$, then $\mathbf{P}_0^{-1} = \mathbf{0}$ Equation (2.10-5) becomes

$$Cov\{\hat{\mathbf{x}}_{0|k}\} \geq CRB(\mathbf{x}_0 | \mathbf{y}_{1:k}) = [\mathcal{F}_{\hat{\mathbf{x}}_{0|k}}]^{-1} = \left[\sum_{i=1}^{k} \left[\mathbf{H}_i \left[\frac{\partial \mathbf{x}_i}{\partial \mathbf{x}_0} \right] \right]^T \mathbf{R}_i^{-1} \left[\mathbf{H}_i \left[\frac{\partial \mathbf{x}_i}{\partial \mathbf{x}_0} \right] \right] \right]^{-1}. \qquad (2.10\text{-}6)$$

Given $\frac{\partial \mathbf{x}_i}{\partial \mathbf{x}_0} = \mathbf{\Phi}_{i,0}$, the above equation is the same as observability condition, $\sum_{i=1}^{k} \mathbf{\Phi}_{i,0}^T \mathbf{H}_i^T \mathbf{H}_i \mathbf{\Phi}_{i,0}$, as in Equation (2.3-6) with $\mathbf{R}_i = \mathbf{I}$. Therefore, it is concluded that when a system is observable, the covariance for any unbiased state estimator exists. The CRB is thus a first step before the attempt to build an unbiased estimator.

A Recursive Algorithm for Computing Cramer–Rao Bound

The computation of $\mathrm{CRB}(\mathbf{x}_k|\mathbf{y}_{1:k})$, Equation (2.10-4), can be greatly simplified with a recursive algorithm. This algorithm will be derived using mathematical induction.

Let $k = 1$, then $i = 1$ only. Using the notation for the Fisher information matrix,

$$\mathcal{F}_{\hat{\mathbf{x}}_{1|1}} = \mathbf{H}_1^T \mathbf{R}_1^{-1} \mathbf{H}_1 + \mathbf{\Phi}_{0,1}^T \mathbf{P}_0^{-1} \mathbf{\Phi}_{0,1}.$$

Let $k = 2$, then $i = 1$ and 2.
For $i = 1$,

$$\mathbf{\Phi}_{1,2}^T \mathbf{H}_1^T \mathbf{R}_1^{-1} \mathbf{H}_1 \mathbf{\Phi}_{1,2} + \mathbf{\Phi}_{1,2}^T \mathbf{\Phi}_{0,1}^T \mathbf{P}_0^{-1} \mathbf{\Phi}_{0,1} \mathbf{\Phi}_{1,2} = \mathbf{\Phi}_{1,2}^T \left[\mathcal{F}_{\hat{\mathbf{x}}_{1|1}} \right] \mathbf{\Phi}_{1,2}.$$

For $i = 2$,

$$\mathbf{H}_2^T \mathbf{R}_2^{-1} \mathbf{H}_2.$$

Combining the above two together yields

$$\mathcal{F}_{\hat{\mathbf{x}}_{2|2}} = \mathbf{\Phi}_{1,2}^T \mathcal{F}_{\hat{\mathbf{x}}_{1|1}} \mathbf{\Phi}_{1,2} + \mathbf{H}_2^T \mathbf{R}_2^{-1} \mathbf{H}_2.$$

One can now easily extend the above results to the general case

$$\mathcal{F}_{\hat{\mathbf{x}}_{k|k}} = \mathbf{\Phi}_{k-1,k}^T \mathcal{F}_{\hat{\mathbf{x}}_{k-1|k-1}} \mathbf{\Phi}_{k-1,k} + \mathbf{H}_k^T \mathbf{R}_k^{-1} \mathbf{H}_k. \tag{2.10-7}$$

This equation gives the recursive evolution of Fisher's information. Its inverse is the Cramer–Rao bound,

$$\mathrm{CRB}(\mathbf{x}_k|\mathbf{y}_{1:k}) = \left[\mathcal{F}_{\hat{\mathbf{x}}_{k|k}} \right]^{-1} = \left[\mathbf{\Phi}_{k-1,k}^T \mathcal{F}_{\hat{\mathbf{x}}_{k-1|k-1}} \mathbf{\Phi}_{k-1,k} + \mathbf{H}_k^T \mathbf{R}_k^{-1} \mathbf{H}_k \right]^{-1}. \tag{2.10-8}$$

The algorithm given in Equations (2.10-7) and (2.10-8) means that one can recursively compute Fisher's information matrix and invert it for the Cramer–Rao bound when it is needed. Note that $\mathcal{F}_{\hat{\mathbf{x}}_{k|k}}$ may be singular for small k, that is, when the number of measurements is small. If $\mathcal{F}_{\hat{\mathbf{x}}_{k|k}}$ is always singular then the system is not observable. As such, there does not exist any unbiased estimator with finite covariance. When $\mathcal{F}_{\hat{\mathbf{x}}_{k|k}}$ becomes nonsingular, the resulting $\mathrm{CRB}(\mathbf{x}_k|\mathbf{y}_{1:k})$ can also be computed recursively. Using the notation $\mathbf{P}_{k|k} = \mathrm{CRB}(\mathbf{x}_k|\mathbf{y}_{1:k})$ and $\mathbf{P}_{k|k-1} = \mathrm{CRB}(\mathbf{x}_k|\mathbf{y}_{1:k-1})$, the recursive algorithm is stated below leaving the derivation to the readers.

$$\mathbf{P}_{k|k} = \left[\mathbf{P}_{k|k-1}^{-1} + \mathbf{H}_k^T \mathbf{R}_k^{-1} \mathbf{H}_k\right]^{-1} = \mathbf{P}_{k|k-1} - \mathbf{P}_{k|k-1} \mathbf{H}_k^T \left[\mathbf{H}_k \mathbf{P}_{k|k-1} \mathbf{H}_k^T + \mathbf{R}_k\right]^{-1} \mathbf{H}_k \mathbf{P}_{k|k-1} \quad (2.10\text{-}9)$$

where

$$\mathbf{P}_{k|k-1} = \mathbf{\Phi}_{k,k-1} \mathbf{P}_{k-1|k-1} \mathbf{\Phi}_{k,k-1}^T. \quad (2.10\text{-}10)$$

Note that the two equations above are the same as the KF covariance equation when the process noise is zero.

2.10.2   For Stochastic Linear Systems

The a posteriori CRB for the nonlinear discrete system developed by Tichavsky [22] is an exact formulation for the solution of Equation (2.10-1). For a nonlinear system, the statistical expectation may need to be approximated (e.g., by Monte Carlo simulation), but for linear systems, it reduces to the KF covariance equation. This approach will be fully described in Section 3.10 of the next chapter, will thus not to be included here. Table 2.4 presents a summary of CRB for linear systems.

Table 2.4   Summary of Cramer–Rao Bound for Linear Systems

Cramer–Rao Bound on the Covariance of Linear State Estimation

Cramer–Rao Bound (CRB) Definition

$$Cov\{\hat{\mathbf{x}}_{1:k|1:k}\} \geq \left[E\left\{ \left[\frac{\partial \ln p(\mathbf{y}_{1:k}; \mathbf{x}_{1:k})}{\partial \mathbf{x}_{1:k}}\right] \left[\frac{\partial \ln p(\mathbf{y}_{1:k}; \mathbf{x}_{1:k})}{\partial \mathbf{x}_{1:k}}\right]^T \right\} \right]^{-1}$$

CRB for Linear Deterministic Systems

$$Cov\{\hat{\mathbf{x}}_{k|k}\} \geq \mathrm{CRB}(\mathbf{x}_k|\mathbf{y}_{1:k}) = [\mathcal{F}_{\hat{\mathbf{x}}_{k|k}}]^{-1} = \left[\sum_{i=1}^{k} \left[\mathbf{H}_i\left[\frac{\partial \mathbf{x}_i}{\partial \mathbf{x}_k}\right]\right]^T \mathbf{R}_i^{-1} \left[\mathbf{H}_i\left[\frac{\partial \mathbf{x}_i}{\partial \mathbf{x}_k}\right]\right] + \left[\frac{\partial \mathbf{x}_0}{\partial \mathbf{x}_k}\right]^T \mathbf{P}_0^{-1} \left[\frac{\partial \mathbf{x}_0}{\partial \mathbf{x}_k}\right] \right]^{-1}$$

The above can be computed recursively exactly the same as KF covariance equation for deterministic systems.

CRB for Linear Stochastic Systems

$$\mathbf{P}_{k|k-1} = \mathbf{\Phi}_{k,k-1} \mathbf{P}_{k-1|k-1} \mathbf{\Phi}_{k,k-1}^T + \mathbf{Q}_{k-1}$$

$$\mathbf{P}_{k|k} = \mathbf{P}_{k|k-1} - \mathbf{P}_{k|k-1} \mathbf{H}_k^T \left[\mathbf{H}_k \mathbf{P}_{k|k-1} \mathbf{H}_k^T + \mathbf{R}_k\right]^{-1} \mathbf{H}_k \mathbf{P}_{k|k-1}$$

This is identical to the KF covariance equation for stochastic systems. See Section 3.10.2 for derivation.

2.11   A Kalman Filter Example

The sinusoidal signal in noise example from Section 1.8 will be revisited. Consider the following ordinary differential equation

$$\ddot{x} = -\omega^2 x \tag{2.11-1}$$

with known ω and initial condition x_0. The solution of this equation is well known,

$$x = a\sin(\omega t + \varphi)$$

where $x_0 = a\sin\varphi$. Equation (2.11-1) can be rewritten with a state vector representation

$$\dot{\mathbf{x}} = \begin{bmatrix} 0 & 1 \\ -\omega^2 & 0 \end{bmatrix} \mathbf{x}, \tag{2.11-2}$$

where $\mathbf{x} = [x_1, x_2]^T$, with $x_1 = x$ and $x_2 = \dot{x}$. Although there is no system noise specified here, in practice, a good KF designer always adds system noise in the covariance equation, which can normally prevent filter divergence (see practical considerations in Chapter 4). In such cases, a modified system equation is considered

$$\dot{\mathbf{x}} = \mathbf{A}\mathbf{x} + \mathbf{B}\xi, \tag{2.11-3}$$

where $\mathbf{A} = \begin{bmatrix} 0 & 1 \\ -\omega^2 & 0 \end{bmatrix}$ and $\mathbf{B} = \begin{bmatrix} 0 \\ 1 \end{bmatrix}$. The system noise term ξ is a scalar and only impacts the higher derivative portion of the state vector, in this case $\dot{x}_2$, with variance σ_q^2. The above is a continuous system while our filtering equations are derived for discrete systems. The discrete equivalent of the continuous system is obtained by using the state transition matrix, as defined in Equation (2.2-6), $\mathbf{\Phi}_{t_k,t_{k-1}} = e^{\mathbf{A}(t_k - t_{k-1})}$. For the system above, a closed form solution for $\mathbf{\Phi}_{t_k,t_{k-1}}$ exists

$$\mathbf{\Phi}_{t_k,t_{k-1}} = \begin{bmatrix} \cos(\omega\Delta t) & \dfrac{\sin(\omega\Delta t)}{\omega} \\ -\omega\sin(\omega\Delta t) & \cos(\omega\Delta t) \end{bmatrix}, \tag{2.11-4}$$

where $\Delta t = t_k - t_{k-1}$. The equivalent noise covariance for the discrete system can be obtained by applying Equation (2.2-9)

$$\mathbf{Q}_{k-1} = \sigma_q^2 \Delta t \begin{bmatrix} \dfrac{2\omega\Delta t - \sin(2\omega\Delta t)}{4\omega^3} & \dfrac{\sin^2(\omega\Delta t)}{2\omega^2} \\ \dfrac{\sin^2(\omega\Delta t)}{2\omega^2} & \dfrac{\Delta t}{2} + \dfrac{\sin(2\omega\Delta t)}{4\omega} \end{bmatrix}.$$

Sampled measurements are taken on the first element of x, x_1, that is,

$$y_k = \mathbf{H}\mathbf{x}_k + v_k, \tag{2.11-5}$$

where $\mathbf{H} = [1\ 0]$ and v_k is white measurement noise with distribution $N(0, \sigma_v^2)$. Let $\hat{\mathbf{x}}_{0|0}$ denote the mean of the initial state of $\mathbf{x}_0$, with covariance $\mathbf{P}_{0|0}$. A proper selection of $\{\hat{\mathbf{x}}_{0|0}, \mathbf{P}_{0|0}\}$ will determine the transient behavior of the filter state estimate, $\hat{\mathbf{x}}_{k|k}$, but will not change its steady state behavior. In practice, $\{\hat{\mathbf{x}}_{0|0}, \mathbf{P}_{0|0}\}$ can be obtained with the first few measurements taken at discrete times before starting the filter. For this example at least two measurements are needed, y_1 and y_2. The derivation for $\hat{\mathbf{x}}_{0|0}$ and $\mathbf{P}_{0|0}$ using y_1 and y_2 is left as a homework problem. The measurements y_1 and y_2 should not be reused in the filter update. We are now ready to build a KF to estimate $\mathbf{x}_k$ given $y_{1:k} = \{y_1, y_2, \dots, y_k\}$. Applying the two steps for discrete KF implementation: predict and update, one obtains:

Predict The predicted state vector is obtained by integrating Equation (2.11-2) from time $k-1$ to k with $\hat{\mathbf{x}}_{k-1|k-1}$ as the initial condition. Because the system represented by Equation (2.11-2) is linear and the exact solution for transition matrix exists, Equation (2.11-4), the propagation of state estimate $\hat{\mathbf{x}}_{k-1|k-1}$ to $\hat{\mathbf{x}}_{k|k-1}$ is computed by using

$$\hat{\mathbf{x}}_{k|k-1} = \mathbf{\Phi}_{t_k, \, t_{k-1}} \hat{\mathbf{x}}_{k-1|k-1}.$$

In the same manner, the covariance prediction is computed as

$$\mathbf{P}_{k|k-1} = \mathbf{\Phi}_{k,k-1} \mathbf{P}_{k-1|k-1} \mathbf{\Phi}_{k,k-1}^T + \mathbf{Q}_{k-1}.$$

Update The update process is a straightforward application of the algorithm

$$\hat{\mathbf{x}}_{k|k} = \hat{\mathbf{x}}_{k|k-1} + \mathbf{P}_{k|k-1} \mathbf{H}^T \left(y_k - \mathbf{H}\hat{\mathbf{x}}_{k|k-1} \right) / \left(\mathbf{H}\mathbf{P}_{k|k-1}\mathbf{H}^T + \sigma_v^2 \right),$$

$$\mathbf{P}_{k|k} = \mathbf{P}_{k|k-1} - \mathbf{P}_{k|k-1} \mathbf{H}^T \mathbf{H}\mathbf{P}_{k|k-1} / \left(\mathbf{H}\mathbf{P}_{k|k-1}\mathbf{H}^T + \sigma_v^2 \right).$$

Note that in this example $\left(\mathbf{H}\mathbf{P}_{k|k-1}\mathbf{H}^T + \sigma_v^2 \right)$ is a scalar and $\mathbf{H}^T\mathbf{H} = \begin{bmatrix} 1 & 0 \\ 0 & 0 \end{bmatrix}$.

Numerical results are presented in Figures 2.4 through 2.8. All these results are for a signal-to-noise ratio (SNR) of 6.667 (or, equivalently, noise variance $\sigma_v^2 = 0.0225$) with exactly the same measurements used in Section 1.8. The parameter values for the sine wave are: $a = 1$, $\omega = 2\pi$, and $\varphi = \pi/6$. Figure 2.4 shows a single Monte Carlo run of 2 sec of observation at a 10 Hz sampling rate and the process noise variance σ_q^2 is set to 0, matching the noise-free truth system. The solid curve is the truth and the data

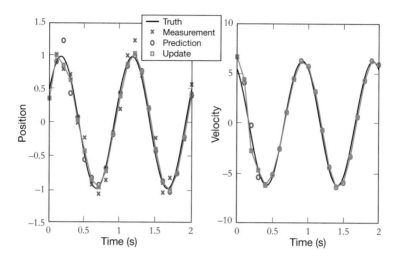

Figure 2.4 Position (x) and velocity ($\dot{x}$) estimates of Equation (2.11-1), for $\sigma_q^2 = 0$.

points are measurements (red crosses), predicted (blue circles), and updated (green squares) estimates, respectively.

Two observations from Figure 2.4 are notable: (a) the updated estimates are made between measurements and prediction [this is true of the position estimate only, since only position is directly measurable, Equation (2.11-4)], and (b) after about 1 sec the predicted estimate and the updated estimate become nearly the same. Figure 2.4 is identical to Figure 1.3, because the filter is solving the exact same problem as the parameter estimation problem. The convergence of the predicted and updated estimate is expected because with the process noise variance σ_q^2 set to zero, the filter gains in both position and velocity have converged to a smaller value in a few samples and approach zero asymptotically (blue curves in Figure 2.6), which results in the updated estimates being less sensitive for the new measurements that follow. The same plots for the system noise variance $\sigma_q^2 = 100$ are shown in Figure 2.5. Note that the updated estimates become closer to the measurements because the process noise opens up filter gains (red curves in Figure 2.6) making the filter more responsive to the new measurements. The filter gains for the system noise variance $\sigma_q^2 = 10$ are also presented in Figure 2.6 for comparison. Higher assumed system noise covariance will lead to a filter design with higher filter gains. How to vary the system noise variance in a filter design to compensate a mismatched system model will be discussed in Section 4.4. This technique is called filter tuning.

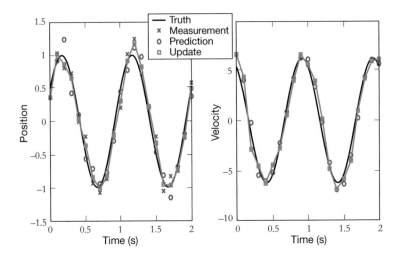

Figure 2.5   Position (x) and velocity ($\dot{x}$) estimates of Equation (2.11-1), for $\sigma_q^2 = 100$.

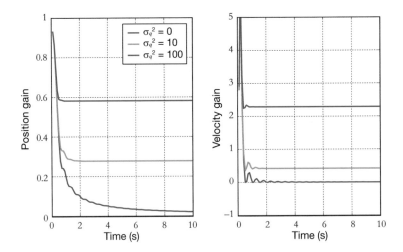

Figure 2.6   Filter gains for position and velocity comparison with three process noise covariance levels.

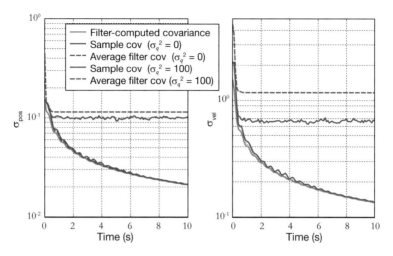

Figure 2.7   Comparison of Monte Carlo RMS errors (red/blue solid line) and average filter computed standard deviations (red/blue dashed line) in position and velocity for filters with difference process noise levels, $\sigma_q^2 = 0$ and 100.

Figure 2.7 presents a comparison of filter-computed covariance, the CRB, and error statistics using Monte Carlo simulations. They are compared as position and velocity root mean square (RMS) errors for filters with different σ_q^2 values. This is for a data window of 10 sec with 1000 Monte Carlo simulations. Note that when the process noise, σ_q^2, is set to zero, the RMS errors and the filter-computed standard deviations are nearly the same and decrease exponentially, representing the case that the filter model matches the truth. With larger process noise variance, $\sigma_q^2 = 100$, the filter RMS errors converge to steady state values, which is bigger than the RMS errors for the filter with $\sigma_q^2 = 0$. This is consistent with Figure 2.6 where the filter gains for $\sigma_q^2 = 100$ are consistently higher than the associated gains for $\sigma_q^2 = 0$ and hence the bandwidths of the filter with $\sigma_q^2 = 100$ are larger and allow more noise to pass through. The RMS errors for the filter with $\sigma_q^2 = 100$ should be higher than the filter with $\sigma_q^2 = 0$.

The obvious question for a filter designer is, which filter is better? Is the filter with $\sigma_q^2 = 0$ better? If so, then why? Which filter has a more accurately computed standard deviation that represents the statistics of the filter output? The average normalized estimation error square (NEES) [23] of both filters are computed and presented in Figure 2.8. The definition of the average NEES will be introduced in Section 4.4, which is a very useful measure of how consistent a filter computed covariance is against sample mean square error. Note that for $\sigma_q^2 = 0$, the filter is actually designed with a matched system process noise level (i.e., process noise free). It is shown in Figure 2.7 that the sample statistics, the average of the filter computed standard deviations, is the same as the CRB in this case. The average NEES values are bounded by both 95%; and 99.99%; regions in Figure 2.8, and therefore the average computed covariance and the mean square error are consistent based on χ^2 statistics. However, the filter with $\sigma_q^2 = 100$ is overly pessimistic in that the average filter computed standard deviations is larger than the RMS error as shown in Figure 2.8. These results illustrate the behavior of a KF for a simple linear system with a known system parameter (ω in this case).

Remarks

Compared with the results of the numerical example in Section 1.8, it can be seen that the WLS estimator for known ω has the same estimation errors in position and velocity, because the predicted and updated estimates at each time step are actually the KF estimates. It should be emphasized that the WLS estimator is optimal because the function governing the system dynamics is completely known. The unknowns are the system parameters (in this case all three: a, ω, φ), which are the parameters being estimated. However, the KF is more robust when the system function is not exactly known and in this case the process noise covariance of the KF plays a crucial role in enabling the filter to work. More exploration on this subject will be covered in a later chapter.

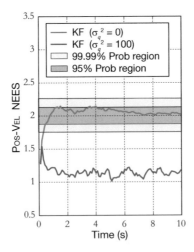

Figure 2.8 Covariance consistence check using average normalized estimation error square (NEES) of filters with two process noise covariance levels, $\sigma_q^2 = 0$ and $\sigma_q^2 = 100$.

Appendix 2.A   Stochastic Processes

A stochastic (random) process $\{\mathbf{x}_t, t \in T\}$ is a family of random vectors[6] indexed by the time parameter in set T. In most cases considered in this book, $\mathbf{x}_t$ is a continuous-valued vector in $\mathbb{R}^n$, and in some special cases, $\mathbf{x}_t$ can be discrete-valued. The time parameter set T can be discrete, such as a set of integers or a continuous interval, for example, $[0, \infty)$. Commonly used terminology will be now be defined in this Appendix [6].

Random Sequence: A continuous valued $\{\mathbf{x}_t, t \in T\}$ is called a random sequence if T is a discrete set. It is also called a discrete time random process. This is the stochastic (random) process used most often in this book.

Stochastic Process or Random Function: A continuous-valued $\{\mathbf{x}_t, t \in T\}$ is called a *(stochastic) process* or *random function*, if T is a continuous interval. It is also called a continuous-time random process. This is the stochastic (random) process describe most of natural phenomena. Instead of a set of random vectors as a sample defined in random sequence, it is a continuous time function as a sample. That is why random function is used for the definition.

Probability Law of a Stochastic Process: Let $\{\mathbf{x}_t, t \in T\}$ be a stochastic process. For any finite set $\{t_1, \dots, t_n, \forall t_i \in T, i = 1, \dots, n\}$, the joint distribution function of the random vector $\mathbf{x}_{t_1}, \dots, \mathbf{x}_{t_n}$ is called the nth order distribution of the stochastic process $\{\mathbf{x}_t, t \in T\}$ evaluated at the finite set of times $\{t_1, \dots, t_n\}$. The stochastic process can be characterized by specifying the finite dimensional distribution

$$F\left(\boldsymbol{x}_{t_1}, \dots, \boldsymbol{x}_{t_n}\right) = \text{Prob}\left\{\mathbf{x}_{t_1} \leq \boldsymbol{x}_{t_1}, \dots, \mathbf{x}_{t_n} \leq \boldsymbol{x}_{t_n}\right\} \text{ for all finite sets } \left\{t_1, \dots, t_n\right\} \subseteq T.$$

Equivalently, the stochastic process can be characterized by specifying the nth order joint density function

$$p\left(\mathbf{x}_{t_1}, \dots, \mathbf{x}_{t_n}\right) \text{ for all finite sets } \left\{t_1, \dots, t_n\right\} \subseteq T.$$

Mean, Correlation, and Covariance Functions of a Random Process: Let $\{\mathbf{x}_t, t \in T\}$ be a stochastic process. The following are important time functions associated with it

$$\boldsymbol{m}_x(t) = E(\mathbf{x}_t) \text{ (mean vector of } \{\mathbf{x}_t, t \in T\})$$

$$\boldsymbol{\gamma}_x(t, \tau) = E\{(\mathbf{x}_t)(\mathbf{x}_\tau)^T\} \text{ (correlation matrix of } \{\mathbf{x}_t, t \in T\})$$

---

6   In this book, lower case letters were used for random variables and lower case *italic* letters for their realizations; and boldface letters for vectors and non-boldface letters for scalars.

$$\Sigma_x(t, \tau) = E\{(\mathbf{x}_t - \mathbf{m}_x(t))(\mathbf{x}_\tau - \mathbf{m}_x(\tau))^T\}$$

$$= \boldsymbol{\gamma}_x(t, \tau) - \mathbf{m}_x(t)\mathbf{m}_x^T(\tau)(\text{covariance matrix of } \{\mathbf{x}_t, t \in T\})$$

Independence and Conditioning of Two Random Processes: Let $\{\mathbf{x}_t, t \in T\}$ and $\{\mathbf{y}_t, t \in T\}$ be two stochastic processes, they are said to be independent, if for all finite sets $\{t_1, \dots, t_n\} \subseteq T$, the random vectors

$$\left(\mathbf{x}_{t_1}^T, \dots, \mathbf{x}_{t_n}^T\right)^T, \left(\mathbf{y}_{t_1}^T, \dots, \mathbf{y}_{t_n}^T\right)^T$$

are independent. The conditional probability density is defined accordingly as in Section 1.3.2

$$p\left(\left(\mathbf{x}_{t_1}^T, \dots, \mathbf{x}_{t_n}^T\right)^T \mid \left(\mathbf{y}_{t_1}^T, \dots, \mathbf{y}_{t_n}^T\right)^T\right) = \frac{p\left(\left(\mathbf{x}_{t_1}^T, \dots, \mathbf{x}_{t_n}^T\right)^T, \left(\mathbf{y}_{t_1}^T, \dots, \mathbf{y}_{t_n}^T\right)^T\right)}{p\left(\left(\mathbf{y}_{t_1}^T, \dots, \mathbf{y}_{t_n}^T\right)^T\right)}$$

$$= \frac{p\left(\left(\mathbf{y}_{t_1}^T, \dots, \mathbf{y}_{t_n}^T\right)^T \mid \left(\mathbf{x}_{t_1}^T, \dots, \mathbf{x}_{t_n}^T\right)^T\right)}{p\left(\left(\mathbf{y}_{t_1}^T, \dots, \mathbf{y}_{t_n}^T\right)^T\right)} p\left(\left(\mathbf{x}_{t_1}^T, \dots, \mathbf{x}_{t_n}^T\right)^T\right)$$

for all finite sets $\{t_1, \dots, t_n\} \subseteq T$.

Continuous-Time Stochastic Process with Independent Increments: A continuous-time stochastic process $\{\mathbf{x}_t, t \in T\}$ has independent increments if, for all finite sets $\{t_1 < t_2 < \dots, < t_n\} \subseteq T$, the random vectors

$$\mathbf{x}_{t_1} - \mathbf{x}_{t_2}, \mathbf{x}_{t_2} - \mathbf{x}_{t_3}, \dots, \mathbf{x}_{t_{n-1}} - \mathbf{x}_{t_n}$$

are independent. It follows directly from Section 2.3.2 that the joint density function

$$p(\mathbf{x}_{t_1} - \mathbf{x}_{t_2}, \mathbf{x}_{t_2} - \mathbf{x}_{t_3}, \dots, \mathbf{x}_{t_{n-1}} - \mathbf{x}_{t_n}) = p(\mathbf{x}_{t_1} - \mathbf{x}_{t_2}) p(\mathbf{x}_{t_2} - \mathbf{x}_{t_3}) \dots p(\mathbf{x}_{t_{n-1}} - \mathbf{x}_{t_n}).$$

The process $\{\mathbf{x}_t, t \in T\}$ has stationary independent increments if, in addition,

$$\mathbf{x}_{t+\tau} - \mathbf{x}_{s+\tau}$$

has the same distribution as $\mathbf{x}_t - \mathbf{x}_s$, for every $t > s \in T$ and $\tau, t + \tau, s + \tau \in T$.

Brownian Motion (Wiener) Process: A continuous-time scalar stochastic process $\{w_t, t \in T\}$ is a Brownian motion process if

1. $\{w_t, t \in T\}$ has stationary independent increments;

2. For every $t \geq 0$, w_t is normally distributed;

3. For every $t \geq 0$, $E\{w_t\} = 0$;

4. $\text{Prob}\{w_0 = 0\} = 1$.

Gaussian Processes: A stochastic process $\{\mathbf{x}_t, t \in T\}$ is a Gaussian (or normal) process if the joint density is Gaussian (normal). Brownian motion is a Gaussian process.

Markov Processes: A stochastic process $\{\mathbf{x}_t, t \in T\}$ is called a Markov process if, for any finite parameter set $\{t_i: t_i < t_{i+1}\} \in T$, and for every $r \in \mathbb{R}^n$,

$$\text{Prob}\{\mathbf{x}_{t_n} \leq r | \mathbf{x}_{t_1}, \ldots, \mathbf{x}_{t_{n-1}}\} = \text{Prob}\{\mathbf{x}_{t_n} \leq r | \mathbf{x}_{t_{n-1}}\}.$$

The above property can be written in terms of density functions

$$p(\mathbf{x}_{t_n} | \mathbf{x}_{t_1}, \ldots, \mathbf{x}_{t_{n-1}}) = p(\mathbf{x}_{t_n} | \mathbf{x}_{t_{n-1}}),$$

where $t_1 < t_2 < \ldots < t_n$. Furthermore, applying Bayes' rule repeatedly yields

$$p(\mathbf{x}_{t_1}, \ldots, \mathbf{x}_{t_n}) = p(\mathbf{x}_{t_n} | \mathbf{x}_{t_{n-1}}) p(\mathbf{x}_{t_{n-1}} | \mathbf{x}_{t_{n-2}}) \ldots p(\mathbf{x}_{t_2} | \mathbf{x}_{t_1}) p(\mathbf{x}_{t_1}).$$

Therefore, the probability law of a Markov process can be specified by $p(\mathbf{x}_{t_1})$ and $p(\mathbf{x}_t | \mathbf{x}_s)$ for all $t > s \in T$. The conditional densities $p(\mathbf{x}_t | \mathbf{x}_s)$ are called the transition probability densities of the Markov process. Brownian motion is a Markov process.

Chapman–Kolmogorov Equation: Let $\{\mathbf{x}_t, t \in T\}$ be a Markov process. The following Chapman–Kolmogorov equation

$$p(\mathbf{x}_{t_n} | \mathbf{x}_{t_{n-2}}) = \int p(\mathbf{x}_{t_n} | \mathbf{x}_{t_{n-1}}) p(\mathbf{x}_{t_{n-1}} | \mathbf{x}_{t_{n-2}}) d\mathbf{x}_{t_{n-1}},$$

can be derived using the definition of marginal density and the Markov process property of $\{\mathbf{x}_t, t \in T\}$

$$p(\mathbf{x}_{t_n} | \mathbf{x}_{t_{n-2}}) = \int p(\mathbf{x}_{t_n}, \mathbf{x}_{t_{n-1}} | \mathbf{x}_{t_{n-2}}) d\mathbf{x}_{t_{n-1}}.$$

White Noise: The random sequence $\{\boldsymbol{\xi}_n, n = 1, 2, \ldots\}$ is called a white random sequence if it is a Markov sequence and

$$p(\boldsymbol{\xi}_k | \boldsymbol{\xi}_j) = p(\boldsymbol{\xi}_k) \text{ for all } k > j,$$

that is, all the $\boldsymbol{\xi}_k$'s are mutually independent. If the $\boldsymbol{\xi}_k$'s are normally distributed, the sequence is called white Gaussian random sequence. A zero-mean, white Gaussian noise sequence is a stochastic process with the following probability properties that the mean

$$E\{\boldsymbol{\xi}_k\} = \mathbf{0}, \text{ for all } k \geq 1,$$

and the covariance matrix

$$E\{(\xi_j - E\{\xi_j\})(\xi_k - E\{\xi_k\})^T\} = \mathbf{Q}_k \delta_{j,k}, \text{ for all } j, k \geq 1,$$

where

$$\delta_{jk} = \begin{cases} 1, j = k \\ 0, j \neq k \end{cases}$$

is the Kronecker delta and $\mathbf{Q}_k$ is a positive semidefinite matrix.

The notion of continuous-time white noise in continuous dynamical systems is more involved. It is beyond the scope of this book to rigorously define it, but some important properties of the process will be stated here formally.

1. A white process $\{\xi_t, t \in T\}$ is a Markov process for which

$$p(\xi_t|\xi_s) = p(\xi_t), \text{ for all } t > s \in T,$$

and if the ξ_t's are normally distributed for each $t \in T$, then the process is a white Gaussian process.

Furthermore, it is a Gaussian process with

$$E\{(\xi_t - E\{\xi_t\})(\xi_s - E\{\xi_s\})^T\} = \mathbf{Q}_t \delta(t - s)$$

where $\mathbf{Q}_t$ is a positive semidefinite covariance matrix, and $\delta(t - s)$ is the Dirac delta function.

2. A scalar white Gaussian process can be approximated by a zero-mean stationary Gaussian process with correlation function

$$\gamma(t + \tau, t) = (\rho/2)\sigma^2 e^{-\rho|\tau|}.$$

For $\rho \to \infty$,

$$\gamma^\infty = \sigma^2 \delta(\tau).$$

3. The power spectral density function of a random process is defined as the Fourier transform of the correlation function

$$f(\omega) = \int_{-\infty}^{\infty} e^{-i\omega\tau} \gamma(t + \tau, t) d\tau.$$

For the Gaussian process in (2), the power spectral density function is

$$f^\rho(\omega) = \frac{\sigma^2}{1 + (\omega/\rho)^2}.$$

With $\rho \to \infty, f^\infty(\omega) = \sigma^2$, a positive constant for all ω, this is the origin of the adjective *white* as in white light.

Homework Problems

1. Let x denote the position of an object traveling along a straight line. The differential equation $\ddot{x} = 0$ denotes this object traveling with CV. The state equation for the motion of this object is

    $$\dot{\mathbf{x}}_t = \mathbf{A}_t \mathbf{x}_t.$$

The solution of the above equation for a given initial condition x_0 gives the trajectory of an object traveling in a straight line. The state vector is $\mathbf{x}_t = [x_1, x_2]^T$, with $x_1 = x$, $x_2 = \dot{x}$, and the system matrix A_t is

$$\mathbf{A}_t = \begin{bmatrix} 0 & 1 \\ 0 & 0 \end{bmatrix}.$$

   a. Derive the discrete equivalent of the system equation for the sampling interval of T.

   b. Repeat the above for $\dddot{x} = 0$. This corresponds to a CA model.

   c. Let (x, y, z) denote the three axes of Euclidean space. For an object traveling in CV, the state vector can be defined as $\mathbf{x}_t = (x, y, z, \dot{x}, \dot{y}, \dot{z})^T$. If motion along each axis is independent, then the state equation is a simple concatenation of three of the above 2D models to a six-dimensional state space. Similarly, for a CA model, this extends to a nine-dimensional state space.

   d. Assume the measurement variables are x, y, z and each has additive white noise $N(0, \sigma^2)$. Derive and compute the CRB of estimating $\mathbf{x}_t$ where t could be 0 or any time along the trajectory. Show that no matter what time is chosen, the covariance propagation equation will advance to any other time with the same result. Any initial condition, $\mathbf{x}_0$, can be chosen to display the result.

   e. Note that the dynamic model used in this problem is the same as the polynomial model in Section 2.2 and Homework Problem 15 of Chapter 1. Show that the results are the same.

2. The above models, whether CV or CA, are not appropriate for the the real world. For instance, a pilot always makes small corrective controls to maintain the airplane to stay on course. Wind turbulence, among other factors, causes

the actual trajectory to vary around a straight line. When one builds a KF that is based strictly on a CV model, the filter will eventually diverge because the filter gain becomes too small, such that the recent measurements are ignored. In experimenting with a model you build, you will see that this situation can be corrected by using process noise covariance $\mathbf{Q}_k$.

a. Use your computer to build a CV trajectory model with a small sinusoidal driving function. Your trajectory model could start at 100 km away with a velocity of 150 m/s. The direction of the velocity could be toward the sensor [located at (0, 0)], or crossing at some angle. Variations of trajectory can be used to compare estimation accuracy due to the geometry difference. Add some driving noise in velocity to simulate random disturbances. Display the results graphically. Assume the object is traveling at a constant altitude, so that you only need to display the trajectory in a 2D Euclidean space (x, y). Add Gaussian noise samples to simulate measurements. You can pick σ to be anywhere from 1 m to 100 m.

b. Build a KF to estimate the state of the trajectory you just built. Experiment with various $\mathbf{Q}_k$ values including zero. Experiment with which state you add the process noise covariance, in velocity or position, or both. Display and observe the results [as in part (a), 2D is fine]. Explain what you have and see if you could come up with some conclusion on how to select $\mathbf{Q}_k$. Repeat the run in Monte Carlo fashion, and compute sample statistics of bias and covariance.

c. Build a piecewise trajectory model, that is, one that starts with CV, switches to CA, and after a while returns to be CV. The CA portion of the trajectory represents an intentional maneuver, such as the pilot is changing a plane's heading, say from heading east to heading north. Display your trajectory in a 2D space (x, y).

d. Apply your CV Kalman filter to the trajectory you built in (c). Experiment with $\mathbf{Q}_k$ values to reduce the filter bias during trajectory switchover.

e. Build a CA Kalman filter, repeat with (b) and (d), explain what you see during the model switches.

f. Compute CRB for CV and CA models, with and without $\mathbf{Q}_k$. Compare them with your sample covariance.

3. The polynomial model (Problem 15 of Chapter 1) can be rewritten in terms of CV or CA models without process noise. The polynomial model is very useful because of its simplicity and the closed form expression on estimate and covariance. It is a valid model when the time window is short, such that ignoring process noise is acceptable. It has been used to estimate initial states and covariance using an initial batch of radar measurements in its three measurement coordinates to obtain position and velocity vectors with covariance. It is thus appropriate for use as a filter initial condition derivation. Derive the recursive system equation from the polynomial model and experiment with the statements above.

4. Complete the derivation of Equations (2.7-2) through (2.7-5), the Bayesian approach for a KF.

5. For the example in Section 2.11, show that the initial state estimate $\hat{\mathbf{x}}_{0|0}$ with covariance $\mathbf{P}_{0|0}$ can be obtained with the first two measurements taken at discrete times t_1 and t_2, that is, y_1 and y_2.

6. Use the fixed-interval smoother (FIS) for a deterministic system, from Equations (2.9-13) and (2.9-14).

   a. Derive the special case when there is no initial condition, or $\mathbf{P}_0^{-1} = \mathbf{0}$;

   b. Derive special cases when k is at the initial time, 0, or end time K;

   c. Show the equivalence with the FIS for the stochastic system while setting the process noise covariance $\mathbf{Q}_i = \mathbf{0}$ for all i.

7. Singer [24] proposed a target model to represent maneuvering of manned aircraft. Letting $\mathbf{x}_t$ be the state vector of an object moving along a straight line, the Singer model is: $\dot{\mathbf{x}}_t = \mathbf{A}_t \mathbf{x}_t + \mathbf{G} a_t$ where $\mathbf{A}_t$ is the same as in problem 1, $\mathbf{G} = \begin{bmatrix} 0 \\ 1 \end{bmatrix}$, a_t is target acceleration represented by a random process with time correlation function: $r_\tau = E\{a_t a_{t+\tau}\} = \sigma_m^2 e^{-\alpha|\tau|}$, σ_m^2 is the magnitude of the maneuver and $1/\alpha$ is the maneuver time constant. For example, when $1/\alpha \cong 60$, it represents a lazy turn, and when $1/\alpha \cong 20$, it represents an evasive maneuver.

   a. Derive the discrete equivalent of the system equation for the sampling interval of T.

   b. Select some values for σ_m^2 and α. Simulate trajectories using this model in a 2D Euclidean space (see Problem 2.a).

   c. Compare results with those obtained in Problem 2.a. Find the parameters that will make these two models come close to each other.

References

1. R.E. Kalman, "A New Approach to Linear Filtering and Prediction Problems," *Transactions of the ASME*, vol. 82, ser. D, pp. 35–45, Mar. 1960.

2. P.M. DeRusso, R.J. Roy, C.M. Close, and A.A. Desrochers, *State Variables for Engineers*, 2nd ed. New York: Wiley, 1998.

3. W.L. Brogan, *Modern Control Theory*. Englewood Cliffs, NJ: Prentice Hall, 1990.

4. A.P. Sage, *Optimum Systems Control*. Englewood Cliffs, NJ: Prentice Hall, 1968.

5. M. Athans and P. Falb, *Optimal Control*. New York: McGraw-Hill, 1966.

6. A.H. Jazwinski, *Stochastic Processes and Filtering Theory*. New York: Academic Press, 1970.

7. A. Gelb (Ed.), *Applied Optimal Estimation*. Cambridge, MA: MIT Press, 1974.

8. F.L. Lewis, *Optimal Estimation*. New York: Wiley, 1986.

9. D. Simon, *Optimal State Estimation*. New York: Wiley, 2006.

10. A. Sage and J. Melsa, *Estimation Theory with Applications to Communications and Control*. New York: McGraw-Hill, 1971.

11. P. Maybeck, *Stochastic Models, Estimation, and Control*, Vol. 1. New York: Academic Press, 1979.

12. T. Kailath, A.H. Sayed, and Babak Hassibi, *Linear Estimation*. Englewood Cliffs, NJ: Prentice Hall, 2000.

13. M. Athans, "The Role and Use of the Stochastic Linear-Quadratic-Gaussian Control Problem in System Design," *IEEE Transactions on Automatic Control*, vol. AC-16, pp. 529–552, Dec. 1971.

14. R.S. Bucy, "Nonlinear Filtering Theory," *IEEE Transactions on Automatic Control*, vol. AC-10, pp. 198–206, Jan. 1965.

15. Y.C. Ho and R.C.K. Lee, "A Bayesian Approach to Problems in Stochastic Estimation and Control," *IEEE Transactions on Automatic Control*, vol. AC-9, pp. 333–339, Oct. 1964.

16. T. Kailath, "An Innovations Approach to Least-Squares Estimation; Part I: Linear Filtering in Additive White Noise," *IEEE Transactions on Automatic Control*, vol. AC-13, pp. 688–706, Dec. 1968.

17. R. Mehra, "On the Identification of Variances and Adaptive Kalman Filtering," *IEEE Transactions on Automatic Control*, vol. AC-15, pp. 175–184, Apr. 1970.

18. I.B. Rhodes, "A Tutorial Introduction to Estimation and Filtering," *IEEE Transactions on Automatic Control*, vol. AC-16, pp. 688–706, Dec. 1971.

19. H.E. Rauch, F. Tung, and C.T. Striebel, "Maximum Likelihood Estimates of Linear Dynamic Systems," *AIAA Journal*, vol. 3, pp. 1445–1450, Aug. 1965.

20. C.B. Chang, R.H. Whiting, L. Youens, and M. Athans, "Application of Fixed-Interval-Smoother to Maneuvering Trajectory Estimation," *IEEE Transactions on Automatic Control*, vol. AC-22, pp. 876–879, Oct. 1977.

21. C.B. Chang, R.H. Whiting, and M. Athans, "On the State and Parameter Estimation for Maneuvering Reentry Vehicles," *IEEE Transactions on Automatic Control*, vol. AC-22, pp. 99–105, Feb. 1977.

22. P. Tichavsky, C.H. Muravchik, and A. Nehorai, "Posterior Cramer–Rao Bounds for Discrete-Time Nonlinear Filtering," *IEEE Transactions on Signal Processing*, vol. SP-46, pp. 1386–1396, May 1998.

23. Y. Bar-Shalom, X. Rong Li, and T. Kirubarajan, *Estimation with Applications to Tracking and Navigation*. New York: Wiley, 2001.

24. R. Singer, "Estimating Optimal Tracking Filter Performance for Manned Maneuvering Targets," *IEEE Transactions on Aerospace and Electronic Systems*, vol. AES-6, pp. 473–483, July 1970.

3

State Estimation for Nonlinear Systems

3.1 Introduction

Most practical problems are nonlinear, that is, the systems dynamics is represented by nonlinear differential (or difference) equations, and/or the measured variables are related to the state variables through nonlinear equations. Up to this point, the book has covered systems with linear dynamics and measurement relations. The only exception was in Section 1.6 where an iterative algorithm was introduced for parameter estimation when the measurement equation is nonlinear. Initially, the basics of estimation algorithms with linear systems are developed because (a) the theory for solving linear problems is rigorous, and (b) most of the algorithms developed for linear systems can be applied to nonlinear problems (as approximated solutions) by substituting the system and measurement matrices with Jacobians of the system and measurement equations. The resulting filter is known as the extended Kalman filter (EKF). The technique used to derive the EKF is referred to as linearization where a first order Taylor series expansion is used to approximate the nonlinear terms.

Mathematically rigorous research in addressing the nonlinear estimation problem took place in the 1960s and 1970s, [see, e.g., 1–7]. Regardless of whether the system is linear or nonlinear, the solution of the state estimation problem is given by the a posteriori density function [1–3]. The temporal evolution of the a posteriori density function for a continuous system can be represented by a differential equation. It has been stated [4–6] that the solution of this differential equation for practical problems was unattainable. Numerous researchers have proposed approximate solutions to the a posteriori density function [e.g., 5–10]. For discrete time systems, this is the density function of the state $\mathbf{x}_k$ conditioned on measurements up to time k, $\mathbf{y}_{1:k} = \{\mathbf{y}_1, \mathbf{y}_2,...,\mathbf{y}_k\}$. Section 2.4 demonstrated that this probability density function can be characterized recursively regardless whether it is a linear or a nonlinear estimation problem [2, 3].

For a linear system with Gaussian initial conditions, process noise and measurement noise terms, the a posteriori density function remains Gaussian for all k. Therefore, it is sufficient to use the conditional mean and covariance to characterize the a posteriori density function. The resulting algorithm for computing the conditional mean and covariance is indeed the Kalman filter (KF). The Gaussian assumption goes away for nonlinear systems, where only approximate solutions can be obtained. This chapter presents estimation algorithms for discrete systems. All the results represent approximate solutions to the conditional mean and covariance. Mean and covariance carry special physical meaning for engineering problems. The conditional mean is the estimate of the state vector that characterizes the dynamics of the system of interest. The covariance is a measure of performance, that is, it represents the accuracy of the state estimation. For this reason, engineers seek solutions of conditional mean and covariance even though they may be approximate solutions. In deriving the answer, Taylor series expansions are applied to the system and measurement equations. The approximate solution for the state and covariance estimates based only on the first order Taylor series expansion is referred to as the EKF. When the second order terms of the Taylor series expansion are retained, the resulting filter is known as the second order filter [10]. Readers interested in filtering applications should consult Gelb [7, 11]. All these filters are presented in this chapter. The Cramer–Rao bound (CRB) for the nonlinear state estimation problem is presented as the last section of this chapter. Its application to practical problems is included in the homework problems. A class of sampling algorithms was developed in parallel with the analytical approaches presented in this chapter that will be discussed separately in Chapter 6, which will include the following topics: A point mass filter was introduced in 1969 [12]. A deterministic sampling scheme utilizing unscented transformation for nonlinear system with additive Gaussian noises was introduced in the late 1990s [13–16]. A much more recent approach to obtain the solution of the a posteriori density function is to use Monte Carlo sampling techniques and is known as the particle filter [17].

3.2   Problem Definition

It is assumed that the state vector $\mathbf{x}$ evolves with time with a known nonlinear discrete stochastic relationship is described as

$$\mathbf{x}_k = \mathbf{f}(\mathbf{x}_{k-1}) + \boldsymbol{\mu}_{k-1} \tag{3.2-1}$$

for $k = 1, 2, \ldots, K$, where K is an arbitrary end time. The driving noise term $\boldsymbol{\mu}_{k-1}$ is a random sequence uncorrelated in time and with known distribution. The initial state

$\mathbf{x}_0$ is random with known distribution and uncorrelated with $\boldsymbol{\mu}_{k-1}$. Similar to the linear case, the system dynamics is usually a continuous process, characterized by a state differential equation. Therefore, Equation (3.2-1) is a time-sampled representation of the continuous process. There is no loss of generality with this model because the estimator is implemented with a computer and measurements are usually taken at discrete times, such that

$$\mathbf{y}_k = \mathbf{h}(\mathbf{x}_k) + \mathbf{v}_k. \tag{3.2-2}$$

The measurement noise sequence $\mathbf{v}_k$ is assumed to be uncorrelated in time and with known distribution. All three random vectors, $\mathbf{x}_0$, $\boldsymbol{\mu}_{k-1}$, and $\mathbf{v}_k$, are mutually independent for all k. Equations (3.2-1) and (3.2-2) define the nonlinear system to be studied in this chapter. The estimation problem is to compute $\mathbf{x}_k$ given all measurements: $\mathbf{y}_{1:k} = \{\mathbf{y}_1, \mathbf{y}_2, \dots, \mathbf{y}_k\}$ for the above nonlinear system Equations (3.2-1) and (3.2-2).

The Bayesian recursive equation for a posteriori probability density functions is reviewed below.

3.3   Bayesian Approach for State Estimation

For the sake of clarity, the following equivalent representation of conditional probabilities introduced in Section 2.4 is restated below

$$p(\mathbf{x}_k|\mathbf{y}_{1:k}) = p(\mathbf{x}_k|\mathbf{y}_k, \mathbf{y}_{1:k-1}). \tag{3.3-1}$$

Applying Bayes' rule of probability to relate priori to posteriori density functions yields

$$p\left(\mathbf{x}_k|\mathbf{y}_{1:k}\right) = \frac{p\left(\mathbf{y}_k|\mathbf{x}_k\right)}{p\left(\mathbf{y}_k|\mathbf{y}_{1:k-1}\right)} \, p\left(\mathbf{x}_k|\mathbf{y}_{1:k-1}\right) \tag{3.3-2}$$

where the density $p(\mathbf{x}_k|\mathbf{y}_{1:k-1})$ can be obtained via

$$p\left(\mathbf{x}_k|\mathbf{y}_{1:k-1}\right) = \int p\left(\mathbf{x}_k|\mathbf{x}_{k-1}\right) p\left(\mathbf{x}_{k-1}|\mathbf{y}_{1:k-1}\right) d\mathbf{x}_{k-1}. \tag{3.3-3}$$

Putting these two together one obtains

$$p\left(\mathbf{x}_k|\mathbf{y}_{1:k}\right) = \frac{p\left(\mathbf{y}_k|\mathbf{x}_k\right)}{p\left(\mathbf{y}_k|\mathbf{y}_{1:k-1}\right)} \int p\left(\mathbf{x}_k|\mathbf{x}_{k-1}\right) p\left(\mathbf{x}_{k-1}|\mathbf{y}_{1:k-1}\right) d\mathbf{x}_{k-1}. \tag{3.3-4}$$

The evolution of the a posteriori density function in Equation (3.3-4) applies to both linear and nonlinear systems. The definition of probability terms and relationship made in Section 2.4 hold equally with nonlinear systems. The process of applying the above relations is as follows. The a priori density $p(\mathbf{x}_{k-1}|\mathbf{y}_{1:k-1})$ is given at time k, and this is possible because at the initial time when $k = 1$, $p(\mathbf{x}_0)$ is given. From this point, $p(\mathbf{x}_1|\mathbf{y}_{1:1})$ is computed and the recursion continues. $p(\mathbf{x}_k|\mathbf{y}_{1:k-1})$ is the one-step predicted density shown on the right-hand side of Equation (3.3-2). A notional way of computing it is shown in Equation (3.3-3). Once the new measurement vector $\mathbf{y}_k$ is obtained, Equation (3.3-4) is applied to complete the computation. The term in the denominator, $p(\mathbf{y}_k|\mathbf{y}_{1:k-1})$, is a normalization factor. Once $p(\mathbf{x}_k|\mathbf{y}_{1:k})$ is obtained, it becomes the a priori density for time $k + 1$. For linear systems with Gaussian initial conditions and Gaussian process and measurement noise sequences, the a posteriori density function $p(\mathbf{x}_k|\mathbf{y}_{1:k})$ remains Gaussian, and it can be completely characterized with mean and covariance. Indeed, the KF can be derived using this relationship as shown in Section 2.7.

For nonlinear systems, even when initial condition and process and measurement noise sequences are Gaussian, the a posteriori density function $p(\mathbf{x}_k|\mathbf{y}_{1:k})$ after the first recursion will no longer be Gaussian. The computation of its mean and covariance can no longer be performed exactly. Furthermore, mean and covariance alone will no longer completely characterize the a posteriori density function. Numerous approaches have been proposed over the years in trying to solve the nonlinear estimation problem, viz. the solution of Equation (3.3-4). Many of the proposed solutions try to approximate the solution of Equation (3.3-4), such as series expansion with moments, or with coefficients of orthogonal series, among other strategies [6–8]. Unfortunately, none of the earlier approaches for obtaining approximated solutions to the a posteriori density function resulted in any practical algorithms ready for implementation. In the remainder of this chapter, more traditional algorithms by means of using Taylor series expansion to approximate the state and measurement equations are derived and discussed.

3.4   Extended Kalman Filter Derivation: As a Weighted Least Squares Estimator

We will apply the techniques used for conditional expectation and the weighted least squares estimator in Chapter 1 to derive the nonlinear filter similar to the derivation of KF in Chapter 2. In order to obtain an explicit solution, the Taylor series expansion is applied to the state and measurement equations, where only the first order term is retained, this results in the EKF.

3.4.1   One-Step Prediction Equation

The one-step prediction equation is used to compute $\hat{\mathbf{x}}_{k|k-1}$ and $\mathbf{P}_{k|k-1}$ given $\hat{\mathbf{x}}_{k-1|k-1}$ and $\mathbf{P}_{k-1|k-1}$. Before we proceed, we first apply the Taylor series expansion to the system equation about $\hat{\mathbf{x}}_{k-1|k-1}$

$$\mathbf{x}_k = \mathbf{f}(\hat{\mathbf{x}}_{k-1|k-1}) + [\mathbf{F}_{\hat{\mathbf{x}}_{k-1|k-1}}](\mathbf{x}_{k-1} - \hat{\mathbf{x}}_{k-1|k-1}) + \boldsymbol{\mu}_{k-1} + \text{HOT}, \qquad (3.4\text{-}1)$$

where HOT stands for higher order terms and $\mathbf{F}_{\hat{\mathbf{x}}_{k-1|k-1}}$ is the Jacobian matrix of $\mathbf{f}(.)$ evaluated at $\hat{\mathbf{x}}_{k-1|k-1}$, that is,

$$\mathbf{F}_{\hat{\mathbf{x}}_{k-1|k-1}} = \left[\frac{\partial \mathbf{f}(\mathbf{x})}{\partial \mathbf{x}}\right]_{\hat{\mathbf{x}}_{k-1|k-1}}.$$

Note that the simplifying assumption that the system noise $\boldsymbol{\mu}_{k-1}$ is additive has been applied. Similar to the KF derivation, the one-step predicted estimate is a conditional mean estimator, that is, $\hat{\mathbf{x}}_{k|k-1} = E\{\mathbf{x}_k | \mathbf{y}_{1:k-1}\}$. Applying the conditional expectation operator over Equation (3.4-1), we obtain

$$\hat{\mathbf{x}}_{k|k-1} = E\{\mathbf{f}(\mathbf{x}_{k-1}) | \mathbf{y}_{1:k-1}\} \approx \mathbf{f}(\hat{\mathbf{x}}_{k-1|k-1}). \qquad (3.4\text{-}2)$$

This is with the assumption that the estimate $\hat{\mathbf{x}}_{k-1|k-1}$ of $\mathbf{x}_{k-1}$ is unbiased, $\boldsymbol{\mu}_{k-1}$ is independent of $\mathbf{y}_{1:k-1}$, and HOT can be ignored. Similarly, applying the definition of covariance and ignoring the HOT of the Taylor series expansion, one obtains an approximate expression for $\mathbf{P}_{k|k-1}$

$$\begin{aligned}
\mathbf{P}_{k|k-1} &\approx E\Big\{\big(\mathbf{f}(\hat{\mathbf{x}}_{k-1|k-1}) + [\mathbf{F}_{\hat{\mathbf{x}}_{k-1|k-1}}](\mathbf{x}_{k-1} - \hat{\mathbf{x}}_{k-1|k-1}) + \boldsymbol{\mu}_{k-1} - \mathbf{f}(\hat{\mathbf{x}}_{k-1|k-1})\big) \\
&\quad \big(\mathbf{f}(\hat{\mathbf{x}}_{k-1|k-1}) + [\mathbf{F}_{\hat{\mathbf{x}}_{k-1|k-1}}](\mathbf{x}_{k-1} - \hat{\mathbf{x}}_{k-1|k-1}) + \boldsymbol{\mu}_{k-1} - \mathbf{f}(\hat{\mathbf{x}}_{k-1|k-1})\big)^T\Big\} \\
&= \mathbf{F}_{\hat{\mathbf{x}}_{k-1|k-1}} \mathbf{P}_{k-1|k-1} \mathbf{F}_{\hat{\mathbf{x}}_{k-1|k-1}}^T + \mathbf{Q}_{k-1}
\end{aligned} \qquad (3.4\text{-}3)$$

This completes the derivation of the one-step prediction equation.

3.4.2   Update Equations

With $\hat{\mathbf{x}}_{k|k-1}$ and $\mathbf{P}_{k|k-1}$ and the new measurement $\mathbf{y}_k$, the weighted least squares derivation is used to obtain the update equation. The updated state estimate $\hat{\mathbf{x}}_{k|k}$ is the $\mathbf{x}_k$ that minimizes the weighted least squares (WLS) performance index

$$J = (\mathbf{y}_k - \mathbf{h}(\mathbf{x}_k))^T \mathbf{R}_k^{-1}(\mathbf{y}_k - \mathbf{h}(\mathbf{x}_k)) + (\mathbf{x}_k - \hat{\mathbf{x}}_{k|k-1})^T \mathbf{P}_{k|k-1}^{-1}(\mathbf{x}_k - \hat{\mathbf{x}}_{k|k-1}). \qquad (3.4\text{-}4)$$

Taking the partial derivative of J with respect to $\mathbf{x}_k$ and setting it to zero yields the necessary condition for minimization

$$-\frac{\partial J}{\partial \mathbf{x}_k} = 2\left\{\left[\frac{\partial \mathbf{h}(\mathbf{x}_k)}{\partial \mathbf{x}_k}\right]^T \left[\mathbf{R}_k^{-1}(\mathbf{y}_k - \mathbf{h}(\mathbf{x}_k))\right] - \mathbf{P}_{k|k-1}^{-1}(\mathbf{x}_k - \hat{\mathbf{x}}_{k|k-1})\right\} = 0 \qquad (3.4\text{-}5)$$

In order to obtain an explicit solution for $\mathbf{x}_k$, taking the Taylor series expansion of $\mathbf{h}(\mathbf{x}_k)$ about the predicted estimate $\hat{\mathbf{x}}_{k|k-1}$ and dropping the HOT results in the approximation

$$\mathbf{h}(\mathbf{x}_k) \approx \mathbf{h}(\hat{\mathbf{x}}_{k|k-1}) + \left[\frac{\partial \mathbf{h}(\mathbf{x}_k)}{\partial \mathbf{x}_k}\right]_{\hat{\mathbf{x}}_{k|k-1}}(\mathbf{x}_k - \hat{\mathbf{x}}_{k|k-1}). \qquad (3.4\text{-}6)$$

Adopting the same notation as $\mathbf{F}_{\hat{\mathbf{x}}_{k-1|k-1}}$ for $\left[\dfrac{\partial \mathbf{h}(\mathbf{x}_k)}{\partial \mathbf{x}_k}\right]_{\hat{\mathbf{x}}_{k|k-1}}$

$$\mathbf{H}_{\hat{\mathbf{x}}_{k|k-1}} = \left[\frac{\partial \mathbf{h}(\mathbf{x}_k)}{\partial \mathbf{x}_k}\right]_{\hat{\mathbf{x}}_{k|k-1}}$$

which is the Jacobian matrix of $\mathbf{h}(\mathbf{x}_k)$. The necessary condition for solving for $\hat{\mathbf{x}}_{k|k}$ is the $\mathbf{x}_k$ that minimizes Equation (3.4-4)

$$\mathbf{H}_{\hat{\mathbf{x}}_{k|k-1}}^T \mathbf{R}_k^{-1}\left(\mathbf{y}_k - \mathbf{h}(\hat{\mathbf{x}}_{k|k-1}) - \mathbf{H}_{\hat{\mathbf{x}}_{k|k-1}}(\mathbf{x}_k - \hat{\mathbf{x}}_{k|k-1})\right) - \mathbf{P}_{k|k-1}^{-1}(\mathbf{x}_k - \hat{\mathbf{x}}_{k|k-1}) = \mathbf{0}. \qquad (3.4\text{-}7)$$

The solution below gives the updated estimate, $\hat{\mathbf{x}}_{k|k}$,

$$\hat{\mathbf{x}}_{k|k} = \hat{\mathbf{x}}_{k|k-1} + \left[\mathbf{P}_{k|k-1}^{-1} + \mathbf{H}_{\hat{\mathbf{x}}_{k|k-1}}^T \mathbf{R}_k^{-1}\mathbf{H}_{\hat{\mathbf{x}}_{k|k-1}}\right]^{-1} \mathbf{H}_{\hat{\mathbf{x}}_{k|k-1}}^T \mathbf{R}_k^{-1}\left(\mathbf{y}_k - \mathbf{h}(\hat{\mathbf{x}}_{k|k-1})\right) \qquad (3.4\text{-}8)$$

and the covariance of $\hat{\mathbf{x}}_{k|k}$ is

$$\begin{aligned}\mathbf{P}_{k|k} &= \left[\mathbf{P}_{k|k-1}^{-1} + \mathbf{H}_{\hat{\mathbf{x}}_{k|k-1}}^T \mathbf{R}_k^{-1}\mathbf{H}_{\hat{\mathbf{x}}_{k|k-1}}\right]^{-1} \\ &= \mathbf{P}_{k|k-1} - \mathbf{P}_{k|k-1}\mathbf{H}_{\hat{\mathbf{x}}_{k|k-1}}^T\left[\mathbf{H}_{\hat{\mathbf{x}}_{k|k-1}}\mathbf{P}_{k|k-1}\mathbf{H}_{\hat{\mathbf{x}}_{k|k-1}}^T + \mathbf{R}_k\right]^{-1}\mathbf{H}_{\hat{\mathbf{x}}_{k|k-1}}\mathbf{P}_{k|k-1}.\end{aligned} \qquad (3.4\text{-}9)$$

This completes the derivation of the EKF.

The EKF algorithm is summarized in Table 3.1.

Remarks

1. The EKF algorithm is identical to the KF in functional form with the linear system and measurement matrices replaced by the first order derivative of the nonlinear system and measurement equations (i.e., their Jacobian matrices).

2. The term $\mathbf{y}_k - \mathbf{h}(\hat{\mathbf{x}}_{k|k-1})$ that appeared in Equation (3.4-9) is the difference between the actual measurement $\mathbf{y}_k$ and the approximation of predicted measurement, $\mathbf{h}(\hat{\mathbf{x}}_{k|k-1})$, both at time k. It has a similar property to the filter residual or the innovation process as defined in Sections 2.6 and 2.8. It is only similar because EKF is a first order solution of the nonlinear estimation problem and its covariance is approximated by $\mathbf{H}_{\hat{\mathbf{x}}_{k|k-1}}\mathbf{P}_{k|k-1}\mathbf{H}^T_{\hat{\mathbf{x}}_{k|k-1}} + \mathbf{R}_k$ where the system and measurement matrices are approximated by the Jacobians of the nonlinear functions. It is the authors' experience in solving many practical problems that it is a good assumption that the residual process approximately holds the property as a white noise process with zero mean and covariance $\mathbf{H}_{\hat{\mathbf{x}}_{k|k-1}}\mathbf{P}_{k|k-1}\mathbf{H}^T_{\hat{\mathbf{x}}_{k|k-1}} + \mathbf{R}_k$ for the EKF. As such, the technique to adaptively adjust $\mathbf{Q}_k$ based upon the behavior of the innovation process can be applied.

3. The KF equation gives the conditional mean and covariance of $\mathbf{x}_k$ given $\mathbf{y}_{1:k}$. Together they completely characterize the a posteriori density function for linear state estimation problems with Gaussian statistics. When the system is nonlinear, the a posteriori density will no longer remain Gaussian even with Gaussian initial conditions and noise processes. Thus, the first two central moments, mean and covariance, can no longer fully characterize the a posteriori density. Furthermore, the state and covariance of EKF are only approximated at the first and second central moments.

4. Although EKF does not compute the a posteriori density, it nevertheless provides a state estimate and its covariance, which are essential for solving practical problems. The error covariance can usually be made consistent with the actual error when the EKF is tuned properly. Due to its simplicity and the fact that it works with many practical problems, we recommend EKF as the first filter to try.

5. The EKF uses only the first order Taylor series expansion to approximate a nonlinear system. It works well in most nonlinear systems when the data rate is high. It tends to perform poorly when the data rate is low and when the first order Taylor series expansion is a poor approximation of the nonlinear system, for example, when the true second central moment is no longer ellipsoid shaped. Means to improve filter performance include (a) numerically seeking a closer solution to minimization (single stage iteration), (b) extension to second order expansion (the second order filter), and (c) applying a sampled approximation to the conditional mean and covariance (the unscented Kalman filter). The first two methods will be discussed later in this chapter and the third method will be discussed in Chapter 6.

Table 3.1   Summary of Extended Kalman Filter Algorithm

Discrete Extended Kalman Filter (EKF)

State and Measurement Equations

$$\mathbf{x}_k = \mathbf{f}(\mathbf{x}_{k-1}) + \boldsymbol{\mu}_{k-1}$$

$$\mathbf{y}_k = \mathbf{h}(\mathbf{x}_k) + \mathbf{v}_k$$

where $\mathbf{x}_k \in \mathbb{R}^n$, $\mathbf{y}_k \in \mathbb{R}^m$, $\mathbf{x}_0 {:} {\sim} N(\hat{\mathbf{x}}_0, \mathbf{P}_0)$, $\boldsymbol{\mu}_k{:} \sim N(\mathbf{0}, \mathbf{Q}_k)$, $\mathbf{v}_k{:} \sim N(\mathbf{0}, \mathbf{R}_k)$, and $\mathbf{f}(.)$ and $\mathbf{h}(.)$ are known functions

Solution Process

Given: $\hat{\mathbf{x}}_{k-1|k-1}$ and $\mathbf{P}_{k-1|k-1}$, they are initial conditions $\hat{\mathbf{x}}_{0|0}$ and $\mathbf{P}_{0|0}$ when $k = 1$

Predict: compute $\hat{\mathbf{x}}_{k|k-1}$ and $\mathbf{P}_{k|k-1}$ using the conditional mean formulation with first order Taylor series approximation of $\mathbf{f}(\mathbf{x}_{k-1})$

Update: compute $\hat{\mathbf{x}}_{k|k}$ and $\mathbf{P}_{k|k}$ with new measurement vector $\mathbf{y}_k$ using weighted least squares estimation formulation with first order Taylor series approximation of $\mathbf{h}(\mathbf{x})$

Jacobian Matrices for First Order Taylor Series Expansion

$$\mathbf{F}_{\hat{\mathbf{x}}_{k-1|k-1}} = \left[\frac{\partial \mathbf{f}(\mathbf{x})}{\partial \mathbf{x}} \right]_{\hat{\mathbf{x}}_{k-1|k-1}}$$

$$\mathbf{H}_{\hat{\mathbf{x}}_{k|k-1}} = \left[\frac{\partial \mathbf{h}(\mathbf{x})}{\partial \mathbf{x}} \right]_{\hat{\mathbf{x}}_{k|k-1}}$$

Solution Equations

Predict:

$$\hat{\mathbf{x}}_{k|k-1} = \mathbf{f}\left(\hat{\mathbf{x}}_{k-1|k-1} \right)$$

$$\mathbf{P}_{k|k-1} = \mathbf{F}_{\hat{\mathbf{x}}_{k-1|k-1}} \mathbf{P}_{k-1|k-1} \mathbf{F}_{\hat{\mathbf{x}}_{k-1|k-1}}^T + \mathbf{Q}_{k-1}$$

If the system dynamics has a continuous representation, the state prediction equation can be replaced with numerical integration from $k - 1$ to k

Update:

$$\hat{\mathbf{x}}_{k|k} = \hat{\mathbf{x}}_{k|k-1} + \mathbf{P}_{k|k-1} \mathbf{H}_{\hat{\mathbf{x}}_{k|k-1}}^T \left[\mathbf{H}_{\hat{\mathbf{x}}_{k|k-1}} \mathbf{P}_{k|k-1} \mathbf{H}_{\hat{\mathbf{x}}_{k|k-1}}^T + \mathbf{R}_k \right]^{-1} \left(\mathbf{y}_k - \mathbf{h}(\hat{\mathbf{x}}_{k|k-1}) \right)$$

$$\mathbf{P}_{k|k} = \mathbf{P}_{k|k-1} - \mathbf{P}_{k|k-1} \mathbf{H}_{\hat{\mathbf{x}}_{k|k-1}}^T \left[\mathbf{H}_{\hat{\mathbf{x}}_{k|k-1}} \mathbf{P}_{k|k-1} \mathbf{H}_{\hat{\mathbf{x}}_{k|k-1}}^T + \mathbf{R}_k \right]^{-1} \mathbf{H}_{\hat{\mathbf{x}}_{k|k-1}} \mathbf{P}_{k|k-1}$$

3.5   Extended Kalman Filter with Single Stage Iteration

The development of the single stage iterative extended Kalman filter (IEKF) is similar to the derivation in Section 1.6; the desire is to derive an iterative algorithm that can improve on the Taylor series expansion on the measurement update step and offer the opportunity to achieve a more accurate estimate than the EKF. The difference between Section 1.6 and this section is that the a priori state and covariance used in Section 1.6 is now the one-step predicted state and covariance, $\hat{\mathbf{x}}_{k-1|k}$ and $\mathbf{P}_{k-1|k}$. The performance index to be minimized remains, Equation (3.4-4), but the state vector used in the Taylor series expansion in Equation (3.4-6) is now changed to an initial guess of the estimate $\hat{\mathbf{x}}_{k|k}$, namely $\hat{\mathbf{x}}_{k|k}^0$. Let $\hat{\mathbf{x}}_{k|k}^0$ be $\hat{\mathbf{x}}_{k|k-1}$ as a logical initial guess, thus

$$\mathbf{h}(\mathbf{x}_k) \approx \mathbf{h}(\hat{\mathbf{x}}_{k|k}^0) + \left[\frac{\partial \mathbf{h}(\mathbf{x}_k)}{\partial \mathbf{x}_k} \right]_{\hat{\mathbf{x}}_{k|k}^0} (\mathbf{x}_k - \hat{\mathbf{x}}_{k|k}^0). \tag{3.5-1}$$

Substituting Equation (3.5-1) into Equation (3.4-5) and solving for $\mathbf{x}_k$ one obtains

$$\hat{\mathbf{x}}_{k|k}^1 = \mathbf{P}_{k|k}^0 \left(\left(\mathbf{H}_{\hat{\mathbf{x}}_{k|k}^0}^T \mathbf{R}_k^{-1} \mathbf{H}_{\hat{\mathbf{x}}_{k|k}^0} \hat{\mathbf{x}}_{k|k}^0 + \mathbf{P}_{k|k-1}^{-1}\hat{\mathbf{x}}_{k|k-1} \right) + \mathbf{H}_{\hat{\mathbf{x}}_{k|k}^0}^T \mathbf{R}_k^{-1} \left(\mathbf{y} - \mathbf{h}(\hat{\mathbf{x}}_{k|k}^0) \right) \right) \tag{3.5-2}$$

$$\mathbf{P}_{k|k}^0 = \left[\mathbf{H}_{\hat{\mathbf{x}}_{k|k}^0}^T \mathbf{R}_k^{-1} \mathbf{H}_{\hat{\mathbf{x}}_{k|k}^0} + \mathbf{P}_{k|k-1}^{-1} \right]^{-1}, \tag{3.5-3}$$

where $\hat{\mathbf{x}}_{k|k}^1$ is the solution of the first iteration. Equation (3.5-2) suggests an iterative algorithm

$$\hat{\mathbf{x}}_{k|k}^{i+1} = \mathbf{P}_{k|k}^i \left(\left(\mathbf{H}_{\hat{\mathbf{x}}_{k|k}^i}^T \mathbf{R}_k^{-1} \mathbf{H}_{\hat{\mathbf{x}}_{k|k}^i} \hat{\mathbf{x}}_{k|k}^i + \mathbf{P}_{k|k-1}^{-1}\hat{\mathbf{x}}_{k|k-1} \right) + \mathbf{H}_{\hat{\mathbf{x}}_{k|k}^i}^T \mathbf{R}_k^{-1} \left(\mathbf{y} - \mathbf{h}(\hat{\mathbf{x}}_{k|k}^i) \right) \right) \tag{3.5-4}$$

$$\mathbf{P}_{k|k}^i = \left[\mathbf{H}_{\hat{\mathbf{x}}_{k|k}^i}^T \mathbf{R}_k^{-1} \mathbf{H}_{\hat{\mathbf{x}}_{k|k}^i} + \mathbf{P}_{k|k-1}^{-1} \right]^{-1}. \tag{3.5-5}$$

Or in a more conventional form,

$$\hat{\mathbf{x}}_{k|k}^{i+1} = \hat{\mathbf{x}}_{k|k}^i + \mathbf{P}_{k|k}^i \left(\mathbf{H}_{\hat{\mathbf{x}}_{k|k}^i}^T \mathbf{R}_k^{-1} \left(\mathbf{y} - \mathbf{h}(\hat{\mathbf{x}}_{k|k}^i) \right) + \mathbf{P}_{k|k-1}^{-1} \left(\hat{\mathbf{x}}_{k|k-1} - \hat{\mathbf{x}}_{k|k}^i \right) \right). \tag{3.5-6}$$

The iteration is stopped when the difference between $\hat{\mathbf{x}}_{k|k}^{i+1}$ and $\hat{\mathbf{x}}_{k|k}^i$ become sufficiently small. Let the final solution be denoted as $\hat{\mathbf{x}}_{k|k}$, the covariance of $\hat{\mathbf{x}}_{k|k}$ is therefore (approximately)

$$\mathbf{P}_{k|k} = \left[\mathbf{H}_{\hat{\mathbf{x}}_{k|k}}^T \mathbf{R}_k^{-1} \mathbf{H}_{\hat{\mathbf{x}}_{k|k}} + \mathbf{P}_{k|k-1}^{-1} \right]^{-1}. \tag{3.5-7}$$

This algorithm is summarized in Table 3.2.

3.6 Derivation of Extended Kalman Filter with Bayesian Approach

The EKF can also be derived using the Bayesian recursive Equation (3.3-2) similar to that for the KF. The difference is that conditional mean and covariance are sufficient to represent the a posteriori density for linear estimation while it only represents an approximate solution for nonlinear estimation. Furthermore, the mean and covariance alone are insufficient to represent the a posteriori density in the nonlinear case. The mean and covariance remain the main objective of estimator design because, as

Table 3.2   Summary of Single Stage Iterative Extended Kalman Filter

---

Single Stage Iterative Extended Kalman Filter (IEKF)
Only applied at the Update step with $\hat{\mathbf{x}}_{k|k}^{0} = \hat{\mathbf{x}}_{k|k-1}$

$$\hat{\mathbf{x}}_{k|k}^{i+1} = \hat{\mathbf{x}}_{k|k}^{i} + \mathbf{P}_{k|k}^{i}\left(\mathbf{H}_{\hat{\mathbf{x}}_{k|k}^{i}}^{T}\mathbf{R}_{k}^{-1}\left(\mathbf{y} - \mathbf{h}\left(\hat{\mathbf{x}}_{k|k}^{i}\right)\right) + \mathbf{P}_{k|k-1}^{-1}\left(\hat{\mathbf{x}}_{k|k-1} - \hat{\mathbf{x}}_{k|k}^{i}\right)\right)$$

$$\mathbf{P}_{k|k}^{i} = \left[\mathbf{H}_{\hat{\mathbf{x}}_{k|k}^{i}}^{T}\mathbf{R}_{k}^{-1}\mathbf{H}_{\hat{\mathbf{x}}_{k|k}^{i}} + \mathbf{P}_{k|k-1}^{-1}\right]^{-1}$$

until $\left\|\hat{\mathbf{x}}_{k|k}^{i+1} - \hat{\mathbf{x}}_{k|k}^{i}\right\|$ becomes sufficiently small, then

$$\hat{\mathbf{x}}_{k|k} = \hat{\mathbf{x}}_{k|k}^{i+1} \text{ and}$$

$$\mathbf{P}_{k|k} = \left[\mathbf{H}_{\hat{\mathbf{x}}_{k|k}}^{T}\mathbf{R}_{k}^{-1}\mathbf{H}_{\hat{\mathbf{x}}_{k|k}} + \mathbf{P}_{k|k-1}^{-1}\right]^{-1}$$

---

stated before, they represent significant physical meaning for solving practical problems. Similarly, a first order Taylor series expansion of the system and measurement equations is applied with the assumption that the a posteriori density functions remain approximately Gaussian. Of the four density functions in Equation (3.3-2), $p(\mathbf{y}_k|\mathbf{y}_{1:k-1})$ is a normalization factor, and therefore only the remaining three need to be defined. Following the development of Section 3.4 yields

1. $p(\mathbf{y}_k|\mathbf{y}_{1:k-1})$

$$E\left\{\mathbf{x}_k|\mathbf{y}_{1:k-1}\right\} = \hat{\mathbf{x}}_{k|k-1} \approx \mathbf{f}\left(\hat{\mathbf{x}}_{k-1|k-1}\right),$$

$$Cov\left\{\mathbf{x}_k|\mathbf{y}_{1:k-1}\right\} = \mathbf{P}_{k|k-1} \approx \mathbf{F}_{\hat{\mathbf{x}}_{k-1|k-1}}\mathbf{P}_{k-1|k-1}\mathbf{F}_{\hat{\mathbf{x}}_{k-1|k-1}}^{T} + \mathbf{Q}_{k-1}.$$

Note that this density function defines the one-step predicted EKF equations.

2. $p(\mathbf{y}_k|\mathbf{x}_k)$

$$E\{\mathbf{y}_k|\mathbf{x}_k\} = \mathbf{h}(\mathbf{x}_k).$$

$$Cov\{\mathbf{y}_k|\mathbf{x}_k\} = \mathbf{R}_k.$$

The density on the left-hand side of Equation (3.4-3) is just the definition of the estimate and covariance.

3. $p(\mathbf{x}_k|\mathbf{y}_{1:k})$

$$E\left\{\mathbf{x}_k|\mathbf{y}_{1:k}\right\} = \hat{\mathbf{x}}_{k|k},$$

$$Cov\{\mathbf{x}_k|\mathbf{y}_{1:k}\} = \mathbf{P}_{k|k}.$$

Expanding out the exponent of $p(\mathbf{x}_k|\mathbf{y}_{1:k})$ yields

$$\left(\mathbf{x}_k - \hat{\mathbf{x}}_{k|k}\right)^{T}\mathbf{P}_{k|k}^{-1}\left(\mathbf{x}_k - \hat{\mathbf{x}}_{k|k}\right) = \mathbf{x}_k^{T}\mathbf{P}_{k|k}^{-1}\mathbf{x}_k + \hat{\mathbf{x}}_{k|k}^{T}\mathbf{P}_{k|k}^{-1}\hat{\mathbf{x}}_{k|k} - \mathbf{x}_k^{T}\mathbf{P}_{k|k}^{-1}\hat{\mathbf{x}}_{k|k} - \hat{\mathbf{x}}_{k|k}^{T}\mathbf{P}_{k|k}^{-1}\mathbf{x}_k.$$

Expanding the exponents of the right-hand side and leaving the density in the denominator constant yields

$$\left(\mathbf{x}_k - \hat{\mathbf{x}}_{k|k-1}\right)^T \mathbf{P}_{k|k-1}^{-1}\left(\mathbf{x}_k - \hat{\mathbf{x}}_{k|k-1}\right) + \left(\mathbf{y}_k - \mathbf{h}\left(\mathbf{x}_k\right)\right)^T \mathbf{R}_k^{-1}\left(\mathbf{y}_k - \mathbf{h}\left(\mathbf{x}_k\right)\right) - \text{const}$$

$$= \mathbf{x}_k^T \mathbf{P}_{k|k-1}^{-1}\mathbf{x}_k + \hat{\mathbf{x}}_{k|k-1}^T \mathbf{P}_{k|k-1}^{-1}\hat{\mathbf{x}}_{k|k-1} - \mathbf{x}_k^T \mathbf{P}_{k|k-1}^{-1}\hat{\mathbf{x}}_{k|k-1} - \hat{\mathbf{x}}_{k|k-1}^T \mathbf{P}_{k|k-1}^{-1}\mathbf{x}_k + \mathbf{y}_k^T \mathbf{R}_k^{-1}\mathbf{y}_k$$

$$+ \mathbf{h}\left(\mathbf{x}_k\right)^T \mathbf{R}_k^{-1}\mathbf{h}\left(\mathbf{x}_k\right) - \mathbf{y}_k^T \mathbf{R}_k^{-1}\mathbf{h}\left(\mathbf{x}_k\right) - \mathbf{h}\left(\mathbf{x}_k\right)^T \mathbf{R}_k^{-1}\mathbf{y}_k - \text{const.}$$

The Taylor series expansion is now applied to $\mathbf{h}(\mathbf{x}_k)$ about the predicted estimate $\hat{\mathbf{x}}_{k|k-1}$, expanded (the details are omitted here), and the terms in the parentheses of $\mathbf{x}_k^T(.)\mathbf{x}_k$ are collected to obtain

$$\mathbf{P}_{k|k}^{-1} = \mathbf{P}_{k|k-1}^{-1} + \mathbf{H}_{\hat{\mathbf{x}}_{k|k-1}}^T \mathbf{R}_k^{-1}\mathbf{H}_{\hat{\mathbf{x}}_{k|k-1}}.$$

An equivalent form is

$$\mathbf{P}_{k|k} = \mathbf{P}_{k|k-1} - \mathbf{P}_{k|k-1}\mathbf{H}_{\hat{\mathbf{x}}_{k|k-1}}^T \left[\mathbf{H}_{\hat{\mathbf{x}}_{k|k-1}}\mathbf{P}_{k|k-1}\mathbf{H}_{\hat{\mathbf{x}}_{k|k-1}}^T + \mathbf{R}_k\right]^{-1}\mathbf{H}_{\hat{\mathbf{x}}_{k|k-1}}\mathbf{P}_{k|k-1}. \qquad (3.6\text{-}1)$$

Similarly, the terms in the parentheses of $\mathbf{x}_k^T(.)\hat{\mathbf{x}}_{k|k-1}$ gives

$$\mathbf{P}_{k|k}^{-1}\hat{\mathbf{x}}_{k|k} = \mathbf{P}_{k|k-1}^{-1}\hat{\mathbf{x}}_{k|k-1} + \mathbf{H}_{\hat{\mathbf{x}}_{k|k-1}}^T \mathbf{R}_k^{-1}\mathbf{y}_k.$$

After standard manipulation one obtains

$$\hat{\mathbf{x}}_{k|k} = \hat{\mathbf{x}}_{k|k-1} + \mathbf{P}_{k|k-1}\mathbf{H}_{\hat{\mathbf{x}}_{k|k-1}}^T \left[\mathbf{H}_{\hat{\mathbf{x}}_{k|k-1}}\mathbf{P}_{k|k-1}\mathbf{H}_{\hat{\mathbf{x}}_{k|k-1}}^T + \mathbf{R}_k\right]^{-1}\left(\mathbf{y}_k - \mathbf{h}\left(\hat{\mathbf{x}}_{k|k-1}\right)\right). \qquad (3.6\text{-}2)$$

Equations (3.6-1) and (3.6-2) are the EKF update equations.

3.7 Nonlinear Filter Equation with Second Order Taylor Series Expansion Retained

An intuitive extension to improve on the EKF is to retain some of the HOT in the Taylor series expansion. A filter derived by retaining up to the second order terms is referred to as the second order filter (SOF). Several researchers have derived filters with this idea, notably Athans, Wishner, and Bertonili [9] and Wishner, Tabaczynski, and Athans [10]. In this section, an SOF is derived. The results of the derivation closely match those of [10].

First, consider the system dynamics. Expanding Equation (3.4-1) to include the second order terms in a Taylor series expansion about $\hat{\mathbf{x}}_{k-1|k-1}$ results in

$$\mathbf{x}_k = \mathbf{f}\left(\hat{\mathbf{x}}_{k-1|k-1}\right) + \mathbf{F}_{\hat{\mathbf{x}}_{k-1|k-1}}\left(\mathbf{x}_{k-1} - \hat{\mathbf{x}}_{k-1|k-1}\right) +$$

$$\frac{1}{2}\sum_{i=1}^{n}\phi_i \left(\mathbf{x}_{k-1} - \hat{\mathbf{x}}_{k-1|k-1}\right)^T \mathbf{I}_{i,\hat{\mathbf{x}}_{k-1|k-1}}\left(\mathbf{x}_{k-1} - \hat{\mathbf{x}}_{k-1|k-1}\right) + \mathbf{\mu}_{k-1} + \text{HOT}, \qquad (3.7\text{-}1)$$

where ϕ_i is an n-element column vector with 0 everywhere except 1 for the ith element, n is the dimension of the state vector, $\mathbf{I}_{i,\hat{\mathbf{x}}_{k-1|k-1}}$ is the matrix of the second derivative of ith element of $\mathbf{f}(\mathbf{x}_{k-1})$ evaluated at $\hat{\mathbf{x}}_{k-1|k-1}$ (known as the Hessian matrix), the (q,r)th element of $\mathbf{I}_{i,\hat{\mathbf{x}}_{k-1|k-1}}$ is

$$\left[\mathbf{I}_{i,\hat{\mathbf{x}}_{k-1|k-1}}\right]_{q,r} = \left[\frac{\partial^2 f_i(\mathbf{x})}{\partial x_q \partial x_r}\right]_{\hat{\mathbf{x}}_{k-1|k-1}},$$

and x_q and x_r are the qth and rth elements of $\mathbf{x}_{k-1}$.

Similarly, the second order Taylor series expansion of the measurement equation is

$$\mathbf{h}(\mathbf{x}_k) = \mathbf{h}(\hat{\mathbf{x}}_{k|k-1}) + \left[\frac{\partial \mathbf{h}(\mathbf{x}_k)}{\partial \mathbf{x}_k}\right]_{\hat{\mathbf{x}}_{k|k-1}} (\mathbf{x}_k - \hat{\mathbf{x}}_{k|k-1})$$

$$+ \frac{1}{2}\sum_{i=1}^{m} \phi_i (\mathbf{x}_k - \hat{\mathbf{x}}_{k|k-1})^T \mathbf{J}_{i,\hat{\mathbf{x}}_{k|k-1}} (\mathbf{x}_k - \hat{\mathbf{x}}_{k|k-1}) + \text{HOT}, \qquad (3.7\text{-}2)$$

where ϕ_i is an m-element column vector with 0 everywhere except 1 for the ith element, m is the dimension of the measurement vector, $\mathbf{J}_{i,\hat{\mathbf{x}}_{k|k-1}}$ is the matrix of the second derivative of the ith element of $\mathbf{h}(\mathbf{x}_k)$ evaluated at $\hat{\mathbf{x}}_{k|k-1}$ (similarly, the Hessian matrix), the (q,r)th element of $\mathbf{J}_{i,\hat{\mathbf{x}}_{k|k-1}}$ is

$$\left[\mathbf{J}_{i,\hat{\mathbf{x}}_{k|k-1}}\right]_{q,r} = \left[\frac{\partial^2 h_i(\mathbf{x})}{\partial x_q \partial x_r}\right]_{\hat{\mathbf{x}}_{k|k-1}},$$

and x_q and x_r are the qth and rth element of $\mathbf{x}_k$.

3.7.1   One-Step Prediction

First, we define two matrix identities that will be used in the derivations to follow.

Matrix Identities

(a) $\mathbf{x}^T \mathbf{A}\mathbf{x} = \text{tr}\left[\mathbf{A}\mathbf{x}\mathbf{x}^T\right]$ and $E\{\mathbf{x}^T \mathbf{A}\mathbf{x}\} = \text{tr}\left[\mathbf{A}\boldsymbol{\Sigma}\right]$

(b) $E\{\text{tr}\left[\mathbf{A}\mathbf{x}\mathbf{x}^T \mathbf{B}\mathbf{x}\mathbf{x}^T\right]\} = E\{\text{tr}\left[\mathbf{A}\mathbf{x}\mathbf{x}^T\right]\text{tr}\left[\mathbf{B}\mathbf{x}\mathbf{x}^T\right]\}$

$\qquad\qquad\qquad\qquad = 2\text{tr}\left[\mathbf{A}\boldsymbol{\Sigma}\mathbf{B}\boldsymbol{\Sigma}\right] + \text{tr}\left[\mathbf{A}\boldsymbol{\Sigma}\right]\text{tr}\left[\mathbf{B}\boldsymbol{\Sigma}\right]$

where $\text{tr}[\mathbf{M}]$ denotes the trace of the enclosed matrix M and $\boldsymbol{\Sigma} = E\{\mathbf{x}\mathbf{x}^T\}$.

These two identities can be proved with tedious but straightforward manipulation. Their derivation is left as a homework problem. Applying the definition

$\hat{\mathbf{x}}_{k|k-1} = E\{\mathbf{x}_k | \mathbf{y}_{1:k-1}\}$ and taking the conditional expectation over Equation (3.7-1), one obtains

$$\hat{\mathbf{x}}_{k|k-1} \approx \mathbf{f}\left(\hat{\mathbf{x}}_{k-1|k-1}\right) + E\left\{\frac{1}{2}\sum_{i=1}^{n}\phi_i\left(\mathbf{x}_{k-1} - \hat{\mathbf{x}}_{k-1|k-1}\right)^T \mathbf{I}_{i,\hat{\mathbf{x}}_{k-1|k-1}}\left(\mathbf{x}_{k-1} - \hat{\mathbf{x}}_{k-1|k-1}\right) | \mathbf{y}_{1:k-1}\right\}. \quad (3.7\text{-}3)$$

Applying the matrix identity (a) to Equation (3.7-3), one obtains the state prediction equation for the second order filter

$$\hat{\mathbf{x}}_{k|k-1} \approx \mathbf{f}\left(\hat{\mathbf{x}}_{k-1|k-1}\right) + \frac{1}{2}\sum_{i=1}^{n}\phi_i \text{tr}\left[\mathbf{I}_{i,\hat{\mathbf{x}}_{k-1|k-1}}\mathbf{P}_{k-1|k-1}\right]. \quad (3.7\text{-}4)$$

Next, derive the one-step prediction equation for the covariance. By applying the covariance definition and using Equation (3.7-1), one obtains

$$\mathbf{P}_{k|k-1} \approx \mathbf{F}_{\hat{\mathbf{x}}_{k-1|k-1}}\mathbf{P}_{k-1|k-1}\mathbf{F}_{\hat{\mathbf{x}}_{k-1|k-1}}^{T} + \mathbf{Q}_{k-1}$$
$$+ \frac{1}{4}\sum_{i=1}^{n}\sum_{j=1}^{n}\phi_i\phi_j^T E\left\{\left(\Delta\mathbf{x}_{k-1}^{T}\mathbf{I}_{i,\hat{\mathbf{x}}_{k-1|k-1}}\Delta\mathbf{x}_{k-1}\right)\left(\Delta\mathbf{x}_{k-1}^{T}\mathbf{I}_{j,\hat{\mathbf{x}}_{k-1|k-1}}\Delta\mathbf{x}_{k-1}\right) | \mathbf{y}_{1:k-1}\right\}. \quad (3.7\text{-}5)$$

Note that simplifying notation $\Delta\mathbf{x}_{k-1}$ for $\mathbf{x}_{k-1} - \hat{\mathbf{x}}_{k-1|k-1}$ has been used here. Let the (i,j)th element of matrix $\mathbf{K}_{\hat{\mathbf{x}}_{k-1|k-1}}$ be defined by the term with the expectation operator in Equation (3.7-5), that is,

$$\left[\mathbf{K}_{\hat{\mathbf{x}}_{k-1|k-1}}\right]_{i,j} = E\left\{\left(\Delta\mathbf{x}_{k-1}^{T}\mathbf{I}_{i,\hat{\mathbf{x}}_{k-1|k-1}}\Delta\mathbf{x}_{k-1}\right)\left(\Delta\mathbf{x}_{k-1}^{T}\mathbf{I}_{j,\hat{\mathbf{x}}_{k-1|k-1}}\Delta\mathbf{x}_{k-1}\right) | \mathbf{y}_{1:k-1}\right\}$$

thus

$$\mathbf{K}_{\hat{\mathbf{x}}_{k-1|k-1}} = \sum_{i=1}^{n}\sum_{j=1}^{n}\phi_i\phi_j^T \left[\mathbf{K}_{\hat{\mathbf{x}}_{k-1|k-1}}\right]_{i,j}.$$

Applying the matrix identities one obtains

$$\left[\mathbf{K}_{\hat{\mathbf{x}}_{k-1|k-1}}\right]_{i,j} = 2\text{tr}\left[\mathbf{I}_{i,\hat{\mathbf{x}}_{k-1|k-1}}\mathbf{P}_{k-1|k-1}\mathbf{I}_{j,\hat{\mathbf{x}}_{k-1|k-1}}\mathbf{P}_{k-1|k-1}\right] + \text{tr}\left[\mathbf{I}_{i,\hat{\mathbf{x}}_{k-1|k-1}}\mathbf{P}_{k-1|k-1}\right]\text{tr}\left[\mathbf{I}_{j,\hat{\mathbf{x}}_{k-1|k-1}}\mathbf{P}_{k-1|k-1}\right]$$
$$(3.7\text{-}6)$$

Equation (3.7-5) can be rewritten using $\mathbf{K}_{\hat{\mathbf{x}}_{k-1|k-1}}$

$$\mathbf{P}_{k|k-1} \approx \mathbf{F}_{\hat{\mathbf{x}}_{k-1|k-1}}\mathbf{P}_{k-1|k-1}\mathbf{F}_{\hat{\mathbf{x}}_{k-1|k-1}}^{T} + \mathbf{Q}_{k-1} + \frac{1}{4}\mathbf{K}_{\hat{\mathbf{x}}_{k-1|k-1}}. \quad (3.7\text{-}7)$$

Equations (3.7-4), (3.7-6), and (3.7-7) are the one step-prediction algorithm for a second order filter.

3.7.2   Update Equations

Normally the update equation would be derived using the WLS approach. Unfortunately, when second order terms are retained, the sufficient condition for minimizing the weighted square error is quadratic in $\mathbf{x}_k$, thus making a closed form solution difficult to obtain. Instead, the idea that the updated estimate is the weighted sum of the one-step predicted estimate and the residual (innovation) vector is applied, that is,

$$\hat{\mathbf{x}}_{k|k} = \mathbf{P}_{k|k}\left(\mathbf{P}_{k|k-1}^{-1}\hat{\mathbf{x}}_{k|k-1} + \mathbf{P}_{\gamma,k|k-1}^{-1}\boldsymbol{\gamma}_k\right) \tag{3.7-8}$$

where $\boldsymbol{\gamma}_k$ is the filter residual process with covariance $\mathbf{P}_{\gamma,k|k-1}$. Applying the second order expansion to the measurement equation and following similar derivation as the development for the one-step prediction equation, one obtains

$$\boldsymbol{\gamma}_k = \mathbf{y}_k - E\{\mathbf{y}_k|\mathbf{y}_{1:k-1}\} = \mathbf{y}_k - \left(\mathbf{h}(\hat{\mathbf{x}}_{k|k-1}) + \frac{1}{2}\sum_{i=1}^{m}\boldsymbol{\phi}_i\mathrm{tr}\left[\mathbf{J}_{i,\hat{\mathbf{x}}_{k|k-1}}\mathbf{P}_{k|k-1}\right]\right). \tag{3.7-9}$$

The covariance of $\boldsymbol{\gamma}_k$ is

$$\mathbf{P}_{\gamma,k|k-1} = \mathbf{H}_{\hat{\mathbf{x}}_{k|k-1}}\mathbf{P}_{k|k-1}\mathbf{H}_{\hat{\mathbf{x}}_{k|k-1}}^{T} + \mathbf{R}_k + \frac{1}{4}\mathbf{L}_{\hat{\mathbf{x}}_{k|k-1}} \tag{3.7-10}$$

where

$$\mathbf{L}_{\hat{\mathbf{x}}_{k|k-1}} = \sum_{i=1}^{n}\sum_{j=1}^{n}\boldsymbol{\phi}_i\boldsymbol{\phi}_j^T E\left\{\left(\Delta\mathbf{x}_k^T\mathbf{J}_{i,\hat{\mathbf{x}}_{k|k-1}}\Delta\mathbf{x}_k\right)\left(\Delta\mathbf{x}_k^T\mathbf{J}_{j,\hat{\mathbf{x}}_{k|k-1}}\Delta\mathbf{x}_k\right)|\mathbf{y}_{1:k-1}\right\} \tag{3.7-11}$$

and $\Delta\mathbf{x}_k = \mathbf{x}_k - \hat{\mathbf{x}}_{k|k-1}$. Applying the same matrix identities, one obtains the following results for the (i,j)th element of $\mathbf{L}_{\hat{\mathbf{x}}_{k|k-1}}$

$$[\mathbf{L}_{\hat{\mathbf{x}}_{k|k-1}}]_{i,j} = 2\mathrm{tr}\left[\mathbf{J}_{i,\hat{\mathbf{x}}_{k|k-1}}\mathbf{P}_{k|k-1}\mathbf{J}_{j,\hat{\mathbf{x}}_{k|k-1}}\mathbf{P}_{k|k-1}\right] + \mathrm{tr}\left[\mathbf{J}_{i,\hat{\mathbf{x}}_{k|k-1}}\mathbf{P}_{k|k-1}\right]\mathrm{tr}\left[\mathbf{J}_{j,\hat{\mathbf{x}}_{k|k-1}}\mathbf{P}_{k|k-1}\right]. \tag{3.7-12}$$

After some algebraic manipulations, one obtains the update equations for the second order filter

$$\hat{\mathbf{x}}_{k|k} = \hat{\mathbf{x}}_{k|k-1} + \mathbf{G}_k\left(\mathbf{y}_k - \mathbf{h}(\hat{\mathbf{x}}_{k|k-1}) - \frac{1}{2}\sum_{i=1}^{m}\boldsymbol{\phi}_i\mathrm{tr}\left[\mathbf{J}_{i,\hat{\mathbf{x}}_{k|k-1}}\mathbf{P}_{k|k-1}\right]\right), \tag{3.7-13}$$

where the filter gain $\mathbf{G}_k$ is

$$\mathbf{G}_k = \mathbf{P}_{k|k}\mathbf{H}_{\hat{\mathbf{x}}_{k|k-1}}^{T}\left[\mathbf{H}_{\hat{\mathbf{x}}_{k|k-1}}\mathbf{P}_{k|k-1}\mathbf{H}_{\hat{\mathbf{x}}_{k|k-1}}^{T} + \mathbf{R}_k + \frac{1}{4}\mathbf{L}_{\hat{\mathbf{x}}_{k|k-1}}\right], \tag{3.7-14}$$

and the updated covariance $\mathbf{P}_{k|k}$ is

$$\mathbf{P}_{k|k} = \mathbf{P}_{k|k-1} - \mathbf{P}_{k|k-1}\mathbf{H}_{\hat{\mathbf{x}}_{k|k-1}}^{T}\left[\mathbf{H}_{\hat{\mathbf{x}}_{k|k-1}}\mathbf{P}_{k|k-1}\mathbf{H}_{\hat{\mathbf{x}}_{k|k-1}}^{T} + \mathbf{R}_{k} + \frac{1}{4}\mathbf{L}_{\hat{\mathbf{x}}_{k|k-1}}\right]^{-1}\mathbf{H}_{\hat{\mathbf{x}}_{k|k-1}}\mathbf{P}_{k|k-1}. \quad (3.7\text{-}15)$$

Equations (3.7-13), (3.7-14), and (3.7-15) are the updated equations of the SOF. Note that if the second order terms are dropped from both the prediction and the update equations, one obtains the EKF equations.

The second order nonlinear filter is summarized in Table 3.3.

3.7.3   A Numerical Example

A numerical example published in [9, 10] is now used to compare the three nonlinear filters described above. A sphere falling through atmosphere can be described by a set of nonlinear differential equations. This object is observed by a radar located at altitude RA above ground and RD distance away from the closest approach of the trajectory, as shown in Figure 3.1. The radar only measures range to target. The three state variables that constitute the state vector are $x_1 =$ altitude, $x_2 =$ rate of change of altitude due to atmospheric drag, and $x_3 =$ the inverse of ballistic coefficient. Their state equations are

$$\dot{x}_1(t) = -x_2(t)$$

$$\dot{x}_2(t) = -K_1 e^{-K_2 x_1(t)} x_2^2(t) x_3(t)$$

$$\dot{x}_3(t) = 0,$$

where $x_3(t) \triangleq \alpha$, an unknown constant to be estimated and $\alpha = 1/\beta$, where β is the ballistic coefficient. K_1 and K_2 are known constants relating to atmosphere. Note that the gravity term is neglected. As seen in Figure 3.1, this simplification should not impact on the observation and conclusion of filter performance comparison.

The range measurement of this radar is a nonlinear relationship with the first state variable, the altitude of the radar, RA, and the distance to the closest approach, RD, to the falling object

$$h_k(\mathbf{x}_k) = \sqrt{(x_1(t_k) - \text{RA})^2 + \text{RD}^2}.$$

The geometry of this problem is illustrated in Figure 3.1.

Table 3.3 Summary of Second Order Nonlinear Filter Algorithm

Second Order Nonlinear Filter

State and Measurement Equations

$$\mathbf{x}_k = \mathbf{f}(\mathbf{x}_{k-1}) + \boldsymbol{\mu}_{k-1}$$

$$\mathbf{y}_k = \mathbf{h}(\mathbf{x}_k) + \mathbf{v}_k$$

where $\mathbf{x}_k \in \mathbb{R}^n$, $\mathbf{y}_k \in \mathbb{R}^m$, $\mathbf{x}_0 \colon \sim N(\hat{\mathbf{x}}_0, \mathbf{P}_0)$, $\boldsymbol{\mu}_k \colon \sim N(0, \mathbf{Q}_k)$, $\mathbf{v}_k \colon \sim N(0, \mathbf{R}_k)$, and $\mathbf{f}(.)$ and $\mathbf{h}(.)$ are known functions

Solution Process

Predict: Compute $\hat{\mathbf{x}}_{k|k-1}$ and $\mathbf{P}_{k|k-1}$ from $\hat{\mathbf{x}}_{k-1|k-1}$ and $\mathbf{P}_{k-1|k-1}$ using the conditional mean formulation with second order Taylor series approximation of $\mathbf{f}(\mathbf{x}_{k-1})$

Update: Compute $\hat{\mathbf{x}}_{k|k}$ and $\mathbf{P}_{k|k}$ with new measurement vector $\mathbf{y}_k$ using weighted least squares formulation with property of innovation process and the second order Taylor series approximation of $\mathbf{h}(\mathbf{x}_k)$

Jacobian Matrices and Hessian Matrices for Second Order Taylor Series Expansion

$$\mathbf{F}_{\hat{\mathbf{x}}_{k-1|k-1}} = \left[\frac{\partial \mathbf{f}(\mathbf{x})}{\partial \mathbf{x}}\right]_{\hat{\mathbf{x}}_{k-1|k-1}}$$

$$[\mathbf{I}_{i,\hat{\mathbf{x}}_{k-1|k-1}}]_{q,r} = \left[\frac{\partial^2 f_i(\mathbf{x})}{\partial x_q \partial x_r}\right]_{\hat{\mathbf{x}}_{k-1|k-1}}, \text{ and } x_q \text{ and } x_r \text{ are the } q\text{th and } r\text{th element of } \mathbf{x}_{k-1}$$

$$\mathbf{H}_{\hat{\mathbf{x}}_{k|k-1}} = \left[\frac{\partial \mathbf{h}(\mathbf{x})}{\partial \mathbf{x}}\right]_{\hat{\mathbf{x}}_{k|k-1}}$$

$$\left[\mathbf{J}_{i,\hat{\mathbf{x}}_{k|k-1}}\right]_{q,r} = \left[\frac{\partial^2 h_i(\mathbf{x})}{\partial x_q \partial x_r}\right]_{\hat{\mathbf{x}}_{k|k-1}}, \text{ and } x_q \text{ and } x_r \text{ are the } q\text{th and } r\text{th element of } \mathbf{x}_k.$$

Solution Equations

Predict:

$$\hat{\mathbf{x}}_{k|k-1} \approx \mathbf{f}\left(\hat{\mathbf{x}}_{k-1|k-1}\right) + \frac{1}{2}\sum_{i=1}^{n}\phi_i \text{tr}\left[\mathbf{I}_{i,\hat{\mathbf{x}}_{k-1|k-1}}\mathbf{P}_{k-1|k-1}\right]$$

$$\mathbf{P}_{k|k-1} \approx \mathbf{F}_{\hat{\mathbf{x}}_{k-1|k-1}}\mathbf{P}_{k-1|k-1}\mathbf{F}_{\hat{\mathbf{x}}_{k-1|k-1}}^T + \mathbf{Q}_{k-1} + \frac{1}{4}\mathbf{K}_{\hat{\mathbf{x}}_{k-1|k-1}}$$

$$\mathbf{K}_{\hat{\mathbf{x}}_{k-1|k-1}} = \sum_{i=1}^{n}\sum_{j=1}^{n}\phi_i\phi_j^T\left[\mathbf{K}_{\hat{\mathbf{x}}_{k-1|k-1}}\right]_{i,j}$$

$$\left[\mathbf{K}_{\hat{\mathbf{x}}_{k-1|k-1}}\right]_{i,j} = 2\text{tr}\left[\mathbf{I}_{i,\hat{\mathbf{x}}_{k-1|k-1}}\mathbf{P}_{k-1|k-1}\mathbf{I}_{j,\hat{\mathbf{x}}_{k-1|k-1}}\mathbf{P}_{k-1|k-1}\right]$$

ϕ_i is a column vector with 0 everywhere except 1 for the ith element. If the system dynamics has a continuous representation, the state prediction equation can be replaced with numerical integration from $k-1$ to k.

Update:

$$\hat{\mathbf{x}}_{k|k} = \hat{\mathbf{x}}_{k|k-1} + \mathbf{G}_k\left(\mathbf{y}_k - \mathbf{h}\left(\hat{\mathbf{x}}_{k|k-1}\right) - \frac{1}{2}\sum_{i=1}^{m}\phi_i\text{tr}\left[\mathbf{J}_{i,\hat{\mathbf{x}}_{k|k-1}}\mathbf{P}_{k|k-1}\right]\right)$$

$$\mathbf{G}_k = \mathbf{P}_{k|k}\mathbf{H}_{\hat{\mathbf{x}}_{k|k-1}}^T\left[\mathbf{H}_{\hat{\mathbf{x}}_{k|k-1}}\mathbf{P}_{k|k-1}\mathbf{H}_{\hat{\mathbf{x}}_{k|k-1}}^T + \mathbf{R}_k + \frac{1}{4}\mathbf{L}_{\hat{\mathbf{x}}_{k|k-1}}\right]$$

$$\mathbf{P}_{k|k} = \mathbf{P}_{k|k-1} - \mathbf{P}_{k|k-1}\mathbf{H}_{\hat{\mathbf{x}}_{k|k-1}}^T\left[\mathbf{H}_{\hat{\mathbf{x}}_{k|k-1}}\mathbf{P}_{k|k-1}\mathbf{H}_{\hat{\mathbf{x}}_{k|k-1}}^T + \mathbf{R}_k + \frac{1}{4}\mathbf{L}_{\hat{\mathbf{x}}_{k|k-1}}\right]^{-1}\times\mathbf{H}_{\hat{\mathbf{x}}_{k|k-1}}\mathbf{P}_{k|k-1}$$

$$[\mathbf{L}_{\hat{\mathbf{x}}_{k|k-1}}]_{i,j} = 2\text{tr}\left[\mathbf{J}_{i,\hat{\mathbf{x}}_{k|k-1}}\mathbf{P}_{k|k-1}\mathbf{J}_{j,\hat{\mathbf{x}}_{k|k-1}}\mathbf{P}_{k|k-1}\right] + \text{tr}\left[\mathbf{J}_{i,\hat{\mathbf{x}}_{k|k-1}}\mathbf{P}_{k|k-1}\right]\text{tr}\left[\mathbf{J}_{j,\hat{\mathbf{x}}_{k|k-1}}\mathbf{P}_{k|k-1}\right]$$

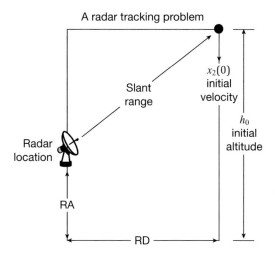

Figure 3.1 The geometry of the radar tracking problem.

Define the filter root-mean-square (RMS) error for each component as

$$\text{RMS Error}_i = \sqrt{\frac{1}{N}\sum_{n=1}^{N}\left(\left(\mathbf{x}(t_k)\right)_i - \left(\hat{\mathbf{x}}_{k|k}^n\right)_i\right)^2}, \text{ for } i = 1,2,3.$$

The RMS error for altitude, velocity, and α estimation are shown in Figures 3.2 and 3.3. The initial conditions for true and estimated states are shown in the figures. The results are obtained for 1 sec measurement update rate and 100 Monte Carlo repetitions.

This particular example offers the following observations.

1. The EKF yields the worst estimation error performance for all three state variables compared to the other filters.

2. The single-stage IEKF performed the best, but requires twice the computation relative to the EKF.

3. The performance of the second order filter lies between the above two filters, but it is much more complicated and requires more computation.

There are more studies, such as linear verse nonlinear, mean and RMS errors, and so on, presented in [9, 10], and interested readers should consult these references for details.

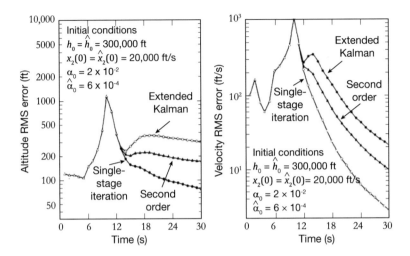

Figure 3.2   Altitude and velocity estimation error comparison.

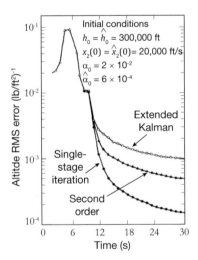

Figure 3.3   Comparison of alpha (inverse of ballistic coefficient) estimation error.

3.8   The Case with Nonlinear but Deterministic Dynamics

Consider the case of a deterministic nonlinear system with nonlinear measurement equations

$$\mathbf{x}_k = \mathbf{f}\left(\mathbf{x}_{k-1}\right), \tag{3.8-1}$$

$$\mathbf{y}_k = \mathbf{h}\left(\mathbf{x}_k\right) + \mathbf{v}_k. \tag{3.8-2}$$

Because the system is deterministic, once the state estimate at any time instance is obtained, the estimate for the entire trajectory is obtained. For the simplicity of notation, consider estimating the state vector at the initial time, that is, $\mathbf{x}_0$ given all measurements $\mathbf{y}_k$, $k = 1,..,K$. The estimate of $\mathbf{x}_0$ is the state vector that minimizes the following performance index

$$J = \sum_{k=1}^{K}\left(\mathbf{y}_k - \mathbf{h}\left(\mathbf{x}_k\right)\right)^T \mathbf{R}_k^{-1}\left(\mathbf{y}_k - \mathbf{h}\left(\mathbf{x}_k\right)\right) + \left(\mathbf{x}_0 - \overline{\mathbf{x}}_0\right)^T \mathbf{P}_0^{-1}\left(\mathbf{x}_0 - \overline{\mathbf{x}}_0\right), \tag{3.8-3}$$

where $\overline{\mathbf{x}}_0$ is the initial condition of Equation (3.8-1) with covariance $\mathbf{P}_0$. It is treated as a priori knowledge that contributes to the estimation of $\mathbf{x}_0$. Taking the derivative of J with respect to $\mathbf{x}_0$ gives the necessary condition for minimization

$$\frac{\partial J}{\partial \mathbf{x}_0} = 2\left(\sum_{k=1}^{K}\left[\frac{\partial \mathbf{h}\left(\mathbf{x}_k\right)}{\partial \mathbf{x}_0}\right]^T \mathbf{R}_k^{-1}\left(\mathbf{y}_k - \mathbf{h}\left(\mathbf{x}_k\right)\right) + \mathbf{P}_0^{-1}\left(\mathbf{x}_0 - \overline{\mathbf{x}}_0\right)\right) = \mathbf{0}. \tag{3.8-4}$$

Equation (3.8-4) can be solved exactly for linear systems. For nonlinear systems, an iterative algorithm is suggested, and is derived below.

An Iterative Least Squares Algorithm

First, $\mathbf{h}(\mathbf{x}_k)$ is approximated with a Taylor series expansion about $\hat{\mathbf{x}}_k^0$, an initial guess for the estimate of $\mathbf{x}_k$, used as a starting point of the iterative algorithm

$$\mathbf{h}\left(\mathbf{x}_k\right) \approx \mathbf{h}\left(\hat{\mathbf{x}}_k^0\right) + \mathbf{H}_{\hat{\mathbf{x}}_k^0}\left(\mathbf{x}_k - \hat{\mathbf{x}}_k^0\right). \tag{3.8-5}$$

Note that the notation $\left[\dfrac{\partial \mathbf{h}\left(\mathbf{x}_k\right)}{\partial \mathbf{x}_k}\right]_{\hat{\mathbf{x}}_k^0} = \mathbf{H}_{\hat{\mathbf{x}}_k^0}$ has been adopted. Because the system is deterministic, Equation (3.8-1) can be repeatedly used to obtain the following relationship

$$\mathbf{x}_k = \mathbf{f}\left(\mathbf{f}\left(\mathbf{f}\left(...\mathbf{f}\left(\mathbf{x}_0\right)...\right)\right)\right) = f_{k,0}\left(\mathbf{x}_0\right) \tag{3.8-6}$$

The expression $f_{k,0}(\mathbf{x}_0)$ is meant to capture the repeated application of the function $\mathbf{f}(.)$. Applying the first order Taylor series expansion of the above equation with respect to $\hat{\mathbf{x}}_0^0$, we obtain

$$\mathbf{x}_k \approx f_{k,0}\left(\hat{\mathbf{x}}_0^0\right) + \left[\frac{\partial f_{k,0}}{\partial \mathbf{x}_0}\right]_{\hat{\mathbf{x}}_0^0} \left(\mathbf{x}_0 - \hat{\mathbf{x}}_0^0\right) \tag{3.8-7}$$

where

$$\left[\frac{\partial f_{k,0}}{\partial \mathbf{x}_0}\right] = \frac{\partial \mathbf{x}_k}{\partial \mathbf{x}_0} = \frac{\partial \mathbf{f}\left(\mathbf{x}_{k-1}\right)}{\partial \mathbf{x}_{k-1}} \frac{\partial \mathbf{f}\left(\mathbf{x}_{k-2}\right)}{\partial \mathbf{x}_{k-2}} \cdots \frac{\partial \mathbf{f}\left(\mathbf{x}_0\right)}{\partial \mathbf{x}_0} = \mathbf{F}_{\mathbf{x}_{k-1}} \mathbf{F}_{\mathbf{x}_{k-2}} \cdots \mathbf{F}_{\mathbf{x}_0}.$$

Note that $\mathbf{F}_{\mathbf{x}_{k-1}}$ is equivalent to the state transition matrix of the linear system, so the notation $\mathbf{F}_{\mathbf{x}_{k-1}} = \mathbf{\Phi}_{k,k-1}$ and $\mathbf{F}_{\mathbf{x}_{k-1}} \mathbf{F}_{\mathbf{x}_{k-2}} \cdots \mathbf{F}_{\mathbf{x}_0} = \mathbf{\Phi}_{k,0}$ is used here. It is further specified that $\mathbf{\Phi}_{k,0}^0$ is the $\mathbf{\Phi}_{k,0}$ evaluated at the initial guess $\hat{\mathbf{x}}_0^0$. Applying these approximations to Equation (3.8-4) after some algebraic manipulation, one obtains

$$\sum\nolimits_{k=1}^{K} \mathbf{\Phi}_{k,0}^0{}^T \mathbf{H}_{\hat{\mathbf{x}}_k^0}{}^T \mathbf{R}_k^{-1}\left(\mathbf{y}_k - \mathbf{h}\left(\hat{\mathbf{x}}_k^0\right) - \mathbf{H}_{\hat{\mathbf{x}}_k^0}\left(f_{k,0}\left(\hat{\mathbf{x}}_0^0\right) + \mathbf{\Phi}_{k,0}^0\left(\mathbf{x}_0 - \hat{\mathbf{x}}_0^0\right) - \hat{\mathbf{x}}_k^0\right)\right) + \mathbf{P}_0^{-1}\left(\overline{\mathbf{x}}_0 - \mathbf{x}_0\right) = \mathbf{0}. \tag{3.8-8}$$

Use $\mathbf{P}_{0|K}^1$ to denote the covariance of $\hat{\mathbf{x}}_0^1$, the estimate of $\mathbf{x}_0$ after the first iteration

$$\mathbf{P}_{0|K}^1 = \left[\sum\nolimits_{k=1}^{K} \mathbf{\Phi}_{k,0}^0{}^T \mathbf{H}_{\hat{\mathbf{x}}_k^0}{}^T \mathbf{R}_k^{-1} \mathbf{H}_{\hat{\mathbf{x}}_k^0} \mathbf{\Phi}_{k,0}^0 + \mathbf{P}_0^{-1}\right]^{-1}. \tag{3.8-9}$$

Then $\hat{\mathbf{x}}_0^1$ is

$$\hat{\mathbf{x}}_0^1 = \mathbf{P}_{0|K}^1\left(\sum\nolimits_{k=1}^{K} \mathbf{\Phi}_{k,0}^0{}^T \mathbf{H}_{\hat{\mathbf{x}}_k^0}{}^T \mathbf{R}_k^{-1} \mathbf{H}_{\hat{\mathbf{x}}_k^0} \mathbf{\Phi}_{k,0}^0 \hat{\mathbf{x}}_0^0 + \mathbf{P}_0^{-1}\overline{\mathbf{x}}_0\right) + \mathbf{P}_{0|K}^1\left(\sum\nolimits_{k=1}^{K} \mathbf{\Phi}_{k,0}^0{}^T \mathbf{H}_{\hat{\mathbf{x}}_k^0}{}^T \mathbf{R}_k^{-1}[\mathbf{y}_k - \mathbf{h}\left(\hat{\mathbf{x}}_k^0\right)]\right). \tag{3.8-10}$$

Finally, the iterative solution to the estimation problem at the $i + 1$ iteration is

$$\hat{\mathbf{x}}_0^{i+1} = \mathbf{P}_{0|K}^{i+1}\left(\sum\nolimits_{k=1}^{K} \mathbf{\Phi}_{k,0}^i{}^T \mathbf{H}_{\hat{\mathbf{x}}_k^i}{}^T \mathbf{R}_k^{-1} \mathbf{H}_{\hat{\mathbf{x}}_k^i} \mathbf{\Phi}_{k,0}^i \hat{\mathbf{x}}_0^i + \mathbf{P}_0^{-1}\overline{\mathbf{x}}_0\right) + \mathbf{P}_{0|K}^{i+1}\left(\sum\nolimits_{k=1}^{K} \mathbf{\Phi}_{k,0}^i{}^T \mathbf{H}_{\hat{\mathbf{x}}_k^i}{}^T \mathbf{R}_k^{-1}\left(\mathbf{y}_k - \mathbf{h}\left(\hat{\mathbf{x}}_k^i\right)\right)\right). \tag{3.8-11}$$

Or in a more conventional form

$$\hat{\mathbf{x}}_0^{i+1} = \hat{\mathbf{x}}_0^i + \mathbf{P}_{0|K}^{i+1}\left(\sum\nolimits_{k=1}^{K} \mathbf{\Phi}_{k,0}^i{}^T \mathbf{H}_{\hat{\mathbf{x}}_k^i}{}^T \mathbf{R}_k^{-1}\left(\mathbf{y}_k - \mathbf{h}\left(\hat{\mathbf{x}}_k^i\right)\right) + \mathbf{P}_0^{-1}\left(\overline{\mathbf{x}}_0 - \hat{\mathbf{x}}_0^i\right)\right) \tag{3.8-12}$$

with covariance

$$\mathbf{P}_{0|K}^{i+1} = \left[\sum_{k=1}^{K} \boldsymbol{\Phi}_{k,0}^{i}{}^{T} \mathbf{H}_{\hat{\mathbf{x}}_{k}^{i}}{}^{T} \mathbf{R}_{k}^{-1} \mathbf{H}_{\hat{\mathbf{x}}_{k}^{i}} \boldsymbol{\Phi}_{k,0}^{i} + \mathbf{P}_{0}^{-1} \right]^{-1}. \tag{3.8-13}$$

Note that $\sum_{k=1}^{K} \boldsymbol{\Phi}_{k,0}^{i}{}^{T} \mathbf{H}_{\hat{\mathbf{x}}_{k}^{i}}{}^{T} \mathbf{R}_{k}^{-1} \mathbf{H}_{\hat{\mathbf{x}}_{k}^{i}} \boldsymbol{\Phi}_{k,0}^{i} + \mathbf{P}_{0}^{-1}$ is identical in functional form to Fisher's information matrix and its inverse in Equation (3.8-13) is functionally identical to the CRB. The difference is that Fisher's information matrix and the CRB are evaluated at true states while the above are evaluated at the converged estimate. The CRB for nonlinear deterministic systems will be developed in Section 3.9.1. When Fisher's information matrix is nonsingular (invertible), the CRB exists, and the system is observable and the estimate is asymptotically efficient.

When there is no a priori knowledge on the initial estimate, it is equivalent to setting $\mathbf{P}_{0}^{-1} = \mathbf{0}$ (there is no information), and one obtains the more familiar form of state and covariance equations

$$\hat{\mathbf{x}}_{0}^{i+1} = \hat{\mathbf{x}}_{0}^{i} + \mathbf{P}_{0|K}^{i+1} \left(\sum_{k=1}^{K} \boldsymbol{\Phi}_{k,0}^{i}{}^{T} \mathbf{H}_{\hat{\mathbf{x}}_{k}^{i}}{}^{T} \mathbf{R}_{k}^{-1} \left(\mathbf{y}_{k} - \mathbf{h}\left(\hat{\mathbf{x}}_{k}^{i}\right) \right) \right) \tag{3.8-14}$$

$$\mathbf{P}_{0|K}^{i+1} = \left[\sum_{k=1}^{K} \boldsymbol{\Phi}_{k,0}^{i}{}^{T} \mathbf{H}_{\hat{\mathbf{x}}_{k}^{i}}{}^{T} \mathbf{R}_{k}^{-1} \mathbf{H}_{\hat{\mathbf{x}}_{k}^{i}} \boldsymbol{\Phi}_{k,0}^{i} \right]^{-1}. \tag{3.8-15}$$

This algorithm is summarized in Table 3.4.

Table 3.4   Summary of an Iterative Least Squares Algorithm for Deterministic System

---

An Iterative Algorithm for Deterministic Systems (A Batch Filter)
Problem Statement
Obtain the weighted least squares estimate of $\mathbf{x}_j$ with all data $\mathbf{y}_k$, $k = 1,..,K$ given by

$$\mathbf{x}_k = \mathbf{f}\left(\mathbf{x}_{k-1}\right)$$

$$\mathbf{y}_k = \mathbf{h}\left(\mathbf{x}_k\right) + \mathbf{v}_k$$

where $\mathbf{x}_k \in \mathbb{R}^n$, $\mathbf{y}_k \in \mathbb{R}^m$, $\mathbf{x}_0 : \sim N(\hat{\mathbf{x}}_0, \mathbf{P}_0)$, $\mathbf{v}_k : \sim N(\mathbf{0}, \mathbf{R}_k)$, and $\mathbf{f}(.)$, $\mathbf{h}(.)$ are known functions.
Algorithm for Obtaining Solution at Initial Time, $\hat{\mathbf{x}}_0$

$$\hat{\mathbf{x}}_0^{i+1} = \hat{\mathbf{x}}_0^i + \mathbf{P}_{\hat{\mathbf{x}}_0^{i+1}}^{i+1} \left(\sum_{k=1}^{K} \boldsymbol{\Phi}_{k,0}^{i}{}^{T} \mathbf{H}_{\hat{\mathbf{x}}_k^i}{}^{T} \mathbf{R}_k^{-1} \left(\mathbf{y}_k - \mathbf{h}\left(\hat{\mathbf{x}}_k^i\right) \right) + \mathbf{P}_0^{-1} \left(\overline{\mathbf{x}}_0 - \hat{\mathbf{x}}_0^i \right) \right)$$

$$\mathbf{P}_K^{i+1} = \left[\sum_{k=1}^{K} \boldsymbol{\Phi}_{k,0}^{i}{}^{T} \mathbf{H}_{\hat{\mathbf{x}}_k^i}{}^{T} \mathbf{R}_k^{-1} \mathbf{H}_{\hat{\mathbf{x}}_k^i} \boldsymbol{\Phi}_{k,0}^{i} + \mathbf{P}_0^{-1} \right]^{-1},$$

where $\hat{\mathbf{x}}_0^0 = \overline{\mathbf{x}}_0$, $\boldsymbol{\Phi}_{k,0} = \mathbf{F}_{\mathbf{x}_{k-1}} \mathbf{F}_{\mathbf{x}_{k-2}} ... \mathbf{F}_{\mathbf{x}_0}$ and $\mathbf{F}_{\mathbf{x}_j} = \left[\dfrac{\partial \mathbf{f}\left(\mathbf{x}_j\right)}{\partial \mathbf{x}_j} \right]$

Since the system is deterministic, finding a solution for $\mathbf{x}_j$ of any j completely characterizes the entire trajectory. One can derive the above algorithm for any time. For the purpose of simplicity, the above is for $\hat{\mathbf{x}}_0$.

Remarks

1. The estimation problem posted in Equations (3.8-1) through (3.8-3) is an optimization problem over the entire history of data, and it is sometimes referred to as a batch filter. This is similar in concept to a smoother (a fixed-interval smoother) but in this case, the system is nonlinear and deterministic. When a new measurement data vector is received, a new estimate is calculated, again with the entire set of data, but not recursively updated as in the KF and EKF.

2. The iterative algorithm presented in Equations (3.8-11) through (3.8-13) is a suggested realization of the solution to the estimation problem posted in Equations (3.8-1) (through 3.8-3). Nothing is said about algorithm convergence, the uniqueness of the solution, or the method for obtaining the initial guess. Depending on the specifics of a particular application, one may choose or design a different algorithm for solving the problem.

3. The above algorithm was successfully applied to a ballistic trajectory estimation problem with angle-only measurements [18].

3.9   Cramer–Rao Bound

Regardless of whether the system is linear or nonlinear [19], the fundamental CRB definition for state estimation remains the same as in Equation (2.10-1), restated below for easy reference

$$Cov\{\hat{\mathbf{x}}_{1:k|1:k}\} \geq \mathrm{CRB}(\mathbf{x}_{1:k}|\mathbf{y}_{1:k}) = \left[E\left\{ \left[\frac{\partial \ln p(\mathbf{y}_{1:k};\mathbf{x}_{1:k})}{\partial \mathbf{x}_{1:k}} \right] \left[\frac{\partial \ln p(\mathbf{y}_{1:k};\mathbf{x}_{1:k})}{\partial \mathbf{x}_{1:k}} \right]^T \right\} \right]^{-1}, \quad (3.9\text{-}1)$$

where the Fisher information matrix for estimating $\mathbf{x}_{1:k}$ is an $nk \times nk$ matrix as shown within the inside of the inversion bracket above

$$\mathcal{F}(\hat{\mathbf{x}}_{1:k|1:k}) \triangleq \left[E\left\{ \left[\frac{\partial \ln p(\mathbf{x}_{1:k};\mathbf{y}_{1:k})}{\partial \mathbf{x}_{1:k}} \right] \left[\frac{\partial \ln p(\mathbf{x}_{1:k};\mathbf{y}_{1:k})}{\partial \mathbf{x}_{1:k}} \right]^T \right\} \right], \quad (3.9\text{-}2)$$

The CRB for a deterministic system will be developed first, followed by the extension to stochastic systems.

3.9.1   For Deterministic Nonlinear Systems

When the system is nonlinear but deterministic, then $\mu_i = 0$ for all $i=1,..,k$. The CRB definition becomes the same as Equation (2.10-2). The derivation of the CRB follows the same steps as in Section 2.10.1 by using the Jacobians of the nonlinear system and measurement equations to replace the linear system and measurement matrices. The result corresponding to Equation (2.10-5) is

$$Cov\{\hat{\mathbf{x}}_{k|k}\} \geq \mathrm{CRB}\left(\mathbf{x}_k | \mathbf{y}_{1:k}\right) = \left[\mathcal{F}_{\hat{\mathbf{x}}_{k|k}}\right]^{-1}$$

$$= \left[\sum_{i=1}^{k} \left[\frac{\partial \mathbf{h}(\mathbf{x}_i)}{\partial \mathbf{x}_k}\right]^T \mathbf{R}_i^{-1} \left[\frac{\partial \mathbf{h}(\mathbf{x}_i)}{\partial \mathbf{x}_k}\right] + \left[\frac{\partial \mathbf{x}_0}{\partial \mathbf{x}_k}\right]^T \mathbf{P}_0^{-1} \left[\frac{\partial \mathbf{x}_0}{\partial \mathbf{x}_k}\right]\right]^{-1}. \qquad (3.9\text{-}3)$$

We now derive the solution for $\dfrac{\partial \mathbf{x}_0}{\partial \mathbf{x}_k}$ and $\dfrac{\partial \mathbf{h}(\mathbf{x}_i)}{\partial \mathbf{x}_k}$ by using the chain rule for derivatives, and following the notation developed in Section 3.8.1

$$\frac{\partial \mathbf{x}_k}{\partial \mathbf{x}_i} = \frac{\partial \mathbf{f}(\mathbf{x}_{k-1})}{\partial \mathbf{x}_{k-1}} \frac{\partial \mathbf{f}(\mathbf{x}_{k-2})}{\partial \mathbf{x}_{k-2}} \cdots \frac{\partial \mathbf{f}(\mathbf{x}_i)}{\partial \mathbf{x}_i} = \mathbf{F}_{\mathbf{x}_{k-1}} \mathbf{F}_{\mathbf{x}_{k-2}} \cdots \mathbf{F}_{\mathbf{x}_i} = \mathbf{F}_{\mathbf{x}_{k-1},\mathbf{x}_i}$$

thus

$$\frac{\partial \mathbf{x}_i}{\partial \mathbf{x}_k} = \mathbf{F}_{\mathbf{x}_i}^{-1} \mathbf{F}_{\mathbf{x}_{i+1}}^{-1} \cdots \mathbf{F}_{\mathbf{x}_{k-1}}^{-1} = \mathbf{F}_{\mathbf{x}_{k-1},\mathbf{x}_i}^{-1},$$

where $\mathbf{F}_{\mathbf{x}_{k-1},\mathbf{x}_i}$ denotes the transition from $\mathbf{x}_i$ to $\mathbf{x}_{k-1}$ and $\mathbf{F}_{\mathbf{x}_{k-1},\mathbf{x}_i}^{-1}$ is its inverse, thus the transition backward in time from $\mathbf{x}_{k-1}$ to $\mathbf{x}_i$. Using the notation employed in Section 3.8, $\mathbf{F}_{\mathbf{x}_{k-1},\mathbf{x}_i}$ can be substituted with $\mathbf{\Phi}_{k,i}$. Similarly, $\mathbf{F}_{\mathbf{x}_{k-1},\mathbf{x}_i}^{-1}$ with $\mathbf{\Phi}_{i,k}$. Therefore

$$\frac{\partial \mathbf{x}_i}{\partial \mathbf{x}_k} = \mathbf{\Phi}_{i,k} \qquad (3.9\text{-}4)$$

when $i=0$, Equation (3.9-4) is $\dfrac{\partial \mathbf{x}_0}{\partial \mathbf{x}_k}$. Similarly,

$$\frac{\partial \mathbf{h}(\mathbf{x}_i)}{\partial \mathbf{x}_k} = \frac{\partial \mathbf{h}(\mathbf{x}_i)}{\partial \mathbf{x}_i} \frac{\partial \mathbf{x}_i}{\partial \mathbf{x}_k} = \mathbf{H}_{\mathbf{x}_i} \mathbf{\Phi}_{i,k} \qquad (3.9\text{-}5)$$

Substituting Equations (3.9-4) and (3.9-5) in Equation (3.9-3), one obtains the final expression of the CRB for nonlinear deterministic systems

$$Cov\{\hat{\mathbf{x}}_{k|k}\} \geq \mathrm{CRB}(\mathbf{x}_k|\mathbf{y}_{1:k}) = [\mathcal{F}_{\hat{\mathbf{x}}_{k|k}}]^{-1} = \left[\sum_{i=1}^{k} [\mathbf{H}_{\mathbf{x}_i}\boldsymbol{\Phi}_{i,k}]^T \mathbf{R}_i^{-1}\mathbf{H}_{\mathbf{x}_i}\boldsymbol{\Phi}_{i,k} + \boldsymbol{\Phi}_{0,k}^T\mathbf{P}_0^{-1}\boldsymbol{\Phi}_{0,k} \right]^{-1}.$$

$$(3.9\text{-}6)$$

Remarks

1. Similar to the linear case, Equation (3.9-6) is related to system observability. However, for nonlinear systems, Equation (3.9-6) does not guarantee uniqueness of solution, it therefore is only a necessary but not sufficient (it is necessary for a solution, but does not guarantee for the solution) condition, see [18, 19] and the cited references for discussions on observability for nonlinear systems.

2. As in the linear case, one can rewrite the above equation in a recursive form and the recursive equation has the same functional form as the covariance equation for the extended Kalman filter without the process noise covariance term. Similar to the comments made in Section 1.7, the main difference between them is in the state used for linearization. The extended Kalman filter covariance equation is evaluated at the state estimate $\hat{\mathbf{x}}$ while the CRB is evaluated at the true state $\mathbf{x}$. This difference can sometimes be significant. Because the filter covariance equation is only an approximation; the CRB will be more meaningful statistically for use in evaluating filter performance.

The CRB for the nonlinear but deterministic system is very useful because it is tight (unlike the case for stochastic systems that will be shown later in this chapter) and easy to compute. This bound was first published by J. H. Taylor in 1979 [20]. It has been used, however, by practicing engineers for some time [e.g., 18, 21].

3.9.2 For Stochastic Nonlinear Systems

The computation of the CRB as defined in Equation (3.9-1) will encounter two difficulties. The first one is in taking the statistical expectation over a nonlinear function. The second one, which has already been mentioned, is the requirement for formulating and inverting a matrix with dimension growing as nk where n is the dimension of state vector and k is the number of measurement times.

There is no general solution to the problem of taking statistical expectation over a nonlinear function. Unless for some specific cases, there does not exist any closed form solution (or algorithm) for computing CRB in the case of nonlinear stochastic systems. An approximate method is to use sampled average, that is, computing the statistical expectation with Monte Carlo sampling.

Two approaches are discussed below for circumventing the issue of growing matrix dimension. The first approach is to treat all process noise sequences as nuisance parameters. In doing so the impact of process noise is accounted for in the statistical expectation operation but not in the formation of the Fisher information matrix. The second method is known as the posteriori CRB (PCRB), in which a recursive algorithm for computing the Fisher information matrix with fixed dimension ($n \times n$) is derived. This method is due to the work of Tichavsky et al. [22]. Both methods will have the same difficulties in computing statistical expectation. The PCRB formulation is mathematically exact. The use of nuisance parameter is an extension of approaches in signal processing. These two approaches are presented individually below.

When Process Noise Is Treated as a Nuisance Parameter Vector

Estimating states for a stochastic system for all time is equivalent to estimating the system noise vector for all time, since once they are determined the corresponding states are determined. The lower bound in the presence of nuisance parameters was introduced in [23]. The lower bound for state estimation in treating the process noise as a nuisance parameter was developed in [24–26]. This approach treats all system noise vectors as nuisance parameters that are averaged out with statistical expectation.

Let $\mathbf{x}_k$ be the state of interest while the system noise prior to time k are assumed given. The joint density function can then be rewritten to incorporate the following conditional density

$$p(\mathbf{y}_{1:k}, \mathbf{x}_0; \mathbf{x}_k | \boldsymbol{\mu}_{1:k-1})$$

where $\mathbf{x}_k$ is the state vector to be estimated, and $\mathbf{y}_{1:k}, \mathbf{x}_0$ represents all measurements of $\mathbf{x}_k$ conditioned on $\boldsymbol{\mu}_{1:k-1}$. The CRB for estimating $\mathbf{x}_k$ can be written as

$$Cov\{\hat{\mathbf{x}}_{k|k}\} \geq \left[E_{\boldsymbol{\mu}_{1:k-1}} \left\{ E_{v_{1:k};\mathbf{x}_0} \left\{ \left[\frac{\partial \ln p(\mathbf{y}_{1:k}, \mathbf{x}_0; \mathbf{x}_k | \boldsymbol{\mu}_{1:k-1})}{\partial \mathbf{x}_k} \right] \left[\frac{\partial \ln p(\mathbf{y}_{1:k}, \mathbf{x}_0; \mathbf{x}_k | \boldsymbol{\mu}_{1:k-1})}{\partial \mathbf{x}_k} \right]^T \right\} \right\} \right]^{-1}.$$

$$(3.9\text{-}7)$$

Note that the subscript r of the operator $E_r\{.\}$ denotes the random variable that the statistical expectation is taken with respect to. With Gaussian assumption on the measurement noise and the initial condition $\mathbf{x}_0$, the log of the density function after dropping additive and multiplicative constants becomes

$$J = \sum_{i=1}^{k} \left(\mathbf{y}_i - \mathbf{h}(\mathbf{x}_i)\right)^T \mathbf{R}_i^{-1} \left(\mathbf{y}_i - \mathbf{h}(\mathbf{x}_i)\right) + \left(\mathbf{x}_0 - \hat{\mathbf{x}}_0\right)^T \mathbf{P}_0^{-1} \left(\mathbf{x}_0 - \hat{\mathbf{x}}_0\right).$$

Similar to all previous CRB derivations, the initial covariance $\mathbf{P}_0$ is included and assumed to be nonsingular. If this initial information is only partially available on certain states, a corresponding Fisher's information matrix can be formulated to replace $\mathbf{P}_0^{-1}$ above. Taking the derivative of J with respect to $\mathbf{x}_k$ yields

$$\frac{\partial J}{\partial \mathbf{x}_k} = \sum_{i=1}^{k} \left[\frac{\partial \mathbf{h}(\mathbf{x}_i)}{\partial \mathbf{x}_k}\right] \mathbf{R}_i^{-1} \left(\mathbf{y}_i - \mathbf{h}(\mathbf{x}_i)\right) + \frac{\partial \mathbf{x}_0}{\partial \mathbf{x}_k} \mathbf{P}_0^{-1} \left(\mathbf{x}_0 - \hat{\mathbf{x}}_0\right).$$

Multiplying $\dfrac{\partial J}{\partial \mathbf{x}_k}$ by its transpose and taking the expectation with respect to $\mathbf{v}_{1:k}$ and $\mathbf{x}_0$ yields

$$E_{\mathbf{v}_{1:k};\mathbf{x}_0}\left\{\left(\frac{\partial J}{\partial \mathbf{x}_k}\right)\left(\frac{\partial J}{\partial \mathbf{x}_k}\right)^T\right\} = E_{\mathbf{v}_{1:k};\mathbf{x}_0}\left\{\sum_{i=1}^{k} \left[\frac{\partial \mathbf{h}(\mathbf{x}_i)}{\partial \mathbf{x}_k}\right]^T \mathbf{R}_i^{-1} \left[\frac{\partial \mathbf{h}(\mathbf{x}_i)}{\partial \mathbf{x}_k}\right] + \left[\frac{\partial \mathbf{x}_0}{\partial \mathbf{x}_k}\right]^T \mathbf{P}_0^{-1} \left[\frac{\partial \mathbf{x}_0}{\partial \mathbf{x}_k}\right]\right\}.$$

$$(3.9\text{-}8)$$

Let $\mathcal{F}_{\mathbf{x}_k|\boldsymbol{\mu}_{1:k-1}}$ denote the Fisher information matrix in estimating $\mathbf{x}_k$ conditioned on $\boldsymbol{\mu}_{1:k-1}$, that is,

$$\mathcal{F}_{\mathbf{x}_k|\boldsymbol{\mu}_{1:k-1}} = E_{\mathbf{v}_{1:k};\mathbf{x}_0}\left\{\left(\frac{\partial J}{\partial \mathbf{x}_k}\right)\left(\frac{\partial J}{\partial \mathbf{x}_k}\right)^T\right\} \tag{3.9-9}$$

Then the CRB for estimating $\mathbf{x}_k$ is

$$Cov\{\hat{\mathbf{x}}_{k|k}\} \geq CRB(\mathbf{x}_k|\mathbf{y}_{1:k}) = E_{\boldsymbol{\mu}_{1:k-1}}\left\{\left[\mathcal{F}_{\mathbf{x}_k|\boldsymbol{\mu}_{1:k-1}}\right]^{-1}\right\}$$

$$= E_{\boldsymbol{\mu}_{1:k-1}}\left\{\left[\sum_{i=1}^{k} \left[\frac{\partial \mathbf{h}(\mathbf{x}_i)}{\partial \mathbf{x}_k}\right]^T \mathbf{R}_i^{-1} \left[\frac{\partial \mathbf{h}(\mathbf{x}_i)}{\partial \mathbf{x}_k}\right] + \left[\frac{\partial \mathbf{x}_0}{\partial \mathbf{x}_k}\right]^T \mathbf{P}_0^{-1} \left[\frac{\partial \mathbf{x}_0}{\partial \mathbf{x}_k}\right]\right]^{-1}\right\}. \tag{3.9-10}$$

Note that when $k=1$, $\mathbf{P}_0^{-1} = \mathbf{0}$, and the dimension of the measurement vector is smaller than that of the state vector, then $\left[\dfrac{\partial \mathbf{h}(\mathbf{x}_1)}{\partial \mathbf{x}_1}\right]^T \mathbf{R}_1^{-1} \left[\dfrac{\partial \mathbf{h}(\mathbf{x}_1)}{\partial \mathbf{x}_1}\right]$ is singular. If the system is observable, $\sum_{i=1}^{k} \left[\dfrac{\partial \mathbf{h}(\mathbf{x}_i)}{\partial \mathbf{x}_k}\right]^T \mathbf{R}_i^{-1} \left[\dfrac{\partial \mathbf{h}(\mathbf{x}_i)}{\partial \mathbf{x}_k}\right]$ is nonsingular for some k.

Given that the system is nonlinear, computing the expectation with a closed form solution or an algorithm is difficult, if not impossible. Some special cases where analytical expressions can be derived are shown in [26], but this is not true in general. An alternative is to compute the expectation by Monte Carlo simulation, that is, by simulating the random trajectory with random samples of $\mu_{1:k-1}$ to compute sample valued $[\mathcal{F}_{\mathbf{x}_k|\mu_{1:k-1}}]^{-1}$, and then statistically averaging over them, is the essence of the alternative approach [25]. The computation of the alternate bound may require a large number of samples; but it is likely to be a manageable task with today's computers.

The Posterior Cramer–Rao Bound

The material presented in this section is due to the work of Tichavsky et al. [22].[1] The fundamental definition of the CRB is restated for continuity.

Cramer–Rao Bound

Given the nonlinear estimation problem defined in Section 3.1, the CRB on the covariance of estimating $\hat{\mathbf{x}}_{1:k\|1:k} \triangleq \{\hat{\mathbf{x}}_{1|k}, ..\hat{\mathbf{x}}_{k|k}\}$ is

$$Cov\{\hat{\mathbf{x}}_{1:k\|1:k}\} \geq \mathrm{CRB}(\mathbf{x}_{1:k}|\mathbf{y}_{1:k}) = \left[E\left\{ \left[\frac{\partial \ln p(\mathbf{y}_{1:k};\mathbf{x}_{1:k})}{\partial \mathbf{x}_{1:k}} \right]\left[\frac{\partial \ln p(\mathbf{y}_{1:k};\mathbf{x}_{1:k})}{\partial \mathbf{x}_{1:k}} \right]^T \right\} \right]^{-1}, \quad (3.9\text{-}11)$$

where the Fisher information matrix for estimating $\mathbf{x}_{1:k}$ is a $nk \times nk$ matrix as shown within the inside of the inversion bracket above

$$\mathcal{F}(\hat{\mathbf{x}}_{1:k\|1:k}) \triangleq \left[E\left\{ \left[\frac{\partial \ln p(\mathbf{x}_{1:k};\mathbf{y}_{1:k})}{\partial \mathbf{x}_{1:k}} \right]\left[\frac{\partial \ln p(\mathbf{x}_{1:k};\mathbf{y}_{1:k})}{\partial \mathbf{x}_{1:k}} \right]^T \right\} \right], \quad (3.9\text{-}12)$$

Let $\mathcal{F}_{\hat{\mathbf{x}}_{k|k}}$ be the Fisher information matrix in estimating $\mathbf{x}_k$, that is, its inverse is the CRB of the covariance of any unbiased estimator for $\hat{\mathbf{x}}_{k|k}$. $\mathcal{F}_{\hat{\mathbf{x}}_{k|k}}$ has a constant dimension $n \times n$. In general, $\mathcal{F}_{\hat{\mathbf{x}}_{k|k}}$ cannot be obtained independent of the prior history of estimating $\mathbf{x}_k$. The CRB for the covariance in estimating $\hat{\mathbf{x}}_{k|k}$ is the right-lower block of the inverse of $[\mathcal{F}(\hat{\mathbf{x}}_{1:k\|1:k})]^{-1}$. However, in order to obtain the CRB for $Cov\{\hat{\mathbf{x}}_{k|k}\}$ it appears necessary to invert a $nk \times nk$ matrix that is growing in dimension with time. A

---

1   The authors are indebted to our reviewers for pointing out this work.

recursive algorithm to circumvent this problem was published in [22] known as posterior CRB. Before PCRB is derived, several relevant mathematical relationships are first presented.

Block Matrix Inversion Identity

Let $\mathbf{M}$ be a positive definite and symmetric matrix with the following block matrix structure

$$\mathbf{M} = \begin{bmatrix} \mathbf{M}_{11} & \mathbf{M}_{12} \\ \mathbf{M}_{21} & \mathbf{M}_{22} \end{bmatrix},$$

where $\mathbf{M}_{21} = \mathbf{M}_{12}^{T}$. The inverse of $\mathbf{M}$ can be defined by its sub-blocks as

$$\mathbf{M}^{-1} = \begin{bmatrix} \left[\mathbf{M}_{11} - \mathbf{M}_{12}\mathbf{M}_{22}^{-1}\mathbf{M}_{21} \right]^{-1} & -\mathbf{M}_{11}^{-1}\mathbf{M}_{12}\left[\mathbf{M}_{22} - \mathbf{M}_{21}\mathbf{M}_{11}^{-1}\mathbf{M}_{12} \right]^{-1} \\ -\mathbf{M}_{22}^{-1}\mathbf{M}_{21}\left[\mathbf{M}_{11} - \mathbf{M}_{12}\mathbf{M}_{22}^{-1}\mathbf{M}_{12} \right]^{-1} & \left[\mathbf{M}_{22} - \mathbf{M}_{21}\mathbf{M}_{11}^{-1}\mathbf{M}_{12} \right]^{-1} \end{bmatrix}.$$

Note that $\mathbf{M}^{-1}$ is also symmetric thus the following must also be an identity

$$-\mathbf{M}_{22}^{-1}\mathbf{M}_{21}\left[\mathbf{M}_{11} - \mathbf{M}_{12}\mathbf{M}_{22}^{-1}\mathbf{M}_{12} \right]^{-1} = \left[-\mathbf{M}_{11}^{-1}\mathbf{M}_{12}\left[\mathbf{M}_{22} - \mathbf{M}_{21}\mathbf{M}_{11}^{-1}\mathbf{M}_{12} \right]^{-1} \right]^{T}.$$

The above identity can be proved by first assuming a matrix $\mathbf{N}$ being the inverse of $\mathbf{M}$ then use $\mathbf{N} \times \mathbf{M} = \mathbf{I}$, where $\mathbf{I}$ is an identity matrix, to solve for the submatrices of $\mathbf{N}$. In the ensuring development, only the lower right side of the submatrix of $\mathbf{M}^{-1}$, $\left[\mathbf{M}_{22} - \mathbf{M}_{21}\mathbf{M}_{11}^{-1}\mathbf{M}_{12} \right]^{-1}$, is of interest to us.

Joint Probability Density Function $p(\mathbf{x}_{1:k}; \mathbf{y}_{1:k})$

With the Markovian assumption of $\mathbf{x}_k$ and the independent assumptions among $\boldsymbol{\mu}_k$ and $\mathbf{v}_k$ for all k, the joint probability density function for $\mathbf{y}_{1:k}$ and $\mathbf{x}_{1:k}$ can be written as

$$p(\mathbf{x}_{1:k}; \mathbf{y}_{1:k}) = p(\mathbf{x}_0)\prod_{i=1}^{k}p(\mathbf{y}_i|\mathbf{x}_i)\prod_{i=1}^{k}p(\mathbf{x}_j|\mathbf{x}_{j-1}).$$

For the purpose of simplification in notation, p_k will be used to denote $p(\mathbf{x}_{1:k}; \mathbf{y}_{1:k})$. The following recursive relationship in p_k,

$$p_k = p_{k-1}p(\mathbf{x}_k|\mathbf{x}_{k-1})p(\mathbf{y}_k|\mathbf{x}_k)$$

is obtained using the chain rule of probability and the Markov process property of $\mathbf{x}_k$.

Notations for First Derivative Operation

Let $\boldsymbol{\alpha}$ and $\boldsymbol{\beta}$ denote two column vectors with dimension n then the first order derivative operator with respect to $\boldsymbol{\alpha}$ or $\boldsymbol{\beta}$ is

$$\nabla_\gamma = \left[\frac{\partial}{\partial \gamma_1}, \ldots, \frac{\partial}{\partial \gamma_n} \right]^T,$$

where $\boldsymbol{\gamma}$ could be $\boldsymbol{\alpha}$ or $\boldsymbol{\beta}$. The outer product of ∇_α and ∇_β is written as

$$\Delta_\beta^\alpha = \nabla_\beta \nabla_\alpha^T.$$

Derivation of PCRB

Using the notations for first derivative operation, the Fisher information matrix $\mathcal{F}(\hat{\mathbf{x}}_{1:k|1:k})$ is written as

$$\mathcal{F}(\hat{\mathbf{x}}_{1:k|1:k}) = \left[E\left\{ \Delta_{\mathbf{x}_{1:k}}^{\mathbf{x}_{1:k}} (\ln p_k) \right\} \right].$$

Let $\mathcal{F}(\hat{\mathbf{x}}_{1:k|1:k})$ be expressed with the submatrices defined as

$$\mathcal{F}(\hat{\mathbf{x}}_{1:k|1:k}) = \begin{bmatrix} E\left\{ \Delta_{\mathbf{x}_{1:k-1}}^{\mathbf{x}_{1:k-1}} (\ln p_k) \right\} & E\left\{ \Delta_{\mathbf{x}_{1:k-1}}^{\mathbf{x}_k} (\ln p_k) \right\} \\ E\left\{ \Delta_{\mathbf{x}_k}^{\mathbf{x}_{1:k-1}} (\ln p_k) \right\} & E\left\{ \Delta_{\mathbf{x}_k}^{\mathbf{x}_k} (\ln p_k) \right\} \end{bmatrix},$$

Clearly the upper left-hand submatrix is $\mathcal{F}(\hat{\mathbf{x}}_{1:k-1|1:k-1})$. Further define the following simplifying notation

$$\mathcal{F}(\hat{\mathbf{x}}_{1:k|1:k}) = \begin{bmatrix} \mathcal{F}(\hat{\mathbf{x}}_{1:k-1|1:k-1}) & \mathbf{B}_k \\ \mathbf{B}_k^T & \mathbf{C}_k \end{bmatrix},$$

The inverse of $\mathcal{F}(\hat{\mathbf{x}}_{1:k|1:k})$ gives the CRB for the covariance of estimating $\mathbf{x}_{1:k}$. The lower right-hand sub-block of $[\mathcal{F}(\hat{\mathbf{x}}_{1:k|1:k})]^{-1}$ is the CRB on the covariance of estimating $\mathbf{x}_k$. Using the sub-block matrix inversion identity, one obtains

$$[\mathrm{CRB}(\mathbf{x}_k|\mathbf{y}_{1:k})]^{-1} = \mathcal{F}_{\hat{\mathbf{x}}_{k|k}} = \mathbf{C}_k - \mathbf{B}_k^T [\mathcal{F}(\hat{\mathbf{x}}_{1:k-1|1:k-1})]^{-1} \mathbf{B}_k. \qquad (3.9\text{-}13)$$

This equation reduces the need for inverting an $nk \times nk$ matrix, $\mathcal{F}(\hat{\mathbf{x}}_{1:k|1:k})$, to the inverting an $n(k-1) \times n(k-1)$ matrix, $\mathcal{F}(\hat{\mathbf{x}}_{1:k-1|1:k-1})$. The goal for a recursive equation is to only require the inversion of $n \times n$ matrices. Following the derivation of [22], extending $\mathcal{F}(\hat{\mathbf{x}}_{1:k\text{II}:k})$ to the next time step $k + 1$ by applying the recursive probability relationship

$$p_{k+1} = p_k p(\mathbf{x}_{k+1}|\mathbf{x}_k) p(\mathbf{y}_{k+1}|\mathbf{x}_{k+1})$$

yields

$$\mathcal{F}(\hat{\mathbf{x}}_{1:k+1\text{II}:k+1}) = \begin{bmatrix} \mathcal{F}(\hat{\mathbf{x}}_{1:k-1\text{II}:k-1}) & \mathbf{B}_k & \mathbf{0} \\ \mathbf{B}_k^T & \mathbf{C}_k + \mathbf{D}_k^{11} & \mathbf{D}_k^{12} \\ \mathbf{0} & \mathbf{D}_k^{21} & \mathbf{D}_k^{22} \end{bmatrix},$$

The new submatrices are,

$$\mathbf{D}_k^{11} = E\{\Delta_{\mathbf{x}_k}^{\mathbf{x}_k}(\ln p(\mathbf{x}_{k+1}|\mathbf{x}_k))\},$$

$$\mathbf{D}_k^{12} = E\{\Delta_{\mathbf{x}_k}^{\mathbf{x}_{k+1}}(\ln p(\mathbf{x}_{k+1}|\mathbf{x}_k))\},$$

$$\mathbf{D}_k^{21} = E\{\Delta_{\mathbf{x}_{k+1}}^{\mathbf{x}_k}(\ln p(\mathbf{x}_{k+1}|\mathbf{x}_k))\} = [\mathbf{D}_k^{12}]^T,$$

$$\mathbf{D}_k^{22} = E\{\Delta_{\mathbf{x}_{k+1}}^{\mathbf{x}_{k+1}}(\ln p(\mathbf{x}_{k+1}|\mathbf{x}_k))\} + E\{\Delta_{\mathbf{x}_{k+1}}^{\mathbf{x}_{k+1}}(\ln p(\mathbf{y}_{k+1}|\mathbf{x}_{k+1}))\},$$

$$\mathbf{0} = \text{a zero matrix.}$$

These submatrices can be readily derived.

The $n \times n$ sub-block of the lower right-hand side of $[\mathcal{F}(\hat{\mathbf{x}}_{1:k+1\text{II}:k+1})]^{-1}$ is the CRB on the covariance of $\hat{\mathbf{x}}_{k+1|k+1}$, or the inverse of this CRB is the Fisher information matrix of estimating $\mathbf{x}_{k+1}$. Applying the block matrix inversion identity by treating the upper left-hand side of four submatrices as one block yields the following relationship

$$\mathcal{F}_{\hat{\mathbf{x}}_{k+1|k+1}} = \mathbf{D}_k^{22} - \begin{bmatrix} \mathbf{0} & \mathbf{D}_k^{21} \end{bmatrix} \begin{bmatrix} \mathcal{F}(\hat{\mathbf{x}}_{1:k-1\text{II}:k-1}) & \mathbf{B}_k \\ \mathbf{B}_k^T & \mathbf{C}_k + \mathbf{D}_k^{11} \end{bmatrix}^{-1} \begin{bmatrix} \mathbf{0} \\ \mathbf{D}_k^{12} \end{bmatrix}$$

$$= \mathbf{D}_k^{22} - \mathbf{D}_k^{21}[\mathbf{C}_k + \mathbf{D}_k^{11} - \mathbf{B}_k^T[\mathcal{F}(\hat{\mathbf{x}}_{1:k-1\text{II}:k-1})]^{-1}\mathbf{B}_k]^{-1}\mathbf{D}_k^{12}.$$

Using $\mathcal{F}_{\hat{\mathbf{x}}_{k|k}} = \mathbf{C}_k - \mathbf{B}_k^T[\mathcal{F}(\hat{\mathbf{x}}_{1:k-1\text{II}:k-1})]^{-1}\mathbf{B}_k$, Equation (3.9-13), we have thus obtained the needed recursive relationship,

$$\mathcal{F}_{\hat{\mathbf{x}}_{k+1|k+1}} = \mathbf{D}_k^{22} - \mathbf{D}_k^{21}[\mathcal{F}_{\hat{\mathbf{x}}_{k|k}} + \mathbf{D}_k^{11}]^{-1}\mathbf{D}_k^{12}. \qquad (3.9\text{-}14)$$

Remarks

1. When the system is linear, one obtains

$$\mathbf{D}_k^{22} = \mathbf{Q}_k + \mathbf{H}_{k+1}{}^T \mathbf{R}_{k+1}^{-1} \mathbf{H}_{k+1},$$

$$\mathbf{D}_k^{12} = \mathbf{\Phi}_{k+1,k}^T \mathbf{Q}_k^{-1},$$

$$\mathbf{D}_k^{21} = \mathbf{Q}_k^{-1} \mathbf{\Phi}_{k+1,k},$$

$$\mathbf{D}_k^{11} = \mathbf{\Phi}_{k+1,k}^T \mathbf{Q}_k^{-1} \mathbf{\Phi}_{k+1,k}.$$

Substituting these terms into the $\mathcal{F}_{\hat{x}_{k+1|k+1}}$ expression above one obtains the KF covariance equation.

2. Thus showing that KF achieves CRB and is an efficient estimator.

3. For nonlinear systems, the difficulty of applying the above set of equations is in taking the statistical expectation $E\{\Delta_\beta^\alpha \ln p(\alpha, \beta)\}$. In general, this cannot be obtained analytically. A Monte Carlo sampling technique may have to be used. The tightness of this CRB formulation will need to be evaluated individually for different applications.

4. Several extensions of PCRB for specific cases are presented in [22]. Interested readers are encouraged to build CRB and PCRB and examine their tightness for their own applications.

The Cramer–Rao bound for nonlinear system is summarized in Table 3.5.

3.10   A Space Trajectory Estimation Problem with Angle Only Measurement and Comparison of Estimation Covariance with Cramer–Rao Bound

Consider an object traveling in space with its trajectory represented by a set of differential equations of motion in an inertial Cartesian coordinate system with origin at the center of the earth. A simplified set of equations of motion can be written as

$$\ddot{x} = -g_0 x R_e^2 / (x^2 + y^2 + z^2)^{3/2},$$

$$\ddot{y} = -g_0 y R_e^2 / (x^2 + y^2 + z^2)^{3/2},$$

$$\ddot{z} = -g_0 z R_e^2 / (x^2 + y^2 + z^2)^{3/2},$$

where g_0 is earth gravity at sea level and R_e is the radius of the earth. This is a very simplified set of equations of motion because of the spherical earth model and the inertial Cartesian coordinate assumption. Many more sophisticated models exist, but

Table 3.5   Summary of Cramer–Rao Bound for Nonlinear Systems

---

Cramer–Rao Bound (CRB) for Nonlinear Systems
CRB Definition, both in Covariance and in Fisher's Information Matrix

$$\mathrm{CRB}(\mathbf{x}_{1:k}|\mathbf{y}_{1:k}) = \left[E\left\{ \left[\frac{\partial \ln p(\mathbf{y}_{1:k};\mathbf{x}_{1:k})}{\partial \mathbf{x}_{1:k}} \right] \left[\frac{\partial \ln p(\mathbf{y}_{1:k};\mathbf{x}_{1:k})}{\partial \mathbf{x}_{1:k}} \right]^T \right\} \right]^{-1},$$

$$\mathcal{F}(\hat{\mathbf{x}}_{1:k|1:k}) \triangleq \left[E\left\{ \left[\frac{\partial \ln p(\mathbf{x}_{1:k};\mathbf{y}_{1:k})}{\partial \mathbf{x}_{1:k}} \right] \left[\frac{\partial \ln p(\mathbf{x}_{1:k};\mathbf{y}_{1:k})}{\partial \mathbf{x}_{1:k}} \right]^T \right\} \right].$$

For Deterministic Systems

$$\mathrm{CRB}(\mathbf{x}_k|\mathbf{y}_{1:k}) = \left[\sum_{i=1}^{k} [\mathbf{H}_{\mathbf{x}_i}\boldsymbol{\Phi}_{i,k}]^T \mathbf{R}_i^{-1} \mathbf{H}_{\mathbf{x}_i}\boldsymbol{\Phi}_{i,k} + \boldsymbol{\Phi}_{0,k}^T \mathbf{P}_0^{-1} \boldsymbol{\Phi}_{0,k} \right]^{-1},$$

where $\boldsymbol{\Phi}_{i,k}$ is defined the same way as in Table 3.4.

For Stochastic Systems
1. When the process noise is treated as nuisance parameter vector

$$\mathrm{CRB}(\mathbf{x}_k|\mathbf{y}_{1:k}) = \left[\mathcal{F}_{\hat{\mathbf{x}}_{k|k}} \right]^{-1}$$

$$\mathcal{F}_{\hat{\mathbf{x}}_{k|k}} = E_{\boldsymbol{\mu}_{1:k-1}} \left\{ \sum_{i=1}^{k} \left[\frac{\partial \mathbf{h}(\mathbf{x}_i)}{\partial \mathbf{x}_k} \right]^T \mathbf{R}_i^{-1} \left[\frac{\partial \mathbf{h}(\mathbf{x}_i)}{\partial \mathbf{x}_k} \right] + \left[\frac{\partial \mathbf{x}_0}{\partial \mathbf{x}_k} \right]^T \mathbf{P}_0^{-1} \left[\frac{\partial \mathbf{x}_0}{\partial \mathbf{x}_k} \right] \right\}.$$

2. Posterior Cramer–Rao Bound (Written in Fisher's information matrix)

$$\mathcal{F}_{\hat{\mathbf{x}}_{k+1|k+1}} = \mathbf{D}_k^{22} - \mathbf{D}_k^{21} \left[\mathcal{F}_{\hat{\mathbf{x}}_{k|k}} + \mathbf{D}_k^{11} \right]^{-1} \mathbf{D}_k^{12}, \quad \mathcal{F}_{\hat{\mathbf{x}}_{0|0}} = \mathbf{P}_0^{-1}.$$

where

$$\mathbf{D}_k^{11} = E\left\{ \Delta_{\mathbf{x}_k}^{\mathbf{x}_k}(\ln p(\mathbf{x}_{k+1}|\mathbf{x}_k)) \right\},$$

$$\mathbf{D}_k^{12} = E\left\{ \Delta_{\mathbf{x}_k}^{\mathbf{x}_{k+1}}(\ln p(\mathbf{x}_{k+1}|\mathbf{x}_k)) \right\},$$

$$\mathbf{D}_k^{21} = E\left\{ \Delta_{\mathbf{x}_{k+1}}^{\mathbf{x}_k}(\ln p(\mathbf{x}_{k+1}|\mathbf{x}_k)) \right\} = [\mathbf{D}_k^{12}]^T,$$

$$\mathbf{D}_k^{22} = E\left\{ \Delta_{\mathbf{x}_{k+1}}^{\mathbf{x}_{k+1}}(\ln p(\mathbf{x}_{k+1}|\mathbf{x}_k)) \right\} + E\left\{ \Delta_{\mathbf{x}_{k+1}}^{\mathbf{x}_{k+1}}(\ln p(\mathbf{y}_{k+1}|\mathbf{x}_{k+1})) \right\},$$

---

the purpose and conclusion of this example remains unchanged. The sensor only measures angles of arrival to the targets. Let the angle measurements be defined by the azimuth (*A*) and elevation (*E*) of the object relative to the sensor

$$A = \tan^{-1}\left(\frac{x}{y} \right),$$

$$E = \tan^{-1}\left(\frac{z}{\sqrt{x^2 + y^2}} \right).$$

Space objects are routinely tracked by radars that measure range and angles, and that are able to locate the object's position in space with each measurement vector.

Angle measurements alone do not provide object position. The estimation of 3D position and velocity is possible because of the known target motion represented by the differential equations shown above. It is expected, however, that the estimation uncertainty along object range will be large. Orbital determination with angle only measurements was solved by Gauss hundreds of years ago. In this example, the problem is revisited using the tools developed in this chapter, namely, the EKF, the iterative least squares (ILS) estimator (Section 3.9.1), and the CRB. The results here are taken from Refs. [18, 21], and interested readers should consult the references for details. Homework Problem 2 in this chapter allows the student to build a set of tools useful in studying the space trajectory estimation problem.

Two cases are considered: (a) The sensor is free falling just as the space object, such as the sensor on a satellite; and (b) The sensor is stationary, equivalent to being ground-based or on an airplane. Figure 3.4 depicts the relationship of target to sensor for a free-falling sensor in a sensor-centered coordinate system. Variable definitions and relationships are self evident in the figure. When the sensor is stationary, that is, free from the influence of gravity, then g_{SR} and g_{SE} are zero. The explicit equations of motion in sensor coordinates are shown in Refs. [18, 21].

The estimation errors using EKF and ILS with comparison to the CRB are shown in Figure 3.5. The explanation on how a priori knowledge impacts estimation error is illustrated in Figure 3.6.

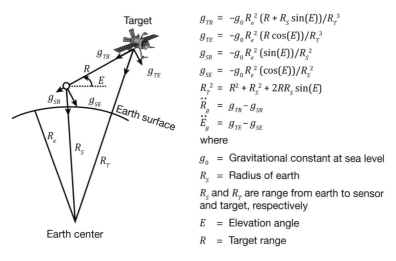

$$g_{TR} = -g_0 R_e^2 (R + R_s \sin(E))/R_T^3$$
$$g_{TE} = -g_0 R_e^2 (R \cos(E))/R_T^3$$
$$g_{SR} = -g_0 R_e^2 (\sin(E))/R_s^2$$
$$g_{SE} = -g_0 R_e^2 (\cos(E))/R_s^2$$
$$R_T^2 = R^2 + R_s^2 + 2RR_s \sin(E)$$
$$\ddot{R}_g = g_{TR} - g_{SR}$$
$$\ddot{E}_g = g_{TE} - g_{SE}$$

where

g_0 = Gravitational constant at sea level

R_s = Radius of earth

R_s and R_T are range from earth to sensor and target, respectively

E = Elevation angle

R = Target range

Figure 3.4 Illustration of target equation of motion in a sensor coordinate for a free-falling sensor. Set g_{SR} and g_{SE} to zero for a stationary sensor.

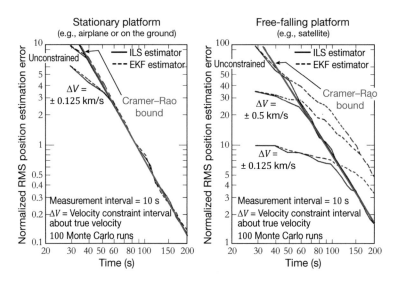

Figure 3.5   Position estimation error and comparison with Cramer–Rao bound.

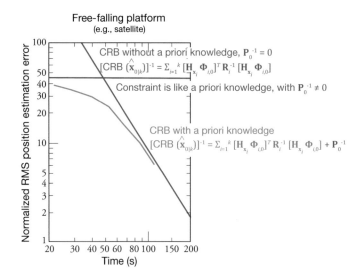

Figure 3.6   Explanation on how a priori knowledge can reduce estimation error.

Remarks

1. The estimation error of a sensor on a stationary platform is much smaller than that of a free-falling platform. This is intuitively correct since angle motion is what enables the estimation of target range. The free-falling platform falls together with the target, resulting in smaller relative angle motion.

2. For the stationary sensor platform case, both the EKF and the ILS estimator achieve the CRB. In the free-falling platform case, only the ILS estimator achieved the CRB, and the EKF diverged. The estimation error is very large in this case because the system is weakly observable (the condition number of the observability Gramian [20] is comparable with the dynamic range of the computer), thus making it difficult for a suboptimal filter such as EKF to converge. On the other hand, the ILS estimator, when converged, is optimal. Achieving the CRB indicates that ILS is efficient.

3. Several levels of velocity constraints are applied. The constraint is equivalent to a priori knowledge; for details see [18, 21]. These constraints help the estimator to achieve more accuracy until the measurement data can provide more information than the constraint. This is explained by the equation shown in Figure 3.5 when the a priori knowledge is incorporated assuming it is a Gaussian distributed random variable.

Homework Problems

1. Following Homework Problem 1 in the last chapter, select the state variable to be represented by a radar centered Cartesian coordinate with x_1 pointing East, x_2 pointing North, and x_3 pointing up (perpendicular to the (x_1, x_2) plane). The basic radar measurements are range and two angles of the target locations, namely range, azimuth, and elevation angles. Let $\mathbf{y}$ denote the measurement vector, thus we have

$$y_1 = \left(x_1^2 + x_2^2 + x_3^2 \right)^{1/2}, \text{range}$$

$$y_2 = \text{atan}\,(x_1/x_2), \text{azimuth}$$

$$y_3 = \text{atan}\left(x_3 / \left(x_1^2 + x_2^2 \right)^{1/2} \right), \text{elevation}$$

a. Using the CV model you built in the previous chapter as a trajectory model, together with the radar measurement variables as defined above, compute and display your CRB with notional radar measurement error as $\sigma_r = 1$ m, $\sigma_{az} = 1$ mrad, $\sigma_{el} = 1$ mrad. Let the target be 100 km from the radar traveling at 150 m/sec. Make the heading any direction of your choice.

b. Build an EKF and compare with CRB.

c. Similar to Homework Problem 2 of Chapter 2, use your EKF to track a CV trajectory perturbed by turbulence.

d. Similarly, experiment with mismatched filters (truth being CA, filter is CV, etc.), and experiment with different $\mathbf{Q}_k$ values.

e. Also as before, let your EKF (could be based upon CV or CA filter model) track your piecewise CV and CA switching trajectory, and experiment with the choice of $\mathbf{Q}_k$ values.

f. The only difference between this problem and the problem in the last chapter is in the measurement equation. Do you see any correlations? Differences? If so, try to offer some explanation.

2. Consider the following simplified ballistic equation of motion (with Coriolis force ignored), equivalent to space objects. The coordinate is chosen as (0,0,0) to be at the center of the earth. Let the three axes of the Cartesian coordinate be denoted as (x, y, z).

$$\ddot{x} = -g_0 \frac{xR_e^2}{\sqrt{x^2 + y^2 + z^2}},$$

$$\ddot{y} = -g_0 \frac{yR_e^2}{\sqrt{x^2 + y^2 + z^2}},$$

$$\ddot{z} = -g_0 \frac{zR_e^2}{\sqrt{x^2 + y^2 + z^2}},$$

where g_0 is gravity at sea level and R_e is the radius of earth. The radar at (x_s, y_s, z_s) measures range, azimuth, and elevation of the object relative to the radar.

a. Build a trajectory model of the above ballistic object. Make sure that the target is at an altitude higher than 100 km so the atmosphere is negligible. Display your trajectory in 3D graphs.

b. Build a CRB for this trajectory estimation problem with notional radar measurement error as $\sigma_r = 1$ m, $\sigma_{az} = 1$ mrad, $\sigma_{el} = 1$ mrad.

c. Build an EKF, run it with Monte Carlo repetition so that you can compare RMS error with the square roots of the filter computed covariance and CRB, respectively.

d. Theoretically, you do not need any process noise covariance for your Kalman filter because the trajectory model for a space object is well understood and has a minimum or nearly zero perturbations. Try to add some $\mathbf{Q}_k$ to your filter, see if the results vary, observe how it varies, and explain why.

You can get more sophisticated by adding a more accurate gravity model and by adding elliptical earth, Coriolis, and centrifugal forces to the trajectory model. Repeat the above exercises. Now you can play with mismatched Kalman filters similar to the CV versus CA models.

3. Consider the following ballistic equation of motion, equivalent to reentry trajectories [21, 27, 28]. Similar to the earlier problems, this is with a flat earth model and without Coriolis and centrifugal forces.

$$\ddot{x} = -\frac{1}{2}\rho V \dot{x}\alpha_d - \frac{1}{2}\rho V^2 \alpha_t \frac{\dot{y}}{V_p} - \frac{1}{2}\rho V \alpha_c \frac{\dot{x}\dot{z}}{V_p},$$

$$\ddot{y} = -\frac{1}{2}\rho V \dot{y}\alpha_d - \frac{1}{2}\rho V^2 \alpha_t \frac{\dot{x}}{V_p} - \frac{1}{2}\rho V \alpha_c \frac{\dot{y}\dot{z}}{V_p},$$

$$\ddot{z} = -\frac{1}{2}\rho V \dot{z}\alpha\alpha_d + \frac{1}{2}\rho V V_p \alpha_c - g,$$

where (x, y, z) are the coordinates of a Cartesian coordinate system with x pointing East, y pointing North, and z pointing up; V is the magnitude of velocity, $\sqrt{\dot{x}^2 + \dot{y}^2 + \dot{z}^2}$; V_p is the magnitude of planar velocity, $\sqrt{\dot{x}^2 + \dot{y}^2}$; α_d is the drag force proportional constant, its inverse is known as the ballistic coefficient, which is aligned with the opposite of the velocity; α_t, α_c are the turn and drag parameters defining the lift force, which is perpendicular to the velocity vector; ρ is air density; and g is the altitude adjusted gravitational constant.

The system dynamic model is illustrated in Figure 3.7.

a. Build a trajectory model of the above ballistic reentry object. You can make the initial altitude to be 100 km usually considered the beginning of reentry. Pick an initial velocity of Mach 7 and a reentry angle of 25°. Use your engineering common sense to select values of the three parameters, α_d, α_t, α_c that vary with time or altitude. Display your trajectory in 3D or in a sequence of 2D graphs.

b. The three parameters, α_d, α_t, α_c are all unknowns. They must be estimated together with the state dynamics. A typical approach is to make them part

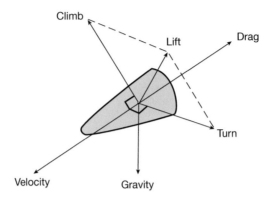

1. Drag is opposite to velocity
2. Lift is perpendicular to drag
3. Turn and climb are components of lift
4. Turn is perpendicular to drag and gravity

Figure 3.7   Geometry of drag and lift forces.

of the state vector with the following model: $\dot{\alpha}_d = \mu_d$, $\dot{\alpha}_t = \mu_t$, $\dot{\alpha}_c = \mu_c$ where they represented a constant parameter models (derivative equals to zero) with μ_d, μ_t, μ_c as process noise terms to represent modeling uncertainties. Use notional radar measurement errors $\sigma_r = 1$ m, $\sigma_{az} = 1$ mrad, $\sigma_{el} = 1$ mrad to simulate radar measurements of this trajectory model. Assume certain values for the corresponding process noise terms, q_d, q_t, and q_c. Build an EKF, run it with Monte Carlo repetition so that you can obtain sample bias and covariance errors, compare with the filter computed covariance. Experiment with various q_d, q_t, q_c values to obtain different random and bias errors of your filter, see if you can find a set that performs the best.

c. Build a CRB for the corresponding model, compare with the results of b.

d. Compare the result of using EKF with CRB.

e. When you are able to find conditions where the CRB closely resembles the sample covariance of EKF, find a way to simulate the state estimate by adding noise samples to the true stats. The noise samples must follow the covariance given by the CRB.

You can get more sophisticated by adding a more accurate gravity model and by adding elliptical earth, Coriolis, and centrifugal forces to the trajectory model.

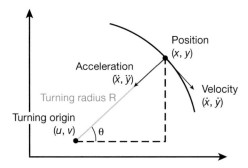

Figure 3.8   Illustration of coordinates for an object turning around a circle.

Repeat the above exercises. Now you can experiment with the case when the dynamics of the true trajectory and the model used in the EKF are different.

4.   Prove the matrix identity in Section 3.7.1

   Matrix Identities

   (1) $\mathbf{x}^T \mathbf{A} \mathbf{x} = \mathrm{tr}\left[\mathbf{A}\mathbf{x}\mathbf{x}^T\right]$ and $E\left\{\mathbf{x}^T \mathbf{A} \mathbf{x}\right\} = \mathrm{tr}\left[\mathbf{A}\Sigma\right]$

   (2) $E\left\{\mathrm{tr}\left[\mathbf{A}\mathbf{x}\mathbf{x}^T \mathbf{B}\mathbf{x}\mathbf{x}^T\right]\right\} = E\left\{\mathrm{tr}\left[\mathbf{A}\mathbf{x}\mathbf{x}^T\right]\mathrm{tr}\left[\mathbf{B}\mathbf{x}\mathbf{x}^T\right]\right\}$
   $$= 2\mathrm{tr}\left[\mathbf{A}\Sigma\mathbf{B}\Sigma\right] + \mathrm{tr}\left[\mathbf{A}\Sigma\right]\mathrm{tr}\left[\mathbf{B}\Sigma\right]$$

   where $\mathrm{tr}[\mathbf{M}]$ denotes the trace of the enclosed matrix $\mathbf{M}$ and $\Sigma = E\left\{\mathbf{x}\mathbf{x}^T\right\}$.

5.   Consider an object traveling with speed v turning around a circle with radius R (see Figure 3.8). The magnitude of the object acceleration is $|\alpha| = v^2/R$. Derive a state space model representing the dynamics of this object in a two-dimensional Cartesian coordinate. (Hint: Consider Figure 3.8 for coordinate and state variable definition.)

References

1.   R.S. Bucy, "Nonlinear Filtering Theory," *IEEE Transactions on Automatic Control*, vol. AC-10, pp. 198–206, Jan. 1965.
2.   R.C.K. Lee, *Optimal Estimation, Identification, and Control*. Cambridge, MA: MIT Press, 1964.

3. Y.C. Ho and R.C.K. Lee, "A Bayesian Approach to Problems in Stochastic Estimation and Control," *IEEE Transactions on Automatic Control*, vol. AC-9, pp. 333–339, Oct. 1964.

4. H.J. Kushner, "On the Differential Equations Satisfied by Conditional Probability Densities of Markov Processes, with Applications," *Journal of SIAM on Control*, vol. 2, pp. 106–119, 1964.

5. H.J. Kushner, "Nonlinear Filtering: The Exact Dynamical Equations Satisfied by the Conditional Mode," *IEEE Transactions on Automatic Control*, vol. AC-12, pp. 262–267, Jun. 1967.

6. H.J. Kushner, "Approximations to Optimal Nonlinear Filters," *IEEE Transactions on Automatic Control*, vol. AC-12, pp. 546–556, Oct. 1967.

7. A.H. Jazwinski, *Stochastic Processes and Filtering Theory*. New York: Academic Press, 1970.

8. H.W. Sorenson and A.R. Stubberud, "Nonlinear Filtering by Approximation of the a Posteriori Density," *International Journal of Control*, vol. 9, pp. 33–51, 1969.

9. M. Athans, R.P. Wishner, and A. Bertolini, "Suboptimal State Estimation for Continuous-Time Nonlinear Systems from Discrete Noisy Measurements," *IEEE Transactions on Automatic Control*, vol. AC-13, pp. 504–514, Oct. 1969.

10. R.P. Wishner, J.A. Tabaczynski, and M. Athans, "A Comparison of Three Nonlinear Filters," *Automatica*, vol. 5, pp. 497–496, 1969.

11. A. Gelb (Ed.), *Applied Optimal Estimation*. Cambridge, MA: MIT Press, 1974.

12. R.S. Bucy, "Bayes Theorem and Digital Realization for Nonlinear Filters," *Journal of Astronautical Sciences*, vol. 17, pp. 80–94, 1969.

13. S.J. Julier and J.K. Uhlmann, "A New Extension of the Kalman Filter to Nonlinear Systems," *AeroSense: 11th International Symposium on Aerospace/Defense Sensing, Simulation and Controls*, pp. 182–193, 1997.

14. S. Julier, J. Uhlmann, and H. F. Durrant-Whyte, "A New Method for the Nonlinear Transformation of Means and Covariances in Filters and Estimators," *IEEE Transactions on Automatic Control*, vol. AC-45, pp. 477–482, Mar. 2000.

15. A. Wan and R. van der Merwe, "The Unscented Kalman Filter for Nonlinear Estimation," in *Proceedings of 2000 AS-SPCC*, pp. 153–158, Alberta, Canada, Oct. 2000.

16. D. Simon, *Optimal State Estimation*. New York: Wiley, 2006.

17. B. Ristic, S. Arulampalam, and N. Gordan, *Beyond the Kalman Filter*. Norwood, MA: Artech House, 2004.

18. C.B. Chang, "Ballistic Trajectory Estimation with Angle-Only Measurements," *IEEE Transactions on Automatic Control*, vol. AC-25, pp. 474–480, Jun. 1980.

19. T.S. Lee, K.P. Dunn, and C.B. Chang, "On Observability and Unbiased Estimation of Nonlinear Systems," *System Modeling and Optimization, Lecture Notes in Control and Information Sciences*, vol. 39, pp. 259–266, New York: Springer, 1982.

20. J.H. Taylor, "The Cramer–Rao Estimation Error Lower Bound Computation for Deterministic Nonlinear Systems," *IEEE Transactions on Automatic Control*, vol. AC-24, pp. 343–344, Apr. 1979.

21. C.B. Chang, "Optimal State Estimation of Ballistic Trajectories with Angle-Only Measurements," MIT Lincoln Laboratory Report TN-1979-1, Jan. 24, 1979.

22. P. Tichavsky, C.H. Muravchik, and A. Nehorai, "Posterior Cramer–Rao Bounds for Discrete-Time Nonlinear Filtering," *IEEE Transactions on Signal Processing*, vol. SP-46, pp. 1386–1396, May 1998.

23. R.W. Miller and C.B. Chang, "A Modified Cramer–Rao Bound and Its Applications," *IEEE Transactions on Information Theory*, vol. IT-24, pp. 399–400, May 1979.

24. C.B. Chang, "Two Lower Bounds on the Covariance for Nonlinear Estimation Problems," *IEEE Transactions on Automatic Control*, vol. AC-26, pp. 1294–1297, Dec. 1981.

25. C.B. Chang, K.P. Dunn, and N.R. Sandell, Jr., "A Cramer–Rao Bound for Nonlinear Filtering Problems with Additive Gaussian Measurement Noise," in *Proceedings of the 19th IEEE Conference on Decision & Control*, pp. 511–512, Dec. 1979.

26. C.B. Chang, "Two Lower Bounds on the Covariance for Nonlinear Filtering Problems," in *Proceedings of the 19th IEEE Conference on Decision & Control*, pp. 374–379, Dec. 1980.

27. C.B. Chang, R.H. Whiting, and M. Athans, "On the State and Parameter Estimation for Maneuvering Reentry Vehicles," *IEEE Transactions on Automatic Control*, vol. AC-22, pp. 99–105, Feb. 1977.

28. C.B. Chang, R.H. Whiting, L. Youens, and M. Athans, "Application of Fixed-Interval-Smoother to Maneuvering Trajectory Estimation," *IEEE Transactions on Automatic Control*, vol. AC-22, pp. 876–879, Oct. 1977.

4

Practical Considerations in Kalman Filter Design

4.1 Model Uncertainty

A Kalman filter (KF) is optimal when certain conditions are met [1]. These conditions are discussed in Chapter 2, and are repeated below for clarity:

1. $\mathbf{\Phi}_{k,k-1}$(or $\mathbf{A}_t$) and $\mathbf{H}_k$ are completely known.

2. The statistics for initial state, system, and measurement noise processes are zero mean Gaussian with known covariance, $\mathbf{x}_0: \sim N(\hat{\mathbf{x}}_0, \mathbf{P}_0)$, $\mathbf{\mu}_k: \sim N(\mathbf{0}, \mathbf{Q}_k)$, $\mathbf{v}_k: \sim N(\mathbf{0}, \mathbf{R}_k)$, and are uncorrelated in time, and with each other.

3. The system is completely observable within the state space of interest.[1]

In most practical problems, these conditions cannot be fully met. When the first two of the above conditions are not met, the resulting filter is suboptimal and referred to as a mismatched KF. It is important to understand this practical problem in order to better conduct filter design, assess filter performance, and to prevent filter divergence. Filter mismatch conditions are further described below.

1. There are unknown parameters in the system dynamics, $\mathbf{\Phi}_{k,k-1}$, and/or measurement device, $\mathbf{H}_k$.

2. The system dynamics and/or measurement device do not fully represent the actual system.

3. Unknown or unmodeled driving functions appear in the system equation.

4. Unknown or unmodeled biases (could be time-varying) appear in the measurement equation.

---

1 A system is observable within the state space of interest when all state in the space can be uniquely defined given measurements (which are observables), for completely observable conditions for linear systems see Section 2.3.1.

5. The system noise is unknown or inappropriately modeled.

6. The measurement noise is unknown or inappropriately modeled.

In this chapter, certain types of model uncertainty are discussed, with approaches offered to deal with them. Before beginning the discussion on model errors and methods for filter compensation, several approaches for filter performance prediction and monitoring are introduced in Section 4.2 to support the ensuing development. A general set of filter error equations with model errors are presented in Section 4.3. A special case when the true system is a subsystem of the model used for filter implementation or vice versa is shown in Section 4.3.2. Steps can be taken during filter design to compensate for the system and filter mismatch issue. Section 4.4 introduces some general concepts on filter compensation. Section 4.5 discusses problems of measurement errors. Section 4.6 details problems for systems that have both unknown system inputs and measurement biases. Section 4.7 presents a special class of state estimation problems in which the system driving functions can make abrupt and unpredictable changes such as system failures. A combination of failure detection and state estimation algorithm is introduced. A treatment to filter divergence caused by ill-conditioning in state covariance computation and a condition called "false observability" in extended Kalman filter (EKF) applications is presented in Section 4.8. In Section 4.9, numerical examples for practical filter design are illustrated with the classical sinusoidal signal in noise problem and filters with unknown constant measurement bias.

4.2   Filter Performance Assessment

Before a filter is designed, it is essential to determine if the system is observable. If the answer is positive, covariance analysis must be conducted to assess the estimation accuracy that a filter can potentially achieve. In this section, two methods, namely the normalized innovation squared method (NIS) and the normalized estimation error squared method (NEES) are introduced. The first, NIS, is suitable for real-time filter performance monitoring when the true state is not known, and the second, NEES, is more appropriate for off-line filter assessment when the true state is known.[2]

4.2.1   Achievable Performance: Cramer–Rao Bound

The Cramer–Rao bound (CRB) provides a performance goal that an unbiased state estimator might be able to achieve. The CRB for estimating the initial state for a

---

2 "Known" true state is possible for a simulation study or for post-mission data analysis using the Best Estimated Trajectory (BET) as ground truth, see Remarks in Section 2.9.5.

deterministic system, Equation (2.10-6), is functionally the same as the system observability condition, Equations (2.3-6) and (2.3-7). For linear systems, existence of CRB answers the question of system observability. If the Fisher information matrix stays singular even with an increasing number of measurements, the system is not observable, and an unbiased estimator with finite covariance does not exist. If there is a priori knowledge about the system in terms of a non-singular $\mathbf{P}_0$, the unbiased state estimate can be computed initially, but there is no guarantee that the covariance will decrease (converge) with increasing number of measurements. For nonlinear systems, the CRB is much more complicated, see Section 3.9. However, for nonlinear but deterministic systems, the CRB can be computed using Jacobian matrices in place of the linear system and measurement matrices and the CRB equation has the same functional form as a necessary condition for nonlinear observability, Section 3.9.1. The topic of nonlinear observability is beyond the scope of this book, interested readers should consult [2] and references cited in it.

An important note here is that for deterministic systems, the covariance of maximum likelihood estimator converges asymptotically to the CRB [3, 4]. When solving practical problems "asymptotical convergence" means higher signal to noise ratio and/or increasing number of measurements. For example, radar range and angle measurement errors derived using CRB shows decreasing error with increasing signal to noise ratio which is shown to be true with real data under proper signal processing conditions. The error covariance of a Maximum Likelihood estimator for orbit determination approaches CRB after measurements over a certain time window have been used, see the example in Section 3.10. It is the author's experience that CRB for deterministic systems is a meaningful measure of filter performance, regardless of whether the system is linear or nonlinear.

4.2.2   Residual Process

The term $\mathbf{y}_k - \mathbf{H}_k\hat{\mathbf{x}}_{k|k-1}$ that appears in the KF update Equation (2.6-5) is the difference between the actual measurement, $\mathbf{y}_k$, and the prediction of that measurement, $\mathbf{H}_k\hat{\mathbf{x}}_{k|k-1}$, at time k. The sequence of these differences is known as the filter residual or the innovation process [5]. When all the KF optimality conditions are met the innovation process is a white noise process with zero mean and covariance $\mathbf{H}_k\mathbf{P}_{k|k-1}\mathbf{H}_k^T + \mathbf{R}_k$. Let $\boldsymbol{\gamma}_k = \mathbf{y}_k - \mathbf{H}_k\hat{\mathbf{x}}_{k|k-1}$, one can readily prove the innovation property by showing $E\{\boldsymbol{\gamma}_k\} = \mathbf{0}$ for all k and $E\{\boldsymbol{\gamma}_k\boldsymbol{\gamma}_\ell^T\} = \mathbf{0}$ for all $k \neq \ell$. With this property, one can monitor whether the filter is performing properly by applying χ^2 statistics on the innovation process. Let

$$\varepsilon_{\gamma_k}^2 = \boldsymbol{\gamma}_k^T \left[\mathbf{H}_k\mathbf{P}_{k|k-1}\mathbf{H}_k^T + \mathbf{R}_k\right]^{-1} \boldsymbol{\gamma}_k, \tag{4.2-1}$$

then $\varepsilon_{\gamma_k}^2$ should follow a χ^2 distribution with degrees of freedom m, where m is the dimension of the measurement vector. Bar-Shalom [6] named $\varepsilon_{\gamma_k}^2$ the NIS.

When the observed sequence, $\varepsilon_{\gamma_k}^2$, follows the χ^2 distribution, one may conclude that the filter's residual process does not show any apparent statistical inconsistency. NIS is a necessary but not sufficient indicator of filter performance. We have observed in application problems that NIS is normal, but large bias errors have already accumulated in the filter. NIS is nevertheless a useful measure since in real time, this is usually the only measure available for filter performance monitoring in real systems. This test can also be conducted with M Monte Carlo trial data when the true state is known $(M > 1)$.

One way to apply NIS for performance monitoring is to set an acceptance interval (probability concentration region) for a hypothesis test, such that

$$\text{Prob}\{\bar{\varepsilon}_{\gamma_k}^2 \in [r_1, r_2]\} = 1 - \alpha, \tag{4.2-2}$$

where

$$\bar{\varepsilon}_{\gamma_k}^2 = \frac{1}{M}\sum_{i=1}^{M} \varepsilon_{\gamma_k}^2 \tag{4.2-3}$$

is the average NIS at time k with M Monte Carlo trials or $M = 1$ with real-time data. If consistent, $M\bar{\varepsilon}_{\gamma_k}^2$ has a χ^2 distribution with mM degrees of freedom (where m is the dimension of measurement vector), the mean is mM, and the variance is $2\ mM$.

The values of r_1 and r_2 can be determined by a χ^2 distribution with mean mM and variance $2\ mM$ [6]. For example, with $M = 1$, $m = 3$, and $\alpha = 0.05$, one has the two-sided interval, $r_1 = 0.22$ and $r_2 = 9.35$; with $M = 50$, $m = 3$, and $\alpha = 0.05$, one has the two-sided interval, $r_1 = 2.36$ and $r_2 = 3.72$. Figure 4.1 shows a typical NIS result for a filter with $M = 1$.

When the γ_k sequence has a nonzero mean, and/or when $\varepsilon_{\gamma_k}^2$ deviates from a χ^2 distribution, it is clear that the filter is not performing as expected. In some cases, the innovation process can be used for adaptive filtering, as will be shown in a later section.

4.2.3   Filter Computed Covariance

When it is determined that the Fisher information matrix is nonsingular and the CRB exists, one may proceed to design the state estimator for a given application. The KF covariance equation represents the filter's knowledge on how well the filter is performing. This is no longer true when the model used to construct the filter is different from

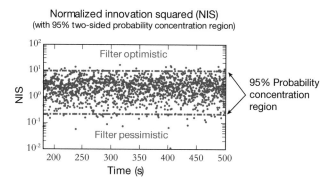

Figure 4.1 NIS illustration.

the actual dynamics, measurement equations, or the associated random processes. A way to determine whether a mismatch exists is by means of simulation. A Monte Carlo study can be conducted using simulated measurements processed by the designed filter. The Monte Carlo repetition allows the designer to obtain sample statistics from which the bias and covariance can be computed. Let $\tilde{\mathbf{x}}^i_{k|k}$ denote the estimation error of the ith trial, thus $\tilde{\mathbf{x}}^i_{k|k} = \mathbf{x}_k - \hat{\mathbf{x}}^i_{k|k}$ where $\hat{\mathbf{x}}^i_{k|k}$ is the estimate obtained from ith Monte Carlo trial. Assuming there are a total of M trials, let $\tilde{\mathbf{x}}_{k|k}$ and $\tilde{\mathbf{P}}_{k|k}$ denote the sample mean and sample mean square errors of $\tilde{\mathbf{x}}^i_{k|k}$, respectively, then,

$$\tilde{\mathbf{x}}_{k|k} = \frac{1}{M} \sum_{i=1}^{M} \tilde{\mathbf{x}}^i_{k|k}, \tag{4.2-4}$$

$$\tilde{\mathbf{P}}_{k|k} = \frac{1}{M} \sum_{i=1}^{M} \left(\tilde{\mathbf{x}}^i_{k|k}\right)\left(\tilde{\mathbf{x}}^i_{k|k}\right)^T. \tag{4.2-5}$$

The square root of the jth diagonal element of $\tilde{\mathbf{P}}_{k|k}$ is called the root mean square (RMS) error of the jth component of the estimator $\hat{\mathbf{x}}_{k|k}$. For an unbiased estimator, one expects $\tilde{\mathbf{x}}_{k|k} \to \mathbf{0}$ as $M \to \infty$. The next question is whether the sample mean square error converges to the CRB, $\mathbf{P}_{k|k}$, computed with the true state at time k, that is, if $\tilde{\mathbf{P}}_{k|k} \to \mathbf{P}_{k|k}$ as $M \to \infty$. To answer this question, we again use χ^2 statistics,

$$\chi^{i^2}_{k|k} = \left(\tilde{\mathbf{x}}^i_{k|k}\right)^T \mathbf{P}^{-1}_{k|k} \left(\tilde{\mathbf{x}}^i_{k|k}\right) \tag{4.2-6}$$

and

$$\overline{\chi^2_{k|k}} = \frac{1}{M} \sum_{i=1}^{M} \chi^{i^2}_{k|k}. \tag{4.2-7}$$

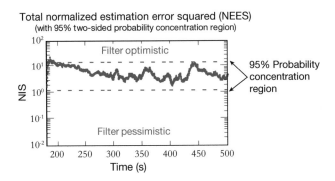

Figure 4.2   NEES illustration.

If $\tilde{\mathbf{x}}_{k|k}^{i}$ is Gaussian with zero mean and with covariance $\mathbf{P}_{k|k}$, then $\chi_{k|k}^{i}{}^{2}$ follows a χ^{2} distribution with n degrees of freedom, where n is the dimension of the state vector $\mathbf{x}_{k}$, and $\overline{\chi_{k|k}^{2}}$ follows χ^{2} distribution with nM degrees of freedom and a mean of nM and a variance of $2\,nM$.

When $\overline{\chi_{k|k}^{2}}$ is within the bounds of the statistical tests, for example, with $M = 1$, $n = 6$, and $\alpha = 0.05$, the two-sided interval is $r_1 = 1.24$ and $r_2 = 14.45$, while if $M = 50$, $n = 6$, and $\alpha = 0.05$, then the two-sided interval is, $r_1 = 5.08$ and $r_2 = 7.0$. Figure 4.2 shows typical filter performance with $M = 1$. For this case it appears that the filter performance is consistent with the CRB predicted covariance, that is, $\mathbf{P}_{k|k}$ does represent the estimation error.

Bar-Shalom [6] named $\overline{\chi_{k|k}^{2}}$ the average NEES. As opposed to NIS from Equation (4.2-1), NEES from Equation (4.2-7) is computed using the true (or best estimated) states. When a filter fails this test, several filter compensation methods discussed in later sections can be tried. The NEES is useful during the filter development phase that the true state, $\mathbf{x}_{k}$, is used for the filter performance prediction. However, it is not useful to check against simulation when the model for simulation is the same as that used for the KF because this is a perfectly but artificially matched case. In some applications, when independent measurements on the true state are available, referred to as the BET, this method is applicable. There are also situations where the true states are generated by sophisticated high fidelity models, or sometimes with the actual hardware as part of the simulation, the so-called hardware in the loop (HWIL) simulation, this method is also applicable.

4.3   Filter Error with Model Uncertainties

Consider the case where the true system, S, and model of the system, S_m, are represented as follows.

S (True system):

$$\mathbf{x}_k = \mathbf{\Phi}\mathbf{x}_{k-1} + \mathbf{\mu}_{k-1}$$

$$\mathbf{y}_k = \mathbf{H}\mathbf{x}_k + \mathbf{v}_k; \qquad (4.3\text{-}1)$$

S_m (Model):

$$\mathbf{x}_k^m = \mathbf{\Phi}^m \mathbf{x}_{k-1}^m + \mathbf{\mu}_{k-1}$$

$$\mathbf{y}_k^m = \mathbf{H}^m \mathbf{x}_k^m + \mathbf{v}_k. \qquad (4.3\text{-}2)$$

For reasons of simplicity, the time variable on both $\mathbf{\Phi}$ and $\mathbf{H}$ has been omitted. The state equation of a KF constructed using S_m is

$$\hat{\mathbf{x}}_{k|k}^m = \mathbf{\Phi}^m \hat{\mathbf{x}}_{k-1|k-1}^m + \mathbf{K}^m (\mathbf{y}_k - \mathbf{H}^m \mathbf{\Phi}^m \hat{\mathbf{x}}_{k-1|k-1}^m), \qquad (4.3\text{-}3)$$

where $\mathbf{K}^m$ is the Kalman gain based on the designer's knowledge of the system, S_m. The filter error for this case is derived next.

4.3.1   Bias and Covariance Equations

Let

$$\tilde{\mathbf{x}}_{k|k}^m \triangleq \mathbf{x}_k - \hat{\mathbf{x}}_{k|k}^m \qquad (4.3\text{-}4)$$

$$\mathbf{\Phi} \triangleq \Delta\mathbf{\Phi} + \mathbf{\Phi}^m \qquad (4.3\text{-}5)$$

$$\mathbf{H} \triangleq \Delta\mathbf{H} + \mathbf{H}^m \qquad (4.3\text{-}6)$$

$$\mathbf{H}\mathbf{\Phi} \triangleq \Delta\mathbf{H}\mathbf{\Phi} + \mathbf{H}^m\mathbf{\Phi}^m \qquad (4.3\text{-}7)$$

where $\Delta\mathbf{\Phi}$, $\Delta\mathbf{H}$, and $\Delta\mathbf{H}\mathbf{\Phi}$ are model errors, then,

$$\tilde{\mathbf{x}}_{k|k}^m = \mathbf{x}_k - \hat{\mathbf{x}}_{k|k}^m = \mathbf{\Phi}\mathbf{x}_{k-1} + \mathbf{\mu}_{k-1} - \left(\mathbf{\Phi}^m \hat{\mathbf{x}}_{k-1|k-1}^m + \mathbf{K}^m \left(\mathbf{y}_k - \mathbf{H}^m\mathbf{\Phi}^m \hat{\mathbf{x}}_{k-1|k-1}^m \right) \right). \qquad (4.3\text{-}8)$$

We expand the innovation process term

$$\mathbf{y}_k - \mathbf{H}^m\mathbf{\Phi}^m \hat{\mathbf{x}}_{k-1|k-1}^m = \mathbf{H}\left(\mathbf{\Phi}\mathbf{x}_{k-1} + \mathbf{\mu}_{k-1} \right) + \mathbf{v}_k - \mathbf{H}^m\mathbf{\Phi}^m \hat{\mathbf{x}}_{k-1|k-1}^m \qquad (4.3\text{-}9)$$

and substitute it back to Equation (4.3-8) to obtain

$$\tilde{\mathbf{x}}_{k|k}^m = \mathbf{\Phi}\mathbf{x}_{k-1} + \boldsymbol{\mu}_{k-1} - \mathbf{\Phi}^m\hat{\mathbf{x}}_{k-1|k-1}^m - \mathbf{K}^m\left(\mathbf{H}\mathbf{\Phi}\mathbf{x}_{k-1} - \mathbf{H}^m\mathbf{\Phi}^m\hat{\mathbf{x}}_{k-1|k-1}^m + \mathbf{H}\boldsymbol{\mu}_{k-1} + \mathbf{v}_k\right). \qquad (4.3\text{-}10)$$

Using system model error term $\Delta\mathbf{\Phi}$ and $\mathbf{\Phi}^m$ defined above to obtain

$$\tilde{\mathbf{x}}_{k|k}^m = \Delta\mathbf{\Phi}\mathbf{x}_{k-1} + \mathbf{\Phi}^m\tilde{\mathbf{x}}_{k-1|k-1}^m + \boldsymbol{\mu}_{k-1}$$
$$- \mathbf{K}^m\left(\mathbf{H}\Delta\mathbf{\Phi}\mathbf{x}_{k-1} + \mathbf{H}\mathbf{\Phi}^m\mathbf{x}_{k-1} - \mathbf{H}^m\mathbf{\Phi}^m\tilde{\mathbf{x}}_{k-1|k-1}^m + \mathbf{H}\boldsymbol{\mu}_{k-1} + \mathbf{v}_k\right). \qquad (4.3\text{-}11)$$

We use the measurement matrix error term defined above to obtain

$$\tilde{\mathbf{x}}_{k|k}^m = \left[\mathbf{I} - \mathbf{K}^m\mathbf{H}^m\right]\mathbf{\Phi}^m\tilde{\mathbf{x}}_{k-1|k-1}^m + \left[\Delta\mathbf{\Phi} - \mathbf{K}^m\left[\mathbf{H}\Delta\mathbf{\Phi} + \Delta\mathbf{H}\mathbf{\Phi}^m\right]\right]\mathbf{x}_{k-1}$$
$$+ \left[\mathbf{I} - \mathbf{K}^m\mathbf{H}\right]\boldsymbol{\mu}_{k-1} - \mathbf{K}^m\mathbf{v}_k. \qquad (4.3\text{-}12)$$

Using the fact that $\mathbf{H}\Delta\mathbf{\Phi} + \Delta\mathbf{H}\mathbf{\Phi}^m = \mathbf{H}\mathbf{\Phi} - \mathbf{H}^m\mathbf{\Phi}^m = \Delta\mathbf{H}\mathbf{\Phi}$, we obtain

$$\tilde{\mathbf{x}}_{k|k}^m = \left[\mathbf{I} - \mathbf{K}^m\mathbf{H}^m\right]\mathbf{\Phi}^m\tilde{\mathbf{x}}_{k-1|k-1}^m + \left[\Delta\mathbf{\Phi} - \mathbf{K}^m\Delta\mathbf{H}\mathbf{\Phi}\right]\mathbf{x}_{k-1} + \left[\mathbf{I} - \mathbf{K}^m\mathbf{H}\right]\boldsymbol{\mu}_{k-1} - \mathbf{K}^m\mathbf{v}_k. \qquad (4.3\text{-}13)$$

Using this expression, it is possible to derive the mean, $\bar{\tilde{\mathbf{x}}}_{k|k}^m$, and covariance, $\tilde{\mathbf{P}}_{k|k}^m$, of $\tilde{\mathbf{x}}_{k|k}^m$. The results are stated below with the derivation omitted.

$$\bar{\tilde{\mathbf{x}}}_{k|k}^m = \left[\mathbf{I} - \mathbf{K}^m\mathbf{H}^m\right]\mathbf{\Phi}^m\bar{\tilde{\mathbf{x}}}_{k-1|k-1}^m + \left[\Delta\mathbf{\Phi} - \mathbf{K}^m\Delta\mathbf{H}\mathbf{\Phi}\right]\bar{\mathbf{x}}_{k-1}, \qquad (4.3\text{-}14)$$

$$\tilde{\mathbf{P}}_{k|k}^m = \left[\mathbf{I} - \mathbf{K}^m\mathbf{H}^m\right]\mathbf{\Phi}^m\tilde{\mathbf{P}}_{k-1|k-1}^m\mathbf{\Phi}^{mT}\left[\mathbf{I} - \mathbf{K}^m\mathbf{H}^m\right]^T + \left[\mathbf{I} - \mathbf{K}^m\mathbf{H}\right]\mathbf{Q}_k\left[\mathbf{I} - \mathbf{K}^m\mathbf{H}\right]^T + \mathbf{K}^m\mathbf{R}\mathbf{K}^{mT}$$
$$+ \left[\Delta\mathbf{\Phi} - \mathbf{K}^m\Delta\mathbf{H}\mathbf{\Phi}\right]\mathbf{P}_{k-1|k-1}\left[\Delta\mathbf{\Phi} - \mathbf{K}^m\Delta\mathbf{H}\mathbf{\Phi}\right]^T + \left[\mathbf{I} - \mathbf{K}^m\mathbf{H}^m\right]\mathbf{\Phi}^m\mathbf{P}_{k-1|k-1}^c\left[\Delta\mathbf{\Phi} - \mathbf{K}^m\Delta\mathbf{H}\mathbf{\Phi}\right]^T$$
$$+ \left[\Delta\mathbf{\Phi} - \mathbf{K}^m\Delta\mathbf{H}\mathbf{\Phi}\right]\mathbf{P}_{k-1|k-1}^{cT}\mathbf{\Phi}^{mT}\left[\mathbf{I} - \mathbf{K}^m\mathbf{H}^m\right]^T \qquad (4.3\text{-}15)$$

$$\mathbf{P}_{k|k}^c = \left[\mathbf{I} - \mathbf{K}^m\mathbf{H}^m\right]\mathbf{\Phi}^m\mathbf{P}_{k-1|k-1}^c\mathbf{\Phi}^T + \left[\Delta\mathbf{\Phi} - \mathbf{K}^m\Delta\mathbf{H}\mathbf{\Phi}\right]\mathbf{P}_{k-1|k-1}^c\mathbf{\Phi}^T + \left[\mathbf{I} - \mathbf{K}^m\mathbf{H}\right]\mathbf{Q}_k, \qquad (4.3\text{-}16)$$

$$\mathbf{P}_{k|k}^x = \mathbf{\Phi}\mathbf{P}_{k-1|k-1}^x\mathbf{\Phi}^T + \mathbf{Q}_k, \qquad (4.3\text{-}17)$$

where $\tilde{\mathbf{P}}_{k|k}^m$ is the covariance of $\tilde{\mathbf{x}}_{k|k}^m$, $\mathbf{P}_{k|k}^c$ is the cross covariance of $\tilde{\mathbf{x}}_{k|k}^m$ and $\mathbf{x}_k$, and $\mathbf{P}_{k|k}^x$ is the covariance of $\mathbf{x}_k$.

These equations are sufficiently general, but are not very useful because the true system is usually not completely known to us. A case where the system designer does have some knowledge of modeling error is now discussed.

4.3.2   Overmodeled and Undermodeled Cases

Consider two systems S_1 and S_2 related as follows:

System S_1:

$$\mathbf{\Phi} = \begin{bmatrix} \mathbf{\Phi}^{11} & 0 \\ 0 & 0 \end{bmatrix}; \mathbf{x} = \begin{bmatrix} \mathbf{x}^1 \\ 0 \end{bmatrix}; \mathbf{H} = \begin{bmatrix} \mathbf{H}^1 & 0 \end{bmatrix}.$$

System S_2:

$$\boldsymbol{\Phi} = \begin{bmatrix} \boldsymbol{\Phi}^{11} & \boldsymbol{\Phi}^{12} \\ \mathbf{0} & \boldsymbol{\Phi}^{22} \end{bmatrix}; \mathbf{x} = \begin{bmatrix} \mathbf{x}^1 \\ \mathbf{x}^2 \end{bmatrix}; \mathbf{H} = \begin{bmatrix} \mathbf{H}^1 & \mathbf{H}^2 \end{bmatrix}.$$

Note that the superscripts refer to positions in the matrix and the relationship between S_1 and S_2. It is clear that S_1 is a subsystem of S_2, and that $\mathbf{x}^2$ is a time-varying input to S_1. If $\boldsymbol{\Phi}^{22} = \mathbf{I}$, then $\mathbf{x}^2$ is a constant vector. If $\boldsymbol{\Phi}^{12} = \mathbf{0}$, then $\mathbf{x}^1$ and $\mathbf{x}^2$ are independent of each other.

Such cases occur in real applications. For example, an object may be traveling with constant velocity or with acceleration. If a filter is designed for a constant velocity motion, then when the object maneuvers (accelerates) without proper compensation, the filter will develop large bias errors. If the filter is designed for an accelerating object while it is traveling at a constant velocity, then the filter that over models the true process will introduce noise in the estimates (estimating accelerations which are actually not there). The application of the above concept is illustrated by the following example.

Consider the following four cases.

Case 1: $S = S_1$; $S_m = S_1$;

Case 2: $S = S_1$; $S_m = S_2$;

Case 3: $S = S_2$; $S_m = S_2$;

Case 4: $S = S_2$; $S_m = S_1$.

Note that Case 1 and Case 3 are both matched cases, that is, the truth and model are the same. Case 2 is an overmodeled case, in which the truth system is a subsystem of the model used for filter design. Case 4 is an undermodeled case, in which the model used for filter design is a subsystem of the truth. Case 1 can be viewed as a subset of Case 3 with $\mathbf{x}^2$, $\boldsymbol{\Phi}^{12}$, $\boldsymbol{\Phi}^{22}$, and $\mathbf{H}^2$ set to zero.

Performance Order

Let $\tilde{\mathbf{P}}_{k|k}^m$ ($S = S_i$; $S_m = S_j$) denote the actual filter covariance from Equation (4.3-15) for the true system, S, to be S_i and the filter model, S_m, to be S_j, then one can obtain the following performance order [7]

$$\tilde{\mathbf{P}}_{k|k}^m \left(S = S_1; \; S_m = S_1 \right) \le \tilde{\mathbf{P}}_{k|k}^m \left(S = S_1; \; S_m = S_2 \right) \le \tilde{\mathbf{P}}_{k|k}^m \left(S = S_2; \; S_m = S_2 \right). \quad (4.3\text{-}18)$$

The matrix ordering is defined as when $\mathbf{A} \ge \mathbf{B}$, then $\mathbf{A} - \mathbf{B}$ is positive semidefinite. The interpretation of this performance order is such that when the true system is a subsystem of the model used by the filter and can at most be equal to the model, then the filter covariance is upper-bounded by the matched case. For example, the covariance for a matched constant velocity object is smaller than that of a filter assuming accelerating object (mismatched), which in turn is smaller than a matched accelerating object.

The derivation of the performance order is straightforward; it is left to the reader as a homework problem. One can also find discussions on this subject in [7, 8]. The use of these equations is now illustrated for both the overmodeled and undermodeled cases.

The Overmodeled Case

With the assumption of Case 2 above, one can show that

$$\left[\Delta\boldsymbol{\Phi} - \mathbf{K}^m \Delta\mathbf{H}\boldsymbol{\Phi}\right]\overline{\mathbf{x}}_{k-1} = \mathbf{0}, \tag{4.3-19}$$

$$\left[\Delta\boldsymbol{\Phi} - \mathbf{K}^m \Delta\mathbf{H}\boldsymbol{\Phi}\right]\mathbf{P}_{k-1|k-1} = \mathbf{0}. \tag{4.3-20}$$

The mean and covariance equations are then simplified to

$$\overline{\tilde{\mathbf{x}}}_{k|k}^m = \left[\mathbf{I} - \mathbf{K}^m \mathbf{H}^m\right]\boldsymbol{\Phi}^m \overline{\tilde{\mathbf{x}}}_{k-1|k-1}^m, \tag{4.3-21}$$

$$\tilde{\mathbf{P}}_{k|k}^m = \left[\mathbf{I} - \mathbf{K}^m \mathbf{H}^m\right]\boldsymbol{\Phi}^m \tilde{\mathbf{P}}_{k-1|k-1}^m \boldsymbol{\Phi}^{mT}\left[\mathbf{I} - \mathbf{K}^m \mathbf{H}^m\right]^T$$
$$+ \left[\mathbf{I} - \mathbf{K}^m \mathbf{H}\right]\mathbf{Q}_k\left[\mathbf{I} - \mathbf{K}^m \mathbf{H}\right]^T + \mathbf{K}^m \mathbf{R}\mathbf{K}^{mT}. \tag{4.3-22}$$

Therefore, in this case, there is no bias issue that causes filter divergence, but there is a filter suboptimality issue due to higher noise or larger covariance.

The Undermodeled Case

This is for Case 4 when $S = S_2$; $S_m = S_1$, in which the model used for filter design is a subsystem of the truth. This corresponds, for example, to a filter based on a constant velocity model, but the object is actually accelerating. The bias and covariance equations are shown in Equations (4.3-14) and (4.3-15) and cannot be simplified. Uncompensated, the filter will diverge. To prevent divergence, any one of the three filter compensation methods described in Section 4.4 can be tried. The magnitude of the object acceleration can be used to preselect the filter process noise covariance. Simulation studies with filter tuning with different process noise covariance levels are required to finalize the filter design. For further discussion along this line with applications, see [9–11].

4.4   Filter Compensation Methods for Mismatched System Dynamics

A most common source of mismatched Kalman filtering is in the system model. As illustrated above, this can be due to the fact that the dynamics of a moving object following a certain order of degrees of freedom while the filter model assumed a different one. It could also be that certain parameters in the system matrix vary with time, and the particular time-varying function is unknown to the filter designer. Another possibility could be that some unknown inputs are driving the system.

Mismatched Kalman filters often lead to filter divergence, taking the form of (a) bias error and/or (b) the computed filter covariance is inconsistent with the actual filter performance. Filter divergence can sometimes be indicated by the NEES and/or NIS defined in Section 4.2.

When there are unknown parameters in the system and/or measurement equations, the filter may diverge or the quality of the estimate (with increasing bias and covariance errors) may be reduced. The parameters are usually included as additional states in the state vector and estimated together with the original state vector. This method is called state augmentation. The unknown parameters could be constants or could vary with time. Their uncertainties are modeled by the covariance of the system input noise or process noise. Process noise is needed to compensate for any unknown and unmodeled system errors. The filter achieves its compensation by virtue of the fact that the filter gain is proportional to the process noise covariance. A higher filter gain allows the filter to place more weight on recent measurements, thus preventing filter divergence.

In the more general case, where the difference between the system model and the true system has no known structure, the state augmentation method will be difficult to apply. Uncompensated, it will cause the filter to diverge with increasing bias error or inconsistent covariance matrix, that is, computed covariance may become much smaller than the actual covariance. Two approaches to potentially address this problem are as follows: (a) increase the system uncertainty in terms of the covariance of the process noise, where it can easily be shown that this will directly increase the filter gain (see numerical examples at the end of this chapter); and (b) limit the memory length of the filter by either using a fixed time window of measurements or by exponentially weighting down the past measurements.

4.4.1   State Augmentation

Let $\mathbf{p}$ denote the parameter vector that consists of all unknown parameters in both the system and measurement matrix. The augmented state vector is now $\mathbf{x}^a = [\mathbf{x}^T, \mathbf{p}^T]^T$ with the new system and measurement equations

$$\mathbf{x}_k^a = \begin{bmatrix} \mathbf{\Phi}_{k,k-1}(\mathbf{p}) & \mathbf{0} \\ \mathbf{0} & \mathbf{I} \end{bmatrix} \mathbf{x}_{k-1}^a + \mathbf{\mu}_{k-1}^a, \qquad (4.4\text{-}1)$$

$$\mathbf{y}_k = [\mathbf{H}_k(\mathbf{p})\,\mathbf{0}]\mathbf{x}_k^a + \mathbf{v}_k, \qquad (4.4\text{-}2)$$

where $\mathbf{\mu}_{k-1}^a = [\mathbf{\mu}_{k-1}^T, \mathbf{\mu}_{k-1}^{p\,T}]^T$ is the new process noise and $\mathbf{\mu}_{k-1}^p$ represents the addition of the process noise corresponding to the parameter vector. Assuming that $\mathbf{\mu}$ and $\mathbf{\mu}^p$ are uncorrelated, the process noise covariance of $\mathbf{\mu}_k^a$ is

$$\mathbf{Q}_k^a = \begin{bmatrix} \mathbf{Q}_k & \mathbf{0} \\ \mathbf{0} & \mathbf{Q}_k^p \end{bmatrix}. \qquad (4.4\text{-}3)$$

where $\mathbf{Q}_k^p$ can be selected to represent the range of uncertainty on the parameter vector $\mathbf{p}$. In the next section, a method is discussed that adaptively adjusts the process noise covariance based upon the observed residual process. It is a technique suitable for real-time implementation, and it is a very widely used technique for jointly estimating the state and the unknown parameters [see 9–12]. This method necessarily turns the linear estimation problem into a nonlinear problem; note that the parameter $\mathbf{p}$ is multiplicative with respect to the state vector. The nonlinear filter design was discussed in the previous chapter.

4.4.2   The Use of Process Noise

The adjustment of process noise covariance to compensate for the apparent filter errors observed in the residual (or innovation) process was a popular subject in the early 1970s [13–17]. These methods are very useful for tuning filters to make it work although this technique is not an exact science [11]. One of the simple methods was due to Jazwinski [17, pp. 305–307]. He suggested that the filter regularity be checked using the χ^2 variable (the NIS),

$$\varepsilon_{\gamma_k}^2 = \mathbf{\gamma}_k^T \left[\mathbf{H}_k \mathbf{P}_{k|k-1} \mathbf{H}_k^T + \mathbf{R}_k \right]^{-1} \mathbf{\gamma}_k. \qquad (4.4\text{-}4)$$

Once determined that $\varepsilon_{\gamma_k}^2$ is out of bounds, the corresponding process noise covariance, $\mathbf{Q}_{k-1}$, is increased so that the one-step predicted covariance, $\mathbf{P}_{k|k-1}$, is increased through

$$\mathbf{P}_{k|k-1} = \mathbf{\Phi}_{k,k-1} \mathbf{P}_{k-1|k-1} \mathbf{\Phi}_{k,k-1}^T + \mathbf{Q}_{k-1}, \qquad (4.4\text{-}5)$$

until $\varepsilon_{\gamma_k}^2$ is within the bound. The computation of the needed $\mathbf{Q}_{k-1}$ could be iterative, through simulation study, or by using a priori knowledge. Further application problems using this method are available [see 9, 10, 12].

The usefulness of adjusting the process noise covariance cannot be overemphasized. Experienced filter designers almost always add some process noise in order to prevent divergence and keep the filter responsive to new measurements. Filter divergence and the use of process noise to prevent overconfidence in the filter design model will be illustrated using a numerical example in Section 4.9.

4.4.3   The Finite Memory Filter

The concept of finite memory filter (FMF) is based upon the assumption that the model used for filter design is a good approximation of the truth trajectory over a limited time span. The FMF will diverge when the time span is too long. The choice of length of the time span is of course an engineering trade-off between bias and random errors. The FMF is synonymous with the windowed estimate in which only measurements within a time window are used for estimation. The FMF shown below is due to Jazwinski [17, pp. 256–258]. Using notation similar to that used in the fixed-interval smoother, let $\hat{\mathbf{x}}_{k|m:k} = E\{\mathbf{x}_k|\mathbf{y}_{m:k}\}$ denote the estimate of $\mathbf{x}_k$ with covariance $\mathbf{P}_{k|m:k}$ based upon all data collected in the time interval $[m, m + 1, ..., k]$, $\mathbf{y}_{m:k}$. With this definition, the time window is $k - m$. After some manipulation, one can show that the $\hat{\mathbf{x}}_{k|m:k}$ and $\mathbf{P}_{k|m:k}$ can be obtained as

$$\hat{\mathbf{x}}_{k|m:k} = \mathbf{P}_{k|m:k}(\mathbf{P}_{k|1:k}^{-1}\hat{\mathbf{x}}_{k|1:k} - \mathbf{P}_{k|1:m}^{-1}\hat{\mathbf{x}}_{k|1:m}), \qquad (4.4\text{-}6)$$

$$\mathbf{P}_{k|m:k}^{-1} = \mathbf{P}_{k|1:k}^{-1} - \mathbf{P}_{k|1:m}^{-1}, \qquad (4.4\text{-}7)$$

where $\hat{\mathbf{x}}_{k|1:k}$ is the estimate of $\mathbf{x}_k$ based upon all data in the interval $[1, ..., k]$, $\hat{\mathbf{x}}_{k|1:m}$ is the predicted estimate of $\mathbf{x}_k$ based upon all data in the interval $[1, ..., m]$, and $\hat{\mathbf{x}}_{k|m:k}$ is the estimate of $\mathbf{x}_k$ based upon the data in the interval $[m, ..., k]$.

The subtraction procedure is intuitively appealing. However, because it involves multiple matrix inversions and subtraction of two positive definite matrices, it is vulnerable to numerical errors, that is, the resulting matrix after subtraction may no longer be positive definite. Computationally efficient and numerically more stable algorithms can be derived [see 17 for details]. The concept of filtering using polynomial models over a limited time window, Homework Problems 15 (Chapter 1) and 5 (Chapter 2), is an FMF. The batch filter concept introduced in Section 3.8 is a generalized FMF.

4.4.4   The Fading Memory Filter

The fading memory filter (sometimes referred to as the aging filter) weights recent data exponentially higher than the past data. A derivation of this filter can be found in the literature [see 12, 17–19]. It is obtained by first applying an exponential weighting to the measurement covariance

$$\mathbf{R}_i^* = \alpha^{k-i}\mathbf{R}_i, \text{ for } i = 1, \ldots, k, \tag{4.4-8}$$

where k is the current time and α is a scalar quantity greater than unity, which is the design parameter for the fading memory filter. Applying the above expression to the KF results in a simple change

$$\mathbf{P}_{k|k-1}^* = \alpha\mathbf{P}_{k|k-1}, \tag{4.4-9}$$

where $\mathbf{P}_{k|k-1}^*$ is to be used in the KF update equation.

4.5   With Uncertain Measurement Noise Model

We now restate the discrete linear system being considered

$$\mathbf{x}_k = \mathbf{\Phi}_{k,k-1}\mathbf{x}_{k-1} + \mathbf{\mu}_{k-1},$$

$$\mathbf{y}_k = \mathbf{H}_k\mathbf{x}_k + \mathbf{v}_k.$$

In applying the KF to the above system, both the process noise vector $\mathbf{\mu}_{k-1}$ and measurement noise vector $\mathbf{v}_k$ are zero mean, white, uncorrelated with each other, and with known covariances. This section discusses the problem where the measurement noise vector contains uncertainty. The problem with unknown system input vector is discussed in the next section.

4.5.1   Unknown Constant Bias

For the problem where the measurement contains unknown constant bias, consider the following measurement equation

$$\mathbf{y}_k = \mathbf{H}_k\mathbf{x}_k + \mathbf{v}_k + \mathbf{b}, \tag{4.5-1}$$

where $\mathbf{b}$ is an unknown constant vector. A typical approach to this problem is to augment the state vector to include $\mathbf{b}$ as $\mathbf{x}^a = [\mathbf{x}^T\mathbf{p}^T]^T$, similar to the approach that was discussed in Section 4.4.1. Now the system and measurement equations are modified to

$$\mathbf{x}_k^a = \begin{bmatrix} \mathbf{\Phi}_{k,k-1} & \mathbf{0} \\ \mathbf{0} & \mathbf{I} \end{bmatrix} \mathbf{x}_{k-1}^a + \mathbf{\mu}_{k-1}^a, \tag{4.5-2}$$

$$\mathbf{y}_k = [\mathbf{H}_k \; \mathbf{0}]\mathbf{x}_k^a + \mathbf{v}_k. \tag{4.5-3}$$

Note that in this case the problem of nonlinearity does not exist (in comparison to the problem in Section 4.4.1). The KF solution to the augmented system is straightforward. However, it may be necessary to examine the observability conditions. The additional part of the augmented state vector, **b**, may not be separable from the original state vector, **x**. This can be checked using the observability criteria. As stated before, computing the CRB will yield the same conclusion.

4.5.2   Residual Bias with Known a Priori Distribution

Bias could be estimated through external means. For example, a sensor can be calibrated before use so that the bias estimates through sensor calibration are used to make corrections on the measurement. Practically speaking, estimated biases will retain uncertainties; this is known as residual bias. In the case that the residual bias follows a known statistic, two methods are commonly used to account for its presence: (a) inflating the measurement covariance according to the known residual bias covariance, and (b) applying the method due to Schmidt [20] known as the Schmidt–Kalman filter (SKF) [17, 21]. Method (a) is mathematically straightforward. It can be shown, however, that it does not always meet the covariance consistency requirement, so Method (a) will not be discussed any further. Method (b) includes the bias in the system equation as a nuisance parameter. Although not explicitly estimated, its impact on covariance propagation is accounted for. It is the practitioner's experience [21] and the example in Section 4.9.2 shows that this method often enables the actual estimation error (RMS error) to be consistent with the filter-computed covariance. The general SKF algorithm accounts for both system and measurement uncertainties [17, pp. 281–286]. Its entirety will be derived later in Section 4.6. Shown below are the SKF equations for measurement bias only. The reader should consult [17] and Section 4.6 for details.

For the SKF formulation, the state and measurement equations are modified to become

$$\mathbf{x}_k = \mathbf{\Phi}_{k,k-1}\mathbf{x}_{k-1} + \mathbf{\mu}_{k-1}, \tag{4.5-4}$$

$$\mathbf{b}_k = \mathbf{b}_{k-1}, \tag{4.5-5}$$

$$\mathbf{y}_k = \mathbf{H}_k\mathbf{x}_k + \mathbf{H}_k^b\mathbf{b}_k + \mathbf{v}_k, \tag{4.5-6}$$

where $\mathbf{b}_k$ is the residual bias that is assumed to be an unknown constant with a priori distribution $\mathbf{b}_{k:} \sim N\,(\mathbf{b}, \mathbf{B})$. In [21], **b** is set to **0**. $\mathbf{H}_k^b$ defines the relationship between the residual bias and the measurement variables. The SKF treats $\mathbf{b}_k$ as a nuisance parameter, and thus not estimated, but its covariance is accounted for. The estimation algorithm is stated below; for derivation see Section 4.6 or [17].

The prediction terms are given by

$$\hat{\mathbf{x}}_{k|k-1} = \boldsymbol{\Phi}_{k,k-1}\hat{\mathbf{x}}_{k-1|k-1}, \tag{4.5-7}$$

$$\mathbf{P}_{k|k-1} = \boldsymbol{\Phi}_{k,k-1}\mathbf{P}_{k-1|k-1}\boldsymbol{\Phi}_{k,k-1}^T + \mathbf{Q}_{k-1}, \tag{4.5-8}$$

$$\mathbf{C}_{k|k-1}^b = \boldsymbol{\Phi}_{k,k-1}\mathbf{C}_{k-1|k-1}^b. \tag{4.5-9}$$

The update terms are

$$\hat{\mathbf{x}}_{k|k} = \hat{\mathbf{x}}_{k|k-1} + \mathbf{K}_k\left(\mathbf{y}_k - \mathbf{H}_k\hat{\mathbf{x}}_{k|k-1}\right), \tag{4.5-10}$$

$$\mathbf{K}_k = \left[\mathbf{P}_{k|k-1}\mathbf{H}_k^T + \mathbf{C}_{k|k-1}^b\mathbf{H}_k^{b^T}\right]\boldsymbol{\Sigma}_k^{-1}, \tag{4.5-11}$$

$$\boldsymbol{\Sigma}_k = \mathbf{H}_k\mathbf{P}_{k|k-1}\mathbf{H}_k^T + \mathbf{R}_k + \mathbf{H}_k^b\mathbf{C}_{k|k-1}^{b^T}\mathbf{H}_k^T + \mathbf{H}_k\mathbf{C}_{k|k-1}^b\mathbf{H}_k^{b^T} + \mathbf{H}_k^b\mathbf{B}\mathbf{H}_k^{b^T}, \tag{4.5-12}$$

$$\mathbf{C}_{k|k}^b = \mathbf{C}_{k|k-1}^b - \mathbf{K}_k\left[\mathbf{H}_k\mathbf{C}_{k|k-1}^b + \mathbf{H}_k^b\mathbf{B}\right], \tag{4.5-13}$$

$$\mathbf{P}_{k|k} = \left[\mathbf{I} - \mathbf{K}_k\mathbf{H}_k\right]\mathbf{P}_{k|k-1}\left[\mathbf{I} - \mathbf{K}_k\mathbf{H}_k\right]^T + \mathbf{K}_k\mathbf{R}_k\mathbf{K}_k^T. \tag{4.5-14}$$

Note that $\mathbf{K}_k$ is the filter gain, $\boldsymbol{\Sigma}_k$ is the covariance of filter residual process, $\mathbf{C}_{k|k}^b$ is the state and bias cross covariance with $\mathbf{C}_{0|0}^b = \mathbf{0}$, and $\mathbf{P}_{k|k}$ is the covariance of $\hat{\mathbf{x}}_{k|k}$.

This algorithm will be shown as a special case of the complete SKF, which will be derived in Section 4.6.

4.5.3   Colored Measurement Noise

Colored noise means that the noise sequence is correlated over time. Consider the modified measurement equation

$$\mathbf{y}_k = \mathbf{H}_k\mathbf{x}_k + \boldsymbol{\varepsilon}_k, \tag{4.5-15}$$

where the time correlation of $\boldsymbol{\varepsilon}_k$ and $\boldsymbol{\varepsilon}_{k-1}$ is characterized by the relationship

$$\boldsymbol{\varepsilon}_k = \boldsymbol{\Psi}_k\boldsymbol{\varepsilon}_{k-1} + \mathbf{v}_{k-1}, \tag{4.5-16}$$

where $\mathbf{v}_{k-1}$ and the system noise $\boldsymbol{\mu}_{k-1}$ are mutually independent uncorrelated time sequences with zero mean Gaussian densities and known covariances. One intuitive approach to solving this problem would be to use state augmentation.

State Augmentation

This results in the following system and measurement equations,

$$\mathbf{x}_k^a = \begin{bmatrix} \boldsymbol{\Phi}_{k,k-1} & \mathbf{0} \\ \mathbf{0} & \boldsymbol{\psi}_k \end{bmatrix} \mathbf{x}_{k-1}^a + \boldsymbol{\mu}_{k-1}^a, \tag{4.5-17}$$

$$\mathbf{y}_k = [\mathbf{H}_k \ \mathbf{I}] \mathbf{x}_k^a, \tag{4.5-18}$$

where $\mathbf{x}^a = [\mathbf{x}^T \boldsymbol{\varepsilon}^T]^T$, $\boldsymbol{\mu}_{k-1}^a = [\boldsymbol{\mu}_{k-1}^T \ \mathbf{v}_{k-1}^T]^T$. Note that the white noise component of the original measurement noise has now been shifted to the augmented process noise. This makes the augmented system measurement of Equation (4.5-18) noise free; thus, an ill-conditioned case. Bryson and Henrikson [22] likened this to solving an $n + m$ dimensional $(\mathbf{x}_k^a \in \mathbb{R}^{n+m})$ estimation problem with only n linear constraints $(\mathbf{x} \in \mathbb{R}^n)$, and consequently, the augmented approach will not work.

Bryson and Henrikson then proceeded to propose a measurement differencing approach, requiring the creation of a derived measurement

$$\mathbf{y}_k^* = \mathbf{y}_{k+1} - \boldsymbol{\psi}_k \mathbf{y}_k. \tag{4.5-19}$$

The measurement noise for the derived measurement $\mathbf{y}_k^*$ is no longer correlated in time, but it is correlated with the process noise $\boldsymbol{\mu}_{k-1}$. They then proceeded to solve this problem, and their approach is rather cumbersome because it involves simultaneously solving a one-step smoothing problem. Note that the derived measurement contains both the future and the present measurement. A more elegant solution was provided by Lambert [23] that is easier to follow, and described below.

Measurement Differencing Approach

To restate the estimation problem, here are the state and measurement equations

$$\mathbf{x}_k = \boldsymbol{\Phi}_{k+1,k} \mathbf{x}_{k-1} + \boldsymbol{\mu}_{k-1}, \tag{4.5-20}$$

$$\mathbf{y}_k = \mathbf{H}_k \mathbf{x}_k + \boldsymbol{\varepsilon}_k, \tag{4.5-21}$$

$$\boldsymbol{\varepsilon}_k = \boldsymbol{\Psi}_k \boldsymbol{\varepsilon}_k + \mathbf{v}_{k-1}. \tag{4.5-22}$$

The derived measurement, $\mathbf{y}_k^*$, is defined as

$$\mathbf{y}_k^* = \mathbf{y}_{k+1} - \boldsymbol{\psi}_k \mathbf{y}_k. \tag{4.5-23}$$

In order to make the derivation clear, the time indices of all matrices and vectors are explicitly noted. Substituting $\mathbf{y}_{k+1}$ and $\mathbf{y}_k$ with their relationship to $\mathbf{x}_k$ and $\boldsymbol{\varepsilon}_k$ yields

$$\mathbf{y}_k^* = [\mathbf{H}_{k+1}\boldsymbol{\Phi}_{k+1,k} - \boldsymbol{\psi}_k\mathbf{H}_k]\mathbf{x}_k + \mathbf{H}_{k+1}\boldsymbol{\mu}_k + \mathbf{v}_k. \tag{4.5-24}$$

Let

$$\mathbf{H}_k^* = \mathbf{H}_{k+1}\boldsymbol{\Phi}_{k+1,k} - \boldsymbol{\psi}_k\mathbf{H}_k, \tag{4.5-25}$$

$$\mathbf{v}_k^* = \mathbf{H}_{k+1}\boldsymbol{\mu}_k + \mathbf{v}_k, \tag{4.5-26}$$

the derived measurement equation is

$$\mathbf{y}_k^* = \mathbf{H}_k^*\mathbf{x}_k + \mathbf{v}_k^*, \tag{4.5-27}$$

where $\mathbf{v}_k^* : \ \sim N(\mathbf{0}, \mathbf{R}_k^*)$ and

$$\mathbf{R}_k^* = \mathbf{H}_{k+1}\mathbf{Q}_k\mathbf{H}_{k+1}^T + \mathbf{R}_k. \tag{4.5-28}$$

Note that the derived measurement noise vector $\mathbf{v}_k^*$ is correlated with the process noise vector, $\boldsymbol{\mu}_k$, and has the following correlation

$$E\{\boldsymbol{\mu}_k\mathbf{v}_k^{*^T}\} = \mathbf{Q}_k\mathbf{H}_{k+1}^T. \tag{4.5-29}$$

In this case, the KF equation cannot be directly applied, so a derived system equation must be defined where the process noise term is uncorrelated with $\mathbf{v}_k^*$. We rewrite the system equation as

$$\mathbf{x}_{k+1} = \boldsymbol{\Phi}_{k+1,k}\mathbf{x}_k + \boldsymbol{\mu}_k + \boldsymbol{\Lambda}_k(\mathbf{y}_k^* - \mathbf{H}_k^*\mathbf{x}_k + \mathbf{v}_k^*), \tag{4.5-30}$$

where $\boldsymbol{\Lambda}_k$ can be any arbitrary $n \times m$ matrix because the vector in the parentheses is equal to $\mathbf{0}$. Rearranging terms yields

$$\mathbf{x}_{k+1} = [\boldsymbol{\Phi}_{k+1,k} - \boldsymbol{\Lambda}_k\mathbf{H}_k^*]\mathbf{x}_k + \boldsymbol{\Lambda}_k\mathbf{y}_k^* + \boldsymbol{\mu}_k^*, \tag{4.5-31}$$

where

$$\boldsymbol{\mu}_k^* = \boldsymbol{\mu}_k - \boldsymbol{\Lambda}_k\mathbf{v}_k^*$$

is the new process noise term. Before the expression for the covariance of $\boldsymbol{\mu}_k^*$, $\mathbf{Q}_k^*$ is derived, the KF assumes that $\boldsymbol{\mu}_k^*$ and $\mathbf{v}_k^*$ are uncorrelated. The correlation matrix is

$$E\{\boldsymbol{\mu}_k^*\mathbf{v}_k^{*^T}\} = \mathbf{Q}_k\mathbf{H}_{k+1}^T - \boldsymbol{\Lambda}_k\mathbf{R}_k^* = \mathbf{0}.$$

Therefore,

$$\mathbf{\Lambda}_k = \mathbf{Q}_k \mathbf{H}_{k+1}^T \mathbf{R}_k^{*^{-1}}. \tag{4.5-32}$$

We now derive the expression for $\mathbf{Q}_k^*$

$$\mathbf{Q}_k^* = Cov\{\mathbf{\mu}_k^*\} = Cov\{\mathbf{\mu}_k - \mathbf{\Lambda}_k(\mathbf{H}_{k+1}\mathbf{\mu}_k + \mathbf{v}_k)\} = \mathbf{Q}_k + \mathbf{\Lambda}_k \mathbf{H}_{k+1} \mathbf{Q}_k \mathbf{H}_{k+1}^T \mathbf{\Lambda}_k^T$$
$$+ \mathbf{\Lambda}_k \mathbf{R}_k \mathbf{\Lambda}_k^T - \mathbf{\Lambda}_k \mathbf{H}_{k+1} \mathbf{Q}_k - \mathbf{Q}_k \mathbf{H}_{k+1}^T \mathbf{\Lambda}_k^T.$$

Note that

$$\mathbf{\Lambda}_k \mathbf{H}_{k+1} \mathbf{Q}_k \mathbf{H}_{k+1}^T \mathbf{\Lambda}_k^T + \mathbf{\Lambda}_k \mathbf{R}_k \mathbf{\Lambda}_k^T = \mathbf{\Lambda}_k \mathbf{R}_k^* \mathbf{\Lambda}_k^T,$$

leading to

$$\mathbf{Q}_k^* = \mathbf{Q}_k + \mathbf{\Lambda}_k \mathbf{R}_k^* \mathbf{\Lambda}_k^T - \mathbf{\Lambda}_k \mathbf{H}_{k+1} \mathbf{Q}_k - \mathbf{Q}_k \mathbf{H}_{k+1}^T \mathbf{\Lambda}_k^T. \tag{4.5-33}$$

Similarly, with the definition of $\mathbf{\Lambda}_k$, $\mathbf{\Lambda}_k \mathbf{R}_k^* - \mathbf{Q}_k \mathbf{H}_{k+1}^T = \mathbf{0}$, the resulting expression for $\mathbf{Q}_k^*$ is

$$\mathbf{Q}_k^* = \mathbf{Q}_k - \mathbf{\Lambda}_k \mathbf{H}_{k+1} \mathbf{Q}_k. \tag{4.5-34}$$

To recap, the system and measurement equations for the colored measurement noise problems are given by

$$\mathbf{x}_{k+1} = [\mathbf{\Phi}_{k+1,k} - \mathbf{\Lambda}_k \mathbf{H}_k^*]\mathbf{x}_k + \mathbf{\Lambda}_k \mathbf{y}_k^* + \mathbf{\mu}_k^* \tag{4.5-35}$$

$$\mathbf{y}_{k+1}^* = \mathbf{H}_{k+1}^* \mathbf{x}_{k+1} + \mathbf{v}_{k+1}^*, \tag{4.5-36}$$

where

$$\mathbf{\mu}_k^*: \sim N(\mathbf{0}, \mathbf{H}_{k+1}\mathbf{Q}_k \mathbf{H}_{k+1}^T + \mathbf{R}_k), \tag{4.5-37}$$

$$\mathbf{v}_k^*: \sim N(\mathbf{0}, \mathbf{Q}_k - \mathbf{\Lambda}_k \mathbf{H}_{k+1} \mathbf{Q}_k), \tag{4.5-38}$$

and $\mathbf{\Lambda}_k$ is as defined in Equation (4.5-32). The KF for this case can be constructed, where the appearance of $\mathbf{y}_k^*$ in the system equation is treated as an input in the prediction process (to go from k to $k + 1$). Another note is that because $\mathbf{y}_k^*$ contains two consecutive measurements, it is necessary to wait for one cycle before initiating the KF.

The derived system and measurement equations suitable for KF application are summarized in Table 4.1.

Table 4.1 Derived System and Measurement Equations for KF Implementation when Measurement Noise Is Time Correlated

---

System and Measurement Equations Derived for KF Application when Measurement Noise Is a Correlated Time Sequence
The Original System and Measurement Equations

$\mathbf{x}_k = \mathbf{\Phi}_{k+1,k}\mathbf{x}_{k-1} + \mathbf{\mu}_{k-1}$,

$\mathbf{y}_k = \mathbf{H}_k\mathbf{x}_k + \mathbf{\varepsilon}_k$,

$\mathbf{\varepsilon}_k = \mathbf{\Psi}_k\mathbf{\varepsilon}_k + \mathbf{v}_{k-1}$.

Where $\mathbf{v}_{k-1}$ and $\mathbf{\mu}_{k-1}$ are mutually independent and uncorrelated time sequences with Gaussian density, zero mean and known covariances.

The Derived System and Measurement Equations Ready for KF Implementation

$\mathbf{x}_{k+1} = [\mathbf{\Phi}_{k+1,k} - \mathbf{\Lambda}_k\mathbf{H}_k^*]\mathbf{x}_k + \mathbf{\Lambda}_k\mathbf{y}_k^* + \mathbf{\mu}_k^*$,

$\mathbf{y}_{k+1}^* = \mathbf{H}_{k+1}^*\mathbf{x}_{k+1} + \mathbf{v}_{k+1}^*$,

Where $\mathbf{\mu}_k^* : \sim N(0, \mathbf{H}_{k+1}\mathbf{Q}_k\mathbf{H}_{k+1}^T + \mathbf{R}_k)$, $\mathbf{v}_k^* : \sim N(0, \mathbf{Q}_k - \mathbf{\Lambda}_k\mathbf{H}_{k+1}\mathbf{Q}_k)$, and $\mathbf{\Lambda}_k = \mathbf{Q}_k\mathbf{H}_{k+1}^T\mathbf{R}_k^{*-1}$.

Remarks

1. Many issues remain: unknown correlation for colored noise, ψ_k, and uncertain colored noise structure, that is, when the real correlation model is not ψ_k.

2. Colored noise is, in a sense, a time-varying bias. Biases among multiple sensors constitute one of the most difficult problems for multiple sensor fusion.

4.6 Systems with Both Unknown System Inputs and Measurement Biases

The SKF that accounts for both unknown system inputs and measurement biases is derived here. This derivation is based on [17, pp. 281–286]. The state and measurement equations accounting for both types of unknowns are modeled as

$$\mathbf{x}_k = \mathbf{\Phi}_{k,k-1}\mathbf{x}_{k-1} + \mathbf{\varphi}_{k,k-1}\mathbf{d}_{k-1} + \mathbf{\mu}_{k-1}, \tag{4.6-1}$$

$$\mathbf{d}_k = \mathbf{d}_{k-1}, \tag{4.6-2}$$

$$\mathbf{b}_k = \mathbf{b}_{k-1}, \tag{4.6-3}$$

$$\mathbf{y}_k = \mathbf{H}_k\mathbf{x}_k + \mathbf{H}_k^b\mathbf{b}_k + \mathbf{v}_k, \tag{4.6-4}$$

where $\mathbf{d}_k$ is the unknown driving vector, assumed to be an unknown constant with a priori distribution $\mathbf{d}_0$: $\sim N$ $(\mathbf{d}, \mathbf{D})$ and where $\mathbf{b}_k$ is the residual bias, assumed to be an unknown constant with a priori distribution $\mathbf{b}_0$: $\sim N$ $(\mathbf{b}, \mathbf{B})$. $\boldsymbol{\varphi}_{k,k-1}$ defines the relationship between the driving vector and the state; and $\mathbf{H}_k^b$ defines the relationship between the residual bias and the measurement vector. Without losing generality, the means for both driving vector and measurement bias are set to $\mathbf{0}$, that is, $\mathbf{d} = \mathbf{0}$ and $\mathbf{b} = \mathbf{0}$ in the ensuring derivation. In solving the estimation problem, one can simply apply the KF by jointly estimating $\mathbf{x}_k$, $\mathbf{d}_k$, and $\mathbf{b}_k$. This can be done through state augmentation. We define the augmented state vector as $\mathbf{x}_k^a = [\mathbf{x}_k, \mathbf{d}_k, \mathbf{b}_k]^T$, which leads to the following state and measurement equations

$$\mathbf{x}_k^a = \begin{bmatrix} \boldsymbol{\Phi}_{k,k-1} & \boldsymbol{\varphi}_{k,k-1} & \mathbf{0} \\ \mathbf{0} & \mathbf{I} & \mathbf{0} \\ \mathbf{0} & \mathbf{0} & \mathbf{I} \end{bmatrix} \mathbf{x}_{k-1}^a + \begin{bmatrix} \boldsymbol{\mu}_{k-1} \\ \mathbf{0} \\ \mathbf{0} \end{bmatrix}, \tag{4.6-5}$$

$$\mathbf{y}_k = \begin{bmatrix} \mathbf{H}_k, \mathbf{0}, \mathbf{H}_k^b \end{bmatrix} \mathbf{x}_k^a + \mathbf{v}_k. \tag{4.6-6}$$

where $\mathbf{I}$ and $\mathbf{0}$ are identity and zero matrices, respectively, with proper dimensions. A KF can be constructed for this system. The covariance of the initial estimates for $\mathbf{d}_0$ and $\mathbf{b}_0$ are set to be $\mathbf{D}$ and $\mathbf{B}$, their a priori knowledge. This approach is straightforward in theory. Should one choose to take this approach, as emphasized in Section 4.4.2, some process noise for both $\mathbf{d}_k$ and $\mathbf{b}_k$ should be added to prevent filter divergence due to the buildup of overconfidence in models[3] for $\mathbf{d}_k$ and $\mathbf{b}_k$.

With concern for excessive computational burden due to increased state dimension, an alternative approach, initially developed by Schmidt [20], is to take into account the effect of the uncertainties on degrading the state estimate, without actually estimating the parameters themselves. This is done by designing suboptimal estimators, $\hat{\mathbf{d}}_{k|k}^*$ and $\hat{\mathbf{b}}_{k|k}^*$, and by ignoring the estimates for $\mathbf{d}_k$ and $\mathbf{b}_k$, while forcing the estimators to satisfy the following conditions

$$\hat{\mathbf{d}}_{k|k}^* \equiv \mathbf{0}, \tag{4.6-7}$$

$$\hat{\mathbf{b}}_{k|k}^* \equiv \mathbf{0}, \tag{4.6-8}$$

and

$$Cov\{\hat{\mathbf{d}}_{k|k}^*\} \equiv \mathbf{D}, \tag{4.6-9}$$

$$Cov\{\hat{\mathbf{b}}_{k|k}^*\} \equiv \mathbf{B}. \tag{4.6-10}$$

---

3   In reality, unknown parameters do not always stay as constants.

Given the above constraints on the a posteriori information, the covariance of the augmented state vector is

$$
\mathbf{P}_{k|k}^a = \begin{bmatrix} \mathbf{P}_{k|k} & \mathbf{C}_{k|k}^d & \mathbf{C}_{k|k}^b \\ \mathbf{C}_{k|k}^{d^T} & \mathbf{D} & \mathbf{0} \\ \mathbf{C}_{k|k}^{b^T} & \mathbf{0} & \mathbf{B} \end{bmatrix},
$$

(4.6-11)

where

$$
\mathbf{C}_{k|k}^d = E\{[\mathbf{x}_k - \hat{\mathbf{x}}_{k|k}^*]\mathbf{d}_k^T\},
$$

(4.6-12)

$$
\mathbf{C}_{k|k}^b = E\{[\mathbf{x}_k - \hat{\mathbf{x}}_{k|k}^*]\mathbf{b}_k^T\},
$$

(4.6-13)

and where $\hat{\mathbf{x}}_{k|k}^*$ is the suboptimal state estimator with the substitution of $\hat{\mathbf{d}}_{k|k}^*$ and $\hat{\mathbf{b}}_{k|k}^*$ for the estimates for $\mathbf{d}_k$ and $\mathbf{b}_k$ in the original augumented filter. The zero matrices in $\mathbf{P}_{k|k}^a$ are due to the assumption that $\mathbf{d}_k$ and $\mathbf{b}_k$ are uncorrelated. Also due to the fact that $Cov\{\hat{\mathbf{d}}_{k|k}^*\}$ and $Cov\{\hat{\mathbf{b}}_{k|k}^*\}$ are forced to remain constant, it is only necessary to compute the evolution of submatrices $\mathbf{P}_{k|k}$, $\mathbf{C}_{k|k}^d$, and $\mathbf{C}_{k|k}^b$.

Using

$$
\mathbf{P}_{k|k-1}^a = \begin{bmatrix} \mathbf{\Phi}_{k,k-1} & \mathbf{\varphi}_{k,k-1} & \mathbf{0} \\ \mathbf{0} & \mathbf{I} & \mathbf{0} \\ \mathbf{0} & \mathbf{0} & \mathbf{I} \end{bmatrix} \begin{bmatrix} \mathbf{P}_{k-1|k-1} & \mathbf{C}_{k-1|k-1}^d & \mathbf{C}_{k-1|k-1}^b \\ \mathbf{C}_{k-1|k-1}^{d^T} & \mathbf{D} & \mathbf{0} \\ \mathbf{C}_{k-1|k-1}^{b^T} & \mathbf{0} & \mathbf{B} \end{bmatrix} \begin{bmatrix} \mathbf{\Phi}_{k,k-1} & \mathbf{\varphi}_{k,k-1} & \mathbf{0} \\ \mathbf{0} & \mathbf{I} & \mathbf{0} \\ \mathbf{0} & \mathbf{0} & \mathbf{I} \end{bmatrix}^T
$$
$$
+ \begin{bmatrix} \mathbf{Q}_{k-1} & \mathbf{0} & \mathbf{0} \\ \mathbf{0} & \mathbf{0} & \mathbf{0} \\ \mathbf{0} & \mathbf{0} & \mathbf{0} \end{bmatrix},
$$

(4.6-14)

with the fact that $\hat{\mathbf{d}}_{k|k}^*$ and $\hat{\mathbf{b}}_{k|k}^*$ are forced to be zero, one obtains the following one-step prediction equations after some manipulation

$$
\hat{\mathbf{x}}_{k|k-1}^* = \mathbf{\Phi}_{k,k-1}\hat{\mathbf{x}}_{k-1|k-1}^*,
$$

(4.6-15)

$$
\mathbf{P}_{k|k-1}^* = \mathbf{\Phi}_{k,k-1}\mathbf{P}_{k-1|k-1}^*\mathbf{\Phi}_{k,k-1}^T + \mathbf{Q}_{k-1} + \mathbf{\Phi}_{k,k-1}\mathbf{C}_{k-1|k-1}^d\mathbf{\varphi}_{k,k-1}^T
$$
$$
+ \mathbf{\varphi}_{k,k-1}\mathbf{C}_{k-1|k-1}^{d^T}\mathbf{\Phi}_{k,k-1}^T + \mathbf{\varphi}_{k,k-1}\mathbf{D}\mathbf{\varphi}_{k,k-1}^T,
$$

(4.6-16)

$$
\mathbf{C}_{k|k-1}^d = \mathbf{\Phi}_{k,k-1}\mathbf{C}_{k-1|k-1}^d + \mathbf{\varphi}_{k,k-1}\mathbf{D},
$$

(4.6-17)

$$
\mathbf{C}_{k|k-1}^b = \mathbf{\Phi}_{k,k-1}\mathbf{C}_{k-1|k-1}^b.
$$

(4.6-18)

Applying the KF update equations yields the following familiar form for the state update equation

$$\hat{\mathbf{x}}_{k|k}^* = \hat{\mathbf{x}}_{k|k-1}^* + \mathbf{K}_k^a (\mathbf{y}_k - \mathbf{H}_k \hat{\mathbf{x}}_{k|k-1}^*), \qquad (4.6\text{-}19)$$

where $\mathbf{K}_k^a$ is the Kalman gain for the augmented system

$$\mathbf{K}_k^a = \mathbf{P}_{k|k-1}^a \mathbf{H}_k^{a^T} \left[\mathbf{H}_k^a \mathbf{P}_{k|k-1}^a \mathbf{H}_k^{a^T} + \mathbf{R}_k \right]^{-1}, \qquad (4.6\text{-}20)$$

and $\mathbf{H}_k^a = [\mathbf{H}_k, \mathbf{0}, \mathbf{H}_k^b]$ is the augmented measurement matrix. Substituting the expression for $\mathbf{H}_k^a$ into the above equation, using the individual submatrices for $\mathbf{P}_{k|k-1}^a$, and after some algebraic manipulation, we obtain the following set of filter update equations

$$\mathbf{K}_k^a = [\mathbf{P}_{k|k-1}^* \mathbf{H}_k^T + \mathbf{C}_{k|k-1}^b \mathbf{H}_k^{b^T}]\Sigma_k^{-1}, \qquad (4.6\text{-}21)$$

$$\Sigma_k = \mathbf{H}_k \mathbf{P}_{k|k-1}^* \mathbf{H}_k^T + \mathbf{R}_k + \mathbf{H}_k^b \mathbf{C}_{k|k-1}^{b^T} \mathbf{H}_k^T + \mathbf{H}_k \mathbf{C}_{k|k-1}^b \mathbf{H}_k^{b^T} + \mathbf{H}_k^b \mathbf{B} \mathbf{H}_k^{b^T}, \qquad (4.6\text{-}22)$$

$$\mathbf{C}_{k|k}^d = \mathbf{C}_{k|k-1}^d - \mathbf{K}_k^a \mathbf{H}_k \mathbf{C}_{k|k-1}^d, \qquad (4.6\text{-}23)$$

$$\mathbf{C}_{k|k}^b = \mathbf{C}_{k|k-1}^b - \mathbf{K}_k^a \left[\mathbf{H}_k \mathbf{C}_{k|k-1}^b + \mathbf{H}_k^b \mathbf{B} \right], \qquad (4.6\text{-}24)$$

$$\mathbf{P}_{k|k}^* = \left[\mathbf{I} - \mathbf{K}_k^a \mathbf{H}_k \right] \mathbf{P}_{k|k-1}^* \left[\mathbf{I} - \mathbf{K}_k^a \mathbf{H}_k \right]^T + \mathbf{K}_k^a \mathbf{R}_k \mathbf{K}_k^{a^T}, \qquad (4.6\text{-}25)$$

where Σ_k is the covariance of the filter residual process. The SKF derived above is summarized in Table 4.2.

Remarks

1. Setting $\varphi_{k,k-1}$ and $\mathbf{D}$ to zero yields the SKF of Section 4.5.2 for residual bias.

2. Recall the target models studied in Homework Problem 1, Chapter 2, and the discussion of Section 4.3 on mismatched systems. The trajectory of a maneuvering object can be modeled with constant acceleration (CA) while an object traveling in a straight line can be modeled with constant velocity (CV). KF built with a CV model can be made to work on a maneuvering target by one of the following two methods: (a) select a process noise covariance that is representative of the target maneuvering level and (b) use the SKF for unknown driving force as shown above.[4] Method (a) is a well-known approach used by practitioners for decades. Method (b), although a subcase of the SKF presented in [17, 20, 21], was republished in 2004 as a reduced state estimator [24]. A paper by Urbano et al. in 2012 [25] pointed out that Method (a) and Method (b) can be tuned to be equivalent when running in steady state.

3. When a target is constantly maneuvering, the filter built with a CA model is a closer representation of the actual object dynamics than the filter with a CV model. The SKF is using a CV model and not estimating the unknown driver (the acceleration in this case), but taking into account its impact through degradation in covariance. While saving computation by not estimating the unknown parameters, it is also suboptimal because it is not estimating the unknown parameter.

4. In reality, objects do not stay in either CV or CA mode. An algorithm to detect when a change occurs and adapt accordingly is the subject of Section 4.7. The approach of building two filters, CV and CA, and allowing the filter outputs to be combined using hypothesis probabilities is the subject of the next chapter, multiple model estimation algorithms (MMEA).

4.7   Systems with Abrupt Input Changes

There is a class of system uncertainty that takes the form of abrupt changes in the system input. In application problems this could be due to a sudden system failure [26–30], or targets making unexpected maneuvers [31–33], among other reasons. When such changes occur, the system designer would like to (a) detect the changes and (b) estimate the changes. The likelihood ratio test is a common statistical technique to detect unknown system failures or changes (hypothesis H_1) against the normal condition (hypothesis H_0) [27]. After the failures or changes are detected, the changes of the parameters in the system can be estimated based on the measurements after the detection. The likelihood function with the estimated parameters is called the generalized likelihood function. The generalized likelihood ratio (GLR) test is designed to solve the detection and estimation problem at the same time. The GLR is used for this problem as shown in [26–33]. The GLR algorithm presented in [32] is derived below.

For the purpose of easy reference, the original estimation problem is restated below. Consider the discrete linear system

$$\mathbf{x}_k = \mathbf{\Phi}_{k,k-1}\mathbf{x}_{k-1} + \mathbf{\mu}_{k-1},$$

$$\mathbf{y}_k = \mathbf{H}_k\mathbf{x}_k + \mathbf{v}_k,$$

where $\mathbf{x}_0 : \sim N(\hat{\mathbf{x}}_0, \mathbf{P}_0)$, $\mathbf{\mu}_k: \sim N(\mathbf{0}, \mathbf{Q}_k)$, $\mathbf{v}_k: \sim N(\mathbf{0}, \mathbf{R}_k)$. The KF algorithm is restated here (see Section 2.6)

---

4   The same equation as above by setting the measurement bias to zero.

Table 4.2   The Schmidt–Kalman Filter for Simultaneous System and Measurement Biases

---

Schmidt–Kalman Filter
System and Measurement Equations

$$\mathbf{x}_k = \boldsymbol{\Phi}_{k,k-1}\mathbf{x}_{k-1} + \boldsymbol{\varphi}_{k,k-1}\mathbf{d}_{k-1} + \boldsymbol{\mu}_{k-1},$$

$$\mathbf{d}_k = \mathbf{d}_{k-1},$$

$$\mathbf{b}_k = \mathbf{b}_{k-1},$$

$$\mathbf{y}_k = \mathbf{H}_k\mathbf{x}_k + \mathbf{H}_k^b\mathbf{b}_k + \mathbf{v}_k,$$

where $\mathbf{d}_k$ is an unknown constant vector with $\mathbf{d}_0$: $\sim N\,(\mathbf{d}, \mathbf{D})$ and $\mathbf{b}_k$ is an unknown constant residual bias vector with a priori distribution $\mathbf{b}_0$: $\sim N\,(\mathbf{b}, \mathbf{B})$. Without losing generality, $\mathbf{d}$ and $\mathbf{b}$ are set to $\mathbf{0}$.

Goal
Estimating $\mathbf{x}_k$ without computing augmented state vector (i.e., without having to estimate $\mathbf{d}_k$ and $\mathbf{b}_k$ together with the state vector) but with the impact of $\mathbf{d}_k$ and $\mathbf{b}_k$ accounted for in the covariance matrix.

Estimation Algorithm
Prediction

$$\hat{\mathbf{x}}_{k|k-1}^* = \boldsymbol{\Phi}_{k,k-1}\hat{\mathbf{x}}_{k-1|k-1}^*,$$

$$\mathbf{P}_{k|k-1}^* = \boldsymbol{\Phi}_{k,k-1}\mathbf{P}_{k-1|k-1}^*\boldsymbol{\Phi}_{k,k-1}^T + \mathbf{Q}_{k-1} + \boldsymbol{\Phi}_{k,k-1}\mathbf{C}_{k-1|k-1}^d\boldsymbol{\varphi}_{k,k-1}^T + \boldsymbol{\varphi}_{k,k-1}\mathbf{C}_{k-1|k-1}^{d^T}\boldsymbol{\Phi}_{k,k-1}^T + \boldsymbol{\varphi}_{k,k-1}\mathbf{D}\boldsymbol{\varphi}_{k,k-1}^T,$$

$$\mathbf{C}_{k|k-1}^d = \boldsymbol{\Phi}_{k,k-1}\mathbf{C}_{k-1|k-1}^d + \boldsymbol{\varphi}_{k,k-1}\mathbf{D},$$

$$\mathbf{C}_{k|k-1}^b = \boldsymbol{\Phi}_{k,k-1}\mathbf{C}_{k-1|k-1}^b.$$

Update

$$\hat{\mathbf{x}}_{k|k}^* = \hat{\mathbf{x}}_{k|k-1}^* + \mathbf{K}_k^a(\mathbf{y}_k - \mathbf{H}_k\hat{\mathbf{x}}_{k|k-1}^*),$$

$$\mathbf{K}_k^a = [\mathbf{P}_{k|k-1}^*\mathbf{H}_k^T + \mathbf{C}_{k|k-1}^b\mathbf{H}_k^{b^T}]\boldsymbol{\Sigma}_k^{-1},$$

$$\boldsymbol{\Sigma}_k = \mathbf{H}_k\mathbf{P}_{k|k-1}^*\mathbf{H}_k^T + \mathbf{R}_k + \mathbf{H}_k^b\mathbf{C}_{k|k-1}^{b^T}\mathbf{H}_k^T + \mathbf{H}_k\mathbf{C}_{k|k-1}^b\mathbf{H}_k^{b^T} + \mathbf{H}_k^b\mathbf{B}\mathbf{H}_k^{b^T},$$

$$\mathbf{C}_{k|k}^d = \mathbf{C}_{k|k-1}^d - \mathbf{K}_k^a\mathbf{H}_k\mathbf{C}_{k|k-1}^d,$$

$$\mathbf{C}_{k|k}^b = \mathbf{C}_{k|k-1}^b - \mathbf{K}_k^a\left[\mathbf{H}_k\mathbf{C}_{k|k-1}^b + \mathbf{H}_k^b\mathbf{B}\right],$$

$$\mathbf{P}_{k|k}^* = [\mathbf{I} - \mathbf{K}_k^a\mathbf{H}_k]\mathbf{P}_{k|k-1}^*[\mathbf{I} - \mathbf{K}_k^a\mathbf{H}_k]^T + \mathbf{K}_k^a\mathbf{R}_k\mathbf{K}_k^{a^T}.$$

where notations $\hat{\mathbf{x}}_{k|k-1}^*$, $\mathbf{P}_{k|k-1}^*$, $\hat{\mathbf{x}}_{k|k}^*$, and $\mathbf{P}_{k|k}^*$ are used to indicate that they are SKF estimates and covariances.

$$\hat{\mathbf{x}}_{k|k-1} = \boldsymbol{\Phi}_{k,k-1}\hat{\mathbf{x}}_{k-1|k-1},$$

$$\hat{\mathbf{x}}_{k|k} = \hat{\mathbf{x}}_{k|k-1} + \mathbf{K}_k\boldsymbol{\gamma}_k,$$

$$\boldsymbol{\gamma}_k = \mathbf{y}_k - \mathbf{H}_k\hat{\mathbf{x}}_{k|k-1},$$

where

$$\mathbf{K}_k = \mathbf{P}_{k|k-1}\mathbf{H}_k^T\left[\mathbf{H}_k\mathbf{P}_{k|k-1}\mathbf{H}_k^T + \mathbf{R}_k\right]^{-1},$$

$$\mathbf{P}_{k|k-1} = \boldsymbol{\Phi}_{k,k-1}\mathbf{P}_{k-1|k-1}\boldsymbol{\Phi}_{k,k-1}^T + \mathbf{Q}_{k-1},$$

$$\mathbf{P}_{k|k} = \mathbf{P}_{k|k-1} - \mathbf{P}_{k|k-1}\mathbf{H}_k^T\left[\mathbf{H}_k\mathbf{P}_{k|k-1}\mathbf{H}_k^T + \mathbf{R}_k\right]^{-1}\mathbf{H}_k\mathbf{P}_{k|k-1},$$

$$\boldsymbol{\Gamma}_k = Cov\{\boldsymbol{\gamma}_k\} = \mathbf{H}_k\mathbf{P}_{k|k-1}\mathbf{H}_k^{\mathrm{T}} + \mathbf{R}_k.$$

Assume that a system change occurred at time k_0, then the residual process is no longer zero mean, that is,

$$\tilde{\boldsymbol{\gamma}}_k = \boldsymbol{\gamma}_k + \mathbf{b}_{k,k_0}, \tag{4.7-1}$$

where $\tilde{\boldsymbol{\gamma}}_k$ is the residual process after changes have occurred, and $\mathbf{b}_{k,k_0}$ is the bias in the residual process that is zero when $k < k_0$. Four types of sudden changes are considered in [30] and include step or impulse inputs to the system or to the measurement device. For example, consider the sudden input change to the system,

$$\mathbf{x}_k = \boldsymbol{\Phi}_{k,k-1}\mathbf{x}_{k-1} + \boldsymbol{\mu}_{k-1} + \mathbf{B}_k\boldsymbol{\delta}_{k,k_0}, \tag{4.7-2}$$

where $\boldsymbol{\delta}_{k,k_0} = \mathbf{0}$ for $k < k_0$ and $\mathbf{B}_k$ is a known input matrix. All four types of system changes considered in [30] can be represented in general form

$$\mathbf{b}_{k,k_0} = \mathbf{G}_{k,k_0}\boldsymbol{\delta}_{k,k_0}.$$

Specific expressions for $\mathbf{G}_{k,k_0}$ of the four types of system changes are given in [27–28]. The GLR algorithm is stated below without specific conditions placed on $\mathbf{G}_{k,k_0}$. Consider the case that $\boldsymbol{\delta}_{k,k_0}$ becomes an unknown constant vector when $k \geq k_0$. Assuming k_0 is given, then $\boldsymbol{\delta}_{k,k_0}$ is estimated, see [32–33], using

$$\hat{\boldsymbol{\delta}}_{k,k_0} = \mathbf{S}_{k,k_0}^{-1}\mathbf{d}_{k,k_0}, \tag{4.7-3}$$

and

$$\mathbf{S}_{k,k_0} = \sum_{n=k_0}^{k}\mathbf{G}_{n,k_0}^T\boldsymbol{\Gamma}_n^{-1}\mathbf{G}_{n,k_0}, \tag{4.7-4}$$

$$\mathbf{d}_{k,k_0} = \sum_{n=k_0}^{k}\mathbf{G}_{n,k_0}^T\boldsymbol{\Gamma}_n^{-1}\tilde{\boldsymbol{\gamma}}_n, \tag{4.7-5}$$

where the estimate of k_0, $\hat{k}_0$, is obtained by selecting the k_0 maximizing the log likelihood ratio ℓ_{k,k_0}

$$\ell_{k,k_0} = \mathbf{d}_{k,k_0}^T \mathbf{S}_{k,k_0}^{-1} \mathbf{d}_{k,k_0} = \mathbf{d}_{k,k_0}^T \hat{\boldsymbol{\delta}}_{k,k_0}. \tag{4.7-6}$$

The sufficient statistics of the GLR is $\ell_{k,\hat{k}_0}$ where $\hat{k}_0$ is obtained by selecting the k_0 that gives the maximum of ℓ_{k,k_0}. It is compared with a threshold λ, and a sudden change at $\hat{k}_0$ is declared when $\ell_{k,\hat{k}_0} > \lambda$. The threshold λ is selected to meet certain detection and false alarm requirements.

Remarks

1. The above input estimation problem assumes that the unknown input vector, $\boldsymbol{\delta}_{k,k_0}$, is a constant, that is, it is a step function with unknown magnitude. The estimation algorithm can be simplified as a KF estimating a constant vector $\boldsymbol{\delta}_{k,k_0}$ with $\tilde{\boldsymbol{\gamma}}_k$ as measurements. The derivation is straightforward. The interested reader can use it as a homework problem or consult [32–33] for details.

2. The input vector can also be time-varying with known transition matrix, such as

$$\boldsymbol{\delta}_{k,k_0} = \boldsymbol{\Phi}_{k,k-1}^{\delta} \boldsymbol{\delta}_{k-1,k_0}.$$

3. It can also be shown that the estimate for $\boldsymbol{\delta}_{k,k_0}$ is again a KF with $\tilde{\boldsymbol{\gamma}}_k$ as measurements.

Once a sudden change is declared, there are two alternatives in obtaining the corrected state estimate.

1. Augment the state vector with $\boldsymbol{\delta}_{k,k_0}$. This results in a larger state dimension, an undesirable situation when computational resource is limited.

2. Continue to estimate $\hat{\boldsymbol{\delta}}_{k,k_0}$ as before, but use $\hat{\boldsymbol{\delta}}_{k,k_0}$ to correct the otherwise biased state estimate. This algorithm is stated below without derivation. The interested reader can consult [32] for details.

In the following, $\hat{\mathbf{x}}_{k|k}$ is used to denote the corrected estimate, and $\tilde{\mathbf{x}}_{k|k}$ is the estimate obtained by assuming $\boldsymbol{\delta}_{k,k_0}$ is zero.

$$\hat{\mathbf{x}}_{k|k} = \tilde{\mathbf{x}}_{k|k} + \mathbf{A}_k \hat{\boldsymbol{\delta}}_{k-1,k_0}, \tag{4.7-7}$$

$$\mathbf{A}_k = [\mathbf{I} - \mathbf{K}_k \mathbf{H}_k][\boldsymbol{\Phi}_{k,k-1}\mathbf{A}_{k-1}\boldsymbol{\Phi}_{k,k-1}^{\delta}{}^{-1} + \mathbf{B}_{k-1}], \tag{4.7-8}$$

$$\hat{\boldsymbol{\delta}}_{k,k_0} = \boldsymbol{\Phi}_{k,k-1}^{\delta}\hat{\boldsymbol{\delta}}_{k-1,k_0} + \mathbf{K}_k^{\delta}\left(\tilde{\boldsymbol{\gamma}}_k - \mathbf{G}_{k,k_0}\boldsymbol{\Phi}_{k,k-1}^{\delta}\hat{\boldsymbol{\delta}}_{k-1,k_0}\right), \tag{4.7-9}$$

$$\mathbf{K}_k^{\delta} = \mathbf{P}_{k|k-1}^{\delta} \mathbf{G}_{k,k_0}^T \left[\mathbf{G}_{k,k_0} \mathbf{P}_{k|k-1}^{\delta} \mathbf{G}_{k,k_0}^T + \mathbf{\Gamma}_k \right]^{-1}. \tag{4.7-10}$$

The covariance of $\hat{\boldsymbol{\delta}}_{k,k_0}$ evolves as

$$\mathbf{P}_{k|k-1}^{\delta} = \mathbf{\Phi}_{k,k-1}^{\delta} \mathbf{P}_{k-1|k-1}^{\delta} \mathbf{\Phi}_{k,k-1}^{\delta}{}^T, \tag{4.7-11}$$

$$\mathbf{P}_{k|k}^{\delta} = \mathbf{P}_{k|k-1}^{\delta} - \mathbf{P}_{k|k-1}^{\delta} \mathbf{G}_{k,k_0}^T \left[\mathbf{G}_{k,k_0} \mathbf{P}_{k|k-1}^{\delta} \mathbf{G}_{k,k_0}^T + \mathbf{\Gamma}_k \right]^{-1} \mathbf{G}_{k,k_0} \mathbf{P}_{k|k-1}^{\delta}. \tag{4.7-12}$$

The covariance of $\hat{\mathbf{x}}_{k|k}$ is now

$$\mathbf{P}_{k|k} = \mathbf{P}_{k|k-1} - \mathbf{P}_{k|k-1} \mathbf{H}_k^T \left[\mathbf{H}_k \mathbf{P}_{k|k-1} \mathbf{H}_k^T + \mathbf{R}_k \right]^{-1} \mathbf{H}_k \mathbf{P}_{k|k-1} + \mathbf{A}_k \mathbf{P}_{k-1|k-1}^{\delta} \mathbf{A}_k^T. \tag{4.7-13}$$

The above algorithm treats the sudden input change, $\boldsymbol{\delta}_{k,k_0}$, as a deterministic system. An experienced algorithm designer will always add some process noise covariance to the $\mathbf{P}_{k|k-1}^{\delta}$ equation to prevent the filter from becoming overconfident. It cannot be emphasized too strongly that adding some process noise covariance is a sound practice for most filter applications.

The above algorithm defines two cascaded Kalman filters, where the estimator for the original state vector is running with the assumption that system input is zero, while the second filter is the change estimator using the residual of the first filter as the measurement. These two estimates are then combined to obtain the final, corrected state estimate.

Due to the fact that the detection process requires estimating $\hat{k}_0$, the above algorithm must be implemented with a growing bank of filters and is computationally intensive. A couple of alternatives come to mind.

1. Always try to estimate input deviations regardless of whether it is expected or not. This results in system overmodeling (see last chapter) in which the overall estimate is noisier than when no system input deviations are modeled.

2. Use a multiple model filter (MMF) to encompass the case with and without input changes. It avoids the need for an explicit detection process, and consequently avoids the apparent transient after detection. However, this approach is also computationally more demanding. Exactly which approach to choose is application dependent and the choice of the system designer. The MMF is the subject of the next chapter.

Algorithms for detecting and estimating the sudden onset of an unknown constant input are summarized in Table 4.3.

Table 4.3 Summary of Sudden Input Detection and Estimation Algorithm

Algorithm to Detect and Estimate Sudden Constant Input
System and Measurement Equations

$$\mathbf{x}_k = \mathbf{\Phi}_{k,k-1}\mathbf{x}_{k-1} + \mathbf{\mu}_{k-1} + \mathbf{B}_k\mathbf{\delta}_{k,k_0},$$

$$\mathbf{y}_k = \mathbf{H}_k\mathbf{x}_k + \mathbf{v}_k,$$

where $\mathbf{x}_0 : \sim N(\hat{\mathbf{x}}_0, \mathbf{P}_0)$, $\mathbf{\mu}_k : \sim N(\mathbf{0}, \mathbf{Q}_k)$, $\mathbf{v}_k : \sim N(\mathbf{0}, \mathbf{R}_k)$, $\mathbf{\delta}_{k,k_0} = \mathbf{0}$ for $k < k_0$ and is an unknown constant vector when $k \geq k_0$. A KF has been running for $k < k_0$. When $k \geq k_0$, the state estimate becomes biased and the bias appears in the residual process as

$$\tilde{\mathbf{\gamma}}_k = (\mathbf{y}_k - \mathbf{H}_k\hat{\mathbf{x}}_{k|k-1}) + \mathbf{G}_{k,k_0}\mathbf{\delta}_{k,k_0},$$

where $\text{Cov}\{\tilde{\mathbf{\gamma}}_k\} = \mathbf{H}_k\mathbf{P}_{k|k-1}\mathbf{H}_k^\mathrm{T} + \mathbf{R}_k$ and $\mathbf{G}_{k,k_0}$ is a known transformation relating input vector to bias in residual.

Problem Statement

1. Detect the onset of $\mathbf{\delta}_{k,k_0}$, which is the same as estimating k_0.
2. Estimate the magnitude of $\mathbf{\delta}_{k,k_0}$.

The GLR Detector

The estimate of k_0, $\hat{k}_0$ is obtained by selecting the k_0 maximizing the generalized log likelihood ratio ℓ_{k,k_0}

$$\ell_{k,k_0} = \mathbf{d}_{k,k_0}^T \mathbf{S}_{k,k_0}^{-1} \mathbf{d}_{k,k_0} = \mathbf{d}_{k,k_0}^T \hat{\mathbf{\delta}}_{k,k_0},$$

where

$$\hat{\mathbf{\delta}}_{k,k_0} = \mathbf{S}_{k,k_0}^{-1} \mathbf{d}_{k,k_0},$$

$$\mathbf{S}_{k,k_0} = \sum_{n=k_0}^{k} \mathbf{G}_{n,k_0}^T \mathbf{\Gamma}_n^{-1} \mathbf{G}_{n,k_0},$$

$$\mathbf{d}_{k,k_0} = \sum_{n=k_0}^{k} \mathbf{G}_{n,k_0}^T \mathbf{\Gamma}_n^{-1} \tilde{\mathbf{\gamma}}_n.$$

The State Estimator with Input Correction

Let $\tilde{\mathbf{x}}_{k|k}$ be the output of a KF assuming $\mathbf{\delta}_{k,k_0}$ is zero and $\hat{\mathbf{x}}_{k|k}$ is the state estimate with input bias correction

$$\hat{\mathbf{x}}_{k|k} = \tilde{\mathbf{x}}_{k|k} + \mathbf{A}_k\hat{\mathbf{\delta}}_{k-1,k_0},$$

$$\hat{\mathbf{\delta}}_{k,k_0} = \mathbf{\Phi}_{k,k-1}^\delta\hat{\mathbf{\delta}}_{k-1,k_0} + \mathbf{K}_k^\delta\left(\tilde{\mathbf{\gamma}}_k - \mathbf{G}_{k,k_0}\mathbf{\Phi}_{k,k-1}^\delta\hat{\mathbf{\delta}}_{k-1,k_0}\right),$$

$$\mathbf{A}_k = [\mathbf{I} - \mathbf{K}_k\mathbf{H}_k]\left[\mathbf{\Phi}_{k,k-1}\mathbf{A}_{k-1}\mathbf{\Phi}_{k,k-1}^{\delta}{}^{-1} + \mathbf{B}_{k-1}\right],$$

$$\mathbf{K}_k^\delta = \mathbf{P}_{k|k-1}^\delta\mathbf{G}_{k,k_0}^T\left[\mathbf{G}_{k,k_0}\mathbf{P}_{k|k-1}^\delta\mathbf{G}_{k,k_0}^T + \mathbf{\Gamma}_k\right]^{-1}.$$

The covariance of $\hat{\mathbf{\delta}}_{k,k_0}$ is

$$\mathbf{P}_{k|k}^\delta = \mathbf{P}_{k|k-1}^\delta - \mathbf{P}_{k|k-1}^\delta\mathbf{G}_{k,k_0}^T\left[\mathbf{G}_{k,k_0}\mathbf{P}_{k|k-1}^\delta\mathbf{G}_{k,k_0}^T + \mathbf{\Gamma}_k\right]^{-1}\mathbf{G}_{k,k_0}\mathbf{P}_{k|k-1}^\delta,$$

$$\mathbf{P}_{k|k-1}^\delta = \mathbf{\Phi}_{k,k-1}^\delta\mathbf{P}_{k-1|k-1}^\delta\mathbf{\Phi}_{k,k-1}^{\delta}{}^T,$$

The covariance of $\hat{\mathbf{x}}_{k|k}$ is

$$\mathbf{P}_{k|k} = \mathbf{P}_{k|k-1} - \mathbf{P}_{k|k-1}\mathbf{H}_k^T\left[\mathbf{H}_k\mathbf{P}_{k|k-1}\mathbf{H}_k^\mathrm{T} + \mathbf{R}_k\right]^{-1}\mathbf{H}_k\mathbf{P}_{k|k-1} + \mathbf{A}_k\mathbf{P}_{k-1|k-1}^\delta\mathbf{A}_k^T.$$

4.8   Ill-Conditioning and False Observability

Numerical problems are the major concern for filter design; in most cases the computed error covariance converges (that seems to imply the system is observable) and yet the state estimation error diverges. This is because in the following state update equation

$$\hat{\mathbf{x}}_{k|k} = \hat{\mathbf{x}}_{k|k-1} + \mathbf{K}_k \left(\mathbf{y}_k - \mathbf{H}_k \hat{\mathbf{x}}_{k|k-1} \right) \tag{4.8-1}$$

the filter gain $\mathbf{K}_k = \mathbf{P}_{k|k-1} \mathbf{H}_k^T \left[\mathbf{H}_k \mathbf{P}_{k|k-1} \mathbf{H}_k^T + \mathbf{R}_k \right]^{-1}$ is proportional to the predicted error covariance $\mathbf{P}_{k|k-1}$, if the computation of $\mathbf{P}_{k|k-1}$, has numerical difficulties such as the need to invert ill-conditioned matrices; for example, in a nearly singular matrix $\left[\mathbf{H}_k^T \mathbf{R}_k^{-1} \mathbf{H}_k + \mathbf{P}_{k|k-1}^{-1} \right]$, the filter gain will not provide a proper projection of the residual for the state update.

As shown in the filter equation derivation, Equation (4.8-1) is equivalent to solving a set of simultaneous linear equations

$$\left[\mathbf{H}_k^T \mathbf{R}^{-1} \mathbf{H}_k + \mathbf{P}_{k|k-1}^{-1} \right] \hat{\mathbf{x}}_{k|k} = \mathbf{H}_k^T \mathbf{R}_k^{-1} \mathbf{y}_k + \mathbf{P}_{k|k-1}^{-1} \hat{\mathbf{x}}_{k|k-1} \tag{4.8-2}$$

for $\hat{\mathbf{x}}_{k|k}$ as for $\mathbf{x}$ of $\mathbf{Ax} = \mathbf{b}$. If we have numerical difficulties to invert $\left[\mathbf{H}_k^T \mathbf{R}^{-1} \mathbf{H}_k + \mathbf{P}_{k|k-1}^{-1} \right]$, the solution of Equation (4.8-2) will not be correct.

Remarks

1. The observability condition in Equation (4.8-1) and the ill-conditioning condition in Equation (4.8-2) are the same. For linear systems, one common practice is to decouple the observable states from unobservable states in the covariance matrix for the gain computation to help in estimating the observable states in Equation (4.8-1), and leaving the unobservable state without measurement update.

2. For a nonlinear system, the observability condition is dependent on the true state as derived in Section 3.9.2. The perceived observability condition in (4.8-1) is dependent on the estimated state through the linearized matrices used in EKF. The decoupling idea may not be easy to implement for a nearly observable system. In [34], several filter coordinate systems were presented for different decoupled EKFs for radar tracking application to avoid ill-conditioned covariance matrices.

4.8.1   False Observability in Radar Tracking Applications

Sometimes, the observability condition for a nonlinear system is not sufficient to guarantee the EKF will converge. A phenomenon has been documented in [34] for long-range radar tracking applications, and is called false cross-range observability. This effect is due to overoptimism on the EKF's covariance matrices computation based on linearized matrices at the estimated states for a weakly observable nonlinear system. Hence, the covariance matrices will converge faster than they should, and the EKF will clamp down the filter gains in Equation (4.8-1), causing the filter to ignore the diverging measurement residuals. The overoptimism comes from the coupling between the observable (range) and unobservable (cross-range) states through the dynamic and measurement equations. The effect is particularly bothersome during the track initialization stage because the initial state of the EKF can be very inaccurate and Jacobian matrices will create an incorrect filter gain in Equation (4.8-1) and direct the filter to an inaccurate direction for the state update, especially when the dynamic system is almost deterministic (i.e., with no or low process noise). This filter behavior is called false observability in [34]. Similar phenomena are observed in other filter designs as overconfidence or particle collapse in particle filtering in Chapter 6.

The false observability problem is discussed next using an example given by [35], and a fix for the radar ballistic object tracking application is recommended. Two simulated ballistic object trajectories are considered in the paper. Time histories of the true target range for a nonorbital ballistic object (e.g., a missile, denoted the near object) and an orbital ballistic object (e.g., a satellite, denoted the far object) are shown in Figure 4.3.

A radar typically has measurements much more accurate in range than angles (or cross-range). Therefore, the cross-range measurements are much poorer for objects at long range than those at close range. The EKF's capability to accurately estimate the cross-range components of the position vector from a sequence of accurate range measurements depends on the known relationship between range and cross-range and their derivatives through the system and measurement equations. The better the velocity estimate, the more accurately the cross-range estimate can be obtained given the ballistic trajectory constraint.

Unfortunately, early in the track when the velocity estimation error is large, the cross coupling between the range and cross-range components in the error covariance matrix (both in position and velocity) can provide a false sense of observability due to the linearization about the estimated trajectory in the EKF. This can give rise to a false sense of direction, resulting in overly optimistic cross-range estimates (false

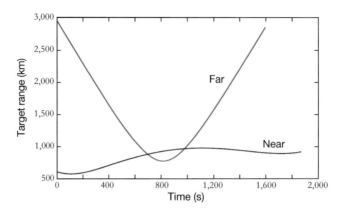

Figure 4.3   True target range for a missile (the near object) and a satellite (the far object).

observability). Therefore, in many EKF designs [34], it is preferable to decouple the range and cross-range dimensions in the error covariance computation during the early track stage to avoid this effect. However, the EKF is then no longer able to benefit from the coupling effect to estimate the cross-range dimensions using accurate range measurement sequences. Over time, due to the assumed target dynamics, the range and cross-range estimates become statistically dependent as a function of the state estimate. However, since the EKF is a suboptimal algorithm, the Kalman gain computed using Equation (4.8-1) tends to over-interpret the statistical dependence between the range and cross-range components of position and velocity because the higher-order terms in the Taylor expansion of the system dynamic and measurement equations are ignored. This peculiar nonlinear effect becomes more severe as the "flattening ratio" (defined as the ratio of the range and cross-range measurement standard deviations) of the position measurement error ellipsoid becomes smaller. How to take advantage of the range-cross-range coupling in the covariance and gain matrix computations while avoiding the false cross-range observability effect in the initial stage and when to switch to a fully coupled filter is an interesting EKF design problem.

4.8.2   Quasi-Decoupling Filter

A method for mitigating the false cross-range observability was originally proposed in [34]. It consists of a quasi-decoupling of range and cross-range components of the predicted covariance matrices when the estimated velocity uncertainty becomes too large. Here, the qualifier *quasi* refers to the fact that the predicted covariance is decoupled, that is, the cross covariance entries of range and cross-range are zeroed only for the computation of the Kalman gain, $\mathbf{K}_k$, as given in Equation (4.8-1), whereas the updated covariance is computed utilizing the following fully coupled version of the Joseph symmetric form [12]

$$\mathbf{P}_{k|k} = [\mathbf{I} - \mathbf{K}_k\mathbf{H}_k]\mathbf{P}_{k|k-1}[\mathbf{I} - \mathbf{K}_k\mathbf{H}_k]^T + \mathbf{K}_k\mathbf{R}_k\mathbf{K}_k^T. \tag{4.8-3}$$

The Joseph symmetric form is the valid representation of the updated covariance for any $\mathbf{K}_k$. It should be used to avoid numerical instabilities rather than Equation (3.4-9) where the Kalman gain was computed with fully coupled covariance, $\mathbf{P}_{k|k-1}$.

It is important that quasi-decoupling be employed when the uncertainty in the velocity estimate is deemed to be large. So how large is large? It can be shown that quasi-decoupling should be applied whenever the following inequality is satisfied

$$\sqrt{\frac{\mathrm{tr}\left[\mathbf{P}_{k|k-1}(4{:}6,\,4{:}6)\right]}{\left\|\hat{\mathbf{v}}_{k|k-1}\right\|^2 - \hat{r}_{k|k-1}^2}} \geq \frac{\sigma_r}{\hat{r}_{k|k-1}\sigma_\theta} \tag{4.8-4}$$

where $\mathrm{tr}[\mathbf{P}_{k|k-1}(4{:}6,\,4{:}6)]$ denotes the trace of the velocity component of the predicted covariance matrix, $\mathbf{P}_{k|k-1}$; $\hat{\mathbf{v}}_{k|k-1}$ is the predicted velocity vector valid at time index k; $\hat{r}_{k|k-1}$, and $\hat{r}_{k|k-1}^2$ are the predicted range and range rate estimate valid at time index k, respectively; and σ_r and σ_θ are the range and angle measurement standard deviation, respectively.

In [35], time histories for the RMS error in estimating the position and velocity components of the target state computed from 10 Monte Carlo trials of trajectories in Figure 4.3 are shown in Figure 4.4. The range measurement standard deviation was assumed to be 3 m, while the azimuth and elevation measurement standard deviations were assumed to be 2.5 mrad each. The results from employing the quasi-decoupling algorithm described above are compared with the results obtained from when false cross-range observability is not accounted for in the implementation of the EKF. It is clear that there is a significant reduction in the estimation error for both the near (missile) and far (satellite) targets—hence, an increase in state estimation accuracy—when the quasi-decoupling algorithm is employed in the formulation of the EKF. Note that the vertical axes of these figures are in logarithmic scales, and the fully coupled portion of the trajectories will experience less fluctuation without the switching effect of the qausi-coupling tests.

Covariance consistency is typically determined using the so-called NEES and NIS metrics (see Section 4.2). The NEES and NIS time histories for the examples considered here are shown in Figure 4.5. Similar to Figure 4.4, the NEES and NIS are computed by averaging over 10 Monte Carlo trials. The regions where the error covariance

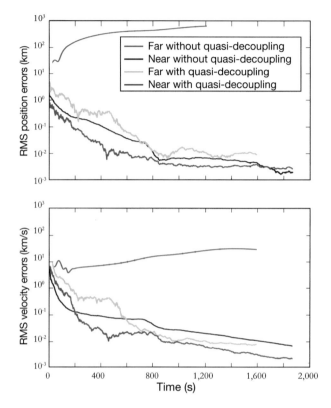

Figure 4.4   Comparison of the RMS errors in position (top) and velocity (bottom) estimates obtained from 10 Monte Carlo trials with/without the quasi-decoupling algorithm.

estimates are deemed *consistent* are shown in gray. These correspond to 95% (two-sided) probability regions for a χ^2 random variable. NEES and NIS values above the gray region correspond to optimistic estimates of the state covariance, while NEES and NIS values below the gray region correspond to pessimistic estimates of the state covariance. Again, the application of the quasi-decoupling algorithm makes the estimated state covariance consistent. When quasi-decoupling is not applied, false cross-range observability can result in the estimated covariance being overly optimistic. The effect is particularly adverse for objects that are farther away (the satellite example in this case) because the measurement error ellipsoid becomes even flatter for objects that are farther away.

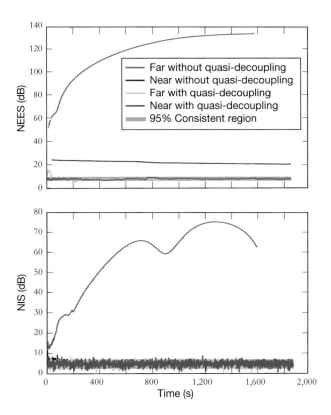

Figure 4.5 Comparison of the NEES (top) and NIS (bottom) obtained from 10 Monte Carlo trials with/without the quasi-decoupling algorithm.

For optimal performance, the quasi-decoupling algorithm of [34] discussed above should be applied whenever the inequality in Equation (4.8-4) is satisfied. For a majority of applications, this inequality is satisfied only during the first few track updates, and subsequently, no decoupling of the state covariance is needed. This is mainly due to the fact that early in track, target state estimates are in general very poor, particularly in the velocity dimension. Of course, at exactly what point in time the decoupling of the initial track update no longer becomes necessary depends on the target trajectory and viewing geometry. In [35], a performance comparison between the EKF applying the quasi-decoupling algorithm based on the inequality in Equation (4.8-4) with the performance of the EKF applying the quasi-decoupling algorithm only during the first six track updates (denoted as a greedy decoupling) is shown in

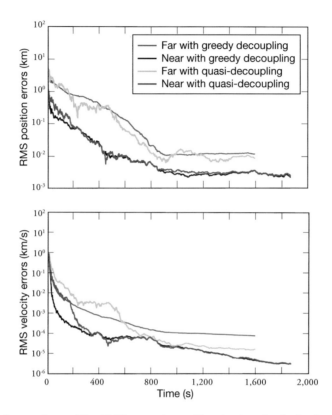

Figure 4.6   Comparison of the RMS errors in position (top) and velocity (bottom) estimates obtained from 10 Monte Carlo trials with the quasi-decoupling algorithm and greedy decoupling algorithm.

Figures 4.6 and 4.7. It is expected that for both versions of EKFs the missile example (i.e., the near object) demonstrates better performance than those obtained for the satellite (the far object). For both trajectories, the performance is better when the quasi-decoupling algorithm is applied based on a continual examination of the inequality in Equation (4.8-4).

4.9   Numerical Examples for Practical Filter Design

In the last section of this chapter, two numerical examples are presented. The first one is the continuation of the sinusoidal signal in noise example presented in Section 1.8

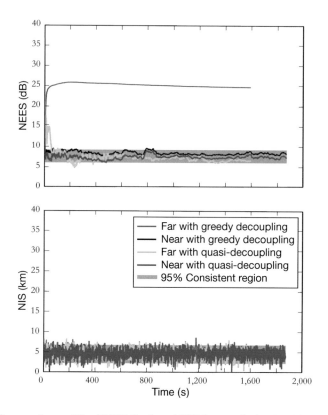

Figure 4.7   Comparison of the NEES (top) and NIS (bottom) obtained from 10 Monte Carlo trials with quasi-decoupling algorithm and greedy decoupling algorithm.

and Section 2.11. The second example is a comparison of three approaches in treating a constant measurement bias with known bias statistics. The purpose of these examples is to illustrate some of the filter design considerations described in this chapter.

4.9.1   Sinusoidal Signal in Noise

It is well known that the solution of the ordinary differential equation

$$\ddot{x} = -\omega^2 x,$$

is the sine wave

$$x = a \sin(\omega t + \varphi)$$

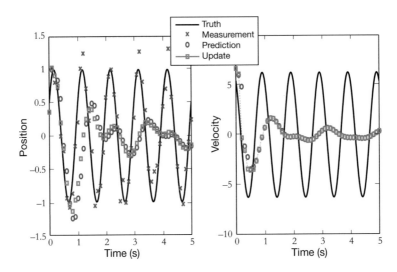

Figure 4.8   Position (x) and velocity ($\dot{x}$) estimates of a mismatched Kalman filter, with $\sigma_q^2 = 0$.

where the phase angle φ defines the initial condition $x_0 = a \sin φ$. The above differential equation can be rewritten in terms of a state vector equation

$$\dot{\mathbf{x}} = \begin{bmatrix} 0 & 1 \\ -\omega^2 & 0 \end{bmatrix} \mathbf{x},$$

where $\mathbf{x} = [x_1, x_2]^T$, with $x_1 = x$ and $x_2 = \dot{x}$. Note that if ω is a known constant, the state equation above is a linear system without process noise and the KF can be constructed as shown in Section 2.11 for $\sigma_q^2 = 0$. If ω is unknown, the KF designer may choose to select a value for ω denoted here as ω_0. When $\omega_0 \neq \omega$, the KF built based upon the assumed value ω_0 is a mismatched filter as discussed in Section 4.3. Filtering results for the two-state KF with $\omega_0 = \pi$ and the process noise $\sigma_q^2 = 0$ are shown in Figure 4.8. Recall the example in Section 1.8 that the true frequency is $\omega = 2\pi$. Because the process noise is set to zero, the filter gain converges to a very small value the same way as shown in Figure 2.6. After about 2 s, the filter updates are no longer influenced by the new measurement. The updated states fall very close to the predicted state, in which the prediction was based upon an incorrectly assumed frequency. This is commonly known as filter divergence or overconfidence. The filter overconfidence issue caused by model mismatch can sometimes be corrected by adding process noise covariance in the filter prediction process, as this will prevent the filter gains from converging to

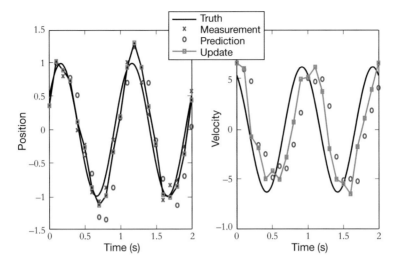

Figure 4.9 Position (x) and velocity ($\dot{x}$) estimates of a mismatched Kalman filter, with $\sigma_q^2 = 250$.

zero, see Figure 2.6. A process noise variance of $\sigma_q^2 = 250$ is added in this example and the result is shown in Figure 4.9. Note that the updated filter states are influenced by the measurements and agree well with the true states. Exactly how to choose the value for σ_q^2 is a black art in KF design, because the true value of the parameter is unknown.

Since adding a process noise covariance improves the filter performance, the next logical question would be whether there is something that can improve performance even more. One alternative is to estimate ω instead of assuming a value for ω. This can be done by augmenting the state vector with ω, as suggested in Section 4.4.1. For simplicity, the new state variable representing frequency is defined as $x_3 = \omega^2$. Using Equation (2.11-2), the augmented state equation becomes

$$\dot{\mathbf{x}} = \begin{bmatrix} 0 & 1 & 0 \\ -x_3 & 0 & 0 \\ 0 & 0 & 0 \end{bmatrix} \mathbf{x}, \tag{4.9-1}$$

where $\mathbf{x} = [x, \dot{x}, \omega^2]^T$.

This is clearly a nonlinear estimation problem because x_3 is multiplicative with the state vector $\mathbf{x}$. In this case, the EKF will be applied to estimate all three state variables. Equation (4.9-1) is a noise-free system. As noted earlier in this chapter, it is always prudent for the filter designer to add some process noise covariance to avoid filter divergence, which is also known as filter overconfidence. To do so, Equation (4.9-1) is modified to

$$\dot{\mathbf{x}} = \mathbf{A}\mathbf{x} + \mathbf{B}\xi, \tag{4.9-2}$$

where $\mathbf{A} = \begin{bmatrix} 0 & 1 & 0 \\ -x_3 & 0 & 0 \\ 0 & 0 & 0 \end{bmatrix}$, $\mathbf{B} = \begin{bmatrix} 0 \\ 0 \\ 1 \end{bmatrix}$, and ξ is the additive process noise term to $\dot{x}_3$. For Equation (4.9-2), a closed form solution for $\mathbf{\Phi}_{t_k,t_{k-1}}$ exists, which is

$$\mathbf{\Phi}_{t_k,t_{k-1}} = \begin{bmatrix} \cos\left(\Delta t \sqrt{\hat{x}_3}\right) & \dfrac{\sin\left(\Delta t \sqrt{\hat{x}_3}\right)}{\sqrt{\hat{x}_3}} & \dfrac{\hat{x}_1\left(\cos\left(\Delta t \sqrt{\hat{x}_3}\right)-1\right)}{\hat{x}_3} \\[2mm] -\sqrt{\hat{x}_3}\cdot\sin\left(\Delta t \sqrt{\hat{x}_3}\right) & \cos\left(\Delta t \sqrt{\hat{x}_3}\right) & \dfrac{-\hat{x}_1\sin\left(\Delta t \sqrt{\hat{x}_3}\right)}{\sqrt{\hat{x}_3}} \\[2mm] 0 & 0 & 1 \end{bmatrix} \tag{4.9-3}$$

where $\Delta t = t_k - t_{k-1}$ and $\hat{x}_1, \hat{x}_3$ are the last update of x_1, x_3 at time step t_{k-1}. The equivalent process noise covariance for the discrete system is obtained by applying Equation (2.2-9)

$$\mathbf{Q}_{k-1} = \sigma_q^2 \Delta t \begin{bmatrix} \dfrac{\hat{x}_1^2\left(\sin\left(2\Delta t\sqrt{\hat{x}_3}\right)-8\sin\left(\Delta t\sqrt{\hat{x}_3}\right)+6\Delta t\sqrt{\hat{x}_3}\right)}{4\left(\sqrt{\hat{x}_3}\right)^5} & \dfrac{2\hat{x}_1^2\sin^4\left(\dfrac{\Delta t\sqrt{\hat{x}_3}}{2}\right)}{\hat{x}_3^2} \\[4mm] \dfrac{2\hat{x}_1^2\sin^4\left(\dfrac{\Delta t\sqrt{\hat{x}_3}}{2}\right)}{\hat{x}_3^2} & \dfrac{\hat{x}_1^2\left(2\Delta t\sqrt{\hat{x}_3}-\sin\left(2\Delta t\sqrt{\hat{x}_3}\right)\right)}{4\left(\sqrt{\hat{x}_3}\right)^3} \\[4mm] \dfrac{\hat{x}_1\left(\sin\left(\Delta t\sqrt{\hat{x}_3}\right)-\Delta t\sqrt{\hat{x}_3}\right)}{\left(\sqrt{\hat{x}_3}\right)^3} & \dfrac{\hat{x}_1\left(\cos\left(\Delta t\sqrt{\hat{x}_3}\right)-1\right)}{\hat{x}_3} \\[4mm] \dfrac{\hat{x}_1\left(\sin\left(\Delta t\sqrt{\hat{x}_3}\right)-\Delta t\sqrt{\hat{x}_3}\right)}{\left(\sqrt{\hat{x}_3}\right)^3} \\[4mm] \dfrac{\hat{x}_1\left(\cos\left(\Delta t\sqrt{\hat{x}_3}\right)-1\right)}{\hat{x}_3} \\[4mm] \Delta t \end{bmatrix} \tag{4.9-4}$$

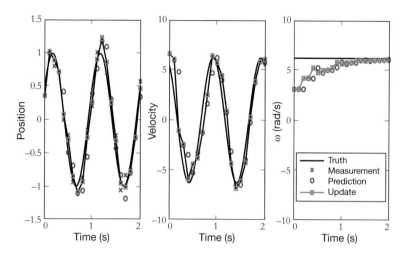

Figure 4.10 Position (x), velocity ($\dot{x}$), and frequency (ω) estimates of Equation (2.11-1), for $\sigma_q^2 = 0.3$.

Sampled measurements are taken on the first element of $\mathbf{x}$, x_1, that is,

$$y_k = \mathbf{H}\mathbf{x}_k + v_k, \tag{4.9-5}$$

where $\mathbf{H} = [1\ 0\ 0]$ and v_k is white measurement noise with distribution $N(0, \sigma_v^2)$. An extended KF can be constructed using the equations shown in Table 3.1 in Section 3.4.

The numerical results for this EKF design are presented in Figures 4.10 and 4.11. All these results are for a signal-to-noise ratio (SNR) of 6.667 (or, equivalently, noise variance $\sigma_v^2 = 0.0225$) using exactly the same measurements used in Section 1.8. The parameter values for the sine wave are $a = 1$, $\omega = 2\pi$, and $\varphi = \pi/6$. Figure 4.10 shows a single Monte Carlo run of 2 s of observation at a 10 Hz sampling rate with $\sigma_q^2 = 0.3$. The solid curve is the true values and the data points are measurements (red crosses), predicted (blue circles) estimates, and updated (green squares) estimates. Compared with Figure 4.9, it can be seen that the three-state EKF performs better than the two-state KF with assumed (incorrectly in this case) frequency and large process noise. With only a very modest level of process noise covariance, $\sigma_q^2 = 0.3$ for the three-state EKF, the estimate of ω converges to a value very close to the true value after about 1 s. The reader should vary different levels of σ_q^2 to see the trade-off of filter convergence rate and the filter random error.

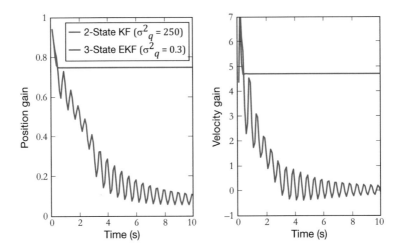

Figure 4.11   Filter gains for position and velocity comparison with a two-state KF with $\sigma_q^2 = 250$ against a 3-state EKF with $\sigma_q^2 = 0.3$.

Figure 4.11 shows the filter gains in position and velocity for the two-state KF (Section 2.11) with $\sigma_q^2 = 250$ and the three-state EKF with $\sigma_q^2 = 0.3$ (from this section). Note that the filter gain for the three-state EKF quickly converges to a much smaller value and in the mean time is able to provide estimates with smaller bias and random errors.

Figure 4.12 presents a comparison of the CRBs for the three-state nonlinear system from Equation (3.11-4), the root of averaged filter-computed covariance, and the RMS errors using Monte Carlo samples, for both two-state KF and three-state EKF. This is for a data window of 10 s with 1000 Monte Carlo repetitions. Note that the CRB for the two-state KF and the three-state EKF are the same because the frequency ω is unknown for both cases. The Monte Carlo RMS errors and the filter-computed covariance for the three-state EKF are very close to the CRB and much lower than those of the two-state KF with larger process noise.

The filter consistency issue is further examined using the averaged NEES [16]. NEES of both filters are computed and presented in Figure 4.13. As in Figure 2.8, the 95% and 99.99% regions are shown with color codes. The NEES of the two-state mismatched KF is oscillatory and overly optimistic.[5] The three-state EKF is generally

---

5   *Optimistic* indicates the actual error is larger than filter covariance, which can cause leakage. *Pessimistic* indicates the actual error is smaller than filter covariance, which can cause false alarms due to noise or interfering signals.

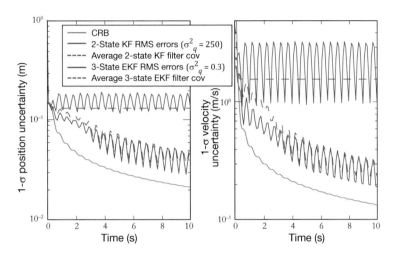

Figure 4.12   Comparison of Monte Carlo RMS errors (red/blue solid) and filter computed standard deviations (red/blue dashed) in position and velocity for two-state KF and three-state EKF.

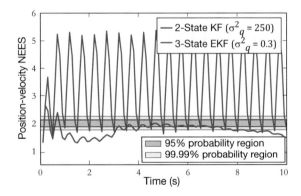

Figure 4.13   Covariance consistence check using average NEES of a two-state KF and a three-state EKF.

within the 99.99% region but slightly pessimistic initially and at the end. This is because of the inclusion of the process noise covariance σ_q^2, which makes the filter computed covariance larger. The above results indicate that a three-state EKF in estimating the frequency is much more accurate and consistent than the two-state KF, even with large process noise.

The fix interval smoother (FIS) presented in Section 2.9.2 is now applied to the sine problem. The results are shown in Figures 4.14 through 4.17. The run conditions such as true parameter values, SNR, and process noise variance are identical to the EKF case considered before. The result of a single Monte Carlo run, corresponding to the results of Figure 4.10, is shown in Figure 4.14. Comparing Figure 4.11 to Figure 4.14, it is clear that the FIS gives much more accurate estimates. This is expected because the smoother is using all data in the time window for all estimates. As shown in Section 2.9.2, the FIS can be implemented as the weighted sum of the forward filtered and one-step backward predicted estimates. The resulting covariance for FIS is the inverse of the sum of the inverses of the covariance of forward and backward estimators, and is lower than the lower of the two. This is illustrated in Figure 4.15.

Figure 4.16 presents the comparison of the root of CRB, filter computed covariance, and RMS errors computed with Monte Carlo samples. The results for forward filter and smoother are shown for comparison. It can be seen that the smoother

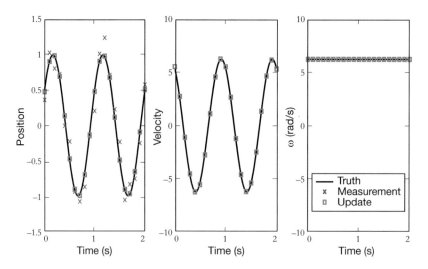

Figure 4.14   Position (x), velocity ($\dot{x}$) and frequency (ω) estimates of Equation (2.11-1) with an FIS over EKF for $\sigma_q^2 = 0.3$.

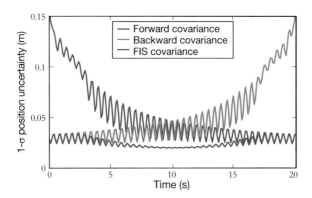

Figure 4.15 Illustration of the smoother's covariance (combined covariance above) as inverse of the sum of inverse of the covariance of forward and backward estimators.

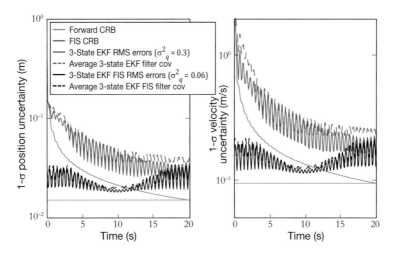

Figure 4.16 Comparison of Monte Carlo RMS errors, filter computed standard deviations, and CRB for the forward filter and the CRB for FIS in position and velocity, all for the three-state nonlinear system.

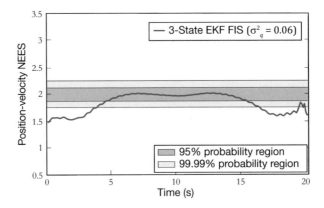

Figure 4.17   Covariance consistence check using average NEES for the three-state FIS of EKF.

produces more accurate estimates than the filters as expected. Better accuracy of smoothed estimates is obtained with the help of data from the future, a luxury that is usually not available for time-critical operations. A smoother is often used to produce the so-called BET. BET is usually computed after a data collection exercise as a reference to ground truth.

Figure 4.17 shows the NEES of the FIS for the sine wave estimation using EKF. The curve falls within the consistency bound for most of the time and is slightly pessimistic during the beginning and the end of the time window.

Summary of Sine Wave Estimation Study

An example of estimating a sine wave based upon noisy measurements has been illustrated using techniques learned from Chapters 1 to 4. The sine wave is defined as $x_k = a\sin(\omega t_k + \varphi)$ with sampled noisy measurements y_k

$$y_k = x_k + v_k = a\sin(\omega t_k + \varphi) + v_k \text{ for } k = 1, 3 \dots N$$

where v_k is the measurement noise.

Four approaches have been introduced for solving this problem and are summarized below.

1. **Weighted Least Squares Estimator (Chapter 1)**

   The three unknown parameters (a, ω, φ) are constants and their relationship to measurements is nonlinear. With Gaussian measurement noise, the weighted least square estimator is theoretically the best estimator to use. The results of this estimator are, however, highly unstable due to the fact that the performance index has many local minimums and the suggested algorithm often fails to find the global minimum. In the case when ω is known, that is, the unknown parameters are limited to (a, φ), the estimator performed well: unbiased and efficient (approaching CRB). The performance index (the weighted norm in measurement space) is periodic with respect to both ω and φ. The phase angle φ can be constrained to be within 2π but ω cannot. This makes the performance index highly ambiguous with periodic behavior with respect to ω. Even when the observability matrix is nonsingular, such ambiguity causes the system to be only locally observable, that is, any of the local peaks of the performance index can satisfy the necessary condition for optimality. The iterative algorithm suggested in Section 1.6 can easily lead the solution to a local minimum. The estimation error becomes very large when one attempts to estimate all three parameters simultaneously.

2. **Kalman Filter (Chapters 2, 4)**

The sine wave is a solution to a linear differential equation $\ddot{x} = -\omega^2 x$ for given ω and x_0 where $x_0 = a\sin\varphi$. A state model can be constructed for this differential equation (DE), which is $\dot{\mathbf{x}} = \begin{bmatrix} 0 & 1 \\ -\omega^2 & 0 \end{bmatrix}\mathbf{x}$, where $\mathbf{x} = [x_1, x_2]^T$ with $x_1 = x$ and $x_2 = \dot{x}$.

A KF build with this system works well when ω is known (Chapter 2). When ω is not known (Chapter 4), then the KF is no longer matched with the true system, and the filter diverges. This filter can be made to work by adding substantial process noise covariance, which is an often-used technique to prevent filter divergence, but the performance is poor.

3. **Extended Kalman Filter (EKF from Chapter 3, example shown in Chapter 4)**

In this case, ω^2 is augmented as the third state, x_3, and the state equation is now $\dot{\mathbf{x}} = \begin{bmatrix} 0 & 1 & 0 \\ -x_3 & 0 & 0 \\ 0 & 0 & 0 \end{bmatrix}\mathbf{x}$. This filter performs very well, and is much more robust than the two-state KF with process noise. The estimation error is close to CRB, and the estimator is nearly consistent.

4. **Extended Kalman Filter with Fixed Interval Smoother (Extension of Chapter 2)**

FIS is applied with EKF by combining the forward filtered and the backward one-step predicted estimates. It was shown that the result of this approach is more accurate than using EKF alone. The FIS is most suitable for post-mission analysis because it produces the best estimate with all data. It is, however, not suitable for real-time operations.

General Observations and Conclusions   Based upon this series of studies, the following broad ranging observations are warranted.

1. A batch filter is theoretically the best, but its realization depends on the robustness of the chosen algorithm when the performance index is ambiguous.

2. A KF works well when there are no unknown parameters. One should always be cautious when applying a KF to a model that may be mismatched with the true system.

3. It is better to augment the unknown parameters and estimate them together with the states even when this makes the estimation problem nonlinear. Although EKF is only an approximated solution to nonlinear estimation, the sine wave estimation example clearly shows that estimating unknown parameters using EKF can achieve very good performance.

4. FIS is very useful when the application is not for real-time application.

Possible Additional Studies on Sine Wave Estimation

There are at least two more approaches that are not explored here, but are left as homework problems for the readers. They are described below.

1. **A Sampling Approach**

    One issue with the particular WLS estimator algorithm of Chapter 1 is that the search direction and step size may cause the subsequent iteration to jump over the optimum solution to a local minimum. An approach to avoid this trap is to exhaustively sample the performance index space with very fine grid of (a, ω, φ). This is a brute force approach. It should work well for sufficiently high SNR. Sufficiency in this case means noise samples will not cause sidelobes of the performance index to become lower than the main valley where the optimal solution resides.

2. **The Classical Approach**

    Take the fast Fourier transform (FFT) over the data. The location of the global minimum and the magnitude of the main valley are estimated as ω and a, respectively. Estimate of the phase angle φ that can be obtained by fitting a linear curve through the phase angles of the frequency samples. Constraints of the sampling filter should be equally applicable here. This approach may not perform as well as the sampling approach, because it is limited by the sensor update rate, but it should be more computationally efficient.

4.9.2   Comparison of Methods in Treating Constant Unknown Biases

The performance of three filters with sensor measurement bias are compared in this section. The bias is an unknown constant with a known covariance. These three filters are (a) KF with a CV model ignoring the measurement bias, (b) the same KV with inflated measurement noise covariance where the amount of inflation is commensurate with the bias covariance, and (c) SKF.

A maneuvering trajectory is used for this example. This is the same trajectory that will be used for examples in Chapters 5 and 7. A top view of a target trajectory in (x, y) Cartesian coordinates (think of it as an airplane flying at a constant altitude) is shown in Figure 4.18. The true trajectory is shown with the black curve. The sensor is a radar making measurements in the range and angle space, the same as illustrated in the Homework Problem 1 of Chapter 3. For a 2D case, the radar only measures one angle. The trajectory corresponding to radar measurements with a 1° bias is shown with the red curve. For illustration purposes, the measured trajectory shown in this figure only contains angular bias error. This target has made two turns, which takes place during $t = 30$ to 45 s and 60 to 72 s. In each of the turns, the radial maneuver force is 5 g. For a target traveling in air (i.e., not a ballistic object), the maneuvers can be simulated by bank-to-turn, or in a simplified model, turning on a circle with a constant radius. The solution to Homework Problem 5 of Chapter 3 will provide the answer on how to simulate objects turning in a circle. Outside of these two time

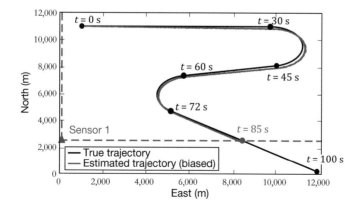

Figure 4.18   Truth trajectory of a maneuvering target (black) and the trajectory corresponding to biased measurements (red). Target maneuverings are taking place during turns.

intervals, the object is flying in a straight line. The sensor location and its field of view (FOV) are also shown. Note that the target is outside of the sensor FOV when it has gone beyond 85 s.

This sensor is taking measurements at a 10 Hz rate with measurement noise standard deviations $\sigma_r = 5$ m, $\sigma_\theta = 1$ mrad. Note that the angle measurement bias of $1°$ (17 mrad) is much bigger than the measurement noise standard deviation. The large bias value is chosen to show the differences in different filtering approaches. Although CV is a straight-line model while the target is maneuvering, a KF built with a CV model can be made to track maneuvering targets. The key is in the choice of process noise covariance, similar to the sinusoidal signal filtering study shown in Section 4.9.1. For a CV filter to track a 5 g target, the process noise covariance is selected with spectral density of 6,000 m^2/s^3, which is equivalent to a discrete-time process noise covariance of 60,000 m^2/s^4 for 0.1 s update period. More on the comparison of various filters in tracking maneuvering targets will be discussed in Chapter 5. Here we only examine the sensor measurement bias compensation problem.

The performances of these three filters are shown in Figures 4.19 and 4.20. Filter computed covariances in position are shown in Figure 4.19. The 1 σ position error is the square root of the trace of the position portion of the total covariance. Recall that the trace of a matrix is the same as the sum of eigenvalues of a matrix where eigenvalues are the semi-major and minor axes of a 1 σ ellipsoid associated with the matrix, thus the square root of the trace of a covariance matrix is a scalar representation of the

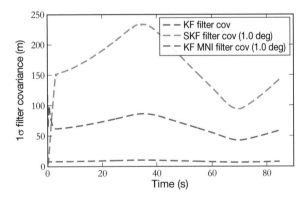

Figure 4.19 Comparison of position estimation covariances of the three filters. The 1 σ covariance is computed as the square-root of the trace of the position components of the covariance.

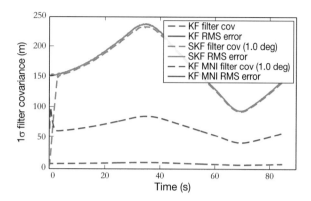

Figure 4.20 Comparison of RMS position estimation errors with the filter computed covariances of the three filters.

total errors in a vector space. The three curves correspond to the three filters: KF, SKF, and KF with measurement noise inflation (MNI) where the amount of inflation is 1° or 17 mrad. The peak and valley of the SKF covariance curve reflects the impact of bias. For example, the target is farthest away from the sensor at about 30 s of the trajectory, thus with the largest position error due to a fixed bias and the SKF covariance peaks at the same time. Similarly, the trajectory is closest to the sensor at about 70 s corresponding to the minimum point of the 1 σ curve. The 1 σ curve for the KF stays flat because it does not contain the error due to bias while the KF with MNI does but the impact is not as dramatic as SKF.

The RMS errors of all three filters computed with 1000 Monte Carlo repetitions are shown in Figure 4.20 together with the 1 σ covariance errors of Figure 4.19. All three filters achieved nearly the same RMS error. All filters have the same estimation error because the dominating error is the measurement bias. The KF ignoring bias is clearly not the filter to choose, because it is too optimistic about its own performance. The filter computed covariance with measurement covariance inflation performs better, but is still much lower than the actual RMS error. The only filter that has computed covariance comparable with the actual RMS error is the SKF; the RMS error curves are overlapping with the SKF computed covariance.

Remarks

1. None of the three methods actually estimates bias. Methods to estimate bias will use state augmentation, as suggested in Sections 4.5.1 and 4.6. A potential issue for state augmentation is observability, that is, the augmented state may not be observable. SKF is not estimating bias, it is a method for properly representing the impact of bias in covariance when not estimating it.

2. SKF requires the bias covariance to be known. The readers are encouraged to investigate sensitivity to the uncertainty in bias covariance. When bias is very large, the SKF will break down. The readers are also encouraged to investigate when will it break down and potential methods of mitigation.

Homework Problems

1. Using the tools and simulations built in the two previous chapters, compute and display the NEES and NIS. If they turn out to be excessively large or small, see what you can do to change them.

2. The exercises of the three previous chapters have already introduced the concept of filter tuning using process noise. Use the model built for CV-CA-CV trajectory and experiment.

   a. Select process noise as driven by the change of filter residual process.

   b. Select a filter memory length to adapt to the unexpected model change.

   c. Select an exponential aging constant to reduce observed biases in the filter residual process.

   d. Build a maneuver detector and use it to change filter model (between CV and CA).

3. Compute the error equation for Equations (4.3-14) and (4.3-15) when the actual system is CV and the model is CA, then reverse the order. Explain your results.

4. Use the tools and results of Problem 3; compare what you obtained to discussions in Section 4.3.2.

5. Using the ballistic target dynamic model, build a maneuver detector when the target is doing each of the following.

   a. Reentering the atmosphere.

   b. Encountering sudden aerodynamic lateral motion.

6. Following the discussion in Section 4.9.1 on the sine wave estimation study, investigate the sampling approach. Obtain results from small to high signal to noise ratio. Compare results with CRB and with the results of the four previous estimators.

7. Similarly, investigate the classical approach (FFT). Compare the results of all six approaches.

8. Following the SKF example of Section 4.9.2.

    a. Investigate the SKF sensitivity when the actual bias and the bias covariance set in SKF are different.

    b. Investigate the performance of SKF with increasing bias value. Observe when it breaks down, and provide an explanation. Investigate ways for mitigation.

References

1. R.E. Kalman, "A New Approach to Linear Filtering and Prediction Problems," *Transactions of the ASME*, vol. 82, ser. D, pp. 35–45, Mar 1960.

2. T.S. Lee, K.P. Dunn, and C.B. Chang, "On Observability and Unbiased Estimation of Nonlinear Systems," *System Modeling and Optimization, Lecture Notes in Control and Information Sciences*, vol. 39, pp. 259–266, New York: Springer, 1982.

3. H.L. Van Trees, *Detection, Estimation, and Modulation Theory, Part 1*. New York: Wiley, 1968.

4. H. Cramer, *Mathematical Methods of Statistics*. Princeton, NJ: Princeton University Press, 1946.

5. T. Kailath, A.H. Sayed, and B. Hassibi, *Linear Estimation*. Englewood Cliffs, NJ: Prentice Hall, 2000.

6. Y. Bar-Shalom, X. Rong Li, and T. Kirubarajan, *Estimation with Applications to Tracking and Navigation*. New York: Wiley, 2001.

7. C.B. Chang and K.P. Dunn, "Kalman Filter Compensation for a Special Class of Systems," *IEEE Transactions on Aerospace and Electronic Systems*, vol. AES-13, pp. 700–706, Nov. 1977.

8. R.J. Fitzgerald, "Divergence of the Kalman Filter," *IEEE Transactions on Automatic Control*, vol. AC-16, pp. 736–747, Dec. 1971.

9. C.B. Chang and J.A. Tabaczynski, "Application of State Estimation to Target Tracking," *IEEE Transactions on Automatic Control*, vol. AC-29, pp. 98–109, Feb. 1984.

10. C.B. Chang, R.H. Whiting, and M. Athans, "On the State and Parameter Estimation for Maneuvering Reentry Vehicles," *IEEE Transactions on Automatic Control*, vol. AC-22, pp. 99–105, Feb. 1977.

11. M. Athans, "The Role and Use of the Stochastic Linear-Quadratic-Gaussian Control Problem in System Design," *IEEE Transactions on Automatic Control*, vol. AC-16, pp. 529–552, Dec. 1971.

12. A. Gelb (Ed.), *Applied Optimal Estimation*. Cambridge, MA: MIT Press, 1974.

13. R. Mehra, "On the Identification of Variances and Adaptive Kalman Filtering," *IEEE Transactions on Automatic Control*, vol. AC-15, pp. 175–184, Apr. 1970.

14. R.K. Mehra, "Approaches to Adaptive Filtering," *IEEE Transactions on Automatic Control*, vol. AC-17, pp. 693–698, Oct. 1971.

15. B. Carew and P.R. Belanger, "Identification of Optimum Filter Steady State Gain for Systems with Unknown Noise Covariances," *IEEE Transactions on Automatic Control*, vol. AC-18, pp. 582–588, Dec. 1973.

16. L. Chin, "Advances in Adaptive Filtering," in *Advances in Control and Dynamic Systems*, vol. 15, C. T. Leondes, Ed., New York: Academic Press, 1979.

17. A.H. Jazwinski, *Stochastic Processes and Filtering Theory*. New York: Academic Press, 1970.

18. S.L. Fagin, "Recursive Linear Regression Theory, Optimal Filter Theory and Error Analysis of Optimal Systems," *IEEE International Convention Record*, pp. 216–240, 1964.

19. R.W. Miller, "Asymptotic Behavior of the Kalman Filter with Exponential Aging," *AIAA Journal*, vol. 9, pp. 537–539, Mar. 1971.

20. G. Schmidt, "Application of State-Space Methods to Navigation Problems," in *Advances in Control Systems*, vol. 3, pp. 293–340, C. Leondes, Ed., New York: Academic Press, 1966.

21. R. Novoselov, S.M. Herman, S. Gadaleta, and A.B. Poor, "Mitigating the Effects of Residual Bias with Schmidt–Kalman Filtering," in *Proceedings of the 8th International Conference on Information Fusion*, July 2005.

22. A.E. Bryson and L.J. Henrikson, "Estimation Using Sampled Data Containing Sequentially Correlated Noise," *Journal of Spacecraft and Rockets*, vol. 5, pp. 662–665, June 1968.

23. H.C. Lambert, "Pre-whitening of Colored Measurement Noise," Private communication, June 25, 2008.

24. P. Mookerjee and F. Reifler, "Reduced State Estimator for Systems with Parametric Inputs," *IEEE Transactions on Aerospace and Electronics Systems*, vol. AES-40, pp. 446–461, Apr. 2004.

25. L.F. Urbano, P. Kalata, and M. Kam, "Optimal Tracking Index Relationship for Random and Deterministic Target Maneuvers," in *Proceedings of IEEE Radar Conference*, May 2012.

26. A.S. Willsky and H.L. Jones, "A Generalized Likelihood Ratio Approach to the Detection and Estimation of Jumps in Linear Systems," *IEEE Transactions on Automatic Control*, vol. AC-21, pp. 108–112, Feb. 1976.

27. H.L. Jones, "Failure Detection in Linear Systems," PhD. dissertation, Dept. of Aeronautics and Astronautics, MIT, Cambridge, MA, Sept. 1973.

28. A.S. Willsky, "A Survey of Design Methods for Failure Detection in Dynamic Systems," *Automatica*, vol. 12, pp. 601–611, Nov. 1976.

29. E.Y. Chow and A.S. Willsky, "Bayesian Design of Detection Rules for Failure Detection," *IEEE Transactions on Aerospace and Electronic Systems*, vol. AES-20, pp. 761–774, Nov. 1984.

30. R. Bueno, E.Y. Chow, K.P. Dunn, S.B. Gersbwin, and A.S. Willsky, "Status Report on the Generalized Likelihood Ratio Failure Detection Technique, with Application to the F-8 Aircraft," in *Proceedings of the 14th IEEE Conference on Decision and Control*, Dec. 1976, pp. 38–47.

31. R.J. McAuLay and E. Denlinger, "A Decision-Directed Adaptive Tracker," *IEEE Transactions on Aerospace and Electronic Systems*, vol. AES-9, pp. 229–236, Mar. 1973.

32. C.B. Chang and K.P. Dunn, "On GLR Detection and Estimation of Unexpected Inputs in Linear Discrete Systems," *IEEE Transactions on Automatic Control*, vol. AC-24, pp. 499–501, June 1979.

33. C.B. Chang and K.P. Dunn, "A Recursive Generalized Likelihood Ratio Test Algorithm for Detecting Sudden Changes in Linear Discrete Systems," in *Proceedings of 17th IEEE Conference on Decision and Control*, Jan. 1979.

34. F. Daum and R. Fitzgerald, "Decoupled Kalman Filters for Phased Array Radar Tracking," *IEEE Transactions on Automatic Control*, vol. AC-28, pp. 269–283, Mar. 1983.

35. H. Lambert and M. Tobias, "False Cross-Range Observability," Private communication, 2008.

5

Multiple Model Estimation Algorithms

5.1 Introduction

Chapter 4 discussed problems with model uncertainties. In this chapter, the case where the model uncertainty is restricted to a finite set of known values is considered. One such example is in the tracking of maneuvering targets. Targets with straight and level flight can be modeled with a dynamic having zero acceleration. When significant target maneuvers begin, the filter must detect the maneuver and estimate its acceleration. The maneuver detection algorithm requires the implementation of a growing bank of filters that could be computationally expensive (see Chapter 4) and detection delays can occur due to filter transients that result when maneuver acceleration estimation is introduced. Because this method declares detection and switches to a different model after the sufficient statistics cross a threshold, it is therefore a "hard" decision method. An alternate approach is to run two filters, one with a maneuver model, and one with an unperturbed flight model. The output of both filters can be combined probabilistically to make the performance of switching between models occur smoothly; thus, this is a "soft" decision or "blending" method. The maneuvering target tracking example is a specific case of a more general multiple model estimation algorithm (MMEA) [1]. Similar to the maneuver target tracking problem, the failure detection problem discussed in Section 4.6 can also be addressed with a multiple model estimator [2].

After providing problem definitions and assumptions in Section 5.2, the MMEA for constant model case (CM^{2}EA) that is applicable to a system with an unknown but constant driving function is developed in Section 5.3. For a maneuvering target or system with unknown switching parameters, a switching model algorithm (SM^{2}EA) is derived in Section 5.4. The SM^{2}EA consists of an exponentially growing bank of filters with each matched to a possible parameter history (for N possible parameters at each time, there could be N^k branches at time k). In practice, a finite-memory switching model is used. The algorithms for finite-memory switching model cases are derived in Section 5.5. A more efficient approximation to the single-step memory filter, the interacting multiple model (IMM) is presented in Section 5.6 and a numerical example is given in Section 5.7.

Similar to the approach used in Chapter 4, the notation for a linear system is used. The results can be easily extended to nonlinear systems using an extended Kalman filter by substituting the Jacobian matrices of the nonlinear equations in place of the linear system and measurement matrices. In Chapter 6, a Monte Carlo sampling scheme is presented to solve this type of problem with random simulations.

5.2   Definitions and Assumptions

As stated several times earlier, the problem is to estimate the state vector of a discrete linear system

$$\mathbf{x}_k = \mathbf{\Phi}_{k,k-1}\mathbf{x}_{k-1} + \mathbf{\mu}_{k-1}, \tag{5.2-1}$$

$$\mathbf{y}_k = \mathbf{H}_k\mathbf{x}_k + \mathbf{v}_k, \tag{5.2-2}$$

with $\mathbf{x}_0 \colon \sim N(\hat{\mathbf{x}}_0, \mathbf{P}_0)$, $\mathbf{\mu}_k \colon \sim N(\mathbf{0}, \mathbf{Q}_k)$, $\mathbf{v}_k \colon \sim N(\mathbf{0}, \mathbf{R}_k)$. The solution to this problem is the well-known Kalman filter (KF). The Bayesian equation [3] for recursive estimation was used in Chapter 2 to derive the KF, and is restated below because it is the foundation for deriving the MMEA.

$$p\left(\mathbf{x}_k | \mathbf{y}_{1:k}\right) = \frac{p\left(\mathbf{y}_k | \mathbf{x}_k\right)}{p\left(\mathbf{y}_k | \mathbf{y}_{1:k-1}\right)} p\left(\mathbf{x}_k | \mathbf{y}_{1:k-1}\right) \tag{5.2-3}$$

and

$$p\left(\mathbf{x}_k | \mathbf{y}_{1:k-1}\right) = \int p\left(\mathbf{x}_k | \mathbf{x}_{k-1}\right) p\left(\mathbf{x}_{k-1} | \mathbf{y}_{1:k-1}\right) d\mathbf{x}_{k-1}, \tag{5.2-4}$$

where $p(\mathbf{x}_k|\mathbf{y}_{1:k})$ is the a posteriori density of $\mathbf{x}_k$ given all measurement from discrete time 1 through k, $p(\mathbf{x}_k|\mathbf{y}_{1:k-1})$ is the a priori density of $\mathbf{x}_k$ given all the measurements from discrete time 1 through $k - 1$, and $p(\mathbf{x}_{k-1}|\mathbf{y}_{1:k-1})$ is the a posteriori density of $\mathbf{x}_{k-1}$ given all measurement from discrete time 1 through $k - 1$. These are used to compute the a priori density for $\mathbf{x}_k$ as shown in the above integral.

Assume that the system under consideration contains an unknown parameter vector. The parameter vector can be modeled in a very general sense, that is, it could appear in the system or measurement matrix, as an input vector to the system or measurement equations, or as an offset for the mean or covariance for any of the noise processes. It could also represent a finite set of models such as the maneuvering versus nonmaneuvering target case. Let us first visit the case where the unknown parameter vector is constant, but can only be selected from a set of known vectors. The MMEA for a constant parameter vector was first addressed by Magill [1]. An example of this is when a target either stays on course or maneuvers during a certain period of time. In reality, a target is more likely to switch back and forth between the maneuvering and nonmaneuvering modes, which leads to the desire to design MMEA algorithms dealing with switching parameters.

The MMEA algorithm for switching (or time-varying) parameters has been considered by many authors [4–14]. Some treated the discrete switching problem [4–5, 10–11], some treated the more theoretical problem of continuous systems [6–7], some extended the results to control systems with applications [2, 7, 8], and some simplified the solution in order to save computation [12–14]. The results discussed in this chapter are based primarily on the work reported in [10, 11] for discrete systems.

5.3   Constant Model Case

Assume that a system has an unknown but constant parameter vector $\mathbf{p}$ that is equal to only one of a known set of N parameter vectors $\{\mathbf{p}_1, \mathbf{p}_2, \dots, \mathbf{p}_N\}$. Let θ be a random variable representing N hypotheses, $\{\theta^1, \theta^2, \dots, \theta^N\}$, such that $\Pr\{\theta^i\}$ is the probability that $\theta = \theta^i$. Define a multiple model system, in which the ith model represents the parameter vector $\mathbf{p}_i$

$$\mathbf{x}_k = \mathbf{\Phi}^i_{k,k-1}\mathbf{x}_{k-1} + \mathbf{\mu}_{k-1} \tag{5.3-1}$$

$$\mathbf{y}_k = \mathbf{H}^i_k\mathbf{x}_k + \mathbf{v}_k. \tag{5.3-2}$$

For the purpose of notational simplicity, the above equations only explicitly show the dependence of the unknown parameter vector in $\mathbf{\Phi}_{k,k-1}$ and/or in $\mathbf{H}_k$, with the same noise statistics. Given the above, the a priori density of θ is

$$p(\theta) = \sum_{i=1}^{N} \Pr\{\theta^i\}\delta(\theta - \theta^i), \tag{5.3-3}$$

where $\delta(.)$ is the Kronecker delta function. Since the true parameter is only equal to one of the N parameter vectors $\{\mathbf{p}_1, \mathbf{p}_2, \dots \mathbf{p}_N\}$, the estimation solution to this problem is referred to as the constant multiple model estimation algorithm (CM²EA). This algorithm is derived below.

Applying the definition of conditional mean to obtain

$$\hat{\mathbf{x}}_{k|k} = E\{\mathbf{x}_k|\mathbf{y}_{1:k}\} = \int \mathbf{x}_k p(\mathbf{x}_k|\mathbf{y}_{1:k})d\mathbf{x}_k. \tag{5.3-4}$$

For the multiple model case, the conditional density can be written in terms of the model hypotheses

$$p(\mathbf{x}_k|\mathbf{y}_{1:k}) = \int p(\mathbf{x}_k, \theta|\mathbf{y}_{1:k})d\theta = \int p(\mathbf{x}_k|\theta, \mathbf{y}_{1:k})p(\theta|\mathbf{y}_{1:k})d\theta.$$

Using $p(\theta|\mathbf{y}_{1:k}) = \sum_{i=1}^{N} \Pr\{\theta|\mathbf{y}_{1:k}\}\delta(\theta - \theta^i)$ in the above equation yields

$$p(\mathbf{x}_k|\mathbf{y}_{1:k}) = \int p(\mathbf{x}_k|\theta, \mathbf{y}_{1:k})\sum_{i=1}^{N} \Pr\{\theta|\mathbf{y}_{1:k}\}\delta(\theta - \theta^i)d\theta$$
$$= \sum_{i=1}^{N} \Pr\{\theta^i|\mathbf{y}_{1:k}\}p(\mathbf{x}_k|\theta^i, \mathbf{y}_{1:k}). \tag{5.3-5}$$

Applying the above equation to the definition of conditional mean estimate and covariance, we obtain the algorithm for state and covariance

$$\hat{\mathbf{x}}_{k|k} = \int \mathbf{x}_k p(\mathbf{x}_k|\mathbf{y}_{1:k})d\mathbf{x}_k = \sum_{i=1}^{N} \Pr\{\theta^i|\mathbf{y}_{1:k}\}\int \mathbf{x}_k p(\mathbf{x}_k|\theta^i, \mathbf{y}_{1:k})d\mathbf{x}_k$$
$$= \sum_{i=1}^{N} \Pr\{\theta^i|\mathbf{y}_{1:k}\}\hat{\mathbf{x}}_{k|k}^i, \tag{5.3-6}$$

$$\mathbf{P}_{k|k} = \int \left[(\mathbf{x}_k - \hat{\mathbf{x}}_{k|k})(\mathbf{x}_k - \hat{\mathbf{x}}_{k|k})^T\right]p(\mathbf{x}_k|\mathbf{y}_{1:k})d\mathbf{x}_k d\mathbf{x}_k$$
$$= \sum_{i=1}^{N} \Pr\{\theta^i|\mathbf{y}_{1:k}\}\int \left[(\mathbf{x}_k - \hat{\mathbf{x}}_{k|k})(\mathbf{x}_k - \hat{\mathbf{x}}_{k|k})^T\right]p(\mathbf{x}_k|\theta^i, \mathbf{y}_{1:k})d\mathbf{x}_k$$
$$= \sum_{i=1}^{N} \Pr\{\theta^i|\mathbf{y}_{1:k}\}\left[\mathbf{P}_{k|k}^i + (\hat{\mathbf{x}}_{k|k}^i - \hat{\mathbf{x}}_{k|k})(\hat{\mathbf{x}}_{k|k}^i - \hat{\mathbf{x}}_{k|k})^T\right], \tag{5.3-7}$$

where $\hat{\mathbf{x}}_{k|k}^i$ is the state estimate generated with a Kalman filter matched to the parameter vector $\mathbf{p}_i$, and $\mathbf{P}_{k|k}^i$ is the covariance of $\hat{\mathbf{x}}_{k|k}^i$. With the conditional mean and covariance definition, we obtain

$$\hat{\mathbf{x}}_{k|k}^i = \int \mathbf{x}_k p(\mathbf{x}_k|\theta^i, \mathbf{y}_{1:k})d\mathbf{x}_k \tag{5.3-8}$$

$$\mathbf{P}_{k|k}^i = \int \left[(\mathbf{x}_k - \hat{\mathbf{x}}_{k|k}^i)(\mathbf{x}_k - \hat{\mathbf{x}}_{k|k}^i)^T\right]p(\mathbf{x}_k|\theta^i, \mathbf{y}_{1:k})d\mathbf{x}_k. \tag{5.3-9}$$

The above development indicates that the solution of the multiple model estimation problems can be obtained via a bank of Kalman filters with each filter matched to a particular parameter vector; the final estimate is obtained with the weighted sums of individual estimates. The weighting factors are the hypothesis probabilities, $\Pr\{\theta^i|\mathbf{y}_{1:k}\}$, matched to each parameter vector $\mathbf{p}_i$, for $i = 1, \ldots, N$. Next, we consider the computation of the hypothesis probability. Bayes' rule for the hypothesis process is

$$p\left(\theta|\mathbf{y}_{1:k}\right) = \frac{p\left(\mathbf{y}_k|\theta, \mathbf{y}_{1:k-1}\right)}{p\left(\mathbf{y}_k|\mathbf{y}_{1:k-1}\right)} p\left(\theta|\mathbf{y}_{1:k-1}\right). \tag{5.3-10}$$

Substituting the discrete density function representation

$$p\left(\theta|\mathbf{y}_{1:k}\right) = \sum_{i=1}^{N} \Pr\left\{\theta^i|\mathbf{y}_{1:k}\right\} \delta\left(\theta - \theta^i\right) \tag{5.3-11}$$

$$p\left(\theta|\mathbf{y}_{1:k-1}\right) = \sum_{i=1}^{N} \Pr\left\{\theta^i|\mathbf{y}_{1:k-1}\right\} \delta\left(\theta - \theta^i\right) \tag{5.3-12}$$

into Equation (5.3-10) yields the recursive updating equation for the hypothesis probabilities

$$\Pr\left\{\theta^i|\mathbf{y}_{1:k}\right\} = \frac{p\left(\mathbf{y}_k|\theta^i, \mathbf{y}_{1:k-1}\right)}{p\left(\mathbf{y}_k|\mathbf{y}_{1:k-1}\right)} \Pr\left\{\theta^i|\mathbf{y}_{1:k-1}\right\}. \tag{5.3-13}$$

where $\Pr\{\theta^i|\mathbf{y}_{1:k}\}$ is the a posteriori probability of θ^i, $\Pr\{\theta^i|\mathbf{y}_{1:k-1}\}$ the a priori probability of θ^i, $p(\mathbf{y}_k|\theta^i, \mathbf{y}_{1:k-1})$ is the density of the residual process of the ith Kalman filter matched to parameter vector $\mathbf{p}_i$, that is,

$$p\left(\mathbf{y}_k|\theta^i, \mathbf{y}_{1:k-1}\right) = C \exp\left\{-\frac{1}{2}\left(\mathbf{y}_k - \mathbf{H}_k^i \hat{\mathbf{x}}_{k|k-1}^i\right)^T \left[\mathbf{H}_k^i \mathbf{P}_{k|k-1}^i \mathbf{H}_k^{iT} + \mathbf{R}_k\right]^{-1} \left(\mathbf{y}_k - \mathbf{H}_k^i \hat{\mathbf{x}}_{k|k-1}^i\right)\right\},$$

where C is the normalization constant for the Gaussian density function, and $p(\mathbf{y}_k|\mathbf{y}_{1:k-1})$ is a normalization factor that is computed as

$$\sum_{i=1}^{N} \Pr\left\{\theta^i|\mathbf{y}_{1:k-1}\right\} p\left(\mathbf{y}_k|\theta^i, \mathbf{y}_{1:k-1}\right).$$

The CM^{2}EA is summarized in Table 5.1.

Table 5.1 Summary of the Multiple Model Estimation Algorithm for a Constant Unknown Parameter Vector

---

CM²EA

State and Measurement Equations

There is a constant and unknown parameter vector $\mathbf{p}$ that is equal to only one of a known set of N parameter vectors $\{\mathbf{p}_1, \mathbf{p}_2, \ldots, \mathbf{p}_N\}$. Let

$$\mathbf{x}_k = \mathbf{\Phi}^i_{k,k-1}\mathbf{x}_{k-1} + \mathbf{\mu}_{k-1},$$

$$\mathbf{y}_k = \mathbf{H}^i_k \mathbf{x}_k + \mathbf{v}_k.$$

where $\mathbf{x} \in \mathbb{R}^n$, $\mathbf{y} \in \mathbb{R}^m$, $\mathbf{x}_0 :\sim N(\hat{\mathbf{x}}_0, \ \mathbf{P}_0)$, $\mathbf{\mu}_k : \sim N(\mathbf{0}, \mathbf{Q}_k)$, and $\mathbf{v}_k : \sim N(\mathbf{0}, \mathbf{R}_k)$ are mutually independent $\forall k$ and $\mathbf{\Phi}^i_{k,k-1}$ and $\mathbf{H}^i_k$ define the system with parameter vector $\mathbf{p}_i$.

Problem Statement

Identify the true $\mathbf{p}_i$ and obtain $\hat{\mathbf{x}}_{k|k}$ and covariance $\mathbf{P}_{k|k}$ given parameter set $\{\mathbf{p}_1, \mathbf{p}_2, \ldots, \mathbf{p}_N\}$ and measurements $\mathbf{y}_{1:k} = \{\mathbf{y}_1, \mathbf{y}_2, \ldots, \mathbf{y}_k\}$.

Solution process

Build a band of N Kalman filters each matched to a $\mathbf{p} \in \{\mathbf{p}_1, \mathbf{p}_2, \ldots, \mathbf{p}_N\}$. Let $\hat{\mathbf{x}}^i_{k|k}$ denote the estimate obtained with a Kalman filter matched to $\mathbf{p}_i$ with covariance $\mathbf{P}^i_{k|k}$, the minimum variance estimate of $\mathbf{x}_k$ (the same as the conditional mean estimate), $\hat{\mathbf{x}}_{k|k}$, with covariance $\mathbf{P}_{k|k}$ are obtained as

$$\hat{\mathbf{x}}_{k|k} = \sum_{i=1}^{N} \Pr\{\theta^i | \mathbf{y}_{1:k}\} \hat{\mathbf{x}}^i_{k|k},$$

$$\mathbf{P}_{k|k} = \sum_{i=1}^{N} \Pr\{\theta^i | \mathbf{y}_{1:k}\} \left[\mathbf{P}^i_{k|k} + \left(\hat{\mathbf{x}}^i_{k|k} - \hat{\mathbf{x}}_{k|k}\right)\left(\hat{\mathbf{x}}^i_{k|k} - \hat{\mathbf{x}}_{k|k}\right)^T \right],$$

where θ is a random variable representing N hypotheses, $\{\theta^1, \theta^2, \ldots, \theta^N\}$, such that $\Pr\{\theta^i | \mathbf{y}_{1:k}\}$ is the probability that $\theta = \theta^i$ given measurements $\mathbf{y}_{1:k}$, and the hypothesis probability evolves recursively as

$$\Pr\{\theta^i | \mathbf{y}_{1:k}\} = \frac{p(\mathbf{y}_k | \theta^i, \mathbf{y}_{1:k-1})}{p(\mathbf{y}_k | \mathbf{y}_{1:k-1})} \Pr\{\theta^i | \mathbf{y}_{1:k-1}\},$$

where $p(\mathbf{y}_k | \theta^i, \mathbf{y}_{1:k-1})$ is the residual density of the KF matched to $\mathbf{p}_i$,

$$p(\mathbf{y}_k | \theta^i, \mathbf{y}_{1:k-1}) = C exp\left\{-\frac{1}{2}\left(\mathbf{y}_k - \mathbf{H}^i_k \hat{\mathbf{x}}^i_{k|k-1}\right)^T \left[\mathbf{H}^i_k \mathbf{P}^i_{k|k-1} \mathbf{H}^{i\,T}_k + \mathbf{R}_k\right]^{-1}\left(\mathbf{y}_k - \mathbf{H}^i_k \hat{\mathbf{x}}^i_{k|k-1}\right)\right\}.$$

Summary and Remarks

1. The CM^{2}EA consists of a bank of N Kalman filters with each matched to a parameter vector $\mathbf{p}_i$ of $\{\mathbf{p}_1, \mathbf{p}_2, \ldots, \mathbf{p}_N\}$.

2. The final conditional mean estimate is obtained as a weighted sum of the output of the filter bank with the a posteriori hypothesis probabilities used as weighting factors.

3. The a posteriori hypothesis probabilities are updated with the density function of the residual process and the a posteriori probability of the previous cycle.

4. Given the problem definition, only one of the models is truly matched to the actual physical process, the output of the Kalman filter for this filter is optimal, with zero bias and consistent covariance. All the other filters will exhibit bias errors with inconsistent filter covariance. It is exactly this property that is manifested through the filter residual process making the hypothesis probability work. The hypothesis probability corresponding to the correct model will converge to one and the rest to zero.

5. It is also clear that when the problem assumption is incorrect, that is, none of the models assumed in CM^{2}EA is equal to the true model and/or the noise statistics are incorrect, the filter performance will become unpredictable. It can converge to the wrong model, bounce back and forth between several different models, and so on. This is the issue facing all practicing engineers. Filter design requires extensive simulation studies anchored by field measurements.

The CM^{2}EA for a two-model case is illustrated in Figure 5.1.

5.4   Switching Model Case

The switching model case assumes that the unknown parameter vector is equal to a different vector of the set over different time intervals. This is referred to as the switch multiple model estimation algorithm (SM^{2}EA). The time evolution of all possible paths for a two parameter case for time up to $k = 3$ is illustrated in Figure 5.2.

Consider the following hypothesis definitions,

θ_k^i is the hypothesis that $\mathbf{p} = \mathbf{p}_i$ at time k. It is referred to as a local hypothesis (in terms of time)

$\theta_{1:k}^j$ is a hypothesis that represents a sequence of parameter vectors from time 1 through time k, thus this is a global hypothesis. A global hypothesis contains a set of local hypotheses such as $\theta_{1:k}^j = \left\{\theta_1^{i_1}, \theta_2^{i_2}, \ldots, \theta_k^{i_k}\right\}$,

where $i_n = 1, \ldots, N$ and $j = 1, \ldots, N^k$ and

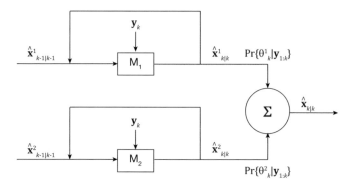

Figure 5.1   Illustration of the constant multiple model estimation algorithm.

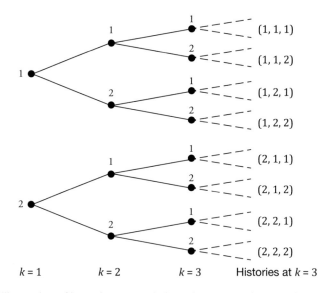

Figure 5.2   Illustration of hypotheses evolution of a two model case for k up to 3.

$\Pr\left\{\theta_k^i | \mathbf{y}_{1:k}\right\}$ is the probability of θ_k^i being true at k, and
$\Pr\left\{\theta_{1:k}^j | \mathbf{y}_{1:k}\right\}$ is the probability of $\theta_{1:k}^j = \left\{\theta_1^{i_1}, \ \theta_2^{i_2}, \ \ldots, \ \theta_k^{i_k}\right\}$ being true at k.

The SM^{2}EA for state vector estimation is now derived,

$$\begin{aligned}
\hat{\mathbf{x}}_{k|k} &= \int \mathbf{x}_k p\left(\mathbf{x}_k | \mathbf{y}_{1:k}\right) d\mathbf{x}_k = \iint \mathbf{x}_k p\left(\mathbf{x}_k | \theta_{1:k}, \mathbf{y}_{1:k}\right) p\left(\theta_{1:k} | \mathbf{y}_{1:k}\right) d\theta_{1:k} d\mathbf{x}_k \\
&= \iint \mathbf{x}_k p\left(\mathbf{x}_k | \theta_{1:k}, \mathbf{y}_{1:k}\right) \sum_{j=1}^{N^k} \Pr\left\{\theta_{1:k}^j | \mathbf{y}_{1:k}\right\} \delta\left(\theta_{1:k} - \theta_{1:k}^j\right) d\theta_{1:k} d\mathbf{x}_k \\
&= \sum_{j=1}^{N^k} \Pr\left\{\theta_{1:k}^j | \mathbf{y}_{1:k}\right\} \int \mathbf{x}_k p\left(\mathbf{x}_k | \theta_{1:k}^j, \mathbf{y}_{1:k}\right) d\mathbf{x}_k = \sum_{j=1}^{N^k} \Pr\left\{\theta_{1:k}^j | \mathbf{y}_{1:k}\right\} \hat{\mathbf{x}}_{k|k}^j. \quad (5.4\text{-}1)
\end{aligned}$$

As expected, the final estimate, $\hat{\mathbf{x}}_{k|k}$, is the weighted sum of all state estimates that match to every possible parameter history, $\hat{\mathbf{x}}_{k|k}^j$ for $j = 1, \ldots, N^k$. The derivation for the covariance of $\hat{\mathbf{x}}_{k|k}$ is similar and the details are left to the reader.

$$\begin{aligned}
\mathbf{P}_{k|k} &= \iint \left(\mathbf{x}_k - \hat{\mathbf{x}}_{k|k}\right)\left(\mathbf{x}_k - \hat{\mathbf{x}}_{k|k}\right)^T p\left(\mathbf{x}_k | \mathbf{y}_{1:k}\right) d\mathbf{x}_k d\mathbf{x}_k \\
&= \sum_{j=1}^{N^k} \Pr\left\{\theta_{1:k}^j | \mathbf{y}_{1:k}\right\} \left[\mathbf{P}_{k|k}^j + \left(\hat{\mathbf{x}}_{k|k}^j - \hat{\mathbf{x}}_{k|k}\right)\left(\hat{\mathbf{x}}_{k|k}^j - \hat{\mathbf{x}}_{k|k}\right)^T\right]. \quad (5.4\text{-}2)
\end{aligned}$$

The a posteriori hypothesis probability equation is now derived by applying the product rule of probability

$$p(\theta_{1:k} | \mathbf{y}_{1:k}) = p(\theta_k | \theta_{1:k-1}, \mathbf{y}_{1:k}) p(\theta_{1:k-1} | \mathbf{y}_{1:k}),$$

using

$$p\left(\theta_k | \theta_{1:k-1}, \mathbf{y}_{1:k}\right) = \frac{p\left(\mathbf{y}_k | \theta_k, \ \theta_{1:k-1}, \ \mathbf{y}_{1:k-1}\right)}{p\left(\mathbf{y}_k | \theta_{1:k-1}, \ \mathbf{y}_{1:k-1}\right)} p\left(\theta_k | \theta_{1:k-1}, \mathbf{y}_{1:k-1}\right)$$

and

$$p\left(\theta_{1:k-1} | \mathbf{y}_{1:k}\right) = \frac{p\left(\mathbf{y}_k | \theta_{1:k-1}, \mathbf{y}_{1:k-1}\right)}{p\left(\mathbf{y}_k | \mathbf{y}_{1:k-1}\right)} p\left(\theta_{1:k-1} | \mathbf{y}_{1:k-1}\right),$$

in the above equations we obtain

$$p\left(\theta_{1:k} | \mathbf{y}_{1:k}\right) = \frac{p\left(\mathbf{y}_k | \theta_k, \theta_{1:k-1}, \mathbf{y}_{1:k-1}\right)}{p\left(\mathbf{y}_k | \mathbf{y}_{1:k-1}\right)} p\left(\theta_k | \theta_{1:k-1}, \mathbf{y}_{1:k-1}\right) p\left(\theta_{1:k-1} | \mathbf{y}_{1:k-1}\right). \quad (5.4\text{-}3)$$

Applying discrete density function representations

$$p\big(\theta_{1:k}|\mathbf{y}_{1:k}\big)=\sum\nolimits_{j=1}^{N^k}\Pr\big\{\theta_{1:k}^{j}|\mathbf{y}_{1:k}\big\}\delta\big(\theta-\theta_{1:k}^{j}\big),$$

$$p\big(\theta_{1:k-1}^{i}|\mathbf{y}_{1:k-1}\big)=\sum\nolimits_{j=1}^{N^{k-1}}\Pr\big\{\theta_{1:k-1}^{j}|\mathbf{y}_{1:k-1}\big\}\delta\big(\theta-\theta_{1:k-1}^{i}\big),$$

yields

$$\Pr\big\{\theta_{1:k}^{j}|\mathbf{y}_{1:k}\big\}=\frac{p\big(\mathbf{y}_{k}|\theta_{k}^{\ell},\theta_{1:k-1}^{i},\mathbf{y}_{1:k-1}\big)}{p\big(\mathbf{y}_{k}|\mathbf{y}_{1:k-1}\big)}\Pr\big\{\theta_{k}^{\ell}|\theta_{1:k-1}^{i},\mathbf{y}_{1:k-1}\big\}\Pr\big\{\theta_{1:k-1}^{j}|\mathbf{y}_{1:k-1}\big\},\quad(5.4\text{-}4)$$

where $\theta_{1:k}^{j}=\big\{\theta_{k}^{\ell},\theta_{1:k-1}^{i}\big\}$, $i=1,\dots,N^{k-1}$; $j=1,\dots,N^{k}$; $\ell=1,\dots,N$. This completes the a posteriori hypothesis probability derivation. See Table 5.2 for algorithm summary.

Summary and Remarks

1. The SM²EA consists of an exponentially growing bank of filters with each matched to a possible parameter history value that is taken from N possibilities at each measurement time. At time k, the total number of branches is N^k.

2. The a posteriori hypothesis probability of a global hypothesis $\theta_{1:k}^{j}$ (denoted as $\big\{\theta_{k}^{\ell},\theta_{1:k-1}^{i}\big\}$) is equal to the multiplication of the normalized residual process of the filter matching to $\big\{\theta_{k}^{\ell},\theta_{1:k-1}^{i}\big\}$ and the transition probability of θ_{k}^{ℓ} given $\theta_{1:k-1}^{i}$ and the a posteriori hypothesis probability of $\theta_{1:k-1}^{j}$ (or as the prior of $\theta_{1:k}^{j}$).

3. Without any prior knowledge, the transition probability of θ_{k}^{ℓ} given $\theta_{1:k-1}^{i}$, $\Pr\big\{\theta_{k}^{\ell}|\theta_{1:k-1}^{i},\mathbf{y}_{1:k-1}\big\}$, is thus unknown. The user will need to specify the entries based upon the property of the physical process. This probability could be learned if there are measurement data to characterize the physical process being modeled, similar to the machine learning process used in artificial intelligence.

4. Similar to the CM²EA case, if all the assumptions for the filter derivation are met, then one of the N^k models is matched to the true physical process. In this case, the filter will exhibit the optimal performance and the probability for this hypothesis being true will converge to unity. In actual applications when the assumptions are not fully met, this often leads to incorrect results. Filter tuning and the use of a priori knowledge for adjustment are important for practical design.

Table 5.2   Summary of the Multiple Model Estimation Algorithm for a Constant Vector Switches Within a Known Set

---

SM²EA

System and Parameter Definition

The unknown parameter vector **p** takes on a different value within the parameter set $\{\mathbf{p}_1, \mathbf{p}_2, \dots \mathbf{p}_N\}$ and measurement time where

$$\mathbf{x}_k = \mathbf{\Phi}^i_{k,k-1}\mathbf{x}_{k-1} + \mathbf{\mu}_{k-1},$$

$$\mathbf{y}_k = \mathbf{H}^i_k\mathbf{x}_k + \mathbf{v}_k.$$

Denote the system taking on the parameter value $\mathbf{p}_i$ at time k. Statistical properties of $\mathbf{x}_0 {:}\sim N(\hat{\mathbf{x}}_0, \ \mathbf{P}_0)$, $\mathbf{\mu}_k {:}\sim N(\mathbf{0}, \mathbf{Q}_k)$, and $\mathbf{v}_k {:}\sim N(\mathbf{0}, \mathbf{R}_k)$ remain the same.

Let θ^i_k be the hypothesis that $\mathbf{p} = \mathbf{p}_i$ at time k and $\theta^j_{1:k}$ be a hypothesis that represents a sequence of parameter vectors from time 1 through time k, thus $\theta^j_{1:k} = \left\{\theta^{i_1}_1, \theta^{i_2}_2, \dots, \theta^{i_n}_k\right\}$, where $i_n = 1, \dots, N$ and $j = 1, \dots, N^k$ and

$\Pr\left\{\theta^i_k | \mathbf{y}_{1:k}\right\}$ is the probability of θ^i_k being true at k,

$\Pr\left\{\theta^j_{1:k} | \mathbf{y}_{1:k}\right\}$ is the probability of $\theta^j_{1:k} = \left\{\theta^{i_1}_1, \theta^{i_2}_2, \dots, \theta^{i_n}_k\right\}$ being true at k, and

$\Pr\left\{\theta^\ell_k | \theta^i_{1:k-1}\right\}$ is the transition probability of θ^ℓ_k being true given the parameter history $\theta^i_{1:k-1}$.

$\Pr\left\{\theta^i_k | \mathbf{y}_{1:k}\right\}$ and $\Pr\left\{\theta^j_{1:k} | \mathbf{y}_{1:k}\right\}$ will be computed and $\Pr\left\{\theta^\ell_k | \theta^i_{1:k-1}\right\}$ is assumed known.

Problem Statement

Obtain $\hat{\mathbf{x}}_{k|k}$ and covariance $\mathbf{P}_{k|k}$ given parameter set $\{\mathbf{p}_1, \mathbf{p}_2, \dots, \mathbf{p}_N\}$, parameter transition probability $\Pr\left\{\theta^\ell_k | \theta^i_{1:k-1}\right\}$, and measurement $\mathbf{y}_{1:k} = (\mathbf{y}_1, \mathbf{y}_2, \dots, \mathbf{y}_k)$.

Solution process

Build an exponentially growing bank of filters, referred to as an elemental filter, with each matched to a possible parameter history. At time k, the total number of elemental filters is N^k. Then

$$\hat{\mathbf{x}}_{k|k} = \sum_{j=1}^{N^k}\Pr\left\{\theta^j_{1:k} | \mathbf{y}_{1:k}\right\}\hat{\mathbf{x}}^j_{k|k},$$

$$\mathbf{P}_{k|k} = \sum_{j=1}^{N^k}\Pr\left\{\theta^j_{1:k} | \mathbf{y}_{1:k}\right\}\left[\mathbf{P}^j_{k|k} + \left(\hat{\mathbf{x}}^j_{k|k} - \hat{\mathbf{x}}_{k|k}\right)\left(\hat{\mathbf{x}}^j_{k|k} - \hat{\mathbf{x}}_{k|k}\right)^T\right], \text{ and}$$

$$\Pr\left\{\theta^j_{1:k} | \mathbf{y}_{1:k}\right\} = \frac{p\left(\mathbf{y}_k | \theta^\ell_k, \theta^i_{1:k-1}, \mathbf{y}_{1:k-1}\right)}{p\left(\mathbf{y}_k | \mathbf{y}_{1:k-1}\right)}\Pr\left\{\theta^\ell_k | \theta^i_{1:k-1}, \mathbf{y}_{1:k-1}\right\}\Pr\left\{\theta^j_{1:k-1} | \mathbf{y}_{1:k-1}\right\}$$

where $\left\{\hat{\mathbf{x}}^j_{k|k}, \mathbf{P}^j_{k|k}\right\}$ for $j = 1, \dots, N^k$ is the state estimate and covariance of the filter matched with jth history and $p\left(\mathbf{y}_k | \theta^\ell_k, \theta^i_{1:k-1}, \mathbf{y}_{1:k-1}\right)$ is the corresponding residual density.

5.5   Finite Memory Switching Model Case

The SM^{2}EA assumes that the current local parameter hypothesis depends on all the complete past history of parameter values. In most physical processes, the current parameter value is only influenced by its recent past. If the assumption is made that the parameter history is limited to a finite duration, then the past history beyond the finite duration could be combined in a weighted average. The averaged state estimate is multimode Gaussian, thus using it for future filter updates invalidates the assumption of using only mean and covariance to represent the a posteriori density. This step therefore makes subsequent derivations heuristic. Nevertheless, this is a sound approach in the physical sense and practical for solving application problems.

The results for the parameter process with a one-step memory will be derived first, then extended to the two-step memory case.

5.5.1   One-Step Model History

In Figure 5.3, a single step history is illustrated for a two-model case. Note that the hypothesis branch grows to N^2 at the next time step. The total number of nodes (or hypotheses) is reduced to N by combining those having the same local hypothesis at the current time. The weighted combination is obtained by averaging over those having different local hypotheses at the previous time.

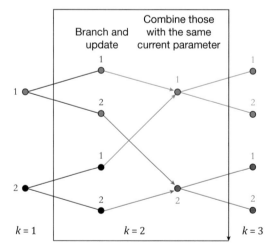

Figure 5.3   Illustration of a single step model memory case for $N = 2$.

With the single stage memory assumption, we obtain the following hypothesis transition probability relationship

$$\Pr\left\{\theta_k^i | \theta_{1:k-1}^m, \mathbf{y}_{1:k-1}\right\} = \Pr\left\{\theta_k^i | \theta_{k-1}^\ell, \theta_{1:k-2}^n, \mathbf{y}_{1:k-1}\right\} = \Pr\left\{\theta_k^i | \theta_{k-1}^\ell\right\},$$

where i, $\ell = 1, \ldots, N$; $m = 1, \ldots, N^{k-1}$; $n = 1, \ldots, N^{k-2}$. This is the hypothesis transition probability. Although there are only N nodes at each time step, the transition to go from $k - 1$ to k is N^2. The desired algorithm is therefore the computation of

$$\hat{\mathbf{x}}_{k|k}^i, \mathbf{P}_{k|k}^i, \Pr\left\{\theta_k^i | \mathbf{y}_{1:k}\right\},$$

given

$$\hat{\mathbf{x}}_{k-1|k-1}^\ell, \mathbf{P}_{k-1|k-1}^\ell, \Pr\left\{\theta_{k-1}^\ell | \mathbf{y}_{1:k-1}\right\},$$

for both i, $\ell = 1, \ldots, N$.

Define the following state and covariance

$\hat{\mathbf{x}}_{k|k}^{\ell i}$: updated state estimate for model i at time k given $\hat{\mathbf{x}}_{k-1|k-1}^\ell$ for model ℓ at $k - 1$,

$\mathbf{P}_{k|k}^{\ell i}$: covariance of $\hat{\mathbf{x}}_{k|k}^{\ell i}$

The state and covariance at time k for the ith model can be obtained by averaging over ℓ,

$$\hat{\mathbf{x}}_{k|k}^i = \sum_{\ell=1}^N \hat{\mathbf{x}}_{k|k}^{\ell i} \Pr\left\{\theta_{k-1}^\ell | \theta_k^i, \mathbf{y}_{1:k}\right\} \tag{5.5-1}$$

$$\mathbf{P}_{k|k}^i = \sum_{\ell=1}^N \Pr\left\{\theta_{k-1}^\ell | \theta_k^i, \mathbf{y}_{1:k}\right\}\left[\mathbf{P}_{k|k}^{\ell i} + \left(\hat{\mathbf{x}}_{k|k}^{\ell i} - \hat{\mathbf{x}}_{k|k}^i\right)\left(\hat{\mathbf{x}}_{k|k}^{\ell i} - \hat{\mathbf{x}}_{k|k}^i\right)^T\right] \tag{5.5-2}$$

The backward time conditional hypothesis probability is computed as

$$\Pr\left\{\theta_{k-1}^\ell | \theta_k^i, \mathbf{y}_{1:k}\right\} = \frac{\Pr\left\{\theta_k^i, \theta_{k-1}^\ell | \mathbf{y}_{1:k}\right\}}{\Pr\left\{\theta_k^i | \mathbf{y}_{1:k}\right\}},$$

$$\Pr\left\{\theta_k^i, \theta_{k-1}^\ell | \mathbf{y}_{1:k}\right\} = \frac{p\left(\mathbf{y}_k | \theta_k^i, \theta_{k-1}^\ell, \mathbf{y}_{1:k-1}\right)}{p\left(\mathbf{y}_k | \mathbf{y}_{1:k-1}\right)} \Pr\left\{\theta_k^i | \theta_{k-1}^\ell\right\} \Pr\left\{\theta_{k-1}^\ell | \mathbf{y}_{1:k-1}\right\},$$

where $p\left(\mathbf{y}_k | \theta_k^i, \theta_{k-1}^\ell, \mathbf{y}_{1:k-1}\right)$ is the residual density of $\hat{\mathbf{x}}_{k|k-1}^{\ell i}$ and $\mathbf{P}_{k|k-1}^{\ell i}$,

$$p\left(\mathbf{y}_k | \theta_k^i, \theta_{k-1}^\ell, \mathbf{y}_{1:k-1}\right) = C exp\left\{-\frac{1}{2}\left(\mathbf{y}_k - \mathbf{H}_k^i \hat{\mathbf{x}}_{k|k-1}^{\ell i}\right)^T\left[\mathbf{H}_k^i \mathbf{P}_{k|k-1}^{\ell i} \mathbf{H}_k^{i\,T} + \mathbf{R}_k\right]^{-1}\left(\mathbf{y}_k - \mathbf{H}_k^i \hat{\mathbf{x}}_{k|k-1}^{\ell i}\right)\right\}.$$

Note that

$$\Pr\left\{\theta_k^i | \mathbf{y}_{1:k}\right\} = \sum_{\ell=1}^N \Pr\left\{\theta_{k-1}^\ell, \theta_k^i | \mathbf{y}_{1:k}\right\}.$$

Substituting $\Pr\{\theta^i_k, \theta^\ell_{k-1}|\mathbf{y}_{1:k}\}$ into the above equation yields the final hypothesis probability update equation

$$\Pr\{\theta^i_k|\mathbf{y}_{1:k}\} = \sum_{\ell=1}^{N} \frac{p\left(\mathbf{y}_k|\theta^i_k, \theta^\ell_{k-1}, \mathbf{y}_{1:k-1}\right)}{p\left(\mathbf{y}_k|\mathbf{y}_{1:k-1}\right)} \Pr\{\theta^i_k|\theta^\ell_{k-1}\} \Pr\{\theta^\ell_{k-1}|\mathbf{y}_{1:k-1}\}. \qquad (5.5\text{-}3)$$

The final estimate $\hat{\mathbf{x}}_{k|k}$ and its covariance $\mathbf{P}_{k|k}$ are obtained by averaging over i

$$\hat{\mathbf{x}}_{k|k} = \sum_{i=1}^{N} \hat{\mathbf{x}}^i_{k|k} \Pr\{\theta^i_k|\mathbf{y}_{1:k}\}, \qquad (5.5\text{-}4)$$

and

$$\mathbf{P}_{k|k} = \sum_{i=1}^{N} \Pr\{\theta^i_k|\mathbf{y}_{1:k}\}\left[\mathbf{P}^i_{k|k} + \left(\hat{\mathbf{x}}_k - \hat{\mathbf{x}}^i_{k|k}\right)\left(\hat{\mathbf{x}}_k - \hat{\mathbf{x}}^i_{k|k}\right)^T\right]. \qquad (5.5\text{-}5)$$

This algorithm, published in [10], is referred to as the generalized pseudo-Bayesian estimator of the second order (GPB2) in [12]. It is summarized in Table 5.3.

Remarks

1. The estimate for a given local hypothesis at time k is obtained by (weighted) averaging of all local hypotheses at time $k - 1$. This is justified because of the one-step parameter history dependency assumption. The number of filters required to implement the algorithm is N^2.

2. If the actual parameter history dependence is longer than one, similar derivation can be applied. The case of a two-step history is shown in the next section.

3. The transition probabilities $\Pr\{\theta^i_k|\theta^\ell_{k-1}\}$ for all $i, \ell = 1, \dots, N$ are dependent on the physical process and assumed to be known to the system designer. In an actual problem, this is usually chosen to be representative of a physical process, such as the probability that an object may be maneuvering during the next step. This probability can be learned should some test data be available, and can be applied similar to machine learning in artificial intelligence.

4. Since the new estimate for each current local hypothesis is obtained as a weighted sum over all previous local hypotheses (Equation 5.5-3) and used in-line for the next measurement update, the a posteriori distribution function of $\mathbf{x}_k$ is multimode Gaussian. Thus, only using mean and covariance, $\{\hat{\mathbf{x}}_{k|k}, \mathbf{P}_{k|k}\}$, is no-longer sufficient to represent all the information.

Table 5.3 Summary of the Multiple Model Estimation Algorithm when the Current Parameter Only Depends on the Immediate Past: Generalized Pseudo-Bayesian Filter

SM^{2}EA with One-Step Model History
System and Parameter Definition
The unknown parameter vector **p** takes on a different value within the parameter set $\{\mathbf{p}_1, \mathbf{p}_2, \dots, \mathbf{p}_N\}$ and measurement time where

$$\mathbf{x}_k = \boldsymbol{\Phi}^i_{k,k-1}\mathbf{x}_{k-1} + \boldsymbol{\mu}_{k-1},$$

$$\mathbf{y}_k = \mathbf{H}^i_k\mathbf{x}_k + \mathbf{v}_k.$$

Denote the system taking on the parameter value $\mathbf{p}_i$ at time k. Statistical properties of $\mathbf{x}_0 :\sim N(\hat{\mathbf{x}}_0,\ \mathbf{P}_0)$, $\boldsymbol{\mu}_k :\sim N(\mathbf{0}, \mathbf{Q}_k)$, and $\mathbf{v}_k :\sim N(\mathbf{0}, \mathbf{R}_k)$ remain the same.
All the assumptions in Table 5.2 apply except that the transition probability is only dependent on the immediate past parameter, not the entire history,

$$\Pr\{\theta^i_k|\theta^m_{1:k-1}\} = \Pr\{\theta^i_k|\theta^\ell_{k-1}\},$$

where $i, \ell = 1, \dots, N; m = 1, \dots, N^{k-1}$.

Problem Statement
Obtain $\hat{\mathbf{x}}_{k|k}$ and covariance $\mathbf{P}_{k|k}$ given parameter set $\{\mathbf{p}_1, \mathbf{p}_2, \dots, \mathbf{p}_N\}$, parameter transition probability $\Pr\{\theta^\ell_k|\theta^\ell_{k-1}\}$, and measurement $\mathbf{y}_{1:k} = \{\mathbf{y}_1, \mathbf{y}_2, \dots, \mathbf{y}_k\}$.

Solution Approach
Obtain $\hat{\mathbf{x}}^i_{k|k}, \mathbf{P}^i_{k|k}, \Pr\{\theta^i_k|\mathbf{y}_{1:k}\}$, with given $\hat{\mathbf{x}}^\ell_{k-1|k-1}, \mathbf{P}^\ell_{k-1|k-1}, \Pr\{\theta^\ell_{k-1}|\mathbf{y}_{1:k-1}\}$, for $i, \ell = 1, \dots, N$.

Solution Process
Compute
$\hat{\mathbf{x}}^{\ell i}_{k|k}$: updated state estimate for model i at time k given $\hat{\mathbf{x}}^\ell_{k-1|k-1}$ at $k-1$, and $\mathbf{P}^{\ell i}_{k|k}$: covariance of $\hat{\mathbf{x}}^{\ell i}_{k|k}$ using a bank of N^2 filters.
Compute

$$\hat{\mathbf{x}}^i_{k|k} = \sum\nolimits_{\ell=1}^{N} \hat{\mathbf{x}}^{\ell i}_{k|k}\Pr\{\theta^\ell_{k-1}|\theta^i_k, \mathbf{y}_{1:k}\}, \text{and}$$

$$\mathbf{P}^i_{k|k} = \sum\nolimits_{\ell=1}^{N}\Pr\{\theta^\ell_{k-1}|\theta^i_k, \mathbf{y}_{1:k}\}\left[\mathbf{P}^{\ell i}_{k|k} + \left(\hat{\mathbf{x}}^{\ell i}_{k|k} - \hat{\mathbf{x}}^i_{k|k}\right)\left(\hat{\mathbf{x}}^{\ell i}_{k|k} - \hat{\mathbf{x}}^i_{k|k}\right)^T\right],$$

$$\Pr\{\theta^i_k|\mathbf{y}_{1:k}\} = \sum\nolimits_{\ell=1}^{N}\frac{p\left(\mathbf{y}_k|\theta^i_k, \theta^\ell_{k-1}, \mathbf{y}_{1:k-1}\right)}{p\left(\mathbf{y}_k|\mathbf{y}_{1:k-1}\right)}\Pr\{\theta^i_k|\theta^\ell_{k-1}\}\Pr = \{\theta^\ell_{k-1}|\mathbf{y}_{1:k-1}\},$$

$$p\left(\mathbf{y}_k|\theta^i_k, \theta^\ell_{k-1}, \mathbf{y}_{1:k-1}\right) = C exp\left\{-\frac{1}{2}\left(\mathbf{y}_k - \mathbf{H}^i_k\hat{\mathbf{x}}^{\ell i}_{k|k-1}\right)^T\left[\mathbf{H}^i_k\mathbf{P}^{\ell i}_{k|k-1}\mathbf{H}^{i\,T}_k + \mathbf{R}_k\right]^{-1}\left(\mathbf{y}_k - \mathbf{H}^i_k\hat{\mathbf{x}}^{\ell i}_{k|k-1}\right)\right\}$$

Solutions

$$\hat{\mathbf{x}}_{k|k} = \sum\nolimits_{i=1}^{N}\hat{\mathbf{x}}^i_{k|k}\Pr\{\theta^i_k|\mathbf{y}_{1:k}\}, \text{and}$$

$$\mathbf{P}_{k|k} = \sum\nolimits_{i=1}^{N}\Pr\{\theta^i_k|\mathbf{y}_{1:k}\}\left[\mathbf{P}^i_{k|k} + \left(\hat{\mathbf{x}}_k - \hat{\mathbf{x}}^i_{k|k}\right)\left(\hat{\mathbf{x}}_k - \hat{\mathbf{x}}^i_{k|k}\right)^T\right].$$

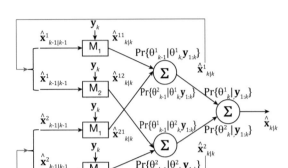

Figure 5.4   A two-model and single-step memory case implementation.

A two filter implementation illustration of the one-step memory MMEA is shown in Figure 5.4.

5.5.2   Two-Step Model History

In this case, at time $k-1$,

$$\hat{\mathbf{x}}_{k-1|k-1}^{\ell i}, \mathbf{P}_{k-1|k-1}^{\ell i}, \Pr\left\{\theta_{k-1}^{i}, \theta_{k-2}^{\ell}|\mathbf{y}_{1:k-1}\right\} \text{ for } \ell, i=1,\ldots,N,$$

and moving to time k one obtains

$$\hat{\mathbf{x}}_{k|k}^{ij}, \mathbf{P}_{k|k}^{ij}, \Pr\left\{\theta_{k}^{j}, \theta_{k-1}^{i}|\mathbf{y}_{1:k}\right\} \text{ for } i, j=1,\ldots,N.$$

Similar to the one-step history case, first expand (ℓ, i) to (ℓ, i, j) for a bank of N^3 filters to obtain

$$\hat{\mathbf{x}}_{k|k}^{\ell ij}, \mathbf{P}_{k|k}^{\ell ij}, \Pr\left\{\theta_{k}^{j}, \theta_{k-1}^{i}, \theta_{k-2}^{\ell}|\mathbf{y}_{1:k}\right\} \text{ for } \ell, i, j=1,\ldots,N,$$

then proceed to average over ℓ (two time steps ago)

$$
\begin{aligned}
\hat{\mathbf{x}}_{k|k}^{ij} &= \int \mathbf{x}_{k} p\left(\mathbf{x}_{k}|\theta_{k}^{j}, \theta_{k-1}^{i}, \mathbf{y}_{1:k}\right) d\mathbf{x}_{k} \\
&= \iint \mathbf{x}_{k} p\left(\mathbf{x}_{k}|\theta_{k}^{j}, \theta_{k-1}^{i}, \theta_{k-2}^{\ell}, \mathbf{y}_{1:k}\right) p\left(\theta_{k-2}^{\ell}|\theta_{k}^{j}, \theta_{k-1}^{i}, \mathbf{y}_{1:k}\right) d\theta_{k-2}^{\ell} d\mathbf{x}_{k} \\
&= \sum_{\ell=1}^{N} \hat{\mathbf{x}}_{k|k}^{\ell ij} \Pr\left\{\theta_{k-2}^{\ell}|\theta_{k}^{j}, \theta_{k-1}^{i}, \mathbf{y}_{1:k}\right\}.
\end{aligned}
\tag{5.5-6}
$$

Similarly, one can derive the equation for $\mathbf{P}_{k|k}^{ij}$ (the details are omitted) [10]

$$\mathbf{P}_{k|k}^{ij} = \sum_{\ell=1}^{N} \Pr\left\{\theta_{k-2}^{\ell} | \theta_{k}^{j}, \theta_{k-1}^{i}, \mathbf{y}_{1:k}\right\} \left[\mathbf{P}_{k|k}^{\ell ij} + \left(\hat{\mathbf{x}}_{k|k}^{ij} - \hat{\mathbf{x}}_{k|k}^{\ell ij}\right)\left(\hat{\mathbf{x}}_{k|k}^{ij} - \hat{\mathbf{x}}_{k|k}^{\ell ij}\right)^{T} \right]. \qquad (5.5\text{-}7)$$

The above computation needs the backward time hypothesis probability $\Pr\left\{\theta_{k-2}^{\ell} | \theta_{k}^{j}, \theta_{k-1}^{i}, \mathbf{y}_{1:k}\right\}$. Starting with the following relationship

$$\Pr\left\{\theta_{k-2}^{\ell} | \theta_{k}^{j}, \theta_{k-1}^{i}, \mathbf{y}_{1:k}\right\} = \frac{\Pr\left\{\theta_{k}^{j}, \theta_{k-1}^{i}, \theta_{k-2}^{\ell} | \mathbf{y}_{1:k}\right\}}{\Pr\left\{\theta_{k}^{j}, \theta_{k-1}^{i} | \mathbf{y}_{1:k}\right\}},$$

where the two terms on the right-hand side are

$$\Pr\left\{\theta_{k}^{j}, \theta_{k-1}^{i}, \theta_{k-2}^{\ell} | \mathbf{y}_{1:k}\right\} = \frac{p\left(\mathbf{y}_{k} | \theta_{k}^{j}, \theta_{k-1}^{i}, \theta_{k-2}^{\ell}, \mathbf{y}_{1:k-1}\right)}{p\left(\mathbf{y}_{k} | \mathbf{y}_{1:k-1}\right)} \Pr\left\{\theta_{k}^{j} | \theta_{k-1}^{i}, \theta_{k-2}^{\ell}\right\} \Pr\left\{\theta_{k-1}^{i}, \theta_{k-2}^{\ell} | \mathbf{y}_{1:k-1}\right\}$$

and

$$\Pr\left\{\theta_{k}^{j}, \theta_{k-1}^{i} | \mathbf{y}_{1:k}\right\} = \int \Pr\left\{\theta_{k}^{j}, \theta_{k-1}^{i}, \theta_{k-2}^{\ell} | \mathbf{y}_{1:k}\right\} p\left(\theta_{k-2}^{\ell}\right) d\theta_{k-2}^{\ell} = \sum_{\ell=1}^{N} \Pr\left\{\theta_{k}^{j}, \theta_{k-1}^{i}, \theta_{k-2}^{\ell} | \mathbf{y}_{1:k}\right\}.$$

The final estimate $\hat{\mathbf{x}}_{k|k}$ and its covariance $\mathbf{P}_{k|k}$ are obtained with double sums

$$\hat{\mathbf{x}}_{k|k} = \sum_{i=1}^{N} \sum_{j=1}^{N} \hat{\mathbf{x}}_{k|k}^{ij} \Pr\left\{\theta_{k}^{j}, \theta_{k-1}^{i} | \mathbf{y}_{1:k}\right\} \qquad (5.5\text{-}8)$$

$$\mathbf{P}_{k|k} = \sum_{i=1}^{N} \sum_{j=1}^{N} \Pr\left\{\theta_{k}^{j}, \theta_{k-1}^{i} | \mathbf{y}_{1:k}\right\} \left[\mathbf{P}_{k|k}^{\ell i} + \left(\hat{\mathbf{x}}_{k|k} - \hat{\mathbf{x}}_{k|k}^{ij}\right)\left(\hat{\mathbf{x}}_{k|k} - \hat{\mathbf{x}}_{k|k}^{ij}\right)^{T} \right]. \qquad (5.5\text{-}9)$$

Remarks

1. The difference to the two-step memory case to the one-step case is that the averaging is taken over two time steps in the past. This is the reason that N^3 filters are required for implementation.

2. Comments on transition probability, $\Pr\left\{\theta_{k}^{j} | \theta_{k-2}^{\ell}, \theta_{k-1}^{i}\right\}$ for all $i, \ell, j = 1, \dots, N$, similar to the one-step parameter history case, also apply here.

3. Again, the resulting state estimate and covariance are no longer sufficient to represent the a posteriori distribution function of $\mathbf{x}_k$, because it is multimode Gaussian.

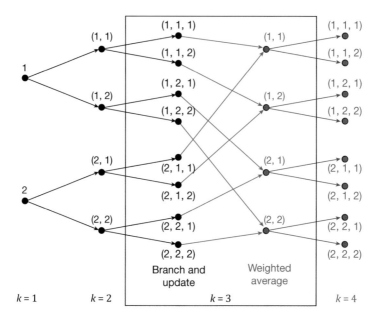

Figure 5.5   Illustration of a two-step model memory case for $N = 2$.

The two-step parameter history process for $N = 2$ is illustrated in Figure 5.5.

The purpose of presenting the algorithm for the two-step model history case is to show the algorithm extensibility. Because the number of elemental filters grows to N^3, it rapidly becomes impractical. For this reason, it will therefore not be summarized in a table.

5.6   Interacting Multiple Model Algorithm

As illustrated, the number of filters required for implementation goes exponentially with the length of the parameter history. As shown in Section 5.5.1, with the simplest case when the memory length is a single step, the number of filters is N^2 where N is the number of model used in the system. An approach to reduce the required filters for implementation to N was first proposed by Blom and Bar-Shalom [13]. This is accomplished by averaging over all local hypotheses of the previous time step before the filter prediction and update processes. This algorithm has achieved wide acceptance in applications and is known as the IMM algorithm. See [12, 14] for a survey on IMM with applications. IMM can be obtained by moving the averaging operation to before filter the update; for comparison, see Section 5.5.1.

Adopting the notation used in the one-step parameter history case, the starting point is

$$\hat{\mathbf{x}}_{k-1|k-1}^{\ell}, \mathbf{P}_{k-1|k-1}^{\ell}, \Pr\{\theta_{k-1}^{\ell}|\mathbf{y}_{1:k-1}\} \text{ for } \ell = 1, \dots, N.$$

The IMM filter uses the premixing algorithm to generate $\hat{\mathbf{x}}_{k-1|k-1}^{0\ell}, \mathbf{P}_{k-1|k-1}^{0\ell}$ with

$$\hat{\mathbf{x}}_{k-1|k-1}^{0\ell} = \sum_{i=1}^{N} \hat{\mathbf{x}}_{k-1|k-1}^{i} \Pr\{\theta_{k-1}^{i}|\theta_{k}^{\ell}, \mathbf{y}_{1:k-1}\}, \quad (5.6\text{-}1)$$

$$\mathbf{P}_{k-1|k-1}^{0\ell} = \sum_{i=1}^{N} \Pr\{\theta_{k-1}^{i}|\theta_{k}^{\ell}, \mathbf{y}_{1:k-1}\} \times \left[\mathbf{P}_{k-1|k-1}^{i} + \left(\hat{\mathbf{x}}_{k-1|k-1}^{0\ell} - \hat{\mathbf{x}}_{k-1|k-1}^{i}\right)\left(\hat{\mathbf{x}}_{k-1|k-1}^{0\ell} - \hat{\mathbf{x}}_{k-1|k-1}^{i}\right)^{T}\right],$$
$$(5.6\text{-}2)$$

$\hat{\mathbf{x}}_{k-1|k-1}^{0\ell}$ and $\mathbf{P}_{k-1|k-1}^{0\ell}$ are used as the a priori estimate for KF predict and update processing with the ℓth model to obtain $\hat{\mathbf{x}}_{k|k}^{\ell}$ and $\mathbf{P}_{k|k}^{\ell}$. For illustration, see Figure 5.6. The backward transition hypothesis probability $\Pr\{\theta_{k-1}^{i}|\theta_{k}^{\ell}, \mathbf{y}_{1:k-1}\}$ is computed by

$$\Pr\{\theta_{k-1}^{i}|\theta_{k}^{\ell}, \mathbf{y}_{1:k-1}\} = \frac{\Pr\{\theta_{k}^{\ell}|\theta_{k-1}^{i}, \mathbf{y}_{1:k-1}\}\Pr\{\theta_{k-1}^{i}|\mathbf{y}_{1:k-1}\}}{\Pr\{\theta_{k}^{\ell}|\mathbf{y}_{1:k-1}\}}$$

and

$$\Pr\{\theta_{k}^{\ell}|\theta_{k-1}^{i}, \mathbf{y}_{1:k-1}\} = \Pr\{\theta_{k}^{\ell}|\theta_{k-1}^{i}\}$$

is the single-step parameter evolution probability.

Next, consider computing $\Pr\{\theta_{k}^{\ell}|\mathbf{y}_{1:k}\}$ given $\Pr\{\theta_{k-1}^{j}|\mathbf{y}_{1:k-1}\}$ and $\mathbf{y}_{k}$ and transition probabilities, $\Pr\{\theta_{k}^{\ell}|\theta_{k-1}^{i}\}$. The basic relationship is Equation (5.5-3)

$$\Pr\{\theta_{k}^{i}|\mathbf{y}_{1:k}\} = \sum_{\ell=1}^{N} \frac{p\left(\mathbf{y}_{k}|\theta_{k}^{i}, \theta_{k-1}^{\ell}, \mathbf{y}_{1:k-1}\right)}{p\left(\mathbf{y}_{k}|\mathbf{y}_{1:k-1}\right)} \Pr\{\theta_{k}^{i}|\theta_{k-1}^{\ell}\}\Pr\{\theta_{k-1}^{\ell}|\mathbf{y}_{1:k-1}\}.$$

Because of the way IMM is implemented for the purpose of saving the number of elemental filters, the residual density $p\left(\mathbf{y}_{k}|\theta_{k}^{i}, \theta_{k-1}^{\ell}, \mathbf{y}_{1:k-1}\right)$ can only be approximated by the residual density of the ℓth filter, that is,

$$p\left(\mathbf{y}_{k}|\theta_{k}^{i}, \theta_{k-1}^{\ell}, \mathbf{y}_{1:k-1}\right) \approx$$
$$\text{Cexp}\left\{-\frac{1}{2}\left(\mathbf{y}_{k} - \mathbf{H}_{k}^{\ell}\hat{\mathbf{x}}_{k|k-1}^{0\ell}\right)^{T}\left[\mathbf{H}_{k}^{\ell}\mathbf{P}_{k|k-1}^{0\ell}\mathbf{H}_{k}^{\ell T} + \mathbf{R}_{k}\right]^{-1}\left(\mathbf{y}_{k} - \mathbf{H}_{k}^{\ell}\hat{\mathbf{x}}_{k|k-1}^{0\ell}\right)\right\}. \quad (5.6\text{-}3)$$

It is an approximation because the dependence on θ_{k}^{i} is only implied by the way in which $\hat{\mathbf{x}}_{k-1|k-1}^{0\ell}, \mathbf{P}_{k-1|k-}^{0\ell}$ is computed. Interested readers should consult Section 11.6.6 of Bar-Shalom's book [12] for discussion.

With $\hat{\mathbf{x}}_{k|k}^{\ell}$ and $\mathbf{P}_{k|k}^{\ell}$, the final estimate and covariance are computed the same way as in Equations (5.5-4) and (5.5-5), which are repeated for the purpose of completeness.

$$\hat{\mathbf{x}}_{k|k} = \sum_{\ell=1}^{N} \hat{\mathbf{x}}_{k|k}^{\ell} \Pr\left\{\theta_k^{\ell} | \mathbf{y}_{1:k}\right\}, \tag{5.6-4}$$

and

$$\mathbf{P}_{k|k} = \sum_{\ell=1}^{N} \Pr\left\{\theta_k^{\ell} | \mathbf{y}_{1:k}\right\} \left[\mathbf{P}_{k|k}^{\ell} + \left(\hat{\mathbf{x}}_k - \hat{\mathbf{x}}_{k|k}^{\ell}\right)\left(\hat{\mathbf{x}}_k - \hat{\mathbf{x}}_{k|k}^{\ell}\right)^T \right] \tag{5.6-5}$$

Equations for IMM are summarized in Table 5.4.

Remarks

1. The advantage of IMM over the single-step memory filter presented in Section 5.4.1 is computation. It has been shown by several researchers [12–14], that the computational saving is significant, with only slight degradation in performance.

2. Like all the MMEA discussed in this chapter, the a priori parameter evolution probability (the history) is critical to filter success. Further research on filter robustness is warranted. This is especially so for IMM because the averaging is conducted before filter update, or before the filters have a chance to learn with the most recent data.

An IMM implementation illustration for $N = 2$ is shown in Figure 5.6.

5.7   Numerical Examples

In this section, two examples for linear systems are presented. The first example uses the algorithm for the constant parameter case although in one of the experiments the true parameter is allowed to switch. The second experiment uses the IMM algorithm for a trajectory estimation problem, where the true target trajectory switches between maneuvering and nonmaneuvering.

Table 5.4 The Interacting Multiple Model Filter

IMM

System and Parameter Definition
All the same definitions and assumptions as in Table 5.3.

Problem Statement
The same as in Table 5.3.

Solution Approach
The same as in Table 5.3.

Solution Process
Instead of building a bank of N^2 filters, IMM uses the premixing algorithm to generate $\hat{\mathbf{x}}_{k-1|k-1}^{0\ell}$, $\mathbf{P}_{k-1|k-1}^{0\ell}$ using the backward transition probabilities $\Pr\{\theta_{k-1}^i|\theta_k^\ell, \mathbf{y}_{1:k-1}\}$, that is,

$$\hat{\mathbf{x}}_{k-1|k-1}^{0\ell} = \sum_{i=1}^{N} \hat{\mathbf{x}}_{k-1|k-1}^i \Pr\{\theta_{k-1}^i|\theta_k^\ell, \mathbf{y}_{1:k-1}\},$$

$$\mathbf{P}_{k-1|k-1}^{0\ell} = \sum_{i=1}^{N} \Pr\{\theta_{k-1}^i|\theta_k^\ell, \mathbf{y}_{1:k-1}\} \left[\mathbf{P}_{k-1|k-1}^i + \left(\hat{\mathbf{x}}_{k-1|k-1}^{0\ell} - \hat{\mathbf{x}}_{k-1|k-1}^i \right)\left(\hat{\mathbf{x}}_{k-1|k-1}^{0\ell} - \hat{\mathbf{x}}_{k-1|k-1}^i \right)^T \right].$$

With $\hat{\mathbf{x}}_{k-1|k-1}^{0\ell}$ and $\mathbf{P}_{k-1|k-1}^{0\ell}$ as the prior estimate and covariance, compute posterior state estimate and covariance, $\hat{\mathbf{x}}_{k|k}^\ell$ $\mathbf{P}_{k|k}^\ell$, and use a KF matched to model ℓ.

Recursive equation for hypothesis probability evolution:

$$\Pr\{\theta_k^i|\mathbf{y}_{1:k}\} = \sum_{\ell=1}^{N} \frac{p\left(\mathbf{y}_k|\theta_k^i, \theta_{k-1}^\ell, \mathbf{y}_{1:k-1}\right)}{p\left(\mathbf{y}_k|\mathbf{y}_{1:k-1}\right)} \Pr\{\theta_k^i|\theta_{k-1}^\ell\} \Pr\{\theta_{k-1}^\ell|\mathbf{y}_{1:k-1}\}.$$

with $p\left(\mathbf{y}_k|\theta_k^i, \theta_{k-1}^\ell, \mathbf{y}_{1:k-1}\right)$ approximated with

$$C\exp\left\{ -\frac{1}{2}\left(\mathbf{y}_k - \mathbf{H}_k^\ell \hat{\mathbf{x}}_{k|k-1}^{0\ell}\right)^T \left[\mathbf{H}_k^\ell \mathbf{P}_{k|k-1}^{0\ell} \mathbf{H}_k^{\ell T} + \mathbf{R}_k \right]^{-1} \left(\mathbf{y}_k - \mathbf{H}_k^\ell \hat{\mathbf{x}}_{k|k-1}^{0\ell}\right) \right\}.$$

Solutions

$$\hat{\mathbf{x}}_{k|k} = \sum_{\ell=1}^{N} \hat{\mathbf{x}}_{k|k}^\ell \Pr\{\theta_k^\ell|\mathbf{y}_{1:k}\},$$

$$\mathbf{P}_{k|k} = \sum_{\ell=1}^{N} \Pr\{\theta_k^\ell|\mathbf{y}_{1:k}\} \left[\mathbf{P}_{k|k}^\ell + \left(\hat{\mathbf{x}}_k - \hat{\mathbf{x}}_{k|k}^\ell \right)\left(\hat{\mathbf{x}}_k - \hat{\mathbf{x}}_{k|k}^\ell \right)^T \right].$$

Example 1[1] Consider a linear system

$$\begin{bmatrix} \dot{x}_1 \\ \dot{x}_2 \end{bmatrix} = \begin{bmatrix} 0 & 1 \\ 0 & \gamma \end{bmatrix}\begin{bmatrix} x_1 \\ x_2 \end{bmatrix} + \begin{bmatrix} 0 \\ 1 \end{bmatrix}\xi + \begin{bmatrix} 0 \\ 1 \end{bmatrix}u,$$

$$y = x_1 + v,$$

1 From Ref. [11].

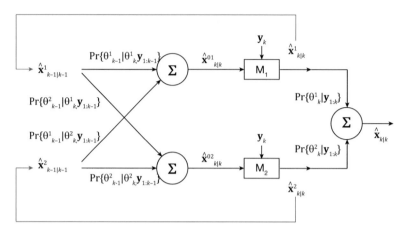

Figure 5.6   Illustration of IMM implementation for a two parameter case.

where ξ and v are process and measurement noise, respectively, both with variance 1. u is a control or excitation input to the system, and its role in this problem will be discussed later. γ is an unknown parameter that can take on the value 0, 0.5, or 1, corresponding to the three hypotheses defined below.

$$H_1: \gamma = 0,$$

$$H_2: \gamma = 0.5,$$

$$H_3: \gamma = 1.$$

The initial states are: $x_1 = 100$, $x_2 = 50$, the sampling rate is 10 Hz. Two experiments were run.

Experiment 1: Three cases, γ set at each of the three values one at a time.

Experiment 2: True parameter value is switched according to

$$\gamma = 0, \text{ for } 0 \leq t < 2,$$

$$\gamma = 0.5, \text{ for } 2 \leq t < 4,$$

$$\gamma = 1, \text{ for } 4 \leq t < 6.$$

The results for the first experiment are expressed in terms of hypothesis probability histories, the second in terms of parameter estimation history, and are shown in Figures 5.7 and 5.8, respectively.

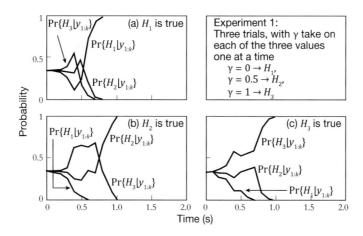

Figure 5.7   Hypothesis probability history for constant parameter case.

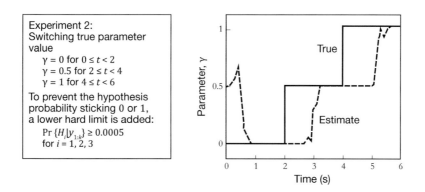

Figure 5.8   Hypothesis probability history for the switching probability case.

Remarks

An example is presented where the true parameters may either stay constant or switch among the three models, while the algorithm used is designed for the constant parameter case. The reason that the constant parameter algorithm works for the second experiment is because of two modifications that are implemented.

1. First, the hypothesis probabilities are hard bounded. This is to prevent any probabilities from converging to zero. If it did, it would be impossible for the probability to move out of zero because any finite number multiplied by zero is zero. The hypothesis problem used in this experiment is lower bounded by (an arbitrary choice)

$\Pr\{H_i|y_{1:k}\} \geq 0.0005$, for all i.

Second, although there is no process noise in the true system, a process noise term[2] with covariance matrix set equal to an identity matrix (also arbitrary, it is just what was used) is applied during the filter design to prevent the filter becoming overconfident in its estimates, thus making it difficult to switch to a different hypothesis. If the process noise is not added, the estimates of a mismatched filter can drift far away from the true state. When the true parameter jumps to a different value, that is, an originally mismatched filter now becomes matched, it would take a long time for the algorithm to identify and settle down to the new true system again. Leaving some process noise level in the filter will allow the mismatched filters to stay close to the true state so that the algorithm can adapt more quickly when the parameter jumps.

2. The control variable u also plays a critical role in this experiment. It represents a persistent excitation to explore differences among these systems. Without it, the state for every model will converge to zero, thus there is no more difference among them. This brings up an interesting question, which is on how to design the input (exciting) function for enabling (or expediting) system identification in using the MMEA.

3. The constant parameter algorithm is optimum when the true parameter stays constant. The modification used in this case to make it work for a switch parameter case is ad hoc but practical.

Example 2   A maneuvering trajectory is used to compare the performance of three filters. This is the same trajectory used in Section 4.9.2. The difference is that the radar measurement bias is set to zero. Three filters are compared: a constant velocity (CV), a constant acceleration (CA), and an IMM which incorporates two elemental

---

2   Recall the discussion in Chapter 4—it is prudent for the system designer to keep the bandwidth of the KF open by including process noise in the filter design.

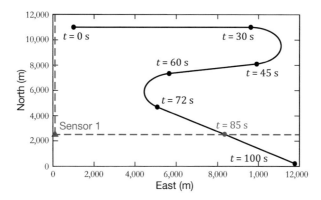

Figure 5.9   Trajectory of a maneuvering target, maneuverings taking place during turns.

models, a CV and a CA. The truth trajectory, which is the same as in Figure 4.18, is repeated in Figure 5.9 for easy reference.

Following the solution to Homework Problems 1 in Chapter 3 and Chapter 4, the mismatched KF (and EKF) discussion, and the example shown in Section 4.9.2, it should be clear to the reader that filters designed using either a CV or CA model[3,4] for the entire trajectory can be made to work; however, it will require that the filter gain be properly tuned using the process noise covariance in the filter prediction process. The tuning parameters (continuous and discrete systems) for a 5 g maneuver are

CV^H: For continuous systems: $6,000 \ m^2/s^3$. For discrete-time systems: $60,000 \ m^2/s^4$, with 0.1 s update period

CA: For continuous systems: $800 \ m^2/s^5$. For discrete-time systems: $8,000 \ m^2/s^6$, with a 0.1 s update period

where CV^H denotes a filter designed assuming a constant velocity, and tuned with high process noise. This is necessary because CV is a reduced order model for the maneuvering target, yet it is being asked to estimate the trajectory of a maneuvering target. The CV filter used in the IMM can be tuned with a much smaller process noise

---

3   For CV and CA models in Cartesian coordinates, see Homework Problem 1 in Chapter 2, on linear estimation. For transformation from Cartesian coordinates to radar measurement coordinates, see Homework Problem 1 in Chapter 3, on nonlinear estimation.

4   Bank to turn is not the same as CA in modeling target maneuvers. Using CA for filter construction is only an approximation. It is nevertheless a simpler model and it is shown here that a CA-based filter achieves fine estimation performance.

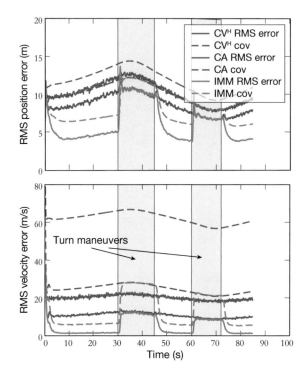

Figure 5.10   Comparison of the root-mean-square errors and covariance of IMM, CV, and CA filters for the maneuvering trajectory shown in Figure 5.9.

because it is only required to accurately estimate the state during the nonmaneuvering portion of the trajectory. In this case the process noise is as follows.

CV^L: For continuous systems: $0.5 \text{ m}^2/\text{s}^3$. For discrete-time systems: $5 \text{ m}^2/\text{s}^4$ with 0.1 s update period

where CV^L denotes the CV filter used in the IMM and tuned with low process noise. The process noise parameter for the CA filter used in IMM remains the same. The results of these filters are shown in Figure 5.10.

   Figure 5.10 compares the performance of the three filters using position and velocity estimation errors (obtained with 1000 Monte Carlo repetitions), and their corresponding error estimates obtained from the computed covariance matrices. The root mean square (RMS) state vector errors are computed by averaging (over the 1000 Monte Carlo repetitions) the square root of the sum of error squares in each coordinate for position and velocity separately. The errors computed using the covariance is the square root of the trace of the computed filter covariance matrix, done separately for position and velocity. The gray shaded areas indicate the periods when the object is maneuvering. All three approaches can track the object, that is, the actual position errors are all contained within the covariances. The IMM gives the smallest error independent of whether the object is maneuvering or not maneuvering. This is because the elemental filters of IMM are matched with these two phases of the trajectory. With smaller error and consistent covariance, IMM is expected to achieve better performance in an environment with multiple and closely spaced objects (Chapter 8). The IMM is able to switch to maneuver mode quickly, but it lags behind for about 4 to 5 s when the object returns to a straight line. This is not a surprise because CA can follow objects traveling in a straight line, that is, CV is a sub-model of CA (see the discussion in Chapter 4 on mismatched models). Shown in Figure 5.11 are mean hypothesis probabilities from Equation (5.6-1) of CV^L and CA over 1000 Monte Carlo repetitions, indicating mode switches between these two filters, which match well with the switching times shown in the error curves of Figure 5.10.

Remarks

1. The above examples clearly indicate that an MMEA can adapt to the model switches of the actual systems with better estimation accuracy.

2. Although the algorithm for estimating the state of a switching system, the IMM in this case, is heuristic, and thus suboptimal, the numerical examples above showed that this algorithm works better than using a single model.

3. There are many additional applications that show the improved performance of multiple model approaches over the single model approach [13, 14].

4. The performance of IMM is only slightly worse than the GPB2 in Section 5.5.1, but the computational savings may be significant [14].

Homework Problems

1. Fill in some of the details in derivation, for example, Equation (5.5-7).

2. Try to derive the IMM algorithm shown in Section 5.6, explain why the filter works even though the averaging step (also referred to as the filter interaction step) does not directly involve filter residuals.

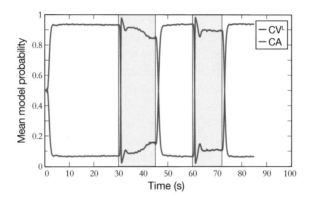

Figure 5.11   Hypothesis probability of CV^L and CA being true, defined in Equation (5.5-1).

3. Build a constant parameter filter (CM²EA) with two models, CA and CV, against a target that is flying strictly either with a CV or a CA trajectory. The CM²EA is a bank of two filters, CV and CA, running in parallel, not switching. Observe the performance of your filter, and comment on the hypothesis probability behavior.

4. Take the simple target maneuvering case of CV-CA-CV, for which you have already built a simulation. Apply your CM²EA filter to it. Adjust the process noise on your filter to see whether it will track this target, paying special attention to the model switch time. Compare with the results you obtained in Homework Problem 1 in Chapter 3.

5. Build an MMEA for a one-step parameter history as shown in Section 5.5.1 and use it to estimate the trajectory of the CV-CA-CV case. Observe how it performs during each phase of the flight and during phase transitions. Adjust the transition probability, $\Pr\{\theta_k^i | \theta_{k-1}^\ell\}$, empirically to observe how the performance changes. Compare with the results of Problem 4 and with the tuned CV or CA only filter which you built for the Homework Problems in Chapter 3.

6. Repeat the same steps for the IMM filter, and compare results.

7. Consider a trajectory going from mid-course to reentry.

    a. Repeat Homework Problems 3 to 6 with a CV or CA filter.

    b. Repeat Homework Problems 3 to 6 for an object switching between ballistic and aerodynamic maneuvers. (This refers to Homework Problem 3 in Chapter 3 for the aerodynamic maneuver model. Note that when the maneuvering coefficients, α_t and α_c, are set to zero, it models a ballistic object.)

References

1. D.T. Magill, "Optimal Adaptive Estimation of Sampled Stochastic Processes," *IEEE Transactions on Automatic Control*, vol. AC-10, pp. 434–439, Oct. 1965.

2. M. Athans, K.-P. Dunn, C.S. Greene, W.H. Lee, N.R. Sandell, I. Segall, and A.S. Willsky, "The Stochastic Control of the F-8C Aircraft Using the Multiple Model Adaptive Control (MMAC) Method," in *Proceedings of the 1975 IEEE Conference on Decision and Control*, pp. 217–228, Dec. 1975.

3. Y.C. Ho and R.C.K. Lee, "A Bayesian Approach to Problems in Stochastic Estimation and Control," *IEEE Transactions on Automatic Control*, vol. AC-9, pp. 333–339, Oct. 1964.

4. G.A. Ackerson and K.S. Fu, "On State Estimation in Switching Environments," *IEEE Transactions on Automatic Control*, vol. AC-15, pp. 10–17, Feb. 1970.

5. J.K. Tugnait and A.H. Haddad, "A Detection-Estimation Scheme for State Estimation in Switching Environments," *Automatica*, vol. 15, pp. 477–481, 1979.

6. D.G. Lainiotis, "Optimal Adaptive Estimation: Structure and Parameter Adaptation," *IEEE Transactions on Automatic Control*, vol. AC-16, pp. 160–170, Apr. 1971.

7. D.G. Lainiotis, "Partitioning: A Unifying Framework for Adaptive Systems. I: Estimation," *Proceedings of the IEEE*, vol. 64, pp. 1126–1143, Aug. 1976.

8. D. Willner, "Observation and Control of Partially Unknown Systems," Ph.D. Thesis, Department of Electrical Engineering, MIT, Cambridge, MA, and MIT Electronic Systems Lab, Rep. ESL-R-496, June 1973.

9. R. Bueno, E.Y. Chow, K.P. Dunn, S.B. Gershwin, and A.S. Willsky, "Status Report on the Generalized Likelihood Ratio Failure Detection Technique, with Application to the F-8 Aircraft," in *Proceedings of the 1976 IEEE Conference on Decision and Control*, Dec. 1976, pp. 38–47.

10. C.B. Chang and M. Athans, "State Estimation for Discrete Systems with Switching Parameters," *IEEE Transactions on Aerospace and Electronic Systems*, vol. AES-14, pp. 418–425, May 1978.

11. M. Athans and C.B. Chang, "Adaptive Estimation and Parameter Identification using Multiple Model Estimation Algorithms," MIT Lincoln Lab., Lexington, MA, Rep.TN-1976-28, June 23, 1976.

12. Y. Bar-Shalom, X. Rong Li, and T. Kirubarajan, *Estimation with Applications to Tracking and Navigation*. New York: Wiley, 2001.

13. H.A.P. Blom and Y. Bar-Shalom, "The Interacting Multiple Model Algorithm for Systems with Markovian Switching Coefficients," *IEEE Transactions on Automatic Control*, vol. AC-33, pp. 780–783, Aug. 1988.

14. E. Mazor, A. Averbuch, Y. Bar-Shalom, and J. Dayan, "Interacting Multiple Model Methods in Target Tracking: A Survey," *IEEE Transactions on Aerospace and Electronic Systems*, AES-34, pp. 103–123, Jan. 1998.

6

Sampling Techniques for State Estimation

6.1 Introduction

When both system and measurement noises are Gaussian, several filters for nonlinear systems have been introduced thus far. All of these have used the Taylor series expansion to approximate the nonlinear system and measurement equations to obtain the statistics of the state estimate (mean and covariance) [1]. Other approximations including the Gaussian sum filter [2], and the second order extended Kalman filter (EKF) [3] (shown in Section 3.7) were proposed in the 1970s. The added computational complexity of the second order EKF has limited its use. In this chapter a different concept is introduced in which the approximated conditional density functions are computed either using systematically selected deterministic samples or statistically selected random samples, and then the statistics (mean and covariance) can be computed numerically.

In 1969, a grid of deterministic sample points (or cells) around the region of interest in the state space was used to compute the conditional density functions for the point-mass filter (PMF) [4]. By propagating these sample points (or cells) according to the nonlinear system equation and computing the likelihood of these samples about the current measurement allowed for the approximation of the updated state conditional density function. In the late 1990s, another numerical approximation technique utilizing the unscented transform in an EKF framework for Gaussian noise cases, known as the unscented Kalman filter (UKF) [5–7], gained some popularity. The conditional density functions are approximated by a set of deterministically selected points (sigma points) from the Gaussian approximation of $p(\mathbf{x}_{k-1}|\mathbf{y}_{1:k-1})$. These points are propagated through the system and measurement, and the mean, $\hat{\mathbf{x}}_{k|k}$ and covariance $\mathbf{P}_{k|k}$ are approximated by the sample statistics of the propagated sample points at time k. This technique is aimed at better approximating mean and covariance after a nonlinear transformation [5–8].

For many application areas, the estimation problems are most likely nonlinear and non-Gaussian, and the optimal solution is intractable. Another class of sampling techniques, referred as particle filtering approximates the solution by random samples to solve the estimation problem [9]. This technique uses a sequential Monte Carlo (SMC) scheme to generate point masses (or "particles") to approximate the probability density functions involved in the estimation problem. There are several Monte Carlo sampling techniques that can be applied to these problems, such as rejection sampling, importance sampling, and Markov chain Monte Carlo (MCMC) [10]. In this chapter, techniques involving importance sampling that was introduced in the field of statistics for estimating properties of a particular distribution back in the 1950s [11] is emphasized. There were a few explorations of these ideas for filtering applications during the 1960s and 1970s, but they were not popular due to the excessive computational requirements and the problems (e.g., particle depletion or degeneration) associated with the early version of the SMC schemes. The major advancement of the SMC for filtering applications was due to availability of faster computers, and the inclusion of a resampling step in the SMC scheme [12]. Since then, research activity in the field has dramatically increased [9, 12–14], resulting in many improvements to the particle filtering approach, as well as a growth in application to a variety of fields that require the computation of conditional probability densities beyond that of the tracking/filtering problems discussed in this book.

In Section 6.2, several computation techniques for the conditional expectation of $g(\mathbf{x})$ given the noisy observation $\mathbf{y} = \mathbf{h}(\mathbf{x}) + \mathbf{v}$ are introduced. These techniques for random vectors $\mathbf{x}$ and $\mathbf{y}$ cover computation strategies for problems that were discussed in the previous chapters as well as the sampling techniques that will be introduced in this chapter. The Bayesian approach to state estimation for nonlinear systems is reviewed in Section 6.3 for easy reference. Two deterministic sampling techniques, namely, the UKF and the PMF, are derived in Section 6.4 and Section 6.5, respectively. Section 6.6 introduces various particle filters based on the Monte Carlo (random) sampling technique. For some cases, these filters have been shown to give better performance than the EKF, UKF, and interacting multiple model (IMM) filters [15, 16].

6.2   Conditional Expectation and Its Approximations

In most parameter or state estimation problems, the following integration is encountered

$$E\{g(\mathbf{x})|\mathbf{y}\} = \int g(\mathbf{x})p(\mathbf{x}|\mathbf{y})d\mathbf{x}, \qquad (6.2\text{-}1)$$

which is the conditional expectation of $g(\mathbf{x})$ given the observation

$$\mathbf{y} = \mathbf{h}(\mathbf{x}) + \mathbf{v} \tag{6.2-2}$$

where $\mathbf{x} \in \mathbb{R}^n$ and $\mathbf{y}, \mathbf{v} \in \mathbb{R}^m$. $\mathbf{y}$ is the measurement of an unknown random vector, $\mathbf{x}$, through Equation (6.2-2) with additive random noise $\mathbf{v}$, which is independent of $\mathbf{x}$. Note that the condition expectation in Equation (6.2-1) is the minimum norm estimator for $g(\mathbf{x})$ given $\mathbf{y}$. The conditional density function $p(\mathbf{x}|\mathbf{y})$ represents the statistical relationship between $\mathbf{x}$ and $\mathbf{y}$. In the following sections, different cases are presented that Equation (6.2-1) can be computed exactly or approximated with analytical formulas or numerical schemes.

6.2.1   Linear and Gaussian Cases

First, considering the linear and Gaussian cases that

$$\mathbf{y} = \mathbf{H}\mathbf{x} + \mathbf{v} \tag{6.2-3}$$

where $\mathbf{x}: \sim N(\bar{\mathbf{x}}, \mathbf{P}_x) \in \mathbb{R}^n$ and $\mathbf{v}: \sim N(\mathbf{0}, \mathbf{R}) \in \mathbb{R}^m$ are two independent Gaussian random vectors. It has been shown in Section 1.5.4 (and in Appendix 1.B "Properties of the Linear Least Squares Estimator") that the conditional density function, $p(\mathbf{x}|\mathbf{y})$, is also Gaussian and its mean and covariance $(\hat{\mathbf{x}}, \mathbf{P})$ can be obtained as

$$\hat{\mathbf{x}} = \bar{\mathbf{x}} + \mathbf{K}(\mathbf{y} - \mathbf{H}\bar{\mathbf{x}}), \tag{6.2-4}$$

$$\mathbf{P} = \left[\mathbf{H}^T\mathbf{R}^{-1}\mathbf{H} + \mathbf{P}_x^{-1}\right]^{-1}, \text{ and} \tag{6.2-5}$$

$$\mathbf{K} = \mathbf{P}\mathbf{H}^T\mathbf{R}^{-1}. \tag{6.2-6}$$

For any continuous function of $\mathbf{x}$, $g(\mathbf{x})$ can be represented by a polynomial of $\mathbf{x}$, the integral in Equation (6.2-1) can be computed exactly as a function of $(\hat{\mathbf{x}}, \mathbf{P})$ (properties of Gaussian random vectors).

6.2.2   Approximated by Taylor Series Expansion

When the measurement is nonlinear

$$\mathbf{y} = \mathbf{h}(\mathbf{x}) + \mathbf{v}$$

where $\mathbf{x}: \sim N(\bar{\mathbf{x}}, \mathbf{P}_x) \in \mathbb{R}^n$ and $\mathbf{v}: \sim N(\mathbf{0}, \mathbf{R}) \in \mathbb{R}^m$ are two independent Gaussian random vectors. The conditional density function, $p(\mathbf{x}|\mathbf{y})$ can be approximated by a Gaussian density function $N(\hat{\mathbf{x}}, \mathbf{P})$ as shown in Section 1.6, with

$$\hat{\mathbf{x}} = \bar{\mathbf{x}} + \mathbf{K}(\mathbf{y} - \mathbf{h}(\bar{\mathbf{x}})), \tag{6.2-7}$$

$$\mathbf{P} = \left[\mathbf{H}_{\bar{\mathbf{x}}}^T \mathbf{R}^{-1} \mathbf{H}_{\bar{\mathbf{x}}} + \mathbf{P}_{\mathbf{x}}^{-1} \right]^{-1}, \tag{6.2-8}$$

$$\mathbf{K} = \mathbf{P} \mathbf{H}_{\bar{\mathbf{x}}}^T \mathbf{R}^{-1}, \tag{6.2-9}$$

where $\mathbf{H}_{\bar{\mathbf{x}}} = \left[\frac{\partial \mathbf{h}(\mathbf{x})}{\partial \mathbf{x}} \right]_{\bar{\mathbf{x}}}$ is the Jacobian of $\mathbf{h}(\mathbf{x})$ evaluated at $\bar{\mathbf{x}}$. Note that the density function $p(\mathbf{x}|\mathbf{y})$ is not Gaussian in general. Depending on the nonlinearity of $\mathbf{h}(\cdot)$, the approximation for Equation (6.2-1) using $N(\hat{\mathbf{x}}, \mathbf{P})$, may not be accurate enough. One can consider a higher order Taylor expansion of $\mathbf{h}(\cdot)$. This will require the computations of the associated Jacobian and Hessian matrices (see Section 3.7) that makes this approach less attractive. The first two moments of $g(\mathbf{x})$ can be approximated by the following equations

$$E\{g(\mathbf{x})|\mathbf{y}\} \approx g(\hat{\mathbf{x}}), \tag{6.2-10}$$

$$E\left\{ (g(\mathbf{x}) - E\{g(\mathbf{x})|\mathbf{y}\})(g(\mathbf{x}) - E\{g(\mathbf{x})|\mathbf{y}\})^T \right\} \approx \mathbf{G}_{\hat{\mathbf{x}}} \mathbf{P} \mathbf{G}_{\hat{\mathbf{x}}}^T, \tag{6.2-11}$$

where $\mathbf{G}_{\hat{\mathbf{x}}} = \left[\frac{\partial g}{\partial \mathbf{x}} \right]_{\hat{\mathbf{x}}}$, the Jacobian of $g(\mathbf{x})$ evaluated at $\hat{\mathbf{x}}$. Note that this transformation was essential for the derivation of the EKF (see Section 3.4).

6.2.3   Approximated by Unscented Transformation

For the cases where both $\mathbf{x} \in \mathbb{R}^n$ and $\mathbf{v} \in \mathbb{R}^m$ are two independent Gaussian random vectors, the integral of Equation (6.2-1) can be approximated by the unscented transformation. This is a method to compute the corresponding mean and covariance of $g(\mathbf{x})$ by averaging the transformations from $\mathbf{x}$ to $g(\mathbf{x})$ of carefully selected sample points in the $\mathbf{x}$ space without the computations of the Jacobian of $g(\mathbf{x})$. First, choose the mean $\bar{\mathbf{x}}$ and $2n$ points[1] ($2n + 1$ deterministic points defined below, also see Figure 6.1) distributed around the mean $\bar{\mathbf{x}}$ on the ellipsoid surface of the covariance matrix, $\bar{\mathbf{P}}$. The mean and the $2n$ points on the surface of the ellipsoids constitute a total of $2n + 1$ sigma points

$$\mathbf{x}^0 = \bar{\mathbf{x}}$$

$$\mathbf{x}^i = \bar{\mathbf{x}} + \left(\sqrt{(n + \kappa)\bar{\mathbf{P}}} \right)_i^T \ \forall \ i = 1, \ldots, n,$$

$$\mathbf{x}^{n+i} = \bar{\mathbf{x}} - \left(\sqrt{(n + \kappa)\bar{\mathbf{P}}} \right)_i^T \ \forall \ i = 1, \ldots, n, \tag{6.2-12}$$

where κ is an arbitrary real number, which is used to fine tune the results to compensate for the fact that higher order moments are not computed for non-Gaussian

---

1   For the simplicity of notation in this chapter, the sample points are deterministic vectors, and the same standard font for random vectors will be used.

distributions. For Gaussian distribution, a useful heuristic selection is $n + \kappa = 3$ [6]. The notation $(\mathbf{A})_i$ denotes the ith row of $\mathbf{A}$ and the matrix square root is defined by $\mathbf{A} = \left[\sqrt{\mathbf{A}}\right]^T \left[\sqrt{\mathbf{A}}\right]$, and can be computed using MATLAB's Cholesky factorization. Let $\mathbf{v}^i = \mathbf{g}(\mathbf{x}^i)$ denote the corresponding sigma points in the $\mathbf{v}$ space, then the mean and covariance of $\mathbf{v}$ are computed as

$$\bar{\mathbf{v}} \approx \sum_{i=0}^{2n} w^i \mathbf{v}^i \tag{6.2-13}$$

and

$$\mathbf{P}_\mathbf{v} \approx \sum_{i=0}^{2n} w^i \left(\mathbf{v}^i - \bar{\mathbf{v}}\right)\left(\mathbf{v}^i - \bar{\mathbf{v}}\right)^T \tag{6.2-14}$$

where the weighting coefficients are defined as

$$w^0 = \frac{\kappa}{(n+\kappa)}, \text{ and } w^i = w^{i+n} = \frac{1}{2(n+\kappa)} \, \forall \, i = 1, \ldots, n. \tag{6.2-15}$$

and

$$E\{\mathbf{g}(\mathbf{x})\} \approx \sum_{i=0}^{2n} w^i \mathbf{g}(\mathbf{x}^i). \tag{6.2-16}$$

One can prove that for $\mathbf{g}(\mathbf{x}) = \mathbf{x}$, then $\bar{\mathbf{v}} = \bar{\mathbf{x}}$ and $\mathbf{P}_\mathbf{v} = \bar{\mathbf{P}}$.

The conditional density function, $p(\mathbf{x}|\mathbf{y})$, can be approximated as a Gaussian density function $N(\hat{\mathbf{x}}, \mathbf{P})$ using the unscented transformation from Equations (6.2-12) to (6.2-15) (see Section 1.5.4), yielding

$$\hat{\mathbf{x}} = \bar{\mathbf{x}} + \mathbf{K}\left(\mathbf{y} - \widehat{\mathbf{h}(\mathbf{x})}\right), \tag{6.2-17}$$

$$\mathbf{P} = \mathbf{P}_\mathbf{x} - \mathbf{P}_{\mathbf{x}\gamma} \mathbf{P}_{\gamma\gamma}^{-1} \mathbf{P}_{\gamma\mathbf{x}}, \tag{6.2-18}$$

$$\mathbf{K} = \mathbf{P}_{\mathbf{x}\gamma} \mathbf{P}_{\gamma\gamma}^{-1}, \tag{6.2-19}$$

where

$$\gamma \triangleq \mathbf{y} - \widehat{\mathbf{h}(\mathbf{x})},$$

$$\widehat{\mathbf{h}(\mathbf{x})} \approx \sum_{i=0}^{2n} w^i \mathbf{h}(\mathbf{x}^i), \tag{6.2-20}$$

$$\mathbf{P}_{\gamma\gamma} \approx \sum_{i=0}^{2n} w^i \left(\mathbf{h}(\mathbf{x}^i) - \widehat{\mathbf{h}(\mathbf{x})}\right)\left(\mathbf{h}(\mathbf{x}^i) - \widehat{\mathbf{h}(\mathbf{x})}\right)^T + \mathbf{R}, \tag{6.2-21}$$

$$\mathbf{P}_{\gamma\mathbf{x}}^T = \mathbf{P}_{\mathbf{x}\gamma} \approx \sum_{i=0}^{2n} w^i \left(\mathbf{x}^i - \bar{\mathbf{x}}\right)\left(\mathbf{h}(\mathbf{x}^i) - \widehat{\mathbf{h}(\mathbf{x})}\right)^T. \tag{6.2-22}$$

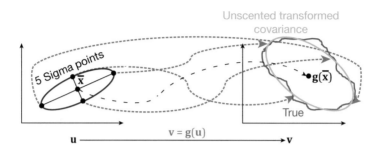

Figure 6.1   Illustration of unscented transformation.

It was argued in [8] that the statistics (mean and covariance) computed with unscented transformation are equivalent to using a Taylor series expansion up to third order, thus providing a better approximation than the method used in Equations (6.2-7) and (6.2-8). An illustration of the unscented transformation is shown in Figure 6.1. The ellipsoids shown in the figure are the 1-sigma ellipsoids with the associate covariance matrices and the red boundary indicates the true 1-sigma extent that may be obtainable with large number of samples using a random sampling approach, which will be illustrated in the next section.

The unscented transformation presented in [8] used a simpler form, as it did not include the mean $\bar{\mathbf{x}}$ in the sample set. This can be obtained simply by selecting $\kappa = 0$. It was argued in [8] that this approach is simpler and the end results are nearly identical. The above method is nevertheless presented for the purpose of completeness.

6.2.4   Approximated by Point Mass Integration

For much more general nonlinear and non-Gaussian cases, we have

$$\mathbf{y} = \mathbf{h}(\mathbf{x}) + \mathbf{v}$$

where $\mathbf{x} \in \mathbb{R}^n$ and $\mathbf{v} \in \mathbb{R}^m$ are two independent random vectors with known density functions $p(\mathbf{x})$ and $p_v(\mathbf{v})$. The integral in Equation (6.2-1) can be computed numerically by a summation similar to the Gaussian quadrature for line integrals as developed in [4, 17]. Let us first divide the region of interest[2] in $\mathbb{R}^n$ into N_S equal-volume hypercubes with associated deterministic centers $\{\mathbf{x}^i, i = 1, \ldots, N_S\}$. The probability density function, $p(\mathbf{x})$ can be approximated as the following summation

---

2   In order to have a reasonable number of hypercubes, N_S, the region of interest has to be selected in advance such that the approximation can be computed accurately and efficiently.

$$p(\mathbf{x}) \approx \sum_{i=1}^{N_S} w_{\mathbf{x}}^i \delta_D(\mathbf{x} - \mathbf{x}^i),\qquad(6.2\text{-}23)$$

where $\delta_D(.)$ is the Dirac delta measure and the weights $\{w_{\mathbf{x}}^i, i = 1, \ldots, N_S\}$ represent the probability of $\mathbf{x}$ in the hypercube centered by $\mathbf{x}^i$, such that

$$\sum_{i=1}^{N_S} w_{\mathbf{x}}^i = 1.$$

The conditional density function, $p(\mathbf{x}|\mathbf{y})$, can be approximated with centers $\{\mathbf{x}^i, i = 1, \ldots, N_S\}$ via Bayes' rule and the density function of $\mathbf{v}$, $p_{\mathbf{v}}(\mathbf{v})$, as follows

$$p(\mathbf{x}|\mathbf{y}) \approx \frac{p_{\mathbf{v}}(\mathbf{y} - \mathbf{h}(\mathbf{x})) \sum_{i=1}^{N_S} w_{\mathbf{x}}^i \delta_D(\mathbf{x} - \mathbf{x}^i)}{\sum_{i=1}^{N_S} w_{\mathbf{x}}^i p_{\mathbf{v}}(\mathbf{y} - \mathbf{h}(\mathbf{x}^i))} = \sum_{i=1}^{N_S} w^i \delta_D(\mathbf{x} - \mathbf{x}^i),\qquad(6.2\text{-}24)$$

where

$$w^i = w_{\mathbf{x}}^i p_{\mathbf{v}}(\mathbf{y} - \mathbf{h}(\mathbf{x}^i)) \Big/ \sum_{j=1}^{N_S} w_{\mathbf{x}}^j p_{\mathbf{v}}(\mathbf{y} - \mathbf{h}(\mathbf{x}^j)),\qquad(6.2\text{-}25)$$

and

$$E\{\boldsymbol{g}(\mathbf{x})|\mathbf{y}\} \approx \sum_{i=1}^{N_S} w^i \boldsymbol{g}(\mathbf{x}^i).\qquad(6.2\text{-}26)$$

To have an accurate approximation, the number of hypercubes, N_S, can be extremely large. If the information about the region of interest can be determined based on $\mathbf{x}$ and $\mathbf{y}$, it can be used to reduce the sample size to some degree, but the computational requirement of this technique to meet a high accuracy level may be excessive as the dimension of $\mathbf{x}$ increases. In the following section, a random sampling technique will be introduced to take advantage of knowing the density function of $\mathbf{x}$, $p(\mathbf{x})$ such that one can approximate the integral with a smaller number of samples at a more important location of $\mathbf{x}$ and create the samples $\{\mathbf{x}^i\}$ more efficiently.

6.2.5   Approximated by Monte Carlo Sampling

The concept of computing the integral in Equation (6.2-1) using random samples generated from the random vector $\mathbf{x}$ with given density function $p(\mathbf{x})$, has been developed in the 1950s for high energy physics [11]. If N_S independent identical distributed (i.i.d.) random sample points, or particles, $\{\mathbf{x}^i: i = 1, \ldots, N_S\}$ can be drawn from $p(\mathbf{x})$, (denoted by $\mathbf{x}^i \sim p(\mathbf{x})$) the density function can be approximated numerically as

$$p(\mathbf{x}) \approx \frac{1}{N_S} \sum_{i=1}^{N_S} \delta_D(\mathbf{x} - \mathbf{x}^i) \qquad (6.2\text{-}27)$$

where $\delta_D(.)$ is the Dirac delta measure. Similar to the point-mass approximation technique in Section 6.2.4, the conditional density function, $p(\mathbf{x}|\mathbf{y})$, can be approximated with $\{\mathbf{x}^i, i = 1, \dots, N_S\}$ via Bayes' rule and the density function of $\mathbf{v}$, $p_v(\mathbf{v})$ as follows

$$p(\mathbf{x}|\mathbf{y}) \approx \frac{p_v(\mathbf{y} - \mathbf{h}(\mathbf{x})) \dfrac{1}{N_S} \sum_{i=1}^{N_S} \delta_D(\mathbf{x} - \mathbf{x}^i)}{\dfrac{1}{N_S} \sum_{i=1}^{N_S} p_v(\mathbf{y} - \mathbf{h}(\mathbf{x}^i))} = \sum_{i=1}^{N_S} w^i \delta_D(\mathbf{x} - \mathbf{x}^i), \qquad (6.2\text{-}28)$$

where

$$w^i = p_v(\mathbf{y} - \mathbf{h}(\mathbf{x}^i)) \Big/ \sum_{j=1}^{N_S} p_v(\mathbf{y} - \mathbf{h}(\mathbf{x}^j)), \qquad (6.2\text{-}29)$$

and

$$E\{\mathbf{g}(\mathbf{x})|\mathbf{y}\} \approx \sum_{i=1}^{N_S} w^i \mathbf{g}(\mathbf{x}^i). \qquad (6.2\text{-}30)$$

In general, sampling directly from $p(\mathbf{x})$ for Equation (6.2-27) may not be practical, but one can sample from another density function $q(\mathbf{x})$ (called the "*importance function*") that has the same support as $p(\mathbf{x})$ [10]. Let $\{\mathbf{x}^i : i = 1, \dots, N_S\}$ be N_S i.i.d. random samples from $q(\mathbf{x})$ (i.e., $\mathbf{x}^i \sim q(\mathbf{x})$) instead of $p(\mathbf{x})$. In order to apply Equation (6.2-30) with $\mathbf{x}^i \sim q(\mathbf{x})$, a correction to Equation (6.2-28) must be made using importance weights, $\{w_q^i : i = 1, \dots, N_S\}$ that are proportional to $p(\cdot)/q(\cdot)$

$$w_q^i \propto \frac{p(\mathbf{x}^i)}{q(\mathbf{x}^i)}. \qquad (6.2\text{-}31)$$

Normalizing the importance weights such that $\sum_{i=1}^{N_S} w_q^i = 1$, Equation (6.2-28) becomes

$$p(\mathbf{x}|\mathbf{y}) \approx \sum_{i=1}^{N_S} w^i \delta_D(\mathbf{x} - \mathbf{x}^i). \qquad (6.2\text{-}32)$$

where

$$w^i = p_v(\mathbf{y} - \mathbf{h}(\mathbf{x}^i)) \frac{p(\mathbf{x}^i)}{q(\mathbf{x}^i)} \Big/ \sum_{j=1}^{N_S} \frac{p_v(\mathbf{y} - \mathbf{h}(\mathbf{x}^j)) p(\mathbf{x}^j)}{q(\mathbf{x}^j)}, \qquad (6.2\text{-}33)$$

and

$$E\{g(\mathbf{x})|\mathbf{y}\} \approx \sum_{i=1}^{N_S} w^i g(\mathbf{x}^i), \qquad (6.2\text{-}34)$$

with $\mathbf{x}^i \sim q(\mathbf{x})$. The choice of the importance density function and the efficiency of the sampling technique has been very well developed over the past decade [2], and is an entirely a new topic that is beyond the scope of this book. It is optional to use nonuniform weighted particles or to resample (or to add particles at the same location) so that all particles have uniform weights with the following procedure.

1. Given samples $\{\mathbf{x}^i : i = 1, \dots, N_S\}$ with weights $\{w^i : i = 1, \dots, N_S\}$, one can eliminate samples associated with negligible weights, and obtain a new sample set $\{\mathbf{x}^j : j = 1, \dots, N_T\}$ with weights $\{w^j : j = 1, \dots, N_T\}$ where $N_T < N_S$. We renormalize $\{w^j\}$ such that $\sum_{j=0}^{N_T} w^j = 1$.

2. A new sample set $\{\bar{\mathbf{x}}^i : i = 1, \dots, N_S\}$ with uniform weights $1/N_S$, is generated such that $\bar{\mathbf{x}}^i = \mathbf{x}^j$ with probability w^j.

A summary of a generic resampling algorithm is presented in Table 6.1.

Table 6.1   Summary of Resampling Algorithm

Resampling Algorithm
Solution Concept
Given random samples and weights $\{\mathbf{x}^i, w^i : i = 1, \dots, N_S\}$ drawn from some density function, there are samples with negligible weights. The purpose of the algorithm is to eliminate those samples with negligible weights and add samples at the same location with larger weights to obtain a new set of samples with N_s samples that has uniform weights, $1/N_S$.

Given samples and weights $\{\mathbf{x}^i, w^i : i = 1, \dots, N_S\}$
Eliminate samples with negligible weights and create a smaller sample set $\{\mathbf{x}^j, w^j : j = 1, \dots, N_T\}$ where $N_T < N_S$. Normalizing $\{w^j\}$ such that $\sum_{j=0}^{N_T} w^j = 1$.

Resampling $\{\mathbf{x}^j, w^j : j = 1, \dots, N_T\}$ **to Obtain a New Sample Set** $\{\bar{\mathbf{x}}^i, 1/N_s, i^j : i = 1, \dots, N_S\}$
Construct a cumulative distribution function (CDF), c^j, from $\{w^j : j = 1, \dots, N_T\}$, where $N_T < N_S$.
For $j = 1$, draw u_1 from a uniform distribution $U[0, 1/N_S]$, and
for $i = 1, \dots, N_S,\ u^i = u^1 + (j-1)/N_S$

$$\bar{\mathbf{x}}^i = \mathbf{x}^j,\ \bar{w}^i = 1/N_S \text{ while } u^i > c^j$$

$$j = j + 1,\ i^j = j \text{ while } u^i \leq c^j$$

where i^j is the index that relates the ith sample, $\bar{\mathbf{x}}^i$, to its parent, the i^jth sample, $\mathbf{x}^{i^j}$.

Remarks

1. Figure 6.2 illustrates the concept of random sampling, importance sampling, and resampling. Note that the deficiency occurred when samples were generated by a proposed importance function $q(x)$. As shown in the top of Figure 6.2, there are more condensed samples around zero of $p(x) = N(0, 1)$, a zero-mean Gaussian with unit standard deviation, while the proposed importance fountain, $q(x) = U[-6, 6]$, a uniform distribution over $[-6, 6]$, has more uniform samples in this interval. The histograms of the two distributions represent the sample statics of the two samples related to their underlying distributions. With large number of samples, one can have the histogram approaching the underlying density. In order to utilize samples from the proposed importance samples, that is, $\{x^i: i = 1, \ldots, N_S\}$, to calculate the integral Equation (6.2-1), one can use Equation (6.2-34). Note that there are samples $p(x^i) \leq 0.001$ (in red) that have negligible weights, w^i, in the calculation of the integral. Those particles can be thrown out and will not affect the approximation in Equation (6.2-34) too much. Therefore, we have smaller usable samples (49 out of 96 [in black] about 50% of the total population), called sample depletion. The resample scheme shown in Table 6.1 will replenish the samples by reusing samples at the same location multiple times to artificially create samples with uniform weight, $w^i = 1/N_S$ as shown in the bottom of Figure 6.2. Although both have approximately the same results for the integral, it is different from the distribution of the original samples shown at the top of Figure 6.2.

2. While this is unnecessary in the random vector case presented here, and resampling would always increase the Monte Carlo variation of the estimators, it is a vital step in a Monte Carlo sampling scheme discussed in the following sections for a random process to avoid the depletion of particles due to degeneracy of the importance weights over time.

In the following sections, a set of filters using sampling techniques is introduced. The sampling technique is aiming at obtaining better approximation of the mean and covariance of state estimate for a nonlinear system or more directly to approximate the a posteriori density function of the state estimates using samples. To set the stage, the Bayesian relationship for state estimation is first reviewed in Section 6.3. The method to sample a few selected points around the a priori covariance matrix then transformed them to the predicted and updated space using the unscented transformation is known as the UKF, Section 6.4. The method to sample the entire state space uniformly for use in filter prediction and update is the PMF, Section 6.5. Both methods sample the space deterministically. Techniques to take random samples in the state space with focused concentration are known as particle filters (PF), which will be covered in Section 6.6. There is a series of development in particle filters each aiming at

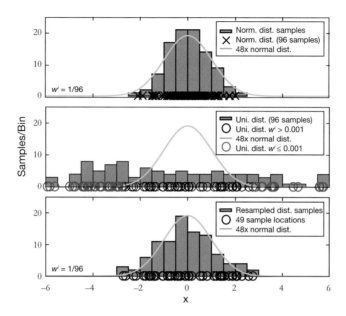

Figure 6.2   Illustration of random sampling, importance sampling, and resampling.

improving the performance or circumventing the deficiencies of earlier filters. The algorithms of all these filters will be described in Section 6.6.

6.3   Bayesian Approach to Nonlinear State Estimation

Following the same notation used in Section 3.2, consider the nonlinear dynamics of a target that are represented as a sequence of state vectors $\{\mathbf{x}_0, \mathbf{x}_1, \dots, \mathbf{x}_k\}$ satisfying the following nonlinear relationship

$$\mathbf{x}_k = \mathbf{f}(\mathbf{x}_{k-1}) + \boldsymbol{\mu}_{k-1}, \tag{6.3-1}$$

for $k = 1, 2 \dots, K$, where $\{\boldsymbol{\mu}_{k-1}, k = 1, 2, \dots, K\}$ is an i.i.d. process noise sequence with distribution $p_{\boldsymbol{\mu}_{k-1}}(\boldsymbol{\mu})$ and end time K. The initial state $\mathbf{x}_0$ is independent of $\{\boldsymbol{\mu}_{k-1}, k = 1, 2, \dots, K\}$ with a known distribution, $p_{\mathbf{x}_0}(\mathbf{x})$. There is no loss of generality with this model because the estimator is implemented with a computer and measurements are usually taken at discrete times represented as

$$\mathbf{y}_k = \mathbf{h}(\mathbf{x}_k) + \mathbf{v}_k \tag{6.3-2}$$

where $\{\mathbf{v}_k, k = 1, 2, \ldots, K\}$ is an i.i.d. measurement noise sequence with known density function, $p_{\mathbf{v}_k}(\mathbf{v})$ and independent of $\mathbf{x}_0$ and $\{\boldsymbol{\mu}_{k-1}, k = 1, 2, \ldots, K\}$. The problem is to estimate $\mathbf{x}_k$ recursively given all measurements $\mathbf{y}_{1:k} = \{\mathbf{y}_1, \mathbf{y}_2, \ldots, \mathbf{y}_k\}$.

In general, the optimal estimator for $\mathbf{x}_k$ given all measurements: $\mathbf{y}_{1:k} = \{\mathbf{y}_1, \mathbf{y}_2, \ldots, \mathbf{y}_k\}$ in the mean square sense, is the conditional mean [1] (see also Appendix 1.B)

$$\hat{\mathbf{x}}_{k|k} = \int \mathbf{x}_k p(\mathbf{x}_k | \mathbf{y}_{1:k}) d\mathbf{x}_k, \qquad (6.3\text{-}3)$$

where $p(\mathbf{x}_k | \mathbf{y}_{1:k})$ is the posterior probability function of $\mathbf{x}_k$ given all measurements $\mathbf{y}_{1:k} = \{\mathbf{y}_1, \mathbf{y}_2, \ldots, \mathbf{y}_k\}$. The challenge in computing $\hat{\mathbf{x}}_{k|k}$ is twofold. The first is to determine the posterior probability density function, and the second is to calculate $\hat{\mathbf{x}}_{k|k}$ with the integral from Equation (6.3-3) (see Section 6.2). To calculate the posterior probability density function $p(\mathbf{x}_k | \mathbf{y}_{1:k})$ formally, one can follow the recursive procedure outlined below.

Step 1: Compute the joint density function

$$p(\mathbf{x}_k, \mathbf{x}_{k-1} | \mathbf{y}_{1:k-1}) = p(\mathbf{x}_k | \mathbf{x}_{k-1}) p(\mathbf{x}_{k-1} | \mathbf{y}_{1:k-1}) \qquad (6.3\text{-}4)$$

given $p(\mathbf{x}_{k-1} | \mathbf{y}_{1:k-1})$ and $p(\mathbf{x}_k | \mathbf{x}_{k-1})$ and Equation (6.3-1).

Step 2: Compute the predicted probability density function of $\mathbf{x}_k$ given $\mathbf{y}_{1:k-1} = \{\mathbf{y}_1, \mathbf{y}_2, \ldots \mathbf{y}_{k-1}\}$ with the Chapman–Kolmogorov integral equation [1]

$$p(\mathbf{x}_k | \mathbf{y}_{1:k-1}) = \int p(\mathbf{x}_k | \mathbf{x}_{k-1}) p(\mathbf{x}_{k-1} | \mathbf{y}_{1:k-1}) d\mathbf{x}_{k-1}. \qquad (6.3\text{-}5)$$

Step 3: Compute the posterior probability density function of $\mathbf{x}_k$ given $\mathbf{y}_{1:k}$, $p(\mathbf{x}_k | \mathbf{y}_{1:k})$ using Bayes' rule

$$p(\mathbf{x}_k | \mathbf{y}_{1:k}) = \frac{p(\mathbf{y}_k | \mathbf{x}_k)}{p(\mathbf{y}_k | \mathbf{y}_{1:k-1})} p(\mathbf{x}_k | \mathbf{y}_{1:k-1}), \qquad (6.3\text{-}6)$$

where $p(\mathbf{y}_k | \mathbf{x}_k)$ is the likelihood function and the normalization constant is given as

$$p(\mathbf{y}_k | \mathbf{y}_{1:k-1}) = \int p(\mathbf{y}_k | \mathbf{x}_k) p(\mathbf{x}_k | \mathbf{y}_{1:k-1}) d\mathbf{x}_k. \qquad (6.3\text{-}7)$$

With the above recursive procedure as the basis, the nonlinear state estimation algorithms for Equations (6.3-1) and (6.3-2) can be derived in the following sections utilizing the sampling techniques presented in Sections 6.2.3 through 6.2.5.

6.4   Unscented Kalman Filter

In this section, attention will be focused on a class of nonlinear estimation problems with the following form

$$\mathbf{x}_k = \mathbf{f}(\mathbf{x}_{k-1}) + \boldsymbol{\mu}_{k-1} \tag{6.4-1}$$

and

$$\mathbf{y}_k = \mathbf{h}(\mathbf{x}_k) + \mathbf{v}_k \tag{6.4-2}$$

where $\mathbf{x}_0: \sim N(\hat{\mathbf{x}}_0, \mathbf{P}_0)$, $\boldsymbol{\mu}_{k-1}: \sim N(\mathbf{0}, \mathbf{Q}_{k-1})$, and $\mathbf{v}_k: \sim N(\mathbf{0}, \mathbf{R}_k)$ are mutually independent and $\boldsymbol{\mu}_{k-1}$ and $\mathbf{v}_k$ are white Gaussian system (process) and measurement noise sequences, respectively. The results of Section 6.2.3 are now used to derive a different form of the EKF for the nonlinear system from Equations (6.4-1) and (6.4-2).

Assume that at time $k-1$, the state estimate can be approximated by a Gaussian random vector $\hat{\mathbf{x}}_{k-1|k-1}: \sim N(\hat{\mathbf{x}}_{k-1|k-1}, \mathbf{P}_{k-1|k-1})$, one will first compute the one-step prediction estimate, $\hat{\mathbf{x}}_{k|k-1}$ and $\mathbf{P}_{k|k-1}$ using unscented transformation, Equations (6.2-13) and (6.2-14). When a new measurement vector, $\mathbf{y}_k$, is obtained at time k, the approximation of the updated state and covariance, $\hat{\mathbf{x}}_{k|k}$ and $\mathbf{P}_{k|k}$, are computed based on Equations (6.2-17) to (6.2-22).

One-Step Prediction

Following Equation (6.2-12), the $2n+1$ sigma points for $\hat{\mathbf{x}}_{k-1|k-1}$ and $\mathbf{P}_{k-1|k-1}$ are defined as[3]

$$\mathbf{x}^0_{k-1|k-1} = \hat{\mathbf{x}}_{k-1|k-1}$$

$$\hat{\mathbf{x}}^i_{k-1|k-1} = \hat{\mathbf{x}}_{k-1|k-1} + \left(\sqrt{(n+\kappa)\mathbf{P}_{k-1|k-1}}\right)^T_i \; \forall \, i = 1, \ldots, n,$$

$$\hat{\mathbf{x}}^{n+i}_{k-1|k-1} = \hat{\mathbf{x}}_{k-1|k-1} - \left(\sqrt{(n+\kappa)\mathbf{P}_{k-1|k-1}}\right)^T_i \; \forall \, i = 1, \ldots, n. \tag{6.4-3}$$

The corresponding one step predicted points are

$$\hat{\mathbf{x}}^i_{k|k-1} = \mathbf{f}\left(\hat{\mathbf{x}}^i_{k-1|k-1}\right) + \boldsymbol{\mu}^i_{k-1} \; \forall \, i = 0, \ldots, 2n. \tag{6.4-4}$$

Note that the system noise is accounted for by using random draws of samples of $\boldsymbol{\mu}^i_{k-1}$ according to the known density function of $\boldsymbol{\mu}_{k-1}$. The predicted state and covariance are computed using sample points $\hat{\mathbf{x}}^i_{k|k-1}$,

---

3   All samples here are deterministic, and the form $\hat{\mathbf{x}}^i_{j|k}$ denotes the ith sample at the time step indicated by the subscript.

$$\hat{\mathbf{x}}_{k|k-1} = \sum_{i=0}^{2n} w^i \hat{\mathbf{x}}_{k|k-1}^i, \text{ and} \tag{6.4-5}$$

$$\mathbf{P}_{k|k-1} = \sum_{i=0}^{2n} w^i \left(\hat{\mathbf{x}}_{k|k-1}^i - \hat{\mathbf{x}}_{k|k-1} \right) \left(\hat{\mathbf{x}}_{k|k-1}^i - \hat{\mathbf{x}}_{k|k-1} \right)^T + \mathbf{Q}_{k-1} \tag{6.4-6}$$

where

$$w^0 = \frac{\kappa}{(n+\kappa)}, \text{ and } w^i = w^{i+n} = \frac{1}{2(n+\kappa)} \ \forall \ i = 1, \dots, n.$$

Measurement Update

To update the filter with a measurement $\mathbf{y}_k$, we use the formula presented in Equations (6.2-17) to (6.2-22) for unscented transformation, yielding

$$\hat{\mathbf{x}}_{k|k} = \hat{\mathbf{x}}_{k|k-1} + \mathbf{K}_k \left(\mathbf{y}_k - \widehat{\mathbf{h}(\mathbf{x})}_{k|k-1} \right) \tag{6.4-7}$$

$$\mathbf{P}_{k|k} = \mathbf{P}_{k|k-1} - \mathbf{K}_k \mathbf{P}_{\gamma_k \gamma_k} \mathbf{K}_k^T \tag{6.4-8}$$

$$\mathbf{K}_k = \mathbf{P}_{\hat{\mathbf{x}}_{k|k-1}, \gamma_k} \mathbf{P}_{\gamma_k \gamma_k}^{-1} \tag{6.4-9}$$

$$\boldsymbol{\gamma}_k = \mathbf{y}_k - \widehat{\mathbf{h}(\mathbf{x})}_{k|k-1}, \tag{6.4-10}$$

where $\mathbf{K}_k$ is known as the filter gain, $\widehat{\mathbf{h}(\mathbf{x})}_{k|k-1}$ is the sample mean, and $\mathbf{P}_{\gamma_k \gamma_k}$ and $\mathbf{P}_{\hat{\mathbf{x}}_{k|k-1}, \gamma_k}$ are the sample covariance matrices, in terms of the sigma points as follows,

$$\widehat{\mathbf{h}(\mathbf{x})}_{k|k-1} = \sum_{i=0}^{2n} w^i \mathbf{h} \left(\hat{\mathbf{x}}_{k|k-1}^i \right), \tag{6.4-11}$$

$$\mathbf{P}_{\gamma_k \gamma_k} = \sum_{i=0}^{2n} w^i \left(\mathbf{h} \left(\hat{\mathbf{x}}_{k|k-1}^i \right) - \widehat{\mathbf{h}(\mathbf{x})}_{k|k-1} \right) \left(\mathbf{h} \left(\hat{\mathbf{x}}_{k|k-1}^i \right) - \widehat{\mathbf{h}(\mathbf{x})}_{k|k-1} \right)^T + \mathbf{R}_k, \tag{6.4-12}$$

$$\mathbf{P}_{\hat{\mathbf{x}}_{k|k-1}, \gamma_k} = \sum_{i=0}^{2n} w^i \left(\hat{\mathbf{x}}_{k|k-1}^i - \hat{\mathbf{x}}_{k|k-1} \right) \left(\mathbf{h} \left(\hat{\mathbf{x}}_{k|k-1}^i \right) - \widehat{\mathbf{h}(\mathbf{x})}_{k|k-1} \right)^T. \tag{6.4-13}$$

where

$$w^0 = \frac{\kappa}{(n+\kappa)}, \text{ and } w^i = w^{i+n} = \frac{1}{2(n+\kappa)} \ \forall \ i = 1, \dots, n.$$

The complete UKF algorithm is summarized in Table 6.2.

Table 6.2   Summary of Unscented Kalman Filter Algorithm

Unscented Kalman Filter (UKF)
Solution Concept
Use sigma points to compute predicted and updated state estimate and covariance matrices for nonlinear systems

Given $\hat{\mathbf{x}}_{k-1|k-1}$ and $\mathbf{P}_{k-1|k-1}$, Obtain their $2n+1$ Sigma Points by

$$\mathbf{x}_{k-1|k-1}^{0} = \hat{\mathbf{x}}_{k-1|k-1}$$

$$\hat{\mathbf{x}}_{k-1|k-1}^{i} = \hat{\mathbf{x}}_{k-1|k-1} + \left(\sqrt{(n+\kappa)\mathbf{P}_{k-1|k-1}}\right)_{i}^{T} \quad \forall \, i = 1, \dots, n,$$

$$\hat{\mathbf{x}}_{k-1|k-1}^{n+i} = \hat{\mathbf{x}}_{k-1|k-1} - \left(\sqrt{(n+\kappa)\mathbf{P}_{k-1|k-1}}\right)_{i}^{T} \quad \forall \, i = 1, \dots, n.$$

Prediction

$$\hat{\mathbf{x}}_{k|k-1}^{i} = \mathbf{f}\left(\hat{\mathbf{x}}_{k-1|k-1}^{i}\right) + \boldsymbol{\mu}_{k-1}^{i} \ \forall \, i = 1, \dots, n,$$

$$\hat{\mathbf{x}}_{k|k-1} = \sum_{i=0}^{2n} w^{i} \hat{\mathbf{x}}_{k|k-1}^{i},$$

$$\mathbf{P}_{k|k-1} = \sum_{i=0}^{2n} w^{i} \left(\hat{\mathbf{x}}_{k|k-1}^{i} - \hat{\mathbf{x}}_{k|k-1}\right)\left(\hat{\mathbf{x}}_{k|k-1}^{i} - \hat{\mathbf{x}}_{k|k-1}\right)^{T} + \mathbf{Q}_{k-1}.$$

where

$$w^{0} = \frac{\kappa}{(n+\kappa)}, \text{ and } w^{i} = w^{i+n} = \frac{1}{2(n+\kappa)} \ \forall \, i = 1, \dots, n.$$

Update
Compute

$$\widehat{\mathbf{h}(\mathbf{x})}_{k|k-1} = \sum_{i=0}^{2n} w^{i} \mathbf{h}\left(\hat{\mathbf{x}}_{k|k-1}^{i}\right),$$

$$\mathbf{P}_{\gamma_{k}\gamma_{k}} = \sum_{i=0}^{2n} w^{i} \left(\mathbf{h}\left(\hat{\mathbf{x}}_{k|k-1}^{i}\right) - \widehat{\mathbf{h}(\mathbf{x})}_{k|k-1}\right)\left(\mathbf{h}\left(\hat{\mathbf{x}}_{k|k-1}^{i}\right) - \widehat{\mathbf{h}(\mathbf{x})}_{k|k-1}\right)^{T} + \mathbf{R}_{k},$$

$$\mathbf{P}_{\hat{\mathbf{x}}_{k|k-1},\gamma_{k}} = \sum_{i=0}^{2n} w^{i} \left(\hat{\mathbf{x}}_{k|k-1}^{i} - \hat{\mathbf{x}}_{k|k-1}\right)\left(\mathbf{h}\left(\hat{\mathbf{x}}_{k|k-1}^{i}\right) - \widehat{\mathbf{h}(\mathbf{x})}_{k|k-1}\right)^{T}.$$

For use in

$$\hat{\mathbf{x}}_{k|k} = \hat{\mathbf{x}}_{k|k-1} + \mathbf{K}_{k}\left(\mathbf{y}_{k} - \widehat{\mathbf{h}(\mathbf{x})}_{k|k-1}\right),$$

$$\mathbf{P}_{k|k} = \mathbf{P}_{k|k-1} - \mathbf{K}_{k}\mathbf{P}_{\gamma_{k}\gamma_{k}}\mathbf{K}_{k}^{T},$$

$$\mathbf{K}_{k} = \mathbf{P}_{\hat{\mathbf{x}}_{k|k-1},\gamma_{k}}\mathbf{P}_{\gamma_{k}\gamma_{k}}^{-1},$$

where

$$w^{0} = \frac{\kappa}{(n+\kappa)}, \text{ and } w^{i} = w^{i+n} = \frac{1}{2(n+\kappa)} \ \forall \, i = 1, \dots, n.$$

Remarks

1. The difference between UKF and EKF are in the covariance computation, as mentioned in Section 6.2.3 the UKF should reduce to KF for a linear system. The advantage of the UKF is the use of sample mean, $\widehat{\mathbf{h}(\mathbf{x})}_{k|k-1}$, and sample covariance, $\mathbf{P}_{\gamma_k \gamma_k}$ and $\mathbf{P}_{\hat{\mathbf{x}}_{k|k-1} \cdot \gamma_k}$, instead of using $\mathbf{h}(\hat{\mathbf{x}}_{k|k-1})$ for $E\{\mathbf{h}(\mathbf{x}_k)|\mathbf{y}_{1:k-1}\}$ and using Equation (6.2-11) for the covariance matrix computation using the Jacobian $\mathbf{H}_{\hat{\mathbf{x}}_{k|k-1}} = \left[\frac{\partial \mathbf{h}(\mathbf{x}_k)}{\partial \mathbf{x}_k}\right]_{\hat{\mathbf{x}}_{k|k-1}}$. While not generating the Jacobian and/or Hessian matrices at each recursive step, the UKF is computationally as efficient as EKF and achieves the same level of accuracy as the second order EKF discussed in Section 4.7; see [5–7].

2. The UKF can easily apply to nonadditive process noise cases for $\mathbf{x}_k = \mathbf{f}(\mathbf{x}_{k-1}, \boldsymbol{\mu}_{k-1})$ by defining the augmented state vector $\mathbf{z}_k = \left(\mathbf{x}_{k-1}^T, \boldsymbol{\mu}_{k-1}^T\right)^T$ as a nonlinear transformation of $\mathbf{x}_{k-1}$ and $\boldsymbol{\mu}_{k-1}$ which has mean and covariance $(\hat{\mathbf{x}}_{k-1}^T, \mathbf{0}^T)^T$ and $\text{diag}[\mathbf{P}_{k-1|k-1}, \mathbf{Q}_{k-1}]$.[4] The unscented transformation can be applied to the random vector $\mathbf{z}_k$ through the nonlinear function $\mathbf{f}$ to obtain the predicted state statistics [see Equations (6.2-12) to (6.2-16)]. The number of sigma points will be $4n + 1$ instead of $2n + 1$. A similar approach can be applied to the nonadditive measurement noise cases.

6.5   The Point-Mass Filter

In this section, attention will be focused on a class of nonlinear estimation problems of the form

$$\mathbf{x}_k = \mathbf{f}(\mathbf{x}_{k-1}) + \boldsymbol{\mu}_{k-1} \qquad (6.5\text{-}1)$$

and

$$\mathbf{y}_k = \mathbf{h}(\mathbf{x}_k) + \mathbf{v}_k \qquad (6.5\text{-}2)$$

where $\mathbf{x}_k, \boldsymbol{\mu}_{k-1} \in \mathbb{R}^n$ and $\mathbf{y}_k, \mathbf{v}_k \in \mathbb{R}^m$; $\mathbf{x}_0, \boldsymbol{\mu}_{k-1}$ and $\mathbf{v}_k$ are mutually independent random vectors with known density functions $p(\mathbf{x}_0)$, $p_{\boldsymbol{\mu}_{k-1}}(\boldsymbol{\mu}_{k-1})$ and $p_{\mathbf{v}_k}(\mathbf{v}_k)$. Treating the state prediction and measurement update as two steps in the computation process and utilizing the sampling technique described in Section 6.2.4 for random vectors with a region of interest in the state space $\mathbf{x}_k$ decomposed into N_s equal-volume hypercubes (or cells) with deterministic centers $\{\mathbf{x}_k^i : i = 1, \ldots, N_s\}$, a PMF (or grid-based Kalman

---

4   diag [A, B] is a shorthand notation of a matrix $\begin{bmatrix} \mathbf{A} & \mathbf{0} \\ \mathbf{0} & \mathbf{B} \end{bmatrix}$.

filter and/or EKF) is formed to approximate the nonlinear estimation problem [4, 9, 17].

Initialization

For $k = 0$

$$p(\mathbf{x}_0) \approx \sum_{i=1}^{N_S} w_0^i \delta_D(\mathbf{x}_0 - \mathbf{x}_0^i), \qquad (6.5\text{-}3)$$

and

$$\sum_{i=1}^{N_S} w_0^i = 1,$$

where $\delta_D(.)$ is the Dirac delta measure.

One-Step Prediction

Assuming the posterior density function at $k-1$ is given by the approximation

$$p(\mathbf{x}_{k-1}|\mathbf{y}_{1:k-1}) \approx \sum_{i=1}^{N_S} w_{k-1|k-1}^i \delta_D(\mathbf{x}_{k-1} - \mathbf{x}_{k-1}^i), \qquad (6.5\text{-}4)$$

and

$$\sum_{i=1}^{N_S} w_{k-1|k-1}^i = 1,$$

where $\delta_D(.)$ is the Dirac delta measure. Then, the one-step prediction density can be approximated as

$$p(\mathbf{x}_k|\mathbf{y}_{1:k-1}) \approx \sum_{i=1}^{N_S} w_{k|k-1}^i \delta_D(\mathbf{x}_k - \mathbf{x}_k^i), \qquad (6.5\text{-}5)$$

where

$$w_{k|k-1}^i \triangleq \sum_{j=1}^{N_S} w_{k-1|k-1}^j p_{\mu_{k-1}}\left(\mathbf{x}_k^i - \mathbf{f}\left(\mathbf{x}_{k-1}^j\right)\right). \qquad (6.5\text{-}6)$$

and

$$E\{\boldsymbol{g}(\mathbf{x}_k)|\mathbf{y}_{1:k-1}\} \approx \sum_{i=1}^{N_S} w_{k|k-1}^i \boldsymbol{g}(\mathbf{x}_k^i). \qquad (6.5\text{-}7)$$

The mean and covariance of $p(\mathbf{x}_k|\mathbf{y}_{1:k-1})$ are the special cases for Equation (6.5-7) with $\boldsymbol{g}(\mathbf{x}_k) = \mathbf{x}_k$ and $\boldsymbol{g}(\mathbf{x}_k) = (\mathbf{x}_k - \hat{\mathbf{x}}_{k|k-1})(\mathbf{x}_k - \hat{\mathbf{x}}_{k|k-1})^T$ with $\hat{\mathbf{x}}_{k|k-1} = E\{\mathbf{x}_k|\mathbf{y}_{1:k-1}\}$.

Measurement Update

Assuming the density function of the measurement noise $\mathbf{v}_k$ is given as $p_{\mathbf{v}_k}(.)$, then the likelihood of $\mathbf{x}_k$ given the measurement $\mathbf{y}_k$ is $p(\mathbf{y}_k|\mathbf{x}_k) = p_{\mathbf{v}_k}(\mathbf{y}_k - \mathbf{h}(\mathbf{x}_k))$. Substituting Equation (6.5-4) into Equation (6.3-6) yields

$$p(\mathbf{x}_k|\mathbf{y}_{1:k}) \approx \frac{p_{\mathbf{v}_k}(\mathbf{y}_k - \mathbf{h}(\mathbf{x}_k)) \sum_{i=1}^{N_S} w_{k|k-1}^i \delta_D(\mathbf{x}_k - \mathbf{x}_k^i)}{\sum_{i=1}^{N_S} w_{k|k-1}^i p_{\mathbf{v}_k}(\mathbf{y}_k - \mathbf{h}(\mathbf{x}_k^i))} = \sum_{i=1}^{N_S} w_{k|k}^i \delta_D(\mathbf{x}_k - \mathbf{x}_k^i), \quad (6.5\text{-}8)$$

where

$$w_{k|k}^i = w_{k|k-1}^i p_{\mathbf{v}_k}(\mathbf{y}_k - \mathbf{h}(\mathbf{x}_k^i)) \Big/ \sum_{i=1}^{N_S} w_{k|k-1}^i p_{\mathbf{v}_k}(\mathbf{y}_k - \mathbf{h}(\mathbf{x}_k^i)). \quad (6.5\text{-}9)$$

and

$$E\{\mathbf{g}(\mathbf{x}_k)|\mathbf{y}_{1:k}\} \approx \sum_{i=1}^{N_S} w_{k|k}^i \mathbf{g}(\mathbf{x}_k^i). \quad (6.5\text{-}10)$$

and $\mathbf{x}_k^i$ denotes the center of the ith cell at time index k. The grid must be sufficiently dense to get a good approximation to the continuous state variable $\mathbf{x}_k$. The mean and covariance of $p(\mathbf{x}_k|\mathbf{y}_{1:k})$ are the special cases for Equation (6.5-10) with $\mathbf{g}(\mathbf{x}_k) = \mathbf{x}_k$ and $\mathbf{g}(\mathbf{x}_k) = (\mathbf{x}_k - \hat{\mathbf{x}}_{k|k})(\mathbf{x}_k - \hat{\mathbf{x}}_{k|k})^T$ with $\hat{\mathbf{x}}_{k|k} = E\{\mathbf{x}_k|\mathbf{y}_{1:k}\}$.

The complete PMF algorithm is summarized in Table 6.3.

Remarks

The advantage of PMF against EKF and UKF is the scheme does not require closed-form expressions for the posterior probability density function. But the penalty to ensure the accuracy of state estimation and covariance is that this technique can become computationally prohibitive when the dimension of $\mathbf{x}_k$ is large. The selection of the N_s equal-volume hypercubes (or cells) with centers $\{\mathbf{x}_k^i : i = 1,\ldots,N_s\}$ can be made by the given statistics $\hat{\mathbf{x}}_{k-1|k-1}$ and $\mathbf{P}_{k-1|k-1}$ at time $k - 1$ to predict the region of interest [17]. An adaptive algorithm utilizing Chebychev's inequality was presented by [18].

For a more general nonlinear and non-Gaussian problem, the Monte Carlo sampling technique discussed in Section 6.2.5 is needed to compute the conditional mean, $\hat{\mathbf{x}}_{k|k}$, because EKF and/or UKF cannot produce adequate results and PMF is computationally prohibitive when the dimension of $\mathbf{x}_k$ is large. This leads us to introduce another sampling technique in the following section.

Table 6.3 Summary of Point Mass Filter Algorithm

Point-Mass Filter (PMF)

Solution Concept

Use equal-volume hypercubes to cover a region of interest in the space of $\mathbf{x}_k$ with center points of these hypercubes, $\{\mathbf{x}_k^i: i = 1, \dots, N_s\}$, and associated weights proportional to the conditional probabilities of the cube to compute the mean and covariance of the predicted and updated state for a nonlinear system.

Given $p\{\mathbf{x}_{k-1}|\mathbf{y}_{1:k-1}\}$ **and its Approximation**

$$p\left(\mathbf{x}_{k-1}^i|\mathbf{y}_{1:k-1}\right) \approx \sum_{i=1}^{N_s} w_{k-1|k-1}^i \delta_D\left(\mathbf{x}_{k-1} - \mathbf{x}_{k-1}^i\right), \text{ and } \sum_{i=1}^{N_s} w_{k-1|k-1}^i = 1,$$

where $\delta_D(.)$ is the Dirac delta measure.

For $k = 1$, $p(\mathbf{x}_0) \approx \sum_{i=1}^{N_s} w_0^i \delta_D\left(\mathbf{x}_0 - \mathbf{x}_0^i\right)$, and $\sum_{i=1}^{N_s} w_0^i = 1$.

Prediction

$$p\left(\mathbf{x}_k|\mathbf{y}_{1:k-1}\right) \approx \sum_{i=1}^{N_s} w_{k|k-1}^i \delta_D\left(\mathbf{x}_k - \mathbf{x}_k^i\right),$$

$$\hat{\mathbf{x}}_{k|k-1} = \sum_{i=1}^{N_s} w_{k|k-1}^i \mathbf{x}_{k-1}^i,$$

$$\mathbf{P}_{k|k-1} = \sum_{i=1}^{N_s} w_{k|k-1}^i \left(\mathbf{x}_{k-1}^i - \hat{\mathbf{x}}_{k|k-1}\right)\left(\mathbf{x}_{k-1}^i - \hat{\mathbf{x}}_{k|k-1}\right)^T,$$

where

$$w_{k|k-1}^i \triangleq \sum_{j=1}^{N_s} w_{k-1|k-1}^j p_{\mu_{k-1}}\left(\mathbf{x}_k^i - \mathbf{f}\left(\mathbf{x}_{k-1}^j\right)\right).$$

Update

Compute the updated weight $\{w_{k|k}^i: i = 1, \dots, N_s\}$ given $\mathbf{y}_k$

$$w_{k|k}^i \triangleq w_{k|k-1}^i p_{\nu_k}\left(\mathbf{y}_k - \mathbf{h}\left(\mathbf{x}_k^i\right)\right) \Big/ \sum_{j=1}^{N_s} w_{k|k-1}^j p_{\nu_k}\left(\mathbf{y}_k - \mathbf{h}\left(\mathbf{x}_k^j\right)\right).$$

Compute the updated state estimate and covariance

$$\hat{\mathbf{x}}_{k|k} = \sum_{i=1}^{N_s} w_{k|k}^i \mathbf{x}_k^i,$$

$$\mathbf{P}_{k|k} = \sum_{i=1}^{N_s} w_{k|k}^i \left(\mathbf{x}_k^i - \hat{\mathbf{x}}_{k|k}\right)\left(\mathbf{x}_k^i - \hat{\mathbf{x}}_{k|k}\right)^T.$$

6.6 Particle Filtering Methods

Let us consider two general random processes[5] at time step k, $\mathbf{x}_{0:k} = \{\mathbf{x}_j: j = 0, \dots, k\}$ and $\mathbf{y}_{1:k} = \{\mathbf{y}_j: j = 1, \dots, k\}$. Assume the posterior probability function $p(\mathbf{x}_{0:k-1}|\mathbf{y}_{1:k-1})$ is given at time $k - 1$, and that it can be approximated as a set of points $\{\mathbf{x}_{0:k-1}^i : i = 1, \dots, N_s\}$ with weights $\{w_{k-1}^i: i = 1, \dots, N_s\}$ such that

---

5 Note that for a general definition of random processes (see Appendix 2.A), $\mathbf{x}_{0:k}$ and $\mathbf{y}_{1:k}$ are treated as two random vectors.

$$p\left(\mathbf{x}_{0:k-1}|\mathbf{y}_{1:k-1}\right) \approx \sum_{i=1}^{N_S} w_{k-1}^i \delta_D\left(\mathbf{x}_{0:k-1} - \mathbf{x}_{0:k-1}^i\right) \tag{6.6-1}$$

where $\delta(.)$ is the Dirac delta measure. In order to compute

$$\hat{\mathbf{x}}_{0:k\Vert:k} = E\left\{\mathbf{x}_{0:k}|\mathbf{y}_{1:k}\right\},$$

one needs to approximate

$$p\left(\mathbf{x}_{0:k}|\mathbf{y}_{1:k}\right) \approx \sum_{i=1}^{N_S} w_k^i \delta_D\left(\mathbf{x}_{0:k} - \mathbf{x}_{0:k}^i\right) \tag{6.6-2}$$

with a set of points $\left\{\mathbf{x}_{0:k}^i: i=1,\ldots,N_S\right\}$ with weights $\left\{w_k^i: i=1,\ldots,N_S\right\}$ where $\sum_{i=1}^{N_S} w_k^i = 1$. Given Equation (6.6-1), with $\mathbf{x}_{0:k}^i = \left\{\mathbf{x}_k^i, \mathbf{x}_{0:k-1}^i\right\}$ for $i=1,\ldots,N_S$ and $\mathbf{y}_{1:k} = \{\mathbf{y}_k, \mathbf{y}_{1:k-1}\}$, the conditional density function, $p(\mathbf{x}_{0:k}|\mathbf{y}_{1:k})$, can be rewritten using Bayes' rule as

$$\begin{aligned}
p\left(\mathbf{x}_{0:k}|\mathbf{y}_{1:k}\right) &= \frac{p\left(\mathbf{y}_k|\mathbf{x}_{0:k}, \mathbf{y}_{1:k-1}\right)}{p\left(\mathbf{y}_k|\mathbf{y}_{1:k-1}\right)} p\left(\mathbf{x}_{0:k}|\mathbf{y}_{1:k-1}\right) \\
&= \frac{p\left(\mathbf{y}_k|\mathbf{x}_{0:k}, \mathbf{y}_{1:k-1}\right) p\left(\mathbf{x}_k|\mathbf{x}_{0:k-1}, \mathbf{y}_{1:k-1}\right)}{p\left(\mathbf{y}_k|\mathbf{y}_{1:k-1}\right)} p\left(\mathbf{x}_{0:k-1}|\mathbf{y}_{1:k-1}\right).
\end{aligned} \tag{6.6-3}$$

In general, the random samples $\left\{\mathbf{x}_{0:k}^i: i=1,\ldots,N_S\right\}$ with density $p(\mathbf{x}_{0:k}|\mathbf{y}_{1:k})$ cannot be drawn directly. Based on the importance sampling technique described in Section 6.2.5, another density function, $q(.)$ is used that $\left\{\mathbf{x}_{0:k}^i: i=1,\ldots,N_S\right\}$ can be easily drawn. When the importance function for $p(.)$, $q(.)$, is selected, the associated weight in Equation (6.6-2) is defined as

$$w_k^i \propto \frac{p\left(\mathbf{x}_{0:k}^i|\mathbf{y}_{1:k}\right)}{q\left(\mathbf{x}_{0:k}^i|\mathbf{y}_{1:k}\right)}. \tag{6.6-4}$$

Let the importance function in Equation (6.6-4) be chosen such that it can be factored as

$$q(\mathbf{x}_{0:k}|\mathbf{y}_{1:k}) = q(\mathbf{x}_k|\mathbf{x}_{0:k-1}, \mathbf{y}_{1:k})q(\mathbf{x}_{0:k-1}|\mathbf{y}_{1:k-1}), \tag{6.6-5}$$

then samples $\mathbf{x}_{0:k}^i$ can be drawn from the density function $q(\mathbf{x}_{0:k}|\mathbf{y}_{1:k})$ by propagating the existing sample $\mathbf{x}_{0:k-1}^i$ from the prior distribution $q(\mathbf{x}_{0:k-1}|\mathbf{y}_{1:k-1})$ with the new state $\mathbf{x}_k^i$ drawn from $q(\mathbf{x}_k|\mathbf{x}_{0:k-1}, \mathbf{y}_{1:k})$. This sampling technique is called sequential important sampling. This is a procedure applicable to any random processes $\mathbf{x}_{0:k}$ and $\mathbf{y}_{1:k}$.

In this book, a special class of random processes, $\mathbf{x}_{0:k} = \{\mathbf{x}_j: j = 0, \ldots, k\}$ and $\mathbf{y}_{1:k} = \{\mathbf{y}_j: j = 1,\ldots,k\}$, are considered, such that they satisfy the following equations

$$\mathbf{x}_k = \mathbf{f}(\mathbf{x}_{k-1}) + \boldsymbol{\mu}_{k-1} \tag{6.6-6}$$

and

$$\mathbf{y}_k = \mathbf{h}(\mathbf{x}_k) + \mathbf{v}_k \tag{6.6-7}$$

where $\mathbf{x}_k, \boldsymbol{\mu}_{k-1} \in \mathbb{R}^n$ and $\mathbf{y}_k, \mathbf{v}_k \in \mathbb{R}^m$; $\mathbf{x}_0, \boldsymbol{\mu}_{k-1}$ and $\mathbf{v}_k$ are mutually independent random vectors with known probability density functions $p(\mathbf{x}_0)$, $p_{\boldsymbol{\mu}_{k-1}}(\boldsymbol{\mu}_{k-1})$ and $p_{\mathbf{v}_k}(\mathbf{v})$. With the independence assumptions on $\mathbf{x}_0$, $\{\boldsymbol{\mu}_k\}$, and $\{\mathbf{v}_k\}$, and the Markovian property of $\{\mathbf{x}_k\}$, the expression in Equation (6.6-3) simplifies to

$$
\begin{aligned}
p(\mathbf{x}_k|\mathbf{y}_{1:k}) &= \frac{p(\mathbf{y}_k|\mathbf{x}_k)\, p(\mathbf{x}_k|\mathbf{x}_{k-1})}{p(\mathbf{y}_k|\mathbf{y}_{1:k-1})} p(\mathbf{x}_{k-1}|\mathbf{y}_{1:k-1}) \\
&\propto p(\mathbf{y}_k|\mathbf{x}_k)\, p(\mathbf{x}_k|\mathbf{x}_{k-1})\, p(\mathbf{x}_{k-1}|\mathbf{y}_{1:k-1}),
\end{aligned}
\tag{6.6-8}
$$

where

$$p(\mathbf{y}_k|\mathbf{x}_k) = p_{\mathbf{v}_k}(\mathbf{y}_k - \mathbf{h}(\mathbf{x}_k))$$

and

$$p(\mathbf{x}_k|\mathbf{x}_{k-1}) = p_{\boldsymbol{\mu}_{k-1}}(\mathbf{x}_k - \mathbf{f}(\mathbf{x}_{k-1})).$$

By substituting Equations (6.6-8) and (6.6-5) into Equation (6.6-3) with given samples and weights $\{\mathbf{x}_{k-1}^i, w_{k-1}^i : i = 1, \ldots, N_S\}$, the associated sample $\mathbf{x}_k^i$ drawn from $q(\mathbf{x}_k|\mathbf{x}_{k-1}, \mathbf{y}_{1:k})$ yields

$$
\begin{aligned}
w_k^i &\propto \frac{p(\mathbf{y}_k|\mathbf{x}_k^i)\, p(\mathbf{x}_k^i|\mathbf{x}_{k-1}^i)\, p(\mathbf{x}_{k-1}^i|\mathbf{y}_{1:k-1})}{q(\mathbf{x}_k^i|\mathbf{x}_{k-1}^i, \mathbf{y}_{1:k})\, q(\mathbf{x}_{k-1}^i|\mathbf{y}_{1:k-1})} = w_{k-1}^i \frac{p(\mathbf{y}_k|\mathbf{x}_k^i)\, p(\mathbf{x}_k^i|\mathbf{x}_{k-1}^i)}{q(\mathbf{x}_k^i|\mathbf{x}_{k-1}^i, \mathbf{y}_{1:k})} \\
&= w_{k-1}^i \frac{p_{\mathbf{v}_k}(\mathbf{y}_k - \mathbf{h}(\mathbf{x}_k^i))\, p_{\boldsymbol{\mu}_{k-1}}(\mathbf{x}_k^i - \mathbf{f}(\mathbf{x}_{k-1}^i))}{q(\mathbf{x}_k^i|\mathbf{x}_{k-1}^i, \mathbf{y}_{1:k})}.
\end{aligned}
\tag{6.6-9}
$$

This is a useful form for most of the particle filter applications considered in this chapter. For a particular case, one could select the importance density function to satisfy the following constraint

$$q(\mathbf{x}_k|\mathbf{x}_{k-1}, \mathbf{y}_{1:k}) = q(\mathbf{x}_k|\mathbf{x}_{k-1}, \mathbf{y}_k),$$

that is, the density function is dependent only on the previous state $\mathbf{x}_{k-1}$, and the last measurement $\mathbf{y}_k$. The modified weight is then

$$w_k^i \propto w_{k-1}^i \frac{p_{\mathbf{v}_k}(\mathbf{y}_k - \mathbf{h}(\mathbf{x}_k^i))\, p_{\boldsymbol{\mu}_{k-1}}(\mathbf{x}_k^i - \mathbf{f}(\mathbf{x}_{k-1}^i))}{q(\mathbf{x}_k^i|\mathbf{x}_{k-1}^i, \mathbf{y}_k)}. \tag{6.6-10}$$

The posterior probability density in Equation (6.6-8) can be approximated as

$$p(\mathbf{x}_k|\mathbf{y}_{1:k}) \approx \sum_{i=1}^{N_S} w_k^i \delta_D(\mathbf{x}_k - \mathbf{x}_k^i)$$

where the weights are defined in Equation (6.6-10). A summary of the generic particle filter is presented in Table 6.4.

The crucial design step of the particle filter (Table 6.4) is the choice of the importance density function, $q(\mathbf{x}_k|\mathbf{x}_{k-1}, \mathbf{y}_k)$. A proper choice of the importance function will lead to a faster convergence of the filter with better performance. The following sections will provide some commonly used approaches, with their advantages as well as shortfalls discussed. The reader should be aware that a larger number of samples will not necessarily yield a filter with better results if the importance function is not chosen properly for the problem.

Table 6.4 Summary of the Generic Particle Filter Algorithm

Generic Particle Filter

Solution Concept

Given $\{\mathbf{x}_{k-1}^i, w_{k-1}^i : i = 1, \dots, N_s\}$ and $\mathbf{y}_k$, draw random samples $\mathbf{x}_k^i \sim q(\mathbf{x}_k|\mathbf{x}_{k-1}^i, \mathbf{y}_k)$.
Obtain $\{w_k^i : i = 1, \dots, N_s\}$ for the approximation of $p(\mathbf{x}_k|\mathbf{y}_{1:k})$ in order to compute the updated state estimate and covariance.

Initialization

For $k = 1$, draw $\mathbf{x}_0^i \sim p(\mathbf{x}_0)$ for $i = 1, \dots, N_s$, such that

$$p(\mathbf{x}_0) \approx \sum_{i=1}^{N_S} w_0^i \delta_D(\mathbf{x}_0 - \mathbf{x}_0^i) \text{ , and } w_0^i = 1/N_s.$$

Prediction

Given $\{\mathbf{x}_{k-1}^i, w_{k-1}^i : i = 1, \dots, N_s\}$ which approximates $p\{\mathbf{x}_{k-1}|\mathbf{y}_{1:k-1}\}$

Draw $\mathbf{x}_k^i \sim q(\mathbf{x}_k|\mathbf{x}_{k-1}^i, \mathbf{y}_k) \ \forall \ i = 1, \dots, N_s$.

Update

Compute the updated weight $\{w_k^i : i = 1, \dots, N_s\}$ given $\mathbf{y}_k$

$$w_k^i = w_{k-1}^i p_{\mathbf{v}_k}(\mathbf{y}_k - \mathbf{h}(\mathbf{x}_k^i)) p_{\mu_{k-1}}(\mathbf{x}_k^i - \mathbf{f}(\mathbf{x}_{k-1}^i))/q(\mathbf{x}_k^i|\mathbf{x}_{k-1}^i, \mathbf{y}_k), \ \forall \ i = 1, \dots, N_s.$$

Normalize $\{w_k^i : i = 1, \dots, N_s\}$, such that $\sum_{i=1}^{N_S} w_k^i = 1$.
Compute the updated state estimate and covariance

$$\hat{\mathbf{x}}_{k|k} = \sum_{i=1}^{N_S} w_k^i \mathbf{x}_k^i,$$

$$\mathbf{P}_{k|k} = \sum_{i=1}^{N_S} w_k^i (\mathbf{x}_k^i - \hat{\mathbf{x}}_{k|k})(\mathbf{x}_k^i - \hat{\mathbf{x}}_{k|k})^T.$$

6.6.1   Sequential Importance Sampling Filter

There are many forms for the importance density function suggested in the literature [9, 12, 13]. The sequential importance sampling (SIS) filter [12–14] approach is known variously as bootstrap filtering [14, 19], the condensation algorithm [20], particle filtering [21], interacting particle approximations [22, 23], and survival of the fittest [24]. The specific choice for the importance density function is [6, 19]

$$q\left(\mathbf{x}_k | \mathbf{x}_{k-1}^i, \mathbf{y}_k\right) = p\left(\mathbf{x}_k | \mathbf{x}_{k-1}^i\right) = p_{\mu_{k-1}}\left(\mathbf{x}_k - \mathbf{f}\left(\mathbf{x}_{k-1}^i\right)\right), \tag{6.6-11}$$

so that Equation (6.6-10) becomes

$$w_k^i \propto w_{k-1}^i p\left(\mathbf{y}_k | \mathbf{x}_k^i\right) = w_{k-1}^i p_{\nu_k}\left(\mathbf{y}_k - \mathbf{h}\left(\mathbf{x}_k^i\right)\right). \tag{6.6-12}$$

The SIS filter algorithm is summarized in Table 6.5.

A common problem with the SIS filter is the degeneracy phenomenon, often in practice showing up after only a few iterations; all particles will have negligible weights with the exception of one. This degeneracy implies that a large computational effort is wasted in updating the particles that have negligible contribution to the computation of $p(\mathbf{x}_k|\mathbf{y}_{1:k})$ and the Monte Carlo integral associated with it. It has been shown in [12] that the variance of the importance weights can only increase over iterations, and the optimal choice for the importance density function that minimizes the variance of $\left\{w_k^i\right\}$ is

$$q\left(\mathbf{x}_k | \mathbf{x}_{k-1}^i, \mathbf{y}_k\right) = p\left(\mathbf{x}_k | \mathbf{x}_{k-1}^i, \mathbf{y}_k\right) = \frac{p\left(\mathbf{y}_k | \mathbf{x}_k, \mathbf{x}_{k-1}^i\right) p\left(\mathbf{x}_k | \mathbf{x}_{k-1}^i\right)}{p\left(\mathbf{y}_k | \mathbf{x}_{k-1}^i\right)}. \tag{6.6-13}$$

Substituting Equation (6.6-13) into Equation (6.6-10) yields

$$\begin{aligned} w_k^i &\propto w_{k-1}^i p\left(\mathbf{y}_k | \mathbf{x}_{k-1}^i\right) = w_{k-1}^i \int p\left(\mathbf{y}_k | \mathbf{x}_k'\right) p(\mathbf{x}_k' | \mathbf{x}_{k-1}^i) d\mathbf{x}_k' \\ &= w_{k-1}^i \int p_{\nu_k}\left(\mathbf{y}_k - \mathbf{h}\left(\mathbf{x}_k'\right)\right) p_{\mu_{k-1}}\left(\mathbf{x}_k' - \mathbf{f}\left(\mathbf{x}_{k-1}^i\right)\right) d\mathbf{x}_k'. \end{aligned} \tag{6.6-14}$$

The choice of importance density is optimal since w_k^i is the same (i.e., $\mathrm{Var}\left\{w_k^i\right\} = 0$) for given $\mathbf{x}_{k-1}^i$ and $\mathbf{y}_k$, no matter what sample is drawn from $q\left(\mathbf{x}_k | \mathbf{x}_{k-1}^i, \mathbf{y}_k\right)$ of Equation (6.6-13). In practice, the optimal choice of importance density suffers a major drawback that is the ability to draw samples $\mathbf{x}_k$ from $p\left(\mathbf{x}_k | \mathbf{x}_{k-1}^i, \mathbf{y}_k\right)$ of Equation (6.6-13) and to evaluate the integral numerically in Equation (6.6-14). Because the sample $\mathbf{x}_k$ is a function of $\mathbf{x}_{k-1}^i$, a random process, the entire trajectory of $\mathbf{x}_{k-1}^i$ needs to be generated.

Table 6.5   Summary of the Sequential Importance Sampling Filter Algorithm

---

Sequential Importance Sampling (SIS) Filter

Solution Concept

Given $\{\mathbf{x}_{k-1}^i, w_{k-1}^i : i = 1, \ldots, N_s\}$ and $\mathbf{y}_k$, draw random samples $\mathbf{x}_k^i \sim q(\mathbf{x}_k | \mathbf{x}_{k-1}^i, \mathbf{y}_k)$, where

$$q(\mathbf{x}_k | \mathbf{x}_{k-1}^i, \mathbf{y}_k) = p(\mathbf{x}_k | \mathbf{x}_{k-1}^i) = p_{\mu_{k-1}}(\mathbf{x}_k - \mathbf{f}(\mathbf{x}_{k-1}^i)).$$

Obtain $\{w_k^i : i = 1, \ldots, N_s\}$ for the approximation of $p(\mathbf{x}_k | \mathbf{y}_{1:k})$ in order to compute the updated state estimate and covariance.

Initialization

For $k = 1$, draw $\mathbf{x}_0^i \sim p(\mathbf{x}_0) \ \forall \ i = 1, \ldots, N_s$, such that

$$p(\mathbf{x}_0) \approx \sum_{i=1}^{N_S} w_0^i \delta_D(\mathbf{x}_0 - \mathbf{x}_0^i), \text{ and } w_0^i = 1/N_s.$$

Prediction

Given $\{\mathbf{x}_{k-1}^i, w_{k-1}^i : i = 1, \ldots, N_s\}$ which approximates $p(\mathbf{x}_{k-1} | \mathbf{y}_{1:k-1})$

Draw $\mathbf{x}_k^i \sim p_{\mu_{k-1}}(\mathbf{x}_k - \mathbf{f}(\mathbf{x}_{k-1}^i)) \ \forall \ i = 1, \ldots, N_s$.

It is equivalent to draw $\boldsymbol{\mu}_{k-1}^i \sim p_{\mu_{k-1}}(\boldsymbol{\mu})$ and compute

$$\mathbf{x}_k^i = \mathbf{f}(\mathbf{x}_{k-1}^i) + \boldsymbol{\mu}_{k-1}^i \ \forall \ i = 1, \ldots, N_s.$$

Update

Compute the updated weight $\{w_k^i : i = 1, \ldots, N_s\}$ given $\mathbf{y}_k$

$$w_k^i = w_{k-1}^i p_{\nu_k}(\mathbf{y}_k - \mathbf{h}(\mathbf{x}_k^i)), \ \forall \ i = 1, \ldots, N_s.$$

Normalize $\{w_k^i : i = 1, \ldots, N_s\}$, such that $\sum_{i=1}^{N_S} w_k^i = 1$.

Compute the updated state estimate and covariance:

$$\hat{\mathbf{x}}_{k|k} = \sum_{i=1}^{N_S} w_k^i \mathbf{x}_k^i,$$

$$\mathbf{P}_{k|k} = \sum_{i=1}^{N_S} w_k^i (\mathbf{x}_k^i - \hat{\mathbf{x}}_{k|k})(\mathbf{x}_k^i - \hat{\mathbf{x}}_{k|k})^T.$$

There are some cases in which the evaluation of $p(\mathbf{x}_k | \mathbf{x}_{k-1}^i, \mathbf{y}_k)$ is possible as shown in [9, 25], for example, for a linear measurement case with Gaussian system and measurement noises.

To detect whether a significant degeneracy has occurred, a measure, N_{eff}, defined as

$$N_{\text{eff}} = \frac{N_s}{1 + \text{Var}(w_k^{*i})}$$

is calculated where $w_k^{*i} = p(\mathbf{x}_k^i | \mathbf{y}_{1:k}) / q(\mathbf{x}_k^i | \mathbf{x}_{k-1}^i, \mathbf{y}_k)$ is referred to as the true weight. N_{eff} cannot be evaluated exactly, but an estimate

$$\hat{N}_{\text{eff}} = \frac{1}{\sum_{i=1}^{N_S}\left(w_k^i\right)^2} \tag{6.6-15}$$

can be obtained with normalized w_k^i computed by Equation (6.6-10). Note that $\hat{N}_{\text{eff}} \leq N_s$, and the algorithm will resample whenever $\hat{N}_{\text{eff}}$ falls below some threshold, $N_{\text{Threshold}}$. The threshold is a design parameter that depends upon the specific problem. A balance between the speed of convergence and the risk of prematurely converging on a wrong answer is something the user must determine empirically.

There are techniques that can be used to reduce the effects of degeneracy. One common technique is called resampling, as described in Table 6.1. This technique for particle filtering will be introduced in the next section.

6.6.2   Sequential Importance Resampling Filter

The SIS algorithm presented in the previous section provides the basis for most particle filters that have been developed in the past [9]. The following presents other related particle filters that may be derived from the SIS algorithm. The sequential importance resampling (SIR) filter [14] can be derived from the SIS algorithm by choosing the importance density $q(\mathbf{x}_k|\mathbf{x}_{k-1}, \mathbf{y}_k)$ to be the state transition density $p\left(\mathbf{x}_k|\mathbf{x}_{k-1}^i\right)$, a set of samples and weights $\left\{\bar{\mathbf{x}}_k^i, w_k^i, i = 1, \ldots, N_s\right\}$ is obtained as SIS in Table 6.5. Eliminate samples with negligible weights. We apply the resampling procedure (see Table 6.1) to the remaining set of degenerated samples and weights at every time step. A new set of samples with uniform weights $\left\{\mathbf{x}_k^i, 1/N_S, i^j : i = 1, \ldots, N_S\right\}$ is obtained where i^j is the index of the parent of $\mathbf{x}_k^i$, $\mathbf{x}_{k-1}^{i^j}$, at time step $k-1$, and $\mathbf{x}_k^i = \bar{\mathbf{x}}_k^{i^j}$. The new approximated $p(\mathbf{x}_k|\mathbf{y}_{1:k})$ becomes

$$p\left(\mathbf{x}_k|\mathbf{y}_{1:k}\right) \approx \sum_{i=1}^{N_S} 1/N_S \delta_D\left(\mathbf{x}_k - \mathbf{x}_k^i\right) \tag{6.6-16}$$

such that

$$\mathbf{x}_k^i = \bar{\mathbf{x}}_k^{i^j},$$

where $\bar{\mathbf{x}}_k^{i^j}$ occurs with probability $w_k^{i^j}$ and $\mathbf{x}_k^i$ occurs with probability $1/N_S$.

The basic idea of resampling is to eliminate particles that have negligible weights and to concentrate on particles with large weights (see Table 6.1). A summary of the algorithm is presented in Table 6.6.

A graphic illustration of the algorithm is shown in Figure 6.3.

Table 6.6   Summary of the Sequential Importance Resampling Filter Algorithm

---

Sequential Importance Resampling (SIR) Filter
Solution Concept
Given $\{\mathbf{x}_{k-1}^i, 1/N_s : i = 1, \ldots, N_s\}$ and $\mathbf{y}_k$. Draw random samples $\overline{\mathbf{x}}_k^i \sim q(\mathbf{x}_k | \mathbf{x}_{k-1}^i, \mathbf{y}_k)$, where

$$q(\mathbf{x}_k | \mathbf{x}_{k-1}^i, \mathbf{y}_k) = p(\mathbf{x}_k | \mathbf{x}_{k-1}^i) = p_{\mu_{k-1}}(\mathbf{x}_k - \mathbf{f}(\mathbf{x}_{k-1}^i)).$$

Obtain $\{w_k^i : i = 1, \ldots, N_s\}$ for the approximation of $p(\mathbf{x}_k | \mathbf{y}_{1:k})$ in order to compute the updated state estimate and covariance. A resampling step is added at the end of each update circle (see the resampling algorithm given in Table 6.1), to obtain a new sample set with uniform weights as $\{\mathbf{x}_k^i, 1/N_s, i^j : i = 1, \ldots, N_s\}$.

Initialization
For $k = 1$, draw $\mathbf{x}_0^i \sim p(\mathbf{x}_0)$ for $i = 1, \ldots, N_s$, such that

$$p(\mathbf{x}_0) \approx \sum_{i=1}^{N_S} w_0^i \delta_D (\mathbf{x}_0 - \mathbf{x}_0^i), \text{ and } w_0^i = 1/N_s.$$

Prediction
Given $\{\mathbf{x}_{k-1}^i, 1/N_s : i = 1, \ldots, N_s\}$ which approximates $p(\mathbf{x}_{k-1} | \mathbf{y}_{1:k-1})$
Draw $\overline{\mathbf{x}}_k^i \sim p_{\mu_{k-1}}(\mathbf{x}_k - \mathbf{f}(\mathbf{x}_{k-1}^i)) \ \forall \ i = 1, \ldots, N_s$. It is equivalent to draw $\mu_{k-1}^i \sim p_{\mu_{k-1}}(\mu)$ and compute

$$\overline{\mathbf{x}}_k^i = \mathbf{f}(\mathbf{x}_{k-1}^i) + \mu_{k-1}^i \ \forall \ i = 1, \ldots, N_s.$$

Update
Compute the updated weight $\{\overline{w}_k^i : i = 1, \ldots, N_s\}$ given $\mathbf{y}_k$

$$\overline{w}_k^i = p_{\mathbf{v}_k}(\mathbf{y}_k - \mathbf{h}(\overline{\mathbf{x}}_k^i)), \ \forall \ i = 1, \ldots, N_s.$$

Normalize $\{\overline{w}_k^i : i = 1, \ldots, N_s\}$, such that $\sum_{i=1}^{N_S} \overline{w}_k^i = 1$.
Compute the updated state estimate and covariance

$$\hat{\mathbf{x}}_{k|k} = \sum_{i=1}^{N_S} \overline{w}_k^i \overline{\mathbf{x}}_k^i,$$

$$\mathbf{P}_{k|k} = \sum_{i=1}^{N_S} \overline{w}_k^i (\overline{\mathbf{x}}_k^i - \hat{\mathbf{x}}_{k|k})(\overline{\mathbf{x}}_k^i - \hat{\mathbf{x}}_{k|k})^T.$$

Resample $\{\overline{\mathbf{x}}_k^i, \overline{w}_k^i : i = 1, \ldots, N_s\}$ (see Table 6.1) to obtain a new set of samples with uniform weights $\{\mathbf{x}_k^i, 1/N_s, i^j : i = 1, \ldots, N_s\}$.

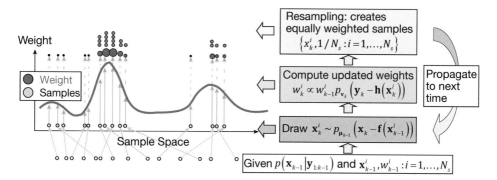

Figure 6.3   Sequential importance resampling (SIR) particle filter illustration.

Remarks

1. The simplicity of an SIR filter is very easy to implement. It can be applied to any system for which the state transition density can be sampled and the likelihood can be computed. But there are issues about the loss of diversity in particles $\{\mathbf{x}_k^i\}$, known as sample impoverishment, because it is drawn from the importance function $p(\mathbf{x}_k|\mathbf{x}_{k-1}^i)$, which is independent of measurement $\mathbf{y}_k$ (see Figure 6.3). These samples $\{\mathbf{x}_k^i\}$ may not have a sufficient number of them with significant weights $w_k^i \propto p(\mathbf{y}_k|\mathbf{x}_k^i)$. Therefore, the samples $\{\mathbf{x}_k^i\}$ do not explore the whole state space for $p(\mathbf{x}_k|\mathbf{y}_{1:k})$ that contain knowledge of $\mathbf{y}_k$. Other importance functions are suggested in the literature [9], and the most common one is discussed in the next section.

2. Sample impoverishment occurs often when the process noise is relatively small. In general, when there is no process noise in the system equation, the use of an SIR filter is not appropriate. In this deterministic case, other estimation techniques, for example the iterative estimator presented in Section 3.8, should be used. There are systematic techniques proposed to solve the sample impoverishment problem, and a good review of these techniques can be found in [9].

3. The resampling step is needed to minimize the effects of degeneracy, and it also makes the exploitation of parallelized computational structure more challenging.

As mentioned earlier, the other deficiency of SIR is that the resampling scheme may cause sample impoverishment or particle collapse, where all N_S particles occupy the same point in the state space resulting in a poor representation of the posterior density $p(\mathbf{x}_k|\mathbf{y}_{1:k})$. The regularized particle filter (RPF) is proposed [26] as a potential solution to this problem. The main idea of this approach is to modify the resampling step of the SIR filter by drawing intermediate samples $\{\overline{\mathbf{x}}_k^i, i = 1, \ldots, N_S\}$ from a continuous approximation of the posterior density $p(\mathbf{x}_k|\mathbf{y}_{1:k})$ instead of the discrete approximation in Equation (6.6-16) as follows

$$p\left(\mathbf{x}_k|\mathbf{y}_{1:k}\right) \approx \sum_{i=1}^{N_S} \overline{w}_k^i K_h\left(\mathbf{x}_k - \overline{\mathbf{x}}_k^i\right) \tag{6.6-17}$$

where

$$K_h(\mathbf{x}) = \frac{1}{h^n} K\left(\frac{\mathbf{x}}{h}\right) \tag{6.6-18}$$

is the rescaled kernel density $K(.)$, $h > 0$ is the kernel bandwidth, n is the dimension of the state vector $\mathbf{x}$, and $\{\overline{w}_k^i, i = 1, \ldots, N_S\}$ are normalized weights. The kernel density is a zero mean bounded symmetric probability density function. There exists an optimal choice for the kernel $K_h(.)$ and bandwidth h. (More design detail about this algorithm can be found in [26].) The key step is, instead of using

$$\mathbf{x}_k^i = \overline{\mathbf{x}}_k^i,$$

replace it by

$$\mathbf{x}_k^i = \overline{\mathbf{x}}_k^i + h\mathbf{D}_k\boldsymbol{\epsilon}^i \tag{6.6-19}$$

with probability $\overline{w}_k^i$ and $w_k^i = 1/N_S$ where $\mathbf{D}_k\mathbf{D}_k^T$ is the sample covariance matrix of $\{\overline{\mathbf{x}}_k^i, \overline{w}_k^i, i = 1, \ldots, N_S\}$ and $\boldsymbol{\epsilon}^i$ is a sample draw from the kernel density $K_h(.)$.

Remarks

1. The disadvantage of the RPF is that its samples are no longer guaranteed to asymptotically approximate samples from the posterior density $p(\mathbf{x}_k|\mathbf{y}_{1:k})$.

2. Another technique to improve the resampling step is called the MCMC move step introduced in [27, 28]. A particular implementation of this step based on the Metropolis–Hastings algorithm [29] is that the resampled particle $\overline{\mathbf{x}}_k^i$ is moved to $\mathbf{x}_k^i$ by Equation (6.6-19) only if $u < \alpha$, where u is a sample drawn from the uniform distribution over [0, 1] and α is the Metropolis–Hastings acceptance probability given by

$$\alpha = \min\left\{1, \frac{p\left(\mathbf{y}_k|\overline{\mathbf{x}}_k^i\right)p\left(\overline{\mathbf{x}}_k^i|\mathbf{x}_{k-1}^i\right)}{p\left(\mathbf{y}_k|\mathbf{x}_k^i\right)p\left(\mathbf{x}_k^i|\mathbf{x}_{k-1}^i\right)}\right\} \tag{6.6-20}$$

6.6.3 Auxiliary Sampling Importance Resampling Filter

There are several strategies to deal with the deficiencies of the SIR algorithm, but there is no silver bullet for all cases. A good survey of research in this area is given in [9]. To circumvent some deficiencies of the SIR sampling scheme for the posterior density, $p(\mathbf{x}_k|\mathbf{y}_{1:k})$, Pitt and Shephard [30] introduced the auxiliary sampling importance resampling (ASIR) filter. An intermediate sample set $\left\{\overline{\mathbf{x}}_k^i, \overline{w}_k^i, i^j: i = 1, \ldots, N_s\right\}$ is drawn from $p\left(\mathbf{x}_k|\mathbf{x}_{k-1}^i\right)$, the SIR. Instead of repeating the same sample $\overline{\mathbf{x}}_k^i$ drawn from $p\left(\mathbf{x}_k|\mathbf{x}_{k-1}^i\right)$ with larger weight $\overline{w}_k^i$ by the SIR and hoping in the next prediction cycle at $k + 1$ that the sample set will be spread to avoid sample impoverishment, alternative sampling approaches are proposed. The following procedure is one of the most popular approaches.

Using Bayes' rule and Equation (6.6-8) yields

$$\begin{aligned} p\left(\mathbf{x}_k, i^j|\mathbf{y}_{1:k}\right) &\propto p\left(\mathbf{y}_k|\mathbf{x}_k\right)p\left(\mathbf{x}_k, i^j|\mathbf{y}_{1:k-1}\right) \\ &= p\left(\mathbf{y}_k|\mathbf{x}_k\right)p\left(\mathbf{x}_k|i^j, \mathbf{y}_{1:k-1}\right)p\left(i^j|\mathbf{y}_{1:k-1}\right) \\ &= p\left(\mathbf{y}_k|\mathbf{x}_k\right)p\left(\mathbf{x}_k|\mathbf{x}_{k-1}^{ij}\right)w_{k-1}^{ij}, \end{aligned} \tag{6.6-21}$$

where i^j is the index of the parent of $\overline{\mathbf{x}}_k^i$, $\mathbf{x}_{k-1}^{ij}$, and $\overline{\mathbf{x}}_k^i = \mathbf{x}_k^{ij}$. One can select an importance density $q(\mathbf{x}_k, i^j|\mathbf{y}_k)$ to satisfy the proportionality

$$q\left(\mathbf{x}_k, i^j|\mathbf{y}_{1:k}\right) \propto p\left(\mathbf{y}_k|\overline{\mathbf{x}}_k^i\right)p\left(\mathbf{x}_k|\mathbf{x}_{k-1}^{ij}\right)w_{k-1}^{ij}. \tag{6.6-22}$$

Here for the ASIR, $\bar{\mathbf{x}}_k^i$ is drawn from $p(\mathbf{x}_k|\mathbf{x}_{k-1}^i)$, with $\mathbf{x}_{k-1}^i = \mathbf{x}_{k-1}^{ij}$. Substituting Equations (6.6-21) and (6.6-22) into Equation (6.6-9), the sample set $\{\bar{\bar{\mathbf{x}}}_k^i : i = 1, \ldots, N_S\}$ is then assigned a weight

$$\bar{w}_k^i \propto \frac{p(\mathbf{x}_k, i^j|\mathbf{y}_{1:k})}{q(\mathbf{x}_k, i^j|\mathbf{y}_{1:k})} = \frac{p(\mathbf{y}_k|\bar{\bar{\mathbf{x}}}_k^i)}{p(\mathbf{y}_k|\bar{\mathbf{x}}_k^i)} = \frac{p_{\mathbf{v}_k}(\mathbf{y}_k - \mathbf{h}(\bar{\bar{\mathbf{x}}}_k^i))}{p_{\mathbf{v}_k}(\mathbf{y}_k - \mathbf{h}(\bar{\mathbf{x}}_k^i))}. \tag{6.6-23}$$

The algorithm is summarized in Table 6.7.

Remarks

1. The advantage of the ASIR over the SIR filter is that it generates points from the sample at time $k - 1$ with the samples at time k, $\{\mathbf{x}_k^i\}$, that are most likely to have generated the current measurement $\mathbf{y}_k$.

2. The ASIR filter can be viewed as resampling at the previous time step, based on some point estimates $\bar{\mathbf{x}}_k^i$ that best characterize $p(\mathbf{x}_k|\mathbf{x}_{k-1}^i)$ given the current measurement $\mathbf{y}_k$ and the likelihood function $p(\mathbf{y}_k|\bar{\mathbf{x}}_k^i)$.

3. If the process noise is small, then the ASIR filter is often not as sensitive to outliers as SIR, but on the other hand, for large process noise, a single point estimate, $\bar{\bar{\mathbf{x}}}_k^i$ does not characterize $p(\mathbf{x}_k|\mathbf{x}_{k-1}^i)$ well and sometimes it may even result in degraded performance.

6.6.4   Extended Kalman Filter Auxiliary Sampling Importance Resampling Filter

There are other popular choices of auxiliary importance sampling functions [19, 15], for example, use samples generated by the EKF [19] and UKF [15] for the systems with Gaussian process and measurement noise, such as $N(\mathbf{0}, \mathbf{Q}_{k-1})$ and $N(\mathbf{0}, \mathbf{R}_k)$, respectively. The importance sample function can be

$$q(\mathbf{x}_k|\mathbf{x}_{k-1}^i, \mathbf{y}_k) = \hat{p}(\mathbf{x}_k|\mathbf{x}_{k-1}^i, \mathbf{y}_k) \tag{6.6-24}$$

where $\hat{p}(\mathbf{x}_k|\mathbf{x}_{k-1}^i, \mathbf{y}_k)$ is the Gaussian density function, $N(\hat{\mathbf{x}}_{k|k}^i, \mathbf{P}_{k|k}^i)$, associated with an EKF or UKF for nonlinear system from Equations (6.6-6) and (6.6-7) with $\hat{\mathbf{x}}_{k-1|k-1}^i = \mathbf{x}_{k-1}^i$ and $\mathbf{P}_{k-1|k-1}^i$. Only the derivation of the extended Kalman filter auxiliary sampling importance resampling (EKF-ASIR) will be shown here. A similar procedure can be applied to the UKF proposed importance function [15]. Let us first create an intermediate sample set $\{\hat{\mathbf{x}}_{k|k}^i, \hat{w}_k^i : i = 1, \ldots, N_s\}$ where

Table 6.7 Summary of the Auxiliary Sampling Importance Resampling Filter Algorithm

Auxiliary Sampling Importance Resampling (ASIR) Filter

Solution Concept

Given $\{\mathbf{x}_{k-1}^i, 1/N_s : i = 1, \ldots, N_s\}$ and $\mathbf{y}_k$ run SIR to obtain $\{\bar{\mathbf{x}}_k^i, 1/N_s, i^j : i = 1, \ldots, N_s\}$ where i^j is the index of the parent of $\bar{\mathbf{x}}_k^i$, $\mathbf{x}_{k-1}^i$ associates with large weight $w_k^{i^j}$. Redraw samples $\bar{\bar{\mathbf{x}}}_k^i \sim p\left(\mathbf{x}_k | \mathbf{x}_{k-1}^{i^j}\right)$ for the importance density

$$q\left(\mathbf{x}_k, i^j | \mathbf{y}_k\right) = p\left(\mathbf{x}_k | \mathbf{x}_{k-1}^{i^j}\right) p\left(\mathbf{y}_k | \bar{\mathbf{x}}_k^{i^j}\right),$$

to obtain the weights $\{\bar{w}_k^i : i = 1, \ldots, N_s\}$ [Equation (6.6-23)] for the approximation of the updated state estimate and covariance for nonlinear estimation. A resampling step is added at the end of each update circle to obtain a new samples with uniform weights as $\{\mathbf{x}_k^i, 1/N_s, i^j : i = 1, \ldots, N_s\}$.

Initialization

For $k = 1$, draw $\mathbf{x}_0^i \sim p(\mathbf{x}_0)$ for $i = 1, \ldots, N_s$, such that

$$p(\mathbf{x}_0) \approx \sum_{i=1}^{N_S} w_0^i \delta_D\left(\mathbf{x}_0 - \mathbf{x}_0^i\right), \text{ and } w_0^i = 1/N_s.$$

Prediction

Given $\{\mathbf{x}_{k-1}^i, 1/N_s : i = 1, \ldots, N_s\}$ which approximates $p(\mathbf{x}_{k-1} | \mathbf{y}_{1:k-1})$

Run SIR (Table 6.6) to obtain $\{\bar{\mathbf{x}}_k^i, 1/N_s, i^j : i = 1, \ldots, N_s\}$ where i^j is the index of the parent of $\bar{\mathbf{x}}_k^i$, $\mathbf{x}_{k-1}^{i^j}$ associates with large weight $w_k^{i^j} \propto p_{\mathbf{v}_k}(\mathbf{y}_k - \mathbf{h}(\bar{\mathbf{x}}_k^i))$.

$$\text{Draw } \bar{\bar{\mathbf{x}}}_k^i \sim p_{\mu_{k-1}}\left(\mathbf{x}_k - \mathbf{f}\left(\mathbf{x}_{k-1}^{i^j}\right)\right) \forall \, i = 1, \ldots, N_s.$$

Update

Compute the updated weight $\{\bar{\bar{w}}_k^i : i = 1, \ldots, N_s\}$ given $\mathbf{y}_k$

$$\bar{\bar{w}}_k^i = p_{\mathbf{v}_k}\left(\mathbf{y}_k - \mathbf{h}\left(\bar{\bar{\mathbf{x}}}_k^i\right)\right) / p_{\mathbf{v}_k}\left(\mathbf{y}_k - \mathbf{h}\left(\bar{\mathbf{x}}_k^i\right)\right), \forall \, i = 1, \ldots, N_s.$$

Normalize $\{\bar{\bar{w}}_k^i : i = 1, \ldots, N_s\}$, such that $\sum_{i=1}^{N_S} \bar{\bar{w}}_k^i = 1$.

Compute the updated state estimate and covariance

$$\hat{\mathbf{x}}_{k|k} = \sum_{i=1}^{N_S} \bar{\bar{w}}_k^i \bar{\bar{\mathbf{x}}}_k^i,$$

$$\mathbf{P}_{k|k} = \sum_{i=1}^{N_S} \bar{\bar{w}}_k^i \left(\bar{\bar{\mathbf{x}}}_k^i - \hat{\mathbf{x}}_{k|k}\right)\left(\bar{\bar{\mathbf{x}}}_k^i - \hat{\mathbf{x}}_{k|k}\right)^T.$$

Resample $\{\bar{\bar{\mathbf{x}}}_k^i, \bar{\bar{w}}_k^i : i = 1, \ldots, N_s\}$ (see Table 6.1) to obtain a new set of samples with uniform weights $\{\mathbf{x}_k^i, 1/N_s : i = 1, \ldots, N_s\}$.

$$\hat{w}_k^i = p_{\boldsymbol{\gamma}_k^i}\left(\mathbf{y}_k - \mathbf{h}\left(\hat{\mathbf{x}}_{k|k-1}^i\right)\right) \quad \forall\, i = 1, \dots, N_s, \text{ with } \sum_{i=1}^{N_S} \hat{w}_k^i = 1$$

and $\hat{\mathbf{x}}_{k|k-1}^i$ is the one-step prediction and $\boldsymbol{\gamma}_k^i$ the residual of $\mathbf{x}_{k-1}^i$, respectively. Then, we resample $\left\{\hat{\mathbf{x}}_{k|k}^i, \hat{w}_k^i: i = 1, \dots, N_s\right\}$ to obtain $\left\{\hat{\mathbf{x}}_k^i, 1/N_s, i^j: i = 1, \dots, N_s\right\}$ where $\hat{\mathbf{x}}_k^i = \hat{\mathbf{x}}_{k|k}^{ij}$ and $\mathbf{P}_k^i = \mathbf{P}_{k|k}^{ij}$.

The sample $\mathbf{x}_k^i \sim N\left(\hat{\mathbf{x}}_k^i, \mathbf{P}_k^i\right)$ and $\mathbf{P}_{k|k}^i = \mathbf{P}_k^i$ with associated weight is given as

$$w_k^i \propto \frac{p\left(\mathbf{y}_k|\mathbf{x}_k^i\right)p\left(\mathbf{x}_k^i|\mathbf{x}_{k-1}^i\right)}{q\left(\mathbf{x}_k^i|\mathbf{x}_{k-1}^i, \mathbf{y}_k\right)} = \frac{p\left(\mathbf{y}_k|\mathbf{x}_k^i\right)}{\hat{p}\left(\mathbf{x}_k^i|\mathbf{x}_{k-1}^i, \mathbf{y}_k\right)} = \frac{p_{\mathbf{v}_k}\left(\mathbf{y}_k - \mathbf{h}\left(\mathbf{x}_k^i\right)\right)}{p_{\mathbf{v}_k}\left(\mathbf{y}_k - \mathbf{h}\left(\hat{\mathbf{x}}_{k|k-1}^{ij}\right) - \mathbf{H}_{\hat{\mathbf{x}}_{k|k-1}^{ij}}\left(\mathbf{x}_k^i - \hat{\mathbf{x}}_{k|k-1}^{ij}\right)\right)}.$$

$$(6.6\text{-}25)$$

The algorithm is summarized in Table 6.8.

Remarks

When the system noise and measurement noise are Gaussian, the EKF-ASIR gives a more accurate approximation of the posterior density function $p(\mathbf{x}_k|\mathbf{y}_{1:k})$. In some cases, the UKF may give better results when the problem is highly nonlinear [15]. The algorithm for the UKF-ASIR is derived from the same principles as the EKF-ASIR.

6.6.5 Sequential Importance Resampling Filter Algorithm for Multiple Model Systems

In this section, the particle filter sampling techniques are applied to the multiple model estimation problems for nonlinear systems [16, 19, 31]. The results presented in Chapter 5 are for linear systems that can be easily extended to nonlinear systems. An outline of these algorithms will be presented here.

A multiple model system can be defined as a mixed random process $\{\mathbf{x}_k, \theta_k\}$, where $\mathbf{x}_k$ is a continue-valued random process and θ_k is a discrete-valued random process that is independent of $\mathbf{x}_0$. θ_k represents N model hypotheses, $\{\theta^1, \theta^2, \dots, \theta^N\}$, such that for $\theta_k = \theta^i$, the hypothesis θ^i is true at time step k and the system and measurement follow the ith model as

$$\mathbf{x}_k = \mathbf{f}^i(\mathbf{x}_{k-1}) + \boldsymbol{\mu}_{k-1} \tag{6.6-26}$$

$$\mathbf{y}_k = \mathbf{h}^i(\mathbf{x}_k) + \mathbf{v}_k, \tag{6.6-27}$$

Table 6.8   Summary of the EKF-Auxiliary Sampling Importance Resampling Filter Algorithm

---

EKF-Auxiliary Sampling Importance Resampling (EKF-ASIR) Filter

Solution Concept

For systems with additive Gaussian noises: $N(\mathbf{0}, \mathbf{Q}_{k-1})$ and $N(\mathbf{0}, \mathbf{R}_k)$.

Given $\{\mathbf{x}_{k-1}^i, \mathbf{P}_{k-1|k-1}^i, 1/N_s : i = 1, \ldots, N_s\}$ draw random samples $\{\mathbf{x}_k^i : i = 1, \ldots, N_s\}$ from an important function

$$q\left(\mathbf{x}_k|\mathbf{x}_{k-1}^i, \mathbf{y}_k\right) = \hat{p}\left(\mathbf{x}_k|\mathbf{x}_{k-1}^i, \mathbf{y}_k\right),$$

where $\hat{p}\left(\mathbf{x}_k|\mathbf{x}_{k-1}^i, \mathbf{y}_k\right)$ is the Gaussian density function, $N\left(\hat{\mathbf{x}}_{k|k}^i, \mathbf{P}_{k|k}^i\right)$, associated with an EKF for nonlinear system from Equations (6.6-6) and (6.6-7) with $\hat{\mathbf{x}}_{k-1|k-1}^i = \mathbf{x}_{k-1}^i$ and $\mathbf{P}_{k-1|k-1}^i$. Obtain the associated weights $\{w_k^i : i = 1, \ldots, N_s\}$ for the approximation of the updated state estimate and covariance. A resampling step is added at the end of each update circle to obtain a new set of samples with uniform weights as $\{\mathbf{x}_k^i, \mathbf{P}_{k|k}^i, 1/N_s : i = 1, \ldots, N_s\}$.

Initialization

For $k = 1$, draw $\mathbf{x}_0^i \sim N(\mathbf{x}_0, \mathbf{P}_0)$ for $i = 1, \ldots, N_s$, such that

$$p(\mathbf{x}_0) = N(\mathbf{x}_0, \mathbf{P}_0) \approx \sum\nolimits_{i=1}^{N_S} w_0^i \delta_D\left(\mathbf{x}_0 - \mathbf{x}_0^i\right), \mathbf{P}_{0|0}^i = \mathbf{P}_0 \text{ and } w_0^i = 1/N_s.$$

Prediction

Given $\{\mathbf{x}_{k-1}^i, \mathbf{P}_{k-1|k-1}^i, 1/N_s : i = 1, \ldots, N_s\}$ which approximates $p(\mathbf{x}_{k-1}|\mathbf{y}_{1:k-1})$.

Run the EKF with $\hat{\mathbf{x}}_{k-1|k-1}^i = \mathbf{x}_{k-1}^i$ and $\mathbf{P}_{k-1|k-1}^i$ and $\mathbf{y}_k$ (see Table 3.1) to obtain $\left\{\hat{\mathbf{x}}_{k|k-1}^i, \mathbf{H}_{\hat{\mathbf{x}}_{k|k-1}^i}, \boldsymbol{\gamma}_k^i, \mathbf{S}_{k|k}^i, \hat{\mathbf{x}}_{k|k}^i, \mathbf{P}_{k|k}^i\right\}$ where

$\mathbf{H}_{\hat{\mathbf{x}}} = \left[\frac{\partial \mathbf{h}(\mathbf{x})}{\partial \mathbf{x}}\right]_{\hat{\mathbf{x}}}$ is the Jacobian of $\mathbf{h}(\mathbf{x})$ evaluated at $\hat{\mathbf{x}}$ and $\mathbf{S}_{k|k}^i$ is the covariance of the residual $\boldsymbol{\gamma}_k^i$.

Generate a sample set $\{\hat{\mathbf{x}}_{k|k}^i, \hat{w}_k^i : i = 1, \ldots, N_s\}$ where

$$\hat{w}_k^i = p_{\boldsymbol{\gamma}_k^i}\left(\mathbf{y}_k - \mathbf{h}\left(\hat{\mathbf{x}}_{k|k-1}^i\right)\right) \forall i = 1, \ldots, N_s.$$

Normalize $\hat{w}_k^i$ such that $\sum_{i=1}^{N_S} \hat{w}_k^i = 1$.

Resample $\{\hat{\mathbf{x}}_{k|k}^i, \hat{w}_k^i : i = 1, \ldots, N_s\}$ to obtain $\{\hat{\mathbf{x}}_k^i, 1/N_s, i^j : i = 1, \ldots, N_s\}$ where $\hat{\mathbf{x}}_k^i = \hat{\mathbf{x}}_{k|k}^{ij}$ and $\mathbf{P}_k^i = \mathbf{P}_{k|k}^{ij}$.

Draw sample $\mathbf{x}_k^i \sim N\left(\hat{\mathbf{x}}_k^i, \mathbf{P}_k^i\right)$ and $\mathbf{P}_{k|k}^i = \mathbf{P}_k^i$.

Update

Compute the updated weight $\{w_k^i : i = 1, \ldots, N_s\}$ given $\mathbf{y}_k$

$$w_k^i = p_{\mathbf{v}_k}\left(\mathbf{y}_k - \mathbf{h}\left(\mathbf{x}_k^i\right)\right) \Big/ p_{\mathbf{v}_k}\left(\mathbf{y}_k - \mathbf{h}\left(\hat{\mathbf{x}}_{k|k-1}^{ij}\right) - \mathbf{H}_{\hat{\mathbf{x}}_{k|k-1}^{ij}}\left(\mathbf{x}_k^i - \hat{\mathbf{x}}_{k|k-1}^{ij}\right)\right) \forall i = 1, \ldots, N_s$$

Normalize w_k^i such that $\sum_{i=1}^{N_S} w_k^i = 1$.

Compute the updated state estimate and covariance

$$\hat{\mathbf{x}}_{k|k} = \sum\nolimits_{i=1}^{N_S} w_k^i \overline{\mathbf{x}}_k^i,$$

$$\mathbf{P}_{k|k} = \sum\nolimits_{i=1}^{N_S} w_k^i \mathbf{P}_{k|k}^i.$$

Resample $\{\mathbf{x}_k^i, w_k^i : i = 1, \ldots, N_s\}$ utilizing the resampling algorithm given in Table 6.1 to obtain a new set of samples with uniform weights as: $\{\mathbf{x}_k^i, \mathbf{P}_{k|k}^i, 1/N_s : i = 1, \ldots, N_s\}$.

---

where $\mathbf{x}_0: \sim N(\hat{\mathbf{x}}_0, \mathbf{P}_0)$, $\boldsymbol{\mu}_{k-1}: \sim N(\mathbf{0}, \mathbf{Q}_{k-1})$ and $\mathbf{v}_k: \sim N(\mathbf{0}, \mathbf{R}_k)$ are mutually independent and $\boldsymbol{\mu}_{k-1}$ and $\mathbf{v}_k$ are white Gaussian system (process) and measurement noise sequences, respectively, and the a priori probability of θ_0 as

$$p(\theta_0) = \sum_{i=1}^{N} \Pr\{\theta_0 = \theta^i\} \delta(\theta_0 - \theta^i), \tag{6.6-28}$$

and $\sum_{i=1}^{N} \Pr\{\theta_0 = \theta^i\} = 1$. For notational simplicity to introduce the concept, we assume all models have the same statistics for $\mathbf{x}_0$, $\boldsymbol{\mu}_{k-1}$, and $\mathbf{v}_k$. Let θ_k^i be a local hypothesis that the ith model is true at time k and $\theta_{1:k}^j$ be a global hypothesis that represents a sequence of hypotheses from time 1 through time k, thus $\theta_{1:k}^j = \{\theta_1^{i_1}, \theta_2^{i_2}, \ldots, \theta_k^{i_k}\}$, where $i_n = 1, \ldots, N$ and $j = 1, \ldots, N^k$ and

$\Pr\{\theta_k^i | \mathbf{y}_{1:k}\}$ is the probability of θ_k^i being true at k, given $\mathbf{y}_{1:k}$

$\Pr\{\theta_{1:k}^j | \mathbf{y}_{1:k}\}$ is the probability of $\theta_{1:k}^j = \{\theta_1^{i_1}, \theta_2^{i_2}, \ldots, \theta_k^{i_k}\}$ being true at k, given $\mathbf{y}_{1:k}$.

Let $\theta_{1:k}^j = \{\theta_k^\ell, \theta_{1:k-1}^j\}$, where $\theta_{1:k-1}^j$ is the parameter history of $\theta_{1:k}^j$ up to time $k-1$. Applying Bayes' rule, the conditional density function, $p(\mathbf{x}_k, \theta_{1:k}^j | \mathbf{y}_{1:k})$ for the system in Equations (6.6-26) and (6.6-27) can be written as

$$p(\mathbf{x}_k, \theta_{1:k}^j | \mathbf{y}_{1:k}) = p(\mathbf{x}_k | \theta_{1:k}^j, \mathbf{y}_{1:k}) \Pr\{\theta_{1:k}^j | \mathbf{y}_{1:k}\}. \tag{6.6-29}$$

Let $\Pr\{\theta_k^\ell | \theta_{1:k-1}^j\}$ be the transition probability of θ_k^ℓ being true given parameter history $\theta_{1:k-1}^j$. For simplicity of notation and ease of conveying the concepts, the single step Markov transition model in Section 5.4.1 is used, that is,

$$\Pr\{\theta_k^\ell | \theta_{1:k-1}^j\} = \Pr\{\theta_k^\ell | \theta_{k-1}^m\}, \tag{6.6-30}$$

which is independent of past history $\theta_{1:k-2}^j$, and θ_{k-1}^m is the hypothesis at time $k-1$. Substituting Equation (6.6-30) into Equation (6.6-29), given θ_k^ℓ and θ_{k-1}^m as the last two hypotheses of $\theta_{1:k}^j$ yields

$$p(\mathbf{x}_k, \theta_k^\ell | \theta_{k-1}^m, \mathbf{y}_{1:k}) \propto p(\mathbf{y}_k | \mathbf{x}_k, \theta_k^\ell) p(\mathbf{x}_k | \mathbf{x}_{k-1}, \theta_k^\ell, \theta_{k-1}^m) \Pr\{\theta_k^\ell | \theta_{k-1}^m\}. \tag{6.6-31}$$

Let the importance function be

$$q(\mathbf{x}_k, \theta_k^\ell | \theta_{k-1}^m, \mathbf{y}_{1:k}) \propto p(\mathbf{y}_k | \mathbf{x}_k, \theta_k^\ell) p(\mathbf{x}_k | \mathbf{x}_{k-1}, \theta_k^\ell, \theta_{k-1}^m) \Pr\{\theta_k^\ell | \theta_{k-1}^m\}. \tag{6.6-32}$$

Treat the multiple models as a mixed random process. Given samples $\{\mathbf{x}_{k-1}^i, \theta_{k-1}^i, 1/N_s : i = 1, \ldots, N_s\}$ and $\mathbf{y}_k$ run SIR for each model hypothesis to obtain NN_s samples $\{\bar{\mathbf{x}}_k^{i\ell}, \theta_k^\ell, \bar{w}_k^{i\ell} : i = 1, \ldots, N_s, \ell = 1, \ldots, N\}$, where

$$\bar{w}_k^{i\ell} \propto p_{\mathbf{v}_k}\left(\mathbf{y}_k - \mathbf{h}^{\theta_k^\ell}(\bar{\mathbf{x}}_k^{i\ell})\right) \Pr\{\theta_k^\ell | \theta_{k-1}^i\} \text{ for } \ell = 1, \ldots, N.$$

Normalize $\overline{w}_k^{i\ell}$ such that $\sum_{i=1}^{N_s} \sum_{\ell=1}^{N} \overline{w}_k^{i,\ell} = 1$.

Resample to obtain $\left\{ \overline{\overline{\mathbf{x}}}_k^i, 1/N_s, i^j : i = 1, \ldots, N_s \right\}$ where i^j is the index of the parent of $\overline{\overline{\mathbf{x}}}_k^i$, $\mathbf{x}_{k-1}^{i^j}$ with model θ_k^ℓ that associates with large weight $\overline{w}_k^{i^j \ell}$. Redraw samples $\mathbf{x}_k^i \sim p\left(\mathbf{x}_k | \mathbf{x}_{k-1}^{i^j}, \theta_k^\ell\right)$ like ASIR for single model to obtain new weights $\left\{ w_k^i : i = 1, \ldots, N_s \right\}$

$$w_k^i \propto \frac{p\left(\mathbf{x}_k, \theta_k^\ell | \theta_{k-1}^m, \mathbf{y}_{1:k}\right)}{q\left(\mathbf{x}_k, \theta_k^\ell | \theta_{k-1}^m, \mathbf{y}_{1:k}\right)} = \frac{p\left(\mathbf{y}_k | \mathbf{x}_k^i, \theta_k^\ell\right)}{p\left(\mathbf{y}_k | \overline{\overline{\mathbf{x}}}_k^i, \theta_k^\ell\right)} = \frac{p_{\mathbf{v}_k}\left(\mathbf{y}_k - \mathbf{h}^{\theta_k^\ell}\left(\mathbf{x}_k^i\right)\right)}{p_{\mathbf{v}_k}\left(\mathbf{y}_k - \mathbf{h}^{\theta_k^\ell}\left(\overline{\overline{\mathbf{x}}}_k^i\right)\right)}. \qquad (6.6\text{-}33)$$

for the approximation of the updated state and covariance for nonlinear estimation. A resampling step is added at the end of each update circle to obtain a new set of samples with uniform weights as $\left\{ \mathbf{x}_k^i, \theta_k^i, 1/N_s : i = 1, \ldots, N_s \right\}$.

A summary of the algorithm is given in Table 6.9.

It was shown in [31] that the multiple model ASIR filter performs better than the EKF and IMM filters. The derivation for the EKF-ASIR or UKF-ASIR for multiple model systems will be very similar to the ASIR for multiple systems, except the procedure for prediction and update should follow that for EKF-ASIR in Table 6.8 for each model.

6.6.6   Particle Filters for Smoothing

In this section, the particle filter framework is extended to the smoothing problems for nonlinear systems [32, 33]. We start with the original filtering problem

$$p\left(\mathbf{x}_{0:k} | \mathbf{y}_{1:k}\right) \approx \sum_{i=1}^{N_s} w_k^i \delta_D \left(\mathbf{x}_{0:k} - \mathbf{x}_{0:k}^i\right), \sum_{i=1}^{N_s} w_k^i = 1, \qquad (6.6\text{-}2)$$

and $\hat{\mathbf{x}}_{k|k} = E\{\mathbf{x}_{0:k} | \mathbf{y}_{1:k}\}$. The smoothed estimate $\hat{\mathbf{x}}_{k|K} = E\{\mathbf{x}_k | \mathbf{y}_{1:K}\}$ for the fixed-interval smoother (FIS, see Section 2.9.2) can be obtained for free, if the whole trajectory for each particle $\mathbf{x}_{0:K}^i$ is stored for the forward path of the particle filter (see Sections 6.6.2–6.6.5; for notational simplicity, the SIR filter is used). Other variations of smoothers discussed in Section 2.9 can be obtained with suitable modifications to the algorithm.

First, assume that particle filtering has been done for $k = 1, 2, \ldots, K$ and the particles and weights $\left\{ \mathbf{x}_k^i, w_k^i : i = 1, \ldots, N_s \right\}$ are stored such that

$$p\left(\mathbf{x}_k | \mathbf{y}_{1:k}\right) \approx \sum_{i=1}^{N_s} w_k^i \delta_D \left(\mathbf{x}_k - \mathbf{x}_k^i\right), \sum_{i=1}^{N_s} w_k^i = 1, \forall k = 1, 2, \ldots, K.$$

Bayes' rule implies

$$p\left(\mathbf{x}_{0:K} | \mathbf{y}_{1:K}\right) = \prod_{k=1}^{K} p\left(\mathbf{x}_k | \mathbf{x}_{k+1:K}, \mathbf{y}_{1:K}\right),$$

Table 6.9   Summary of the ASIR Filter Algorithm for Multiple Model Systems

---

ASIR Filter Algorithm for Multiple Model Systems

Solution Concept

Treat the multiple model systems as a mixed random process $\{\mathbf{x}_k, \theta_k\}$. Given $\{\mathbf{x}_{k-1}^i, \theta_{k-1}^i, 1/N_s : i = 1, \ldots, N_s\}$ and $\mathbf{y}_k$. Run SIS for each model hypothesis to obtain NN_s samples $\{\overline{\mathbf{x}}_k^{i,\ell}, \theta_k^\ell, \overline{w}_k^{i,\ell} : i = 1, \ldots, N_s, \ell = 1, \ldots, N\}$ where

$$\overline{w}_k^{i,\ell} \propto p_{\mathbf{v}_k}\left(\mathbf{y}_k - \mathbf{h}^{\theta_k^\ell}\left(\overline{\mathbf{x}}_k^{i,\ell}\right)\right)\Pr\{\theta_k^\ell | \theta_{k-1}^i\} \text{ for } \ell = 1, \ldots, N.$$

Resample to obtain $\{\overline{\overline{\mathbf{x}}}_k^i, 1/N_s, i^j : i = 1, \ldots, N_s\}$ where i^j is the index of the parent of $\overline{\overline{\mathbf{x}}}_k^i$ $\mathbf{x}_{k-1}^i$ with model θ_k^ℓ that associates with large weight $\overline{w}_k^{i,\ell}$. Redraw samples $\mathbf{x}_k^i \sim p\left(\mathbf{x}_k | \mathbf{x}_{k-1}^{i^j}, \theta_k^\ell\right)$ like ASIR for a single model to obtain new weights $\{w_k^i : i = 1, \ldots, N_s\}$ [Equation (6.6-33)] for the approximation of the updated state and covariance for nonlinear estimation. A resampling step is added at the end of each update circle to obtain a new set of samples with uniform weights as: $\{\mathbf{x}_k^i, \theta_k^\ell, 1/N_s, i^j : i = 1, \ldots, N_s\}$.

Initialization

For $k = 1$, draw $\mathbf{x}_0^i, \theta_0^i \sim p(\mathbf{x}_0, \theta_0)$ for $i = 1, \ldots, N_s$, such that

$$p(\mathbf{x}_0, \theta_0) \approx \sum_{i=1}^{N_S} w_0^i \Pr\{\theta_0 = \theta_0^i\}\delta_D\left(\mathbf{x}_0 - \mathbf{x}_0^i\right)\delta\left(\theta_0 - \theta_0^i\right), \text{ and } w_0^i = 1/N_s.$$

Prediction

Given $\{\mathbf{x}_{k-1}^i, \theta_{k-1}^i, 1/N_s : i = 1, \ldots, N_s\}$ which approximates $p\left(\mathbf{x}_{k-1}, \theta_{k-1} | \mathbf{y}_{1:k-1}\right)$

Run SIS (Table 6.5) with each hypothesis such that $\overline{\mathbf{x}}_k^i \sim p\left(\mathbf{x}_k, \theta_k^\ell | \mathbf{x}_{k-1}^i, \theta_{k-1}^i\right)$ for $\ell = 1, \ldots, N$ to obtain $\{\overline{\mathbf{x}}_k^{i,\ell}, \theta_k^\ell, \overline{w}_k^{i,\ell} : i = 1, \ldots, N_s, \ell = 1, \ldots, N\}$ with the associated weight

$$\overline{w}_k^{i,\ell} \propto p_{\mathbf{v}_k}\left(\mathbf{y}_k - \mathbf{h}^{\theta_k^\ell}\left(\overline{\mathbf{x}}_k^{i,\ell}\right)\right)\Pr\{\theta_k^\ell | \theta_{k-1}^i\} \text{ for } \ell = 1, \ldots, N.$$

Normalize $\overline{w}_k^{i,\ell}$, such that $\sum_{i=1}^{N_S}\sum_{\ell=1}^{N}\overline{w}_k^{i,\ell} = 1$. Resample $\{\overline{\mathbf{x}}_k^{i,\ell}, \theta_k^\ell, \overline{w}_k^{i,\ell} : i = 1, \ldots, N_s, \ell = 1, \ldots, N\}$ to obtain N_s samples $\{\overline{\overline{\mathbf{x}}}_k^i, 1/N_s, i^j : i = 1, \ldots, N_s\}$ where i^j is the index of the parent of $\overline{\overline{\mathbf{x}}}_k^i$, $\mathbf{x}_{k-1}^i$ with model θ_k^ℓ that associates with large weight $\overline{w}_k^{i,\ell}$.

Redraw samples $\mathbf{x}_k^i \sim p\left(\mathbf{x}_k | \mathbf{x}_{k-1}^{i^j}, \theta_k^\ell\right)$ like ASIR for a single model.

Update

Compute the updated weight $\{w_k^i : i = 1, \ldots, N_s\}$ given $\mathbf{y}_k$:

$$w_k^i \propto p_{\mathbf{v}_k}\left(\mathbf{y}_k - \mathbf{h}^{\theta_k^\ell}\left(\mathbf{x}_k^i\right)\right)\Big/ p_{\mathbf{v}_k}\left(\mathbf{y}_k - \mathbf{h}^{\theta_k^\ell}\left(\overline{\overline{\mathbf{x}}}_k^i\right)\right) \forall i = 1, \ldots, N_s.$$

Normalize $\{w_k^i : i = 1, \ldots, N_s\}$, such that $\sum_{i=1}^{N_S} w_k^i = 1$.

Compute the updated state estimate and covariance

$$\hat{\mathbf{x}}_{k|k} = \sum_{i=1}^{N_S} w_k^i \mathbf{x}_k^i,$$

$$\mathbf{P}_{k|k} = \sum_{i=1}^{N_S} w_k^i \left(\mathbf{x}_k^i - \hat{\mathbf{x}}_{k|k}\right)\left(\mathbf{x}_k^i - \hat{\mathbf{x}}_{k|k}\right)^T.$$

Resample $\{\mathbf{x}_k^i, \theta_k^\ell, w_k^i : i = 1, \ldots, N_s\}$ (see Table 6.1) to obtain a new sample with uniform weights $\{\mathbf{x}_k^i, \theta_k^\ell, 1/N_s, i^j : i = 1, \ldots, N_s\}$.

---

where, with the assumptions that $\mathbf{x}_k$ is a Markov process that satisfies Equation (6.6-6)

$$p(\mathbf{x}_k|\mathbf{x}_{k+1:K}, \mathbf{y}_{1:K}) \propto p(\mathbf{x}_k|\mathbf{y}_{1:K})p(\mathbf{x}_{k+1}|\mathbf{x}_k). \tag{6.6-34}$$

A recursive algorithm can be constructed based on the factorization in Equation (6.6-34) with the time index running in reverse, $k = K, K - 1, \ldots, 2, 1$, as follows

1. Resample $\{\mathbf{x}_K^i, w_K^i : i = 1, \ldots, N_s\}$ to obtain $\{\overline{\mathbf{x}}_K^i, 1/N_s, i^j : i = 1, \ldots, N_s\}$, where i^j is the index of the parent of $\overline{\mathbf{x}}_K^i$ with large w_K^i and $\overline{\mathbf{x}}_{K-1}^i = \mathbf{x}_{K-1}^{i^j}$.

2. At time step $K - 1$, compute $w_{K-1|K}^i \propto w_{K-1}^{i^j} p_{\mu_{K-1}}\left(\overline{\mathbf{x}}_K^i - \mathbf{f}\left(\overline{\mathbf{x}}_{K-1}^i\right)\right)$, and normalize $w_{K-1|K}^i$ such that $\sum_{i=1}^{N_S} w_{K-1|K}^i = 1$.

The approximation of $p(\mathbf{x}_{K-1}|\mathbf{y}_{1:K})$ becomes

$$p\left(\mathbf{x}_{K-1}|\mathbf{y}_{1:K}\right) \approx \sum_{i=1}^{N_S} w_{K-1|K}^i \delta_D\left(\mathbf{x}_{K-1} - \overline{\mathbf{x}}_{K-1}^i\right).$$

The smoothed state and covariance can be computed as

$$\hat{\mathbf{x}}_{K-1|K} = \sum_{i=1}^{N_S} w_{K-1|K}^i \overline{\mathbf{x}}_{K-1}^i$$

and

$$\mathbf{P}_{K-1|K} = \sum_{i=1}^{N_S} w_{K-1|K}^i \left(\overline{\mathbf{x}}_{K-1}^i - \hat{\mathbf{x}}_{K-1|K}\right)\left(\overline{\mathbf{x}}_{K-1}^i - \hat{\mathbf{x}}_{K-1|K}\right)^T.$$

3. Realign the filtered samples $\{\mathbf{x}_{K-1}^i\}$ with $\{\overline{\mathbf{x}}_{K-1}^i\}$; the parent of $\overline{\mathbf{x}}_{K-1}^i$ is located as $\overline{\mathbf{x}}_{K-2}^i = \mathbf{x}_{K-2}^{i^j}$. Resample $\{\overline{\mathbf{x}}_{K-1}^i, w_{K-1|K}^i : i = 1, \ldots, N_s\}$ to obtain $\{\overline{\mathbf{x}}_{K-1}^i, 1/N_s, i^j : i = 1, \ldots, N_s\}$, where i^j is the index of the parent of $\overline{\mathbf{x}}_{K-1}^i$ with large weight $w_{K-1|K}^{i^j}$ and $\overline{\mathbf{x}}_{K-2}^i = \mathbf{x}_{K-2}^{i^j}$ to form $\overline{\mathbf{x}}_{K-1:K}^i = \{\overline{\mathbf{x}}_{K-1}^i, \overline{\mathbf{x}}_K^i\}$ and compute $w_{K-2|K}^i$ to form $\{\mathbf{x}_{K-2}^i, w_{K-2|K}^i : i = 1, \ldots, N_s\}$ as in Step 2.

4. Repeat the above process for $k = K - 2, \ldots, 2, 1$, to form $\overline{\mathbf{x}}_{0:K}^i = \{\overline{\mathbf{x}}_0^i, \overline{\mathbf{x}}_1^i, \overline{\mathbf{x}}_2^i, \ldots, \overline{\mathbf{x}}_K^i\}$ as sample of $\mathbf{x}_{0:K}$ for $p(\mathbf{x}_{0:K}|\mathbf{y}_{1:K})$ and

$$\hat{\mathbf{x}}_{k|K} = (1/N_s)\sum_{i=1}^{N_S} \overline{\mathbf{x}}_k^i$$

and

$$\mathbf{P}_{k|K} = (1/N_s)\sum_{i=1}^{N_S} \left(\overline{\mathbf{x}}_k^i - \hat{\mathbf{x}}_{k|K}\right)\left(\overline{\mathbf{x}}_k^i - \hat{\mathbf{x}}_{k|K}\right)^T \text{ for } k = 1, 2, \ldots, K.$$

A summary of the algorithm is given in Table 6.10.

Table 6.10   Summary of the SIR Filter Algorithm for Smoothing

---

SIR Filter Algorithm for Smoothing

Solution Concept

Assume that particle filtering for nonlinear system from Equations (6.6-2) and (6.6-3) has been done for $k = 1, \ldots, K$, and the particles and weights $\{\mathbf{x}_k^i, w_k^i : i = 1, \ldots, N_s\}$ are stored. Start with the random samples and weights $\{\mathbf{x}_K^i, w_K^i : i = 1, \ldots, N_s\}$ at step K and using the following relationship

$$p(\mathbf{x}_k | \mathbf{x}_{k+1:K}, \mathbf{y}_{1:K}) \propto p(\mathbf{x}_k | \mathbf{y}_{1:K}) p(\mathbf{x}_{k+1} | \mathbf{x}_k) = p(\mathbf{x}_k | \mathbf{y}_{1:K}) p_{\mu_k}(\mathbf{x}_{k+1} - \mathbf{f}(\mathbf{x}_k)),$$

to obtain the samples and weights $\{\overline{\mathbf{x}}_k^i, w_{k|K}^i : i = 1, \ldots, N_s\}$ recursively for the approximation of the smoothed state and covariance for a nonlinear system.

Initialization

Run a particle filtering for nonlinear system from Equations (6.6-2) and (6.6-3) for $k = 1, \ldots, K$, and store the particles and weights $\{\mathbf{x}_k^i, w_k^i : i = 1, \ldots, N_s\}$.

Resample $\{\mathbf{x}_K^i, w_K^i : i = 1, \ldots, N_s\}$ to obtain $\{\overline{\mathbf{x}}_K^i, 1/N_s, i^j : i = 1, \ldots, N_s\}$, where i^j is the index of the parent of $\overline{\mathbf{x}}_K^i$ and $\overline{\mathbf{x}}_{K-1}^i = \mathbf{x}_{K-1}^{i^j}$ with weight $w_{K-1}^{i^j}$. Compute $w_{K-1|K}^i \propto w_{K-1}^{i^j} p_{\mu_{K-1}}(\overline{\mathbf{x}}_K^i - \mathbf{f}(\overline{\mathbf{x}}_{K-1}^i))$, and normalize $w_{K-1|K}^i$ such that $\sum_{i=1}^{N_s} w_{K-1|K}^i = 1$.

At Time Step $k + 1$

Given $\{\mathbf{x}_k^i, w_k^i : i = 1, \ldots, N_s\}$ and $\{\mathbf{x}_{k+1}^i, w_{k+1}^i : i = 1, \ldots, N_s\}$ stored from the particle filter, realign $\{\mathbf{x}_{k+1}^i\}$ with $\{\overline{\mathbf{x}}_{k+1}^i\}$, such that the parent of $\overline{\mathbf{x}}_{k+1}^i$ is located as $\overline{\mathbf{x}}_k^i = \mathbf{x}_k^{i^j}$.

Resample $\{\overline{\mathbf{x}}_{k+1}^i, w_{k+1|K}^i : i = 1, \ldots, N_s\}$ to obtain $\{\overline{\mathbf{x}}_{k+1}^i, 1/N_s, i^j : i = 1, \ldots, N_s\}$, where i^j is the index of the parent of $\overline{\mathbf{x}}_{k+1}^i$ with large weight $w_{k+1|K}^{i^j}$.

Smoothing at Time Step k

Compute the updated weight $\{w_{k|K}^i : i = 1, \ldots, N_s\}$

$$w_{k|K}^i = w_k^{i^j} p_{\mu_k}(\overline{\mathbf{x}}_{k+1}^i - \mathbf{f}(\overline{\mathbf{x}}_k^i)) \Big/ \sum_{i=1}^{N_s} w_k^{i^j} p_{\mu_{k+1}}(\overline{\mathbf{x}}_{k+1}^i - \mathbf{f}(\overline{\mathbf{x}}_k^i)), \ \forall \ i = 1, \ldots, N_s.$$

Compute the smoothed state estimate and covariance

$$\hat{\mathbf{x}}_{k|K} = \sum_{i=1}^{N_s} w_{k|K}^i \overline{\mathbf{x}}_k^i,$$

$$\mathbf{P}_{k|K} = \sum_{i=1}^{N_s} w_{k|K}^i (\overline{\mathbf{x}}_k^i - \hat{\mathbf{x}}_{k|K})(\overline{\mathbf{x}}_k^i - \hat{\mathbf{x}}_{k|K})^T.$$

Resample $\{\mathbf{x}_k^i, w_{k|K}^i : i = 1, \ldots, N_s\}$ to obtain $\{\overline{\mathbf{x}}_k^i, 1/N_s, i^j : i = 1, \ldots, N_s\}$, where i^j is the index of the parent of $\overline{\mathbf{x}}_k^i$ and $\overline{\mathbf{x}}_{k-1}^i = \mathbf{x}_{k-1}^{i^j}$.

6.7   Summary

In this chapter, sampling techniques to approximate the mean and covariance or to compute the posterior density functions are introduced. The steps to compute the mean and covariance of Gaussian densities that approximate the probability densities given in Equations (6.3-4) to (6.3-6) for a class of problems with additive Gaussian noise were presented in previous chapters. As discussed in Chapter 3, the techniques based on algorithms that compute the mean and covariance of Gaussian approximations to Equations (6.3-4) to (6.3-6) may not work well because the nonlinearities of the system and measurement functions can cause the approximated mean and covariance to be very inaccurate, or in some cases, the Gaussian distribution is simply not a good representation of the true distribution. Two deterministic sampling approaches were introduced in Sections 6.4 and 6.5, with deterministic samples used to compute the mean and covariance of the filter estimate. Between the two sampling approaches, the UKF is limited to additive Gaussian cases. UKF is more popular, easier to implement, and computationally efficient; on the other hand the PMF is less restrictive but requires a lot of cells (or sample points) to cover the interested region in the state space. A better strategy of selecting important sample points to compute the approximation of the conditional density functions is needed for highly nonlinear and non-Gaussian cases.

Monte Carlo sampling techniques were introduced in this chapter to solve difficult problems for nonlinear and non-Gaussian state estimation. Since the density functions in Equations (6.3-4) to (6.3-6) are approximated using random samples as in Section 6.2.5, all related conditional expectations in Equation (6.2-1), such as the conditional mean, $\hat{\mathbf{x}}$, can be approximated using the Monte Carlo integral in Equation (6.2-30). The challenge is to generate samples that can closely represent the associated density functions for Equation (6.2-28). Section 6.5 describes the Monte Carlo techniques known as particle filters. Two key steps of a generic particle filter are (a) sequential importance sampling, and (b) resampling. Generating samples for the pairs $\{\mathbf{x}_k^i, \mathbf{x}_{k-1}^i, i = 1, \ldots, N_S\}$ can be accomplished by propagating $\mathbf{x}_{k-1}^i$ through the system Equation (6.3-1) with a sample of process noise $\boldsymbol{\mu}_{k-1}$. However, completing the recursive cycle for Equations (6.3-4) to (6.3-6) is nontrivial. The density functions in Equations (6.3-5) and (6.3-6) can be evaluated with a given measurement $\mathbf{y}_k$, but to draw samples from the posterior density function $p(\mathbf{x}_k|\mathbf{y}_{1:k})$ is not straightforward. The concept of importance sampling is used to sample the posterior density, but the sequential algorithm using suboptimal important function leads to the degeneracy problem very quickly. The resampling step was added to the particle filter algorithm

Table 6.11   Summary of All Filter Algorithms

Filters	Systems	Noises	Approaches
KF	Linear	Additive Gaussian	Exact formula
EKF	Nonlinear	Additive Gaussian	Approximate formula
UKF	Nonlinear	Additive Gaussian	$2n + 1$ Deterministic sample points (sigma points)
MPF	Nonlinear	No restriction	Sample on large Uniform volume hypercubes
PF	Nonlinear	No restriction	Large statistical meaningful random samples

to address the degeneracy problem [14]. The SIR then became the most popular, simple, and intuitive particle filter implementation. Yet deficiencies in the SIR filters still remain, mainly in poor sample diversity and overconfidence or particle collapse for some applications. There is a large body of work [9, 10, 12–16, 25–33] devoted to these topics—how to improve the resampling step for generating samples from the posterior density function, $p(\mathbf{x}_k|\mathbf{y}_{1:k})$. Several algorithms were introduced in Section 6.6 for the readers to sample the field.

There are additional algorithms that have been developed to improve performance and computational efficiency. Included among these are those that take advantage of using part of the state space over which the density functions in Equations (6.3-4) to (6.3-6) can be calculated directly using KF or approximated closely by EKF or UKF. One can separate the state space into states using KF (EKF or UKF) and the rest of the states using a particle filter for state estimation. This type of approach is called Rao–Blackwellization for SIS in [12, 29]. The PF has been extended to multiple model problems [16, 19, 31] and for the smoothing problems [12, 32, 33].

Since the particle filters are very expensive in terms of computational requirements [34], one should use them only when problems are too difficult for the application of the conventional Kalman filter (and EKF/UHF) techniques. Table 6.11 summarizes all filters discussed in this book. To the authors' knowledge, partial filters work well in some of these cases.

Homework Problems

1.   Consider a nonlinear case

$$y = \sin(x) + v$$

where $x: \sim N\left(0, \sigma_x^2\right)$ and $v: \sim N\left(0, \sigma_v^2\right)$ are two independent Gaussian random variables. Compute the approximated conditional density function, $p(x|y)$ and $E\{g(x)|y\}$, using the techniques in Section 6.2 by (a) Taylor series expansion, (b) unscented transformation, (c) point mass integration, and (d) Monte Carlo sampling for

a. $g(x) = \frac{\lambda}{2} e^{-\lambda|x|}$, (Laplace),

b. $g(x) = \frac{z}{\lambda} e^{-(z/\lambda)^2/2}$, (Rayleigh).

Run Monte Carlo samples of y for a given sample of x (the "truth") and plot the sample of the estimate $\hat{x}$. Compare its sample mean and covariance against the mean and covariance computed by the approximated $p(x|y)$ for each technique, respectively. Compare the sample statistics of $E\{g(x)|y\}$ with the true value.

2.   Considering Homework Problem 3 of Chapter 3, build a UKF and compare results with CRB and EKF. You can get more sophisticated by adding a more accurate gravity model and by adding elliptical earth, Coriolis, and centrifugal forces to the trajectory model. Now you can experiment with the case when the dynamics of the true trajectory and the model used in the UKF are different.

3.   Consider a 2D true trajectory to be either CV or CA with linear measurements (Homework Problems 1 and 2 in Chapter 2 without the z dimension), and build particle filters to compute their a posteriori densities. Can you see whether these densities are Gaussian for Gaussian measurement noise? (Hint: Compare the first and second central moments computed using your density estimate with that of KF.)

4.   For the same CV or CA trajectory, consider nonlinear measurements as defined in Homework Problem 1 in Chapter 3, and build particle filters to compute their a posteriori densities. Compute the first and second central moments and compare them with the results of EKF.

References

1.   H. Jazwinski, *Stochastic Processes and Filtering Theory*. New York: Academic Press, 1990.

2.   D.L. Alspach and H.W. Sorenson, "Nonlinear Bayesian Estimation using Gaussian Sum Approximation," *IEEE Transactions on Automatic Control*, vol. 17, pp. 439–447, 1972.

3. R.P. Wishner, J.A. Tabaczynski, and M. Athans, "A Comparison of Three Non-Linear Filters," *Automatica*, vol. 5, pp. 487–496, 1969.

4. R.S. Bucy, "Bayes Theorem and Digital Realization for Nonlinear Filters," *Journal of Astronautical Sciences*, vol. 17, pp. 80–94, 1969.

5. S. Julier and J. Uhlmann, "A New Extension of the Kalman Filter to Nonlinear Systems," in *Proceedings of AeroSense: 11th International Symposium on Aerospace/Defense Sensing, Simulation and Controls*, 1997.

6. S. Julier, J. Uhlmann, and H.F. Durrant-Whyte, "A New Method for the Non-linear Transformation of Means and Covariances in Filters and Estimators," *IEEE Transactions on Automatic Control*, vol. AC-45, pp. 477–482, Mar. 2000.

7. E.A. Wan and R. van der Merwe, "The Unscented Kalman Filter for Nonlinear Estimation," in *Proceedings of AS-SPCC 2000*, pp. 153–158, Oct. 2000.

8. D. Simon, *Optimal State Estimation*. New York: Wiley, 2006.

9. S. Arulampalam, S.R. Maskell, N.J. Gordon, and T. Clapp, "A Tutorial on Particle Filters for On-line Nonlinear/Non-Gaussian Bayesian Tracking," *IEEE Transactions on Signal Processing*, vol. 50, pp. 174–188, 2002.

10. P. Robert and G. Casella, *Monte Carlo Statistical Methods*. New York: Springer-Verlag, 2004.

11. J.M. Hammersley and K.W. Morton, "Poor Man's Monte Carlo," *Journal of the Royal Statistical Society B*, vol. 16, pp. 23–38, 1954.

12. A. Doucet, S. Godsill, and C. Andrieu, "On Sequential Monte Carlo Sampling Methods for Bayesian Filtering," *Statistics and Computing*, vol. 10, pp. 197–208, 2000.

13. A. Doucet, N. de Freitas, and N. Gordon (Eds.), *Sequential Monte Carlo Methods in Practice*. New York: Springer-Verlag, 2001.

14. N.J. Gordon, D.J. Salmond, and A.F.M. Smith, "Novel Approach to Nonlinear/Non-Gaussian Bayesian State Estimation," *IEEE Proceedings*, vol. 140, pp. 107–113, 1993.

15. R. van der Merwe, A. Doucet, N. de Freitas, and E.A. Wan, "The Unscented Particle Filter," in *Proceedings of NIPS 2000*, Dec. 2000.

16. R. Karlsson and N. Bergman, "Auxiliary Particle Filters for Tracking a Maneuvering Target," *Proceedings of the 39th IEEE Conference on Decision and Control*, pp. 3891–3895, Dec. 2000.

17. R.S. Bucy and K.D. Senne, "Digital Synthesis of Non-Linear Filters," *Automatica*, vol. 7, pp. 287–298, 1970.

18.   S. Challa, "Nonlinear State Estimation and Filtering with Applications to Target Tracking Problems," Ph.D. Thesis, Queensland University of Technology, 1998.

19.   S. Challa, M.R. Morelande, D. Musicki, and R.J. Evans, *Fundamentals of Object Tracking*. Cambridge, UK: Cambridge University Press, 2011.

20.   J. MacCormick and A. Blake, "A Probabilistic Exclusion Principle for Tracking Multiple Objects," *Proceedings of the International Journal of Computer Vision*, vol. 39, pp. 57–71, Aug. 2000.

21.   J. Carpenter, P. Clifford, and P. Fearnhead, "Improved Particle Filter for Nonlinear Problems," *IEEE Proceedings on Radar and Sonar Navigation*, vol. 146, pp. 2–7, 1999.

22.   Crisan, P. Del Moral, and T.J. Lyons, "Nonlinear Filtering Using Branching and Interacting Particle Systems," *Markov Processes and Related Fields*, vol. 5, pp. 293–319, 1999.

23.   P. Del Moral, "Nonlinear Filtering: Interacting Particle Solution," *Markov Processes and Related Fields*, vol. 2, pp. 555–580, 1996.

24.   K. Kanazawa, D. Koller, and S.J. Russell, "Stochastic Simulation Algorithms for Dynamic Probabilistic Networks," in *Proceedings of the 11th UAI Conference*, pp. 346–351, 1995.

25.   A. Doucet, N. Gordon, and V. Krishnamurthy, "Particle Filters for State Estimation of Jump Markov Linear Systems," *IEEE Transactions on Signal Processing*, vol. 49, pp. 613–624, 2001.

26.   C. Musso, N. Oudjane, and F. LeGland, "Improving Regularized Particle Filters," in *Sequential Monte Carlo Methods in Practice*, A. Doucet, J.F. de Freitas, and N.J. Gordon, Eds. New York: Springer-Verlag, 2001.

27.   W. Gilks and C. Berzuini, "Following a Moving Target; Monte Carlo Inference for Dynamic Bayesian Models," *Journal of the Royal Statistical Society, B*, vol. 63, pp. 127–146, 2001.

28.   C. Berzuini and W. Gilks, "RESAMPLE-MOVE Filtering with Cross-Model Jumps," in *Sequential Monte Carlo Methods in Practice*, A. Doucet, J.F. de Freitas, and N.J. Gordon, Eds. New York: Springer-Verlag, 2001.

29.   G. Castella and C. Robert, "Rao-Blackwellization of Sampling Scheme," *Biometrika*, vol. 83, pp. 81–94, 1996.

30.   M.K. Pitt and N. Shephard, "Filtering via Simulation: Auxiliary Particle Filters," *Journal of the American Statistical Association*, vol. 94 (446), pp. 590–591, 1999.

31.   M.F. Bugallo, S. Xu, and P.M. Djuric, "Performance Comparison of EKF and Particle Filtering Methods for Maneuvering Targets," *Digital Signal Processing*, vol. 17, pp. 774–786, 2007.

32.   W. Fong, S. Godsill, A. Doucet, and M. West, "Monte Carlo Smoothing with Application to Audio Signal Enhancement," *IEEE Transactions on Signal Processing*, vol. 50, pp. 438–449, 2002.

33.   S. Godsill, A. Doucet, and M. West, "Monte Carlo Smoothing for Nonlinear Time Series," *Journal of the American Statistical Association*, vol. 99 (465), pp. 156–168, 2004.

34.   F. Daum and J. Huang, "Curse of Dimensionality and Particle Filters," in *Proceedings of the IEEE Aerospace Conference*, 2003.

7

State Estimation with Multiple Sensor Systems

7.1 Introduction

There are many applications where multiple sensors are networked to observe and track the same set of objects. For example, multiple air traffic control radars cover an overlapping geographical area of interest. In some applications, there may be radars or IR sensors deployed on the surface, in the air, and in space to share surveillance coverage. There are many reasons for employing multiple sensors: improved accuracy through geometric diversity, improved coverage of shared surveillance areas resulting in higher probability of detection, continuous coverage of contiguous regions resulting in better track continuity and object identification, viewing of the same object complex in different aspect angles with possibly different frequencies (different sensors operating at different wave bands) to allow for the exploitation of phenomenology differences, and so on. A thorough discussion of these benefits is beyond the scope of this book, but the topic of improvement in target location accuracy (thus estimation) is of interest here and illustrated in Figure 7.1. A typical radar system measures range more accurately than angle, resulting in a wide and narrow error ellipsoid, as shown in Figure 7.1(a). Using two radars, the intersection of the two individual error ellipsoids results in a smaller joint error ellipsoid, as shown in Figure 7.1(b). Exactly how much improvement one might obtain is geometry and radar parameter dependent, and can be quantified with covariance analysis and/or simulation. An example showing estimation accuracy improvement with two sensors is given in a later section.

In this chapter, the basic algorithm for state estimation with measurements from a single sensor is extended to the case where measurements are taken from multiple sensors. Only the case where there is no ambiguity in assigning measurements to a unique track is considered in this chapter. This corresponds to the case where the

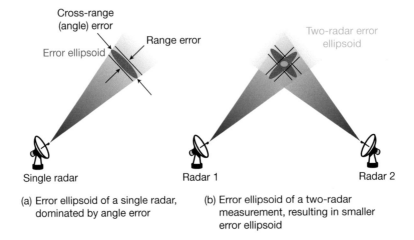

Figure 7.1   Illustration of covariance reduction with two sensors.

physical separation between objects is large relative to measurement uncertainty, so that the new measurements can be assigned unambiguously to existing tracks. For clarity, only the case of a single object being tracked by multiple sensors is considered in this chapter. The problem of multiple targets with either a single sensor or with multiple sensors where ambiguity may arise is considered in Chapters 8–10.

Combining multiple sensor measurements to obtain a final state estimate is called fusion. Two architectures for performing fusion in a multisensor system are illustrated in Figure 7.2. The first one sends measurements directly to fusion processing, as shown in Figure 7.2(a), and is referred to as measurement fusion. Because the state estimate obtained in this way uses the measurements of all sensors, it is referred to as a global estimate. In the second case, in Figure 7.2(b), the state estimate is computed first for each individual sensor, then the sensor estimates are sent to the fusion process. This approach is referred to as state fusion. In contrast to the global estimate, state estimates obtained at the individual sensors are referred to as local estimates. Local estimates can also be fused to obtain a state estimate that encompasses all the measurements [the joint estimate in Figure 7.2(b)]. The joint estimate obtained in this way is usually suboptimal.[1] Measurement fusion produces the optimal estimate while state fusion usually does not, and is discussed later in this chapter. These two architectures are

---

1   A rigorous proof is shown in the Appendix of this chapter.

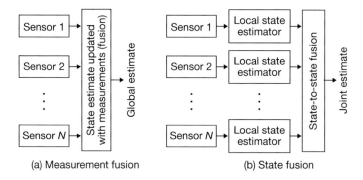

Figure 7.2   Measurement and state fusion.

discussed separately in Sections 7.2 and 7.3. For a general discussion on multiple sensor tracking systems consult [1, Chapters 8–9].

For the purpose of simplicity, linear system notation is used. The results can be easily extended to nonlinear systems using the extended Kalman filter (EKF) with the substitution of system and measurement matrices by Jacobian matrices of the nonlinear system.

7.2   Problem Definition

Here the same discrete system used earlier in the book is considered

$$\mathbf{x}_k = \mathbf{\Phi}_{k,k-1}\mathbf{x}_{k-1} + \mathbf{\mu}_{k-1} \tag{7.2-1}$$

where $\mathbf{\Phi}_{k,k-1}$ is the state transition matrix and $\mathbf{\mu}_{k-1}$ is the system or process noise representing random disturbances or the uncertainty of the state model from Equation (7.2-1) in the actual system dynamics. It is assumed to be Gaussian with zero mean and known covariance, $\mathbf{\mu}_k : \sim N(\mathbf{0}, \mathbf{Q}_k)$. The measurements on $\mathbf{x}_k$ are taken by multiple sensors denoted as

$$\mathbf{y}_{k_i}^i = \mathbf{H}_{k_i}^i \mathbf{x}_{k_i} + \mathbf{v}_{k_i}^i \tag{7.2-2}$$

where the superscript and subscript i denotes the individual sensors $i = 1, 2, ..., I$. Note that different sensors may be of different types (radar, IR, etc.) and may be at different locations, therefore $\mathbf{H}_{k_i}^i$ and $\mathbf{H}_{k_j}^j$ may be different for $i \neq j$. Note that the measurement times for different sensors may also be different, thus a subscript index is added on k. Using the same indexing notation, the measurement noise is white Gaussian with zero mean and known covariance, $\mathbf{v}_{k_i}^i : \sim N(\mathbf{0}, \mathbf{R}_{k_i}^i)$. Further, note that measurements made by different sensors are statistically independent; therefore, the measurement noise for sensor i and sensor j are uncorrelated, that is, $E\left\{\mathbf{v}_{k_i}^i \mathbf{v}_{k_j}^{j^T}\right\} = \mathbf{0}$ for $i \neq j$.

7.3 Measurement Fusion

Two cases of measurement fusion are considered. The first case considers time synchronous measurements in which all sensor systems take measurements on the same object at the same time. This results in $k_i = k_j = k$, for all i, j. In the second case each sensor runs on its own clock, taking measurements nonsynchronously. Most practical multisensor systems are of the second type. Time adjustment by means of extrapolation/interpolation of measurements can make them synchronized and this is often done in practice. The main reason for time synchronization is to simplify the process for state to measurement association. This topic will be discussed in Chapter 8. Time aligned measurements can be processed as in the synchronous measurement case using the estimation algorithms described below.

7.3.1 Synchronous Measurement Case

In the synchronous measurement case, the subscript on k is no longer necessary, that is, the measurement equation becomes

$$\mathbf{y}_k^i = \mathbf{H}_k^i \mathbf{x}_k + \mathbf{v}_k^i \quad i = 1, 2, \ldots, I. \tag{7.3-1}$$

Three approaches to solve this problem are discussed below, namely measurement vector concatenation, sequential processing, and data compression. It can be shown [2, 3] that all three algorithms, although they may look different, are mathematically identical. The choice as to which one to use is implementation dependent.

Measurement Vector Concatenation: Parallel Filter

Since measurements of all sensors take place at time k, $\mathbf{y}_k^i$ can be concatenated for all i to create a single larger dimension measurement vector

$$\mathbf{y}_k = \left[\mathbf{y}_k^{1^T}, \mathbf{y}_k^{2^T}, \ldots, \mathbf{y}_k^{I^T} \right]^T$$

and

$$\mathbf{H}_k = \left[\mathbf{H}_k^{1^T}, \mathbf{H}_k^{2^T}, \ldots, \mathbf{H}_k^{I^T} \right]^T$$

where the measurement noise covariance for the higher dimension measurement vector $\mathbf{y}_k$ is

$$\mathbf{R}_k = \begin{bmatrix} \mathbf{R}_k^1 & \cdots & \mathbf{0} \\ \vdots & \ddots & \vdots \\ \mathbf{0} & \cdots & \mathbf{R}_k^I \end{bmatrix}.$$

Because measurements of all sensors are taken in the same time, measurement vector concatenation is possible, and is conceptually the same as processing all measurement vectors in parallel, hence the name parallel filter. Note that the new measurement covariance matrix $\mathbf{R}_k$ is a block diagonal matrix. This makes it possible to break down the Kalman filter update equation into steps. Applying the Kalman filter equation to the large measurement vector yields

$$\hat{\mathbf{x}}_{k|k} = \hat{\mathbf{x}}_{k|k-1} + \mathbf{K}_k \left(\mathbf{y}_k - \mathbf{H}_k \hat{\mathbf{x}}_{k|k-1} \right),$$

$$\mathbf{K}_k = \mathbf{P}_{k|k} \mathbf{H}_k^T \mathbf{R}_k^{-1},$$

$$\mathbf{P}_{k|k}^{-1} = \mathbf{P}_{k|k-1}^{-1} + \mathbf{H}_k^T \mathbf{R}_k^{-1} \mathbf{H}_k.$$

Substituting in the expressions for $\mathbf{y}_k$, $\mathbf{H}_k$, and $\mathbf{R}_k$, and after some manipulation the following filter update equations are obtained

$$\hat{\mathbf{x}}_{k|k} = \hat{\mathbf{x}}_{k|k-1} + \sum_{i=1}^{I} \mathbf{K}_k^i \left(\mathbf{y}_k^i - \mathbf{H}_k^i \hat{\mathbf{x}}_{k|k-1} \right), \tag{7.3-2}$$

$$\mathbf{K}_k^i = \mathbf{P}_{k|k} \mathbf{H}_k^{i^T} \mathbf{R}_k^{i^{-1}}, \tag{7.3-3}$$

$$\mathbf{P}_{k|k}^{-1} = \mathbf{P}_{k|k-1}^{-1} + \sum_{i=1}^{I} \mathbf{H}_k^{i^T} \mathbf{R}_k^{i^{-1}} \mathbf{H}_k^i. \tag{7.3-4}$$

The one-step predicted state and covariance, $\hat{\mathbf{x}}_{k|k-1}$ and $\mathbf{P}_{k|k-1}$, are computed the same way as before since there is only one state equation. The inverse covariance form of the Kalman filter has been selected here, and it can be easily shown that the other forms of Kalman gain and covariance equations also give the same result. The time-synchronization enables all the measurement vectors to be handled in parallel. The parallel filter concept is illustrated in Figure 7.3.

Sequential Processing

Since all measurements occur at the same instant of time, one can process these measurements sequentially by treating the state vector in each update with zero prediction

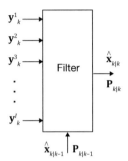

Figure 7.3   Parallel filter illustration.

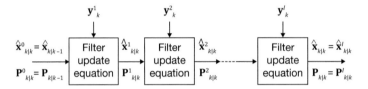

Figure 7.4   Sequential filter.

time. Let $\hat{\mathbf{x}}_{k|k-1}$ be the one-step predicted estimate from $\hat{\mathbf{x}}_{k-1|k-1}$, state estimate and covariance with each sensor measurement updated sequentially as

$$\hat{\mathbf{x}}_{k|k}^{i} = \hat{\mathbf{x}}_{k|k}^{i-1} + \mathbf{K}_k^i \left(\mathbf{y}_k^i - \mathbf{H}_k^i \hat{\mathbf{x}}_{k|k}^{i-1} \right), \tag{7.3-5}$$

$$\mathbf{K}_k^i = \mathbf{P}_{k|k}^i \mathbf{H}_k^{i^T} \mathbf{R}_k^{i^{-1}}, \tag{7.3-6}$$

$$\mathbf{P}_{k|k}^{i^{-1}} = \mathbf{P}_{k|k}^{i-1^{-1}} + \mathbf{H}_k^{i^T} \mathbf{R}_k^{i^{-1}} \mathbf{H}_k^i, \tag{7.3-7}$$

where $i = 1, 2, \ldots, I$. The above process starts by setting $i = 1$ where

$$\hat{\mathbf{x}}_{k|k}^0 = \hat{\mathbf{x}}_{k|k-1},$$

$$\mathbf{P}_{k|k}^0 = \mathbf{P}_{k|k-1}.$$

When the last sensor is reached, $i = I$, then

$$\hat{\mathbf{x}}_{k|k} = \hat{\mathbf{x}}_{k|k}^I,$$

$$\mathbf{P}_{k|k} = \mathbf{P}_{k|k}^I.$$

The sequential filter is illustrated in Figure 7.4.

Remarks on Measurement Vector Concatenation and Sequential Processing

Consider the case of a single sensor system such that all elements of the measurement vector are statistically independent. This configuration is the same as a measurement time synchronized multisensor system in which each element of the measurement vector is equivalent to an independent sensor system. The filter update Equations (7.3-2) through (7.3-4) and (7.3-5) through (7.3-7) can thus be applied as updating the Kalman filter with one measurement element at a time. The significance of this approach is that the inversion of the measurement covariance matrix $\mathbf{R}$ is replaced by the summation of a sequence of scalar inverses. This is an important finding because eliminating the matrix inversion reduces numerical errors.

Data Compression Across Multiple Sensors

Measurements across multiple sensors can be preprocessed to obtain a pseudo-state estimate valid for those measurements. The resulting pseudo-state and covariance can then be used to update the global estimate. The objective is to find the $\mathbf{x}_k$ that minimizes a weighted sum across all sensors

$$J = \sum_{i=1}^{I} \left(\mathbf{y}_k^i - \mathbf{H}_k^i \mathbf{x}_k \right)^T \mathbf{R}_k^{i^{-1}} \left(\mathbf{y}_k^i - \mathbf{H}_k^i \mathbf{x}_k \right) \tag{7.3-8}$$

where the index i denotes the ith sensor, and subscript k denotes that measurements taken at time t_k. In this formulation, estimating the state vector for only one instant of time becomes a parameter estimation problem. The reader should now be able to readily solve this problem to obtain the following answer

$$\hat{\mathbf{x}}_k^I = \left[\sum_{i=1}^{I} \mathbf{H}_k^{i^T} \mathbf{R}_k^{i^{-1}} \mathbf{H}_k^i \right]^{-1} \left(\sum_{i=1}^{I} \mathbf{H}_k^{i^T} \mathbf{R}_k^{i^{-1}} \mathbf{y}_k^i \right), \tag{7.3-9}$$

$$\mathbf{P}_k^I = \left[\sum_{i=1}^{I} \mathbf{H}_k^{i^T} \mathbf{R}_k^{i^{-1}} \mathbf{H}_k^i \right]^{-1}. \tag{7.3-10}$$

The notation $\hat{\mathbf{x}}_k^I$ is used to represent the estimate of $\mathbf{x}_k$ based on the measurements of I sensors taken at the same time t_k, and $\mathbf{P}_k^I$ is the covariance of $\hat{\mathbf{x}}_k^I$. Note that a necessary condition for this method is that $\sum_{i=1}^{I} \mathbf{H}_k^{i^T} \mathbf{R}_k^{i^{-1}} \mathbf{H}_k^i$ be nonsingular. In the case, when the total number of measurements is greater than the number of states, this condition is likely to be satisfied. This method, where a state estimate for a given instant of time obtained by combining all sensor measurements, is sometimes referred to as data compression, measurement preprocessing, or measurement compression.

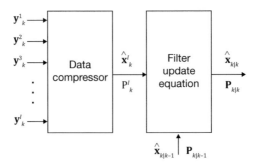

Figure 7.5 Data compression filter.

The Kalman filter uses the compressed data, $\hat{\mathbf{x}}_k^I$ with covariance $\mathbf{P}_k^I$ and the final updated estimate and covariance are given by

$$\hat{\mathbf{x}}_{k|k} = \hat{\mathbf{x}}_{k|k-1} + \mathbf{K}_k\left(\hat{\mathbf{x}}_k^I - \hat{\mathbf{x}}_{k|k-1}\right), \tag{7.3-11}$$

$$\mathbf{K}_k = \mathbf{P}_{k|k}\mathbf{P}_k^{I^{-1}}, \tag{7.3-12}$$

$$\mathbf{P}_{k|k} = \left[\mathbf{P}_{k|k-1}^{-1} + \mathbf{P}_k^I\right]^{-1}. \tag{7.3-13}$$

The flow of the data compression filter is illustrated in Figure 7.5.

The above three methods for updating a state estimate with multiple sensor measurements taken at the same time can be shown to yield the same result. Interested readers can prove this or consult [2, 3] for details. The differences in computational requirements for each approach are not significant with today's computing machines (a count of number of multiplications for each algorithm is shown in [2, 3]), and consequently the choice of an algorithm is dependent on application and ease of implementation.

Although simultaneous measurement by all sensors is generally not true, it is a close approximation for sensors having a high data rate. Furthermore, measurements can be time-aligned (or synchronized) by extrapolation or interpolation.

7.3.2 Asynchronous Measurement Case

The case when sensors in a net do not take measurements at identical times is referred to as the asynchronous measurement case. In applications when the system designer chooses not to time-align the measurements, the algorithm in this section applies. The system and measurement equations are represented by

$$\mathbf{x}_k = \boldsymbol{\Phi}_{k,k-1}\mathbf{x}_{k-1} + \boldsymbol{\mu}_{k-1}, \qquad (7.3\text{-}14)$$

$$\mathbf{y}^i_{k_i} = \mathbf{H}^i_{k_i}\mathbf{x}_{k_i} + \mathbf{v}^i_{k_i}, \qquad (7.3\text{-}15)$$

and the time index k is defined as follows. The actual sampling time, denoted as t_k, is not necessarily uniformly spaced within a time interval. Since sensors are no longer sampled at the same time, k_i is not equal to k_j for $i \neq j$, where k_i and k_j are the measurement time of ith and jth sensors, respectively. A system time index, illustrated in Figure 7.6, is introduced to facilitate implementation. For example, measurements of Sensor 1 will fall on the time step (1, 3, 6, 9, 12, 14), Sensor 2 will have the time step (2, 5, 7, 11, 15), and so on. Given this convention, the Kalman filter for asynchronous measurements is just a straightforward application of the Kalman equation whenever a new measurement, which may come from any of the I sensors, occurs. This is represented by the next equations in this section.

Sequential Updates

Let $k - 1$ be the time of the last update with the system time index, and let k_i be the time that the next measurement is available for filter update from sensor i. Using the scheme in Figure 7.6, k_i is k of the system time index. The Kalman filter for this case simply propagates the state and covariance from the last update $k - 1$ to k_i, $\{\hat{\mathbf{x}}_{k_i|k-1}, \mathbf{P}_{k_i|k-1}\}$, and uses the filter update equation with measurements represented by $\mathbf{y}^i_{k_i}$ with covariance $\mathbf{R}^i_{k_i}$

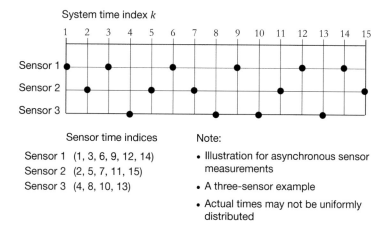

Figure 7.6   Illustration of time indexing for asynchronous measurements.

$$\hat{\mathbf{x}}_{k_i|k_i} = \hat{\mathbf{x}}_{k_i|k-1} + \mathbf{K}_{k_i}\left(\mathbf{y}_{k_i}^i - \mathbf{H}_{k_i}^i \hat{\mathbf{x}}_{k_i|k-1}\right) \tag{7.3-16}$$

$$\mathbf{K}_{k_i} = \mathbf{P}_{k_i|k_i}\mathbf{H}_{k_i}^{i\ T} \mathbf{R}_{k_i}^{i\ -1} \tag{7.3-17}$$

$$\mathbf{P}_{k_i|k_i}^{-1} = \mathbf{P}_{k_i|k-1}^{-1} + \mathbf{H}_{k_i}^{i\ T} \mathbf{R}_{k_i}^{i\ -1}\mathbf{H}_{k_i}^{i}. \tag{7.3-18}$$

For example, using Figure 7.6, only the measurement of Sensor 1 is used for filter update at $k = 1$, only measurements from Sensor 2 are used for the filter update at $k = 2$, and so on. The updated state and covariance using $\mathbf{y}_{k_i}^i$ shown in Equations (7.3-16) through (7.3-18) can be rewritten as $\hat{\mathbf{x}}_{k|k}$ and $\mathbf{P}_{k|k}$ with the system time index k.

Relationship with the Synchronous Measurement Case

The key to implementing the sequential update algorithm, Equations (7.3-16) through (7.3-18), is proper management of the time indices k and k_i. Once organized, the indices function the same way as in the fundamental Kalman filter equation. The same remarks can be made for the sequential update algorithm for the synchronous measurement system case, as in Equations (7.3-5) through (7.3-7). The key challenge is integrating the time indices of all sensors into a system time index. One can show that when all k_i are equal, the asynchronous measurement case reduces to the synchronized measurement case.

7.3.3   Measurement Preprocessing for a Given Sensor to Reduce Data Exchange Rate

The data rate of some sensors can be very high: a radar entering the precision track mode can take many measurements in a second. A tracking IR sensor can operate at a 20–40 Hz frame rate, or even higher. Some netted sensor systems share measurement data from all sensors, resulting in the need for large data communication bandwidth. When such high rates become an issue or when the system simply does not require a high track update rate, a measurement preprocessing technique can be employed to retain the information obtained while reducing the rate of data exchange. This approach is different from the data compression method discussed in Section 7.3.1 where data from different sensors are combined into a single state vector before using it to update the filter. The method discussed here is to combine measurements from the same sensor obtained over a time window. The combined measurement can be in either the synchronous or asynchronous sensor system case, and the concept is illustrated in Figure 7.7. The state estimates obtained over a small time window are sometimes referred to as tracklets.

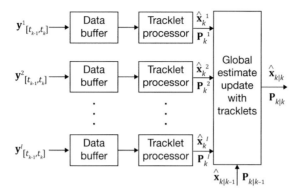

Figure 7.7   Measurement preprocessing (tracklets) before multisensor fusion.

The fixed interval smoother (FIS) for measurement preprocessing is used here for the deterministic system case. This approach is selected because the impact of process noise is small for a short time interval.[2] Let $[t_{k-1}, t_k]$ denote the data time interval where a set of measurements will be used to obtain a tracklet for sensor i at time step k. One can rewrite the FIS smoother equations (Section 2.10.5) for a deterministic system for sensor i as

$$\hat{\mathbf{x}}_k^i = \left[\sum_{k_i=t_{k-1}}^{t_k} \boldsymbol{\Phi}_{k_i,t_k}^T \mathbf{H}_{k_i}^{i\,T} \mathbf{R}_{k_i}^{i\,-1} \mathbf{H}_{k_i}^i \boldsymbol{\Phi}_{k_i,t_k} \right]^{-1} \left(\sum_{k_i=t_{k-1}}^{t_k} \boldsymbol{\Phi}_{k_i,t_k}^T \mathbf{H}_{k_i}^{i\,T} \mathbf{R}_{k_i}^{i\,-1} \mathbf{y}_{k_i}^i \right), \quad (7.3\text{-}19)$$

$$\mathbf{P}_k^i = \left[\sum_{i=t_{k-1}}^{t_k} \boldsymbol{\Phi}_{i,t_k}^T \mathbf{H}_{k_i}^{i\,T} \mathbf{R}_{k_i}^{i\,-1} \mathbf{H}_{k_i}^i \boldsymbol{\Phi}_{i,t_k} \right]^{-1}. \quad (7.3\text{-}20)$$

Comparing the above equations to the FIS equation for deterministic system, Equations (2.9-13) and (2.9-14), note that the initial state and covariance $\hat{\mathbf{x}}_0$ and $\mathbf{P}_0$ are dropped here. This is because the state estimates over different time windows must be statistically independent.

The above equation can be applied to all sensors requiring preprocessing. In this process, the state estimates over the limited time window for each time step k are then used to update the global state estimate using any of the algorithms presented in Section 7.3.1 as pseudo-measurements from each sensor, either in parallel or sequentially. This processing architecture is illustrated in Figure 7.7.

---

2   This concept can be extended to the stochastic case; however, the construction of a Kalman filter for such a case will be more complicated.

7.3.4   Update with Out-of-Sequence Measurements

Due to communication delays, some past measurements may arrive at a processing node after estimates have been updated with more recent measurements. Let t_α denote the time of the delayed measurement and t_k the current time,[3] then $t_\alpha < t_k$. The measurement at time t_α is referred to as an out-of-sequence measurement (OOSM). If the estimated state and covariance immediately before t_α and all measurements between t_α and t_k have been stored, then the estimate that is updated with the delayed measurement can be obtained by reprocessing all stored measurements between t_α and t_k with the delayed measurements in the correct time sequence.

In many real systems, past measurements and estimates are not available for reprocessing. One therefore needs an algorithm to update the estimate with the delayed measurement at time t_α. The algorithm shown below is due to the work of Bar-Shalom [4, 5].

For the purpose of simplicity, indexing for different sensors will be omitted and $t_{k-1} < t_\alpha < t_k$ (one-step lag OOSM [4]).[4] As defined before, $\hat{\mathbf{x}}_{k|k-1}$, $\mathbf{P}_{k|k-1}$, $\hat{\mathbf{x}}_{k|k}$, and $\mathbf{P}_{k|k}$ are the predicted and updated state and covariance at time t_k given $\mathbf{y}_{1:k} = \{\mathbf{y}_1, \mathbf{y}_2, ..., \mathbf{y}_k,\}$ without the OOSM at time t_α. New notation used with the addition of $\mathbf{y}_\alpha$ at time t_α includes

$\hat{\mathbf{x}}_{\alpha|k}$: estimate of $\mathbf{x}_\alpha$ based on $\hat{\mathbf{x}}_{k|k}$ in backward prediction[5] without using $\mathbf{y}_\alpha$,

$\hat{\mathbf{x}}_{k|k,\alpha}$: estimate of $\mathbf{x}_k$ updated with the OOSM $\mathbf{y}_\alpha$.

Covariance of $\hat{\mathbf{x}}_{\alpha|k}$ and $\hat{\mathbf{x}}_{k|k,\alpha}$ are $\mathbf{P}_{\alpha|k}$ and $\mathbf{P}_{k|k,\alpha}$, respectively.

The Approach

1. Compute $\hat{\mathbf{x}}_{\alpha|k}$ and $\mathbf{P}_{\alpha|k}$ using $\hat{\mathbf{x}}_{k|k}$ and $\mathbf{P}_{k|k}$.

2. Compute $\hat{\mathbf{x}}_{k|k,\alpha}$ and $\mathbf{P}_{k|k,\alpha}$ using $\hat{\mathbf{x}}_{\alpha|k}$, $\mathbf{P}_{\alpha|k}$, and $\mathbf{y}_\alpha$.

The Algorithm

Retrodiction Processing   Following Equation (7.2-1), the state vector relationship from time t_α to time t_k is

$$\mathbf{x}_k = \mathbf{\Phi}_{k,\alpha}\mathbf{x}_\alpha + \mathbf{\mu}_{k,\alpha}, \tag{7.3-21}$$

---

3   It is assumed that all measurements are time-tagged, and that the delayed measurement at t_α is out of sequence.

4   For multistep lag OOSM, that is, $t_{k-\ell} < t_\alpha < t_{k-\ell+1}$, the derivation follows the same concept as the one-step lag OOSM. Readers interested in the derivation should refer to [4].

5   Referred to as retrodiction in [4, 5].

$$\mathbf{y}_\alpha = \mathbf{H}_\alpha \mathbf{x}_\alpha + \mathbf{v}_\alpha, \tag{7.3-22}$$

where $\boldsymbol{\mu}_{k,\alpha}$ is the effect of the process noise over the interval between time t_α and t_k and assumed to be white with zero mean and covariance $\mathbf{Q}_{k,\alpha}$ and $\mathbf{v}_\alpha$: $\sim N(\mathbf{0}, \mathbf{R}_\alpha)$ is the OOSM noise which is independent to the process and measurement noises. Given that $\mathbf{x}_k$ and $\mathbf{y}_\alpha$ are two Gaussian random vectors, the state estimates giving measurements $\mathbf{y}_{1:k}$ at time t_α without the OOSM, $\mathbf{y}_\alpha$ Equation (7.3-22), are the conditional mean of the states $\mathbf{x}_k$, $\mathbf{x}_\alpha$ and the process noise $\boldsymbol{\mu}_{k,\alpha}$, that is, $\hat{\mathbf{x}}_{k|k} = E\{\mathbf{x}_k | \mathbf{y}_{1:k}\}$, $\hat{\mathbf{x}}_{\alpha|k} = E\{\mathbf{x}_\alpha | \mathbf{y}_{1:k}\}$ and $\hat{\boldsymbol{\mu}}_{\alpha|k} = E\{\boldsymbol{\mu}_{k,\alpha} | \mathbf{y}_{1:k}\}$, respectively. Taking the conditional expectation over Equation (7.3-21) given $\mathbf{y}_{1:k}$ and applying the new notations we obtain [4]

$$\hat{\mathbf{x}}_{\alpha|k} = \boldsymbol{\Phi}_{\alpha,k}\left(\hat{\mathbf{x}}_{k|k} - \hat{\boldsymbol{\mu}}_{\alpha|k}\right), \tag{7.3-23}$$

$$\mathbf{P}_{\alpha|k} = \boldsymbol{\Phi}_{\alpha,k}\left[\mathbf{P}_{k|k} + \mathbf{P}_{\mu_\alpha} - \mathbf{P}_{\mathbf{x}_k,\mu_\alpha} - \mathbf{P}_{\mathbf{x}_k,\mu_\alpha}^T\right]\boldsymbol{\Phi}_{\alpha,k}^T, \tag{7.3-24}$$

and

$$\hat{\boldsymbol{\mu}}_{\alpha|k} = \mathbf{Q}_{k,\alpha}\mathbf{H}_k^T\boldsymbol{\Gamma}_k^{-1}\boldsymbol{\gamma}_k, \tag{7.3-25}$$

$$\boldsymbol{\gamma}_k = \mathbf{y}_k - \mathbf{H}_k\hat{\mathbf{x}}_{k|k-1}. \tag{7.3-26}$$

$$\boldsymbol{\Gamma}_k = Cov\{\boldsymbol{\gamma}_k\} = \mathbf{H}_k\mathbf{P}_{k|k-1}\mathbf{H}_k^T + \mathbf{R}_k, \tag{7.3-27}$$

$$\mathbf{P}_{\mu_\alpha} = \mathbf{Q}_{k,\alpha} - \mathbf{Q}_{k,\alpha}\mathbf{H}_k^T\boldsymbol{\Gamma}_k^{-1}\mathbf{H}_k\mathbf{Q}_{k,\alpha}, \tag{7.3-28}$$

$$\mathbf{P}_{\mathbf{x}_k,\mu_\alpha} = \mathbf{Q}_{k,\alpha} - \mathbf{P}_{k|k-1}\mathbf{H}_k^T\boldsymbol{\Gamma}_k^{-1}\mathbf{H}_k\mathbf{Q}_{k,\alpha}. \tag{7.3-29}$$

Updated State with OOSM   Following Equations (7.3-21) and (7.3-22), the least squares linear estimator, $\hat{\mathbf{x}}_{k|k,\alpha}$, given $\mathbf{y}_{1:k}$ and $\mathbf{y}_\alpha$ can be derived (see Appendix 1.B, Proposition 3) as follows, going backward in time,

$$\hat{\mathbf{x}}_{k|k,\alpha} = \hat{\mathbf{x}}_{k|k} + \mathbf{P}_{\mathbf{x}_k,\mathbf{y}_\alpha}\mathbf{P}_{\mathbf{y}_\alpha,\mathbf{y}_\alpha}^{-1}\left(\mathbf{y}_\alpha - \hat{\mathbf{y}}_{\alpha|k}\right), \tag{7.3-30}$$

$$\mathbf{P}_{k|k,\alpha} = \mathbf{P}_{k|k} - \mathbf{P}_{\mathbf{x}_k,\mathbf{y}_\alpha}\mathbf{P}_{\mathbf{y}_\alpha,\mathbf{y}_\alpha}^{-1}\mathbf{P}_{\mathbf{x}_k,\mathbf{y}_\alpha}^T, \tag{7.3-31}$$

where $\hat{\mathbf{y}}_{\alpha|k}$ is the backward predicted measurement, expressed as

$$\hat{\mathbf{y}}_{\alpha|k} = \mathbf{H}_\alpha\boldsymbol{\Phi}_{\alpha,k}\left(\hat{\mathbf{x}}_{k|k} - \mathbf{Q}_{k,\alpha}\mathbf{H}_\alpha^T\boldsymbol{\Gamma}_\alpha^{-1}\boldsymbol{\gamma}_k\right), \tag{7.3-32}$$

and

$$\boldsymbol{\Gamma}_\alpha = \mathbf{P}_{\mathbf{y}_\alpha,\mathbf{y}_\alpha} = \mathbf{H}_\alpha\mathbf{P}_{\alpha|k}\mathbf{H}_\alpha^T + \mathbf{R}_\alpha, \tag{7.3-33}$$

The cross covariance $\mathbf{P}_{\mathbf{x}_k,\mathbf{y}_\alpha}$ in Equation (7.3-30) is given by

$$\mathbf{P}_{\mathbf{x}_k,\mathbf{y}_\alpha} = \left[\mathbf{P}_{k|k} - \mathbf{P}_{\mathbf{x}_k,\mu_\alpha}\right]\mathbf{\Phi}_{\alpha,k}^T\mathbf{H}_\alpha^T \tag{7.3-34}$$

The OOSM updating algorithm is summarized in Table 7.1.

This above OOSM updating algorithm is optimal in the conditional expectation sense. Two simplified algorithms (thus suboptimal) were presented in [4, 5] with all three algorithms compared using numerical results. It was shown that one of the two simplified algorithms achieved nearly the same performance as the optimal algorithm. The OOSM updating algorithm was also extended to the interacting multiple model (IMM) filter in [5].

Table 7.1   Summary of Algorithm for Updating Out-of-Sequence Measurements

---

Algorithm for Updating Out-of-Sequence Measurements

Problem Statement

At time t_k, given state estimate and covariance, $\hat{\mathbf{x}}_{k|k}$ and $\mathbf{P}_{k|k}$ and late measurement $\mathbf{y}_\alpha$ where $t_\alpha < t_k$, update the state and covariance without having to reprocess all the old measurements.

Retrodiction Processing

$$\hat{\mathbf{x}}_{\alpha|k} = \mathbf{\Phi}_{\alpha,k}\left(\hat{\mathbf{x}}_{k|k} - \mathbf{Q}_{k,\alpha}\mathbf{H}_k^T\mathbf{\Gamma}_k^{-1}\mathbf{\gamma}_k\right),$$

$$\mathbf{P}_{\alpha|k} = \mathbf{\Phi}_{\alpha,k}\left[\mathbf{P}_{k|k} + \mathbf{P}_{\mu_\alpha} - \mathbf{P}_{\mathbf{x}_k,\mu_\alpha} - \mathbf{P}_{\mathbf{x}_k,\mu_\alpha}^T\right]\mathbf{\Phi}_{\alpha,k}^T,$$

where

$$\mathbf{\gamma}_k = \mathbf{y}_k - \mathbf{H}_k\hat{\mathbf{x}}_{k|k-1},$$

$$\mathbf{\Gamma}_k = Cov\{\mathbf{\gamma}_k\} = \mathbf{H}_k\mathbf{P}_{k|k-1}\mathbf{H}_k^T + \mathbf{R}_k$$

$$\mathbf{P}_{\mu_\alpha} = \mathbf{Q}_{k,\alpha} - \mathbf{Q}_{k,\alpha}\mathbf{H}_k^T\mathbf{\Gamma}_k^{-1}\mathbf{H}_k\mathbf{Q}_{k,\alpha},$$

$$\mathbf{P}_{\mathbf{x}_k,\mu_\alpha} = \mathbf{Q}_{k,\alpha} - \mathbf{P}_{k|k-1}\mathbf{H}_k^T\mathbf{\Gamma}_k^{-1}\mathbf{H}_k\mathbf{Q}_{k,\alpha}.$$

$\mathbf{P}_{k|1:k}$ and $\mathbf{P}_{k|1:k-1}$ are the covariance of $\hat{\mathbf{x}}_{k|1:k}$ and $\hat{\mathbf{x}}_{k|1:k-1}$, respectively, which are computed using KF before the OOSM is received.

Update Processing

$$\hat{\mathbf{x}}_{k|k,\alpha} = \hat{\mathbf{x}}_{k|k} + \mathbf{K}_\alpha\left(\mathbf{y}_\alpha - \mathbf{H}_\alpha\hat{\mathbf{x}}_{\alpha|k}\right),$$

$$\mathbf{P}_{k|k,\alpha} = \mathbf{P}_{k|k} - \mathbf{P}_{\mathbf{x}_k,\mathbf{y}_\alpha}\mathbf{\Gamma}_\alpha^{-1}\mathbf{P}_{\mathbf{x}_k,\mathbf{y}_\alpha}^T.$$

where

$$\mathbf{K}_\alpha = \mathbf{P}_{\mathbf{x}_k,\mathbf{y}_\alpha}\mathbf{\Gamma}_\alpha^{-1},$$

$$\mathbf{P}_{\mathbf{x}_k,\mathbf{y}_\alpha} = \left[\mathbf{P}_{k|k} - \mathbf{P}_{\mathbf{x}_k,\mu_\alpha}\right]\mathbf{\Phi}_{\alpha,k}^T\mathbf{H}_\alpha^T,$$

$$\mathbf{\Gamma}_\alpha = \mathbf{H}_\alpha\mathbf{P}_{\alpha|k}\mathbf{H}_\alpha^T + \mathbf{R}_\alpha.$$

7.4 State Fusion

A drawback of measurement fusion is that the measurements of all sensors must be transmitted to a single processing node, resulting in a demanding communication requirement. An alternative is to let each sensor compute its local state estimate, and send the local estimate to the central processor at a frequency that is much lower than the measurement rate. One such possible architecture is illustrated in Figure 7.8. In this architecture, the joint estimate is a result of fusing local estimates and the fusion takes place at a rate lower than the measurement update of individual sensors (for the sake of saving communication bandwidth, the lower rate of transmitting local estimates fulfills this goal). This architecture also indicates that the joint estimate of the current time is not used to update estimates of future time. This is because updates with new measurements have already taken place within local tracks of individual sensors.

The joint estimate obtained in this way, as shown in Figure 7.8, takes two processing steps: (a) generation of a local estimate using only the data of the local sensor, and (b) joining all local estimates to obtain the final estimate. The local estimate can be thought of as preprocessing a subset of measurements, or applying a transformation of the measurements to a different vector space before obtaining the final estimate. The subject of estimation with transformed measurements is discussed in Appendix 7.A, where an operator theoretic approach is applied to compare the covariance of joint and global estimates.

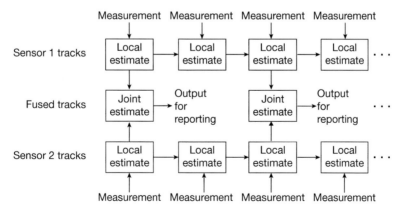

Figure 7.8 A state fusion architecture.

The individual local estimate obtained in Figure 7.8 may be statistically correlated. The correlation is due to the fact that (a) the system dynamics is driven by the same process noise, and (b) the initial estimate of all sensors shares the same prior distribution. Assuming cross covariances are available, the optimum joint estimate, in the sense of weighted least squares (WLS) or maximum likelihood (for Gaussian distribution) can be derived and the result will be as shown in Section 7.4.1. The joint estimate will generally have larger covariance than the global estimate, see Appendix 7.A. However, if the system is deterministic, state fusion is optimal and yields the same results as measurement fusion.

7.4.1   The Fundamental State Fusion Algorithm

Let $\hat{\mathbf{x}}_{k|k}^i, \mathbf{P}_{k|k}^i$ denote the state estimate and covariance of $\mathbf{x}_k$ based upon measurements from the ith sensor, $\mathbf{y}_j^i$ for $j = 1, \ldots k$. $\hat{\mathbf{x}}_{k|k}^i, \mathbf{P}_{k|k}^i$ are referred to as local estimates because they depend only on the measurements from a single local sensor. The local state estimate $\hat{\mathbf{x}}_{k|k}^i$ can be viewed as a measurement of $\mathbf{x}_k$ with measurement covariance $\mathbf{P}_{k|k}^i$, that is,

$$\hat{\mathbf{x}}_{k|k}^i = \mathbf{x}_k + \bar{\bar{\mathbf{v}}}_k^i$$

where $\bar{\bar{\mathbf{v}}}_k^i : \sim N\left(\mathbf{0}, \mathbf{P}_{k|k}^i\right)$. Let $\mathbf{z}_k$ and $\mathbf{v}_k$ denote the concatenation of all $\hat{\mathbf{x}}_{k|k}^i$ and $\bar{\bar{\mathbf{v}}}_k^i$ vectors, respectively, then

$$\mathbf{z}_k = \left[\hat{\mathbf{x}}_{k|k}^{1^T}, \hat{\mathbf{x}}_{k|k}^{2^T}, \ldots, \hat{\mathbf{x}}_{k|k}^{l^T} \right]^T,$$
$$\mathbf{v}_k = \left[\bar{\bar{\mathbf{v}}}_k^{1^T}, \bar{\bar{\mathbf{v}}}_k^{2^T}, \ldots, \bar{\bar{\mathbf{v}}}_k^{l^T} \right]^T,$$

and

$$\mathbf{z}_k = \bar{\bar{\mathbf{H}}}\mathbf{x}_k + \mathbf{v}_k,$$

where

$$\bar{\bar{\mathbf{H}}} = \begin{bmatrix} \mathbf{I} \\ \cdot \\ \cdot \\ \mathbf{I} \end{bmatrix},$$

I is an identity matrix with dimension n, $\mathbf{v}_k \colon \sim N\left(\mathbf{0}, \bar{\bar{\mathbf{R}}}_{k|k}\right)$, and

$$\bar{\bar{\mathbf{R}}}_{k|k} = \begin{bmatrix} \mathbf{P}_{k|k}^1 & \cdots & \mathbf{P}_{k|k}^{1,I} \\ \vdots & \ddots & \vdots \\ \mathbf{P}_{k|k}^{I,1} & \cdots & \mathbf{P}_{k|k}^I \end{bmatrix}.$$

The notation $\mathbf{P}_{k|k}^{i,j}$ denotes the cross covariance of $\hat{\mathbf{x}}_{k|k}^i$ and $\hat{\mathbf{x}}_{k|k}^j$, or $\bar{\bar{\mathbf{v}}}_k^i$ and $\bar{\bar{\mathbf{v}}}_k^j$, which are the same. Due to symmetric property of covariance matrices, $\mathbf{P}_{k|k}^{i,j} = \mathbf{P}_{k|k}^{j,i\,T}$. Given $\hat{\mathbf{x}}_{k|k}^i, \mathbf{P}_{k|k}^i$ for $i = 1, \ldots, I$, the estimate of $\mathbf{x}_k$ denoted as $\bar{\bar{\mathbf{x}}}_{k|k}, \bar{\bar{\mathbf{P}}}_{k|k}$, can be obtained via the WLS estimator formulation developed in Chapter 1. This yields

$$\left[\bar{\bar{\mathbf{H}}}^T \bar{\bar{\mathbf{R}}}_{k|k}^{-1} \bar{\bar{\mathbf{H}}}\right] \bar{\bar{\mathbf{x}}}_{k|k} = \bar{\bar{\mathbf{H}}}^T \bar{\bar{\mathbf{R}}}_{k|k}^{-1} \mathbf{z}_k, \tag{7.4-1}$$

or,

$$\bar{\bar{\mathbf{x}}}_{k|k} = \left[\bar{\bar{\mathbf{H}}}^T \bar{\bar{\mathbf{R}}}_{k|k}^{-1} \bar{\bar{\mathbf{H}}}\right]^{-1} \bar{\bar{\mathbf{H}}}^T \bar{\bar{\mathbf{R}}}_{k|k}^{-1} \mathbf{z}_k, \tag{7.4-2}$$

with covariance

$$\bar{\bar{\mathbf{P}}}_{k|k} = \left[\bar{\bar{\mathbf{H}}}^T \bar{\bar{\mathbf{R}}}_{k|k}^{-1} \bar{\bar{\mathbf{H}}}\right]^{-1}. \tag{7.4-3}$$

The resulting $\bar{\bar{\mathbf{x}}}_{k|k}, \bar{\bar{\mathbf{P}}}_{k|k}$ are referred to as joint estimates because they are obtained by joining (in the WLS sense) all local estimates. Notations $\bar{\bar{\mathbf{x}}}_{k|k}, \bar{\bar{\mathbf{P}}}_{k|k}$ are used to distinguish them from the global estimate $\hat{\mathbf{x}}_{k|k}, \mathbf{P}_{k|k}$.

The State Fusion Equations for Two Sensors

Consider a two-sensor case. Let $\hat{\mathbf{x}}_{k|k}^1, \mathbf{P}_{k|k}^1$ and $\hat{\mathbf{x}}_{k|k}^2, \mathbf{P}_{k|k}^2$ denote the local state estimate and covariance of $\mathbf{x}_k$ generated by the two sensors and $\mathbf{P}_{k|k}^{1,2}$ the cross-covariance of $\hat{\mathbf{x}}_{k|k}^1$ and $\hat{\mathbf{x}}_{k|k}^2$. Equation (7.4-1) becomes

$$\left[\begin{bmatrix}\mathbf{I} & \mathbf{I}\end{bmatrix}\begin{bmatrix}\mathbf{P}_{k|k}^1 & \mathbf{P}_{k|k}^{1,2} \\ \mathbf{P}_{k|k}^{2,1} & \mathbf{P}_{k|k}^2\end{bmatrix}^{-1}\begin{bmatrix}\mathbf{I} \\ \mathbf{I}\end{bmatrix}\right]\bar{\bar{\mathbf{x}}}_{k|k} = \begin{bmatrix}\mathbf{I} & \mathbf{I}\end{bmatrix}\begin{bmatrix}\mathbf{P}_{k|k}^1 & \mathbf{P}_{k|k}^{1,2} \\ \mathbf{P}_{k|k}^{2,1} & \mathbf{P}_{k|k}^2\end{bmatrix}^{-1}\begin{bmatrix}\hat{\mathbf{x}}_{k|k}^1 \\ \hat{\mathbf{x}}_{k|k}^2\end{bmatrix},$$

or

$$\bar{\bar{\mathbf{x}}}_{k|k} = \left[\begin{bmatrix}\mathbf{I} & \mathbf{I}\end{bmatrix}\begin{bmatrix}\mathbf{P}_{k|k}^1 & \mathbf{P}_{k|k}^{1,2} \\ \mathbf{P}_{k|k}^{2,1} & \mathbf{P}_{k|k}^2\end{bmatrix}^{-1}\begin{bmatrix}\mathbf{I} \\ \mathbf{I}\end{bmatrix}\right]^{-1}\begin{bmatrix}\mathbf{I} & \mathbf{I}\end{bmatrix}\begin{bmatrix}\mathbf{P}_{k|k}^1 & \mathbf{P}_{k|k}^{1,2} \\ \mathbf{P}_{k|k}^{2,1} & \mathbf{P}_{k|k}^2\end{bmatrix}^{-1}\begin{bmatrix}\hat{\mathbf{x}}_{k|k}^1 \\ \hat{\mathbf{x}}_{k|k}^2\end{bmatrix},$$

$$\bar{\bar{\mathbf{P}}}_{k|k} = \left[\begin{bmatrix}\mathbf{I} & \mathbf{I}\end{bmatrix}\begin{bmatrix}\mathbf{P}_{k|k}^1 & \mathbf{P}_{k|k}^{1,2} \\ \mathbf{P}_{k|k}^{2,1} & \mathbf{P}_{k|k}^2\end{bmatrix}^{-1}\begin{bmatrix}\mathbf{I} \\ \mathbf{I}\end{bmatrix}\right]^{-1}.$$

Using the block matrix inversion identity shown in Section 3.9.2, we obtain the following set of equations,

$$\bar{\bar{\mathbf{P}}}_{k|k}^{-1}\bar{\bar{\mathbf{x}}}_{k|k} = \bar{\bar{\mathbf{P}}}_{k|k}^{1^{-1}}\hat{\mathbf{x}}_{k|k}^{1} + \bar{\bar{\mathbf{P}}}_{k|k}^{2^{-1}}\hat{\mathbf{x}}_{k|k}^{2} \tag{7.4-4}$$

where $\bar{\bar{\mathbf{P}}}_{k|k}$ is the covariance of $\bar{\bar{\mathbf{x}}}_{k|k}$,

$$\bar{\bar{\mathbf{P}}}_{k|k} = \left[\bar{\bar{\mathbf{P}}}_{k|k}^{1^{-1}} + \bar{\bar{\mathbf{P}}}_{k|k}^{2^{-1}}\right]^{-1} \tag{7.4-5}$$

and

$$\bar{\bar{\mathbf{P}}}_{k|k}^{1^{-1}} = \left[\mathbf{P}_{k|k}^{1} - \mathbf{P}_{k|k}^{1,2}\mathbf{P}_{k|k}^{2^{-1}}\mathbf{P}_{k|k}^{2,1}\right]^{-1} - \mathbf{P}_{k|k}^{2^{-1}}\mathbf{P}_{k|k}^{2,1}\left[\mathbf{P}_{k|k}^{1} - \mathbf{P}_{k|k}^{1,2}\mathbf{P}_{k|k}^{2^{-1}}\mathbf{P}_{k|k}^{2,1}\right]^{-1}$$

$$\bar{\bar{\mathbf{P}}}_{k|k}^{2^{-1}} = \left[\mathbf{P}_{k|k}^{2} - \mathbf{P}_{k|k}^{2,1}\mathbf{P}_{k|k}^{1^{-1}}\mathbf{P}_{k|k}^{1,2}\right]^{-1} - \mathbf{P}_{k|k}^{1^{-1}}\mathbf{P}_{k|k}^{1,2}\left[\mathbf{P}_{k|k}^{2} - \mathbf{P}_{k|k}^{2,1}\mathbf{P}_{k|k}^{1^{-1}}\mathbf{P}_{k|k}^{1,2}\right]^{-1}.$$

Similar to the Kalman filter (KF) equations, these alternative forms for fused state estimate $\bar{\bar{\mathbf{x}}}_{k|k}$ and covariance $\bar{\bar{\mathbf{P}}}_{k|k}$ for the two-sensor case can be obtained by applying the matrix inversion lemma in Appendix A. After some tedious manipulations, we obtain

$$\bar{\bar{\mathbf{x}}}_{k|k} = \hat{\mathbf{x}}_{k|k}^{i} + \left[\mathbf{P}_{k|k}^{i} - \mathbf{P}_{k|k}^{i,j}\right]\left[\mathbf{P}_{k|k}^{1} + \mathbf{P}_{k|k}^{2} - \mathbf{P}_{k|k}^{1,2} - \mathbf{P}_{k|k}^{2,1}\right]^{-1}\left(\hat{\mathbf{x}}_{k|k}^{j} - \hat{\mathbf{x}}_{k|k}^{i}\right), \tag{7.4-6}$$

$$\bar{\bar{\mathbf{P}}}_{k|k} = \mathbf{P}_{k|k}^{i} - \left[\mathbf{P}_{k|k}^{i} - \mathbf{P}_{k|k}^{i,j}\right]\left[\mathbf{P}_{k|k}^{1} + \mathbf{P}_{k|k}^{2} - \mathbf{P}_{k|k}^{1,2} - \mathbf{P}_{k|k}^{2,1}\right]^{-1}\left[\mathbf{P}_{k|k}^{i} - \mathbf{P}_{k|k}^{j,i}\right]. \tag{7.4-7}$$

Using either ($i = 1, j = 2$) or ($i = 2, j = 1$) will result in the same $\bar{\bar{\mathbf{x}}}_{k|k}$ and $\bar{\bar{\mathbf{P}}}_{k|k}$.

Equations (7.4-4) and (7.4-5) show that in state fusion, the cross correlation of the local estimates must be subtracted from the covariance of the local estimates. If ignored, the resulting computed covariance will be inconsistent with the actual state estimation error.

Alternatives to State Fusion Architectures and Algorithms   The basic state fusion architecture is as shown in Figure 7.8. Note that this is a feed forward architecture and the joint estimates are not used in generating future fused estimates. The fusion equations shown in Equations (7.4-2) and (7.4-3) give the optimal joint estimate, giving local estimates. The main difficulty in implementing this algorithm is in the need to compute the cross correlation covariances, $\mathbf{P}_{k|k}^{i,j}$ for all $i, j = 1,\dots I$. Because of this difficulty, and because state fusion is sometimes a necessary choice for multisensor systems, many researchers have developed state fusion algorithms and addressed their performance issues [6-16]. Refs. [12, 13] provide a summary of architectures and algorithms for state fusion. The following discussion follows the presentation of [12].

State Fusion Using Information Decorrelation

Without losing generality, a two-sensor case will be used for most of the following discussions. For the purpose of simplicity, the time index k will be dropped. Let $\hat{\mathbf{x}}^{i}$ and

$\mathbf{P}^i$ denote the local estimate of the *i*th sensor and let $\mathbf{y}^i$ denote its corresponding measurement vector with measurement matrix $\mathbf{H}^i$ and measurement noise covariance $\mathbf{R}^i$ where *i* could indicate either 1 or 2. Assume that the prior of $\mathbf{x}$ that is common to both sensors follows a Gaussian distribution with mean and covariance $\overline{\mathbf{x}}$ and $\overline{\mathbf{P}}$. Further assume that all noise processes are Gaussian, independent of each other, and uncorrelated in time, then the local estimates can be optimally computed with a KF. Using the information filter expression (in which the inverse covariance is the information matrix) we have the following expression

$$\mathbf{P}^{i^{-1}}\hat{\mathbf{x}}^i = \overline{\mathbf{P}}^{-1}\overline{\mathbf{x}} + \mathbf{H}^{i^T}\mathbf{R}^{i^{-1}}\mathbf{y}^i, \tag{7.4-8}$$

$$\mathbf{P}^{i^{-1}} = \overline{\mathbf{P}}^{-1} + \mathbf{H}^{i^T}\mathbf{R}^{i^{-1}}\mathbf{H}^i, \tag{7.4-9}$$

for obtaining the local estimate $\hat{\mathbf{x}}^i$ and covariance $\mathbf{P}^i$, where *i* is equal to either 1 or 2. The estimate $\hat{\mathbf{x}}$ with covariance $\mathbf{P}$ using both measurement vectors $\mathbf{y}^1$ and $\mathbf{y}^2$ can be obtained with the sequential update algorithm of measurement fusion

$$\mathbf{P}^{-1}\hat{\mathbf{x}} = \overline{\mathbf{P}}^{-1}\overline{\mathbf{x}} + \mathbf{H}^{1^T}\mathbf{R}^{1^{-1}}\mathbf{y}^1 + \mathbf{H}^{2^T}\mathbf{R}^{2^{-1}}\mathbf{y}^2, \tag{7.4-10}$$

$$\mathbf{P}^{-1} = \overline{\mathbf{P}}^{-1} + \mathbf{H}^{1^T}\mathbf{R}^{1^{-1}}\mathbf{H}^1 + \mathbf{H}^{2^T}\mathbf{R}^{2^{-1}}\mathbf{H}^2. \tag{7.4-11}$$

Combining Equation (7.4-8) with Equation (7.4-10) and Equation (7.4-9) with Equation (7.4-11), we obtain the following state fusion equation

$$\mathbf{P}^{-1}\hat{\mathbf{x}} = \mathbf{P}^{1^{-1}}\hat{\mathbf{x}}^1 + \mathbf{P}^{2^{-1}}\hat{\mathbf{x}}^2 - \overline{\mathbf{P}}^{-1}\overline{\mathbf{x}}, \tag{7.4-12}$$

$$\mathbf{P}^{-1} = \mathbf{P}^{1^{-1}} + \mathbf{P}^{2^{-1}} - \overline{\mathbf{P}}^{-1}. \tag{7.4-13}$$

This is often referred to as the state fusion with information decorrelation because the prior state and covariance are subtracted in the fusion processing. This development follows that of [12], in which the Bayesian distributed fusion was first derived using Bayes' probability argument. The information decorrelation method is the basis for several state fusion architectures that will be discussed next.

Hierarchical Architecture

Figure 7.9 depicts a two-sensor case for the hierarchical architecture shown in [12]. The solid squares are sensor measurement nodes and the hollow squares are processing nodes. The top flow is Sensor 1, the bottom flow is Sensor 2, and the middle row is fuse processing for two sensors. In addition to fuse local estimates, the middle row is also responsible for propagating fused and local estimates for time alignment in the asynchronous measurement case.

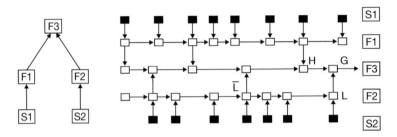

Figure 7.9   Hierarchical architecture for state fusion.

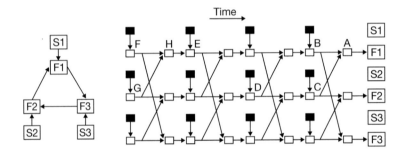

Figure 7.10   Cyclic architecture for state fusion.

Consider fusion at Node G. In this node the state vector previously fused at Node H is to be fused with the new local estimate from Sensor 2 generated at Node L. Node H fused a local estimate of Sensor 1 and the previously fused information done at Node $\bar{L}$, the information of Node $\bar{L}$ becomes the prior for fusion at Node G. In this case, the state fusion equations are

$$\mathbf{P}^{G^{-1}}\hat{\mathbf{x}} = \mathbf{P}^{H^{-1}}\hat{\mathbf{x}}^{H} + \mathbf{P}^{L^{-1}}\hat{\mathbf{x}}^{L} - \mathbf{P}^{\bar{L}^{-1}}\mathbf{x}^{\bar{L}},\qquad(7.4\text{-}14)$$

$$\mathbf{P}^{G^{-1}} = \mathbf{P}^{H^{-1}} + \mathbf{P}^{L^{-1}} - \mathbf{P}^{\bar{L}^{-1}},\qquad(7.4\text{-}15)$$

Cyclic Architecture

Now consider a three-sensor case with a more complicated fusion architecture, named cyclic architecture, as shown in Figure 7.10. Each of the three sensors is represented by a flow of data going from left to right. Each sensor is communicating its local and fused estimates to its neighbor in a cyclic fashion, depicted by cross arrows and illustrated in the left-hand diagram of Figure 7.10. The fusion processing nodes are located inline with sensor processing nodes and depicted by white boxes right before the sensor processing nodes for local measurements.

Consider the case that information from Nodes B and C is to be fused at Node A. State estimates and covariance of D, E, and H, feed into B and C. They are the prior of B and C and must therefore be subtracted. This results in the following equations

$$\mathbf{P}^{A^{-1}}\hat{\mathbf{x}}^A = \mathbf{P}^{B^{-1}}\hat{\mathbf{x}}^B + \mathbf{P}^{C^{-1}}\hat{\mathbf{x}}^C - \mathbf{P}^{D^{-1}}\hat{\mathbf{x}}^D - \mathbf{P}^{E^{-1}}\hat{\mathbf{x}}^E - \mathbf{P}^{H^{-1}}\hat{\mathbf{x}}^H, \qquad (7.4\text{-}16)$$

$$\mathbf{P}^{A^{-1}} = \mathbf{P}^{B^{-1}} + \mathbf{P}^{C^{-1}} - \mathbf{P}^{D^{-1}} - \mathbf{P}^{E^{-1}} - \mathbf{P}^{H^{-1}}, \qquad (7.4\text{-}17)$$

Note that in these two architectures, especially the cyclic architecture, previous relevant fusion information must be kept and used in subsequent fusion. When this is too difficult and/or too complicated to manage, suboptimal approaches have been used in practical systems, some of them are briefly described without derivation. Interested readers can consult [12, 13] and the references cited in them.

Suboptimal Fusion Algorithms

1. **Naïve Fusion**

   In this approach, the prior information is just ignored, resulting in

   $$\mathbf{P}^{-1}\hat{\mathbf{x}} = \mathbf{P}^{1^{-1}}\hat{\mathbf{x}}^1 + \mathbf{P}^{2^{-1}}\hat{\mathbf{x}}^2, \qquad (7.4\text{-}18)$$

   $$\mathbf{P}^{-1} = \mathbf{P}^{1^{-1}} + \mathbf{P}^{2^{-1}}. \qquad (7.4\text{-}19)$$

As mentioned before, ignoring prior (or similarly, the cross covariance) information, the resulting fused covariance does not represent the actual estimation error (for instance, the fused state estimate may be biased). The method of so-called covariance inflation is sometimes used to account for the additional errors.

2. **Channel Filter Fusion**

   The Channel filter is a first order approximation of the information decorrelation filter because it assumes that the most significant information is from the most recent past, thus the only one to be used for subtraction. In the case of the example for the cyclic architecture, the fusion equations become

   $$\mathbf{P}^{A^{-1}}\hat{\mathbf{x}}^A = \mathbf{P}^{B^{-1}}\hat{\mathbf{x}}^B + \mathbf{P}^{C^{-1}}\hat{\mathbf{x}}^C - \mathbf{P}^{D^{-1}}\mathbf{x}^D, \qquad (7.4\text{-}20)$$

   $$\mathbf{P}^{A^{-1}} = \mathbf{P}^{B^{-1}} + \mathbf{P}^{C^{-1}} - \mathbf{P}^{D^{-1}}. \qquad (7.4\text{-}21)$$

Node D is chosen because it is the immediate past fusion nodes of A, B, and C.

3. Chernoff Fusion

Chernoff fusion is naïve fusion with weights, that is,

$$\mathbf{P}^{-1}\hat{\mathbf{x}} = w\mathbf{P}^{1^{-1}}\hat{\mathbf{x}}^1 + (1-w)\mathbf{P}^{2^{-1}}\hat{\mathbf{x}}^2, \tag{7.4-22}$$

$$\mathbf{P}^{-1} = w\mathbf{P}^{1^{-1}} + (1-w)\mathbf{P}^{2^{-1}}, \tag{7.4-23}$$

where $w \in [0\ 1]$. There is no established rule for determining w. In designing practical systems, the engineers usually conduct studies and tests to find the best w.

4. Bhattacharyya Fusion

This is a special case of Chernoff fusion for $w = 0.5$. It is referred to as Bhattacharyya fusion because it can be derived using the square root of the product of two probabilities, similar to the derivation of the Bhattacharyya bound. The resulting fusion equations are

$$\mathbf{P}^{-1}\hat{\mathbf{x}} = \frac{1}{2}\left(\mathbf{P}^{1^{-1}}\hat{\mathbf{x}}^1 + \mathbf{P}^{2^{-1}}\hat{\mathbf{x}}^2\right), \tag{7.4-24}$$

$$\mathbf{P}^{-1} = \frac{1}{2}\left(\mathbf{P}^{1^{-1}} + \mathbf{P}^{2^{-1}}\right). \tag{7.4-25}$$

The fused information is the average of the local information. Both Chernoff and Bhattacharyya fusion are trying to reduce the fused information; they are thus in the category of covariance inflation.

Summary

1. Measurement fusion is optimal. The reason for not considering measurement fusion is due to its need in sending all measurement and measurement covariance to a centralized processing node. When this is not a constraint, measurement fusion is often used.

2. The state fusion is generally less accurate, or with higher covariance. The WLS estimator for state fusion is optimal given local estimates, but it requires cross covariance among local sensors. If the cross covariances are ignored, the joint estimate becomes suboptimal. Ref [8] provides a numerical example to quantitatively assess the effect when cross covariance is neglected.

3. State fusion achieves the same result as measurement fusion when the system is deterministic. One such example is in estimating the trajectory of space objects. This is because the dynamics for objects flying though space is known to within negligible errors.

4. A variety of state fusion algorithms have been derived and compared, see [6–16] and the references cited therein. A brief survey of these methods was presented in this section, which primarily follows the development presented in references [12–13].

7.5   Cramer–Rao Bound

It is a straightforward exercise to show that the Cramer–Rao bound (CRB) for the multisensor case is the same as the very general form of CRB derived in Section 3.9.2. Using the asynchronous measurement case with sequential updates, measurements from different sensors differ only in terms of the coordinate system, thus one can apply the CRB equations directly as

$$Cov\{\hat{\mathbf{x}}_{k|k}\} = \left[\mathbf{P}_{k|k-1}^{-1} + \mathbf{H}_{\mathbf{x}_k}^T \mathbf{R}_k^{-1} \mathbf{H}_{\mathbf{x}_k}\right]^{-1} \tag{7.5-1}$$

where the different coordinate transformations are captured in $\mathbf{H}_{\mathbf{x}_k}$. $\mathbf{P}_{k|k-1}$ is computed from $\mathbf{P}_{k-1|k-1}$ using Equation (3.5-3). Under such conditions the CRB for the multiple sensor and nonlinear system is the same as the covariance equation for the extended Kalman filter with multiple sensors.

7.6   A Numerical Example

Using the same trajectory as in Section 4.9.2, a second sensor is added as shown in Figure 7.11. The angle coverage of these two sensors partially overlaps. This is selected for the purpose of illustrating the improvement in estimation error during the overlapping region. The two elemental filters used for the IMM filter are constant velocity low process noise (CV^L) and constant acceleration (CA) (see Section 5.7 Example 2).

The results are shown in Figure 7.12. All the discussions on filter performance comparisons made for the second example of Section 5.6 apply here. An additional

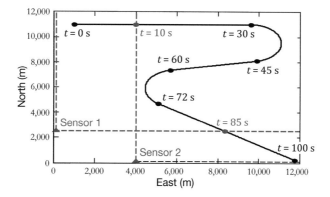

Figure 7.11   The same trajectory as shown in Figure 5.9 with the addition of the second sensor (blue).

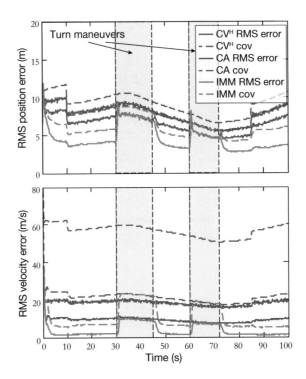

Figure 7.12 Comparison of the root mean square errors and covariance of IMM, CV, and CA filters for the maneuvering trajectory shown in Figure 7.11.

observation applies to the sensor overlap region (t = 10 to 85 s) where a change in estimation error occurs at those boundaries. The improvement in estimation error is due to (a) the increased data rate (two sensors contributing), and (b) the diverse viewing geometry. It is important to emphasize that the improvement of estimation error will diminish in the presence of unknown and unaccounted for intersensor biases. In the case where the bias is known and modeled statistically, such as a Gaussian distribution with known mean and covariance, the Schmidt-Kalman filter (SKF; shown in Section 4.5.2 with numerical example shown in Section 4.9.2) can be used to maintain filter consistency (i.e., the filter computed error matches the actual statistical error). In the Homework Problems, the student is asked to expand the results of this example to include intersensor bias and to experiment with the SKF.

Appendix 7.A   Estimation with Transformed Measurements

Data preprocessing is often used in practical problems in order to reduce the vast amount of data that must be handled. The goal is to reduce communication and storage requirements while incurring minimal loss of information. The idea of state fusion has the same goal, to reduce communication bandwidth while trading off estimation performance. The purpose of this appendix is to apply an operator-theoretic approach to prove that the covariance of the joint estimate is larger than the covariance of the global estimate. The material in this appendix is based upon the work of [17].

7.A.1   Problem Definition

Let $\mathbf{y}$ be a measurement vector of an unknown parameter vector $\mathbf{x}$,

$$\mathbf{y} = \mathbf{Hx} + \boldsymbol{v}$$

where $\boldsymbol{v} = N(\mathbf{0}, \mathbf{R})$. The WLS of $\mathbf{x}$ is[6]

$$\hat{\mathbf{x}} = \left[\mathbf{H}^*\mathbf{R}^{-1}\mathbf{H}\right]^{-1}\left(\mathbf{H}^*\mathbf{R}^{-1}\mathbf{y}\right)$$

with covariance

$$\mathbf{P} = [\mathbf{H}^*\mathbf{R}^{-1}\mathbf{H}]^{-1}.$$

Consider the case where the measurement vector $\mathbf{y}$ is first transformed to a different space, that is,

$$\mathbf{z} = \mathbf{Ty} = \mathbf{THx} + \mathbf{T}\boldsymbol{v}$$

The estimate of $\mathbf{x}$ given $\mathbf{z}$ is

$$\tilde{\mathbf{x}} = \left[\mathbf{H}^*\mathbf{T}^*[\mathbf{TRT}^*]^{-1}\mathbf{TH}\right]^{-1}\left(\mathbf{H}^*\mathbf{T}^*[\mathbf{TRT}^*]^{-1}\mathbf{z}\right)$$

with covariance

$$\tilde{\mathbf{P}} = \left[\mathbf{H}^*\mathbf{T}^*[\mathbf{TRT}^*]^{-1}\mathbf{TH}\right]^{-1}.$$

---

6   Superscript * of a transformation denotes adjoint operator, which is the same as transpose of a matrix in finite dimensional linear vector spaces. Here we use the more general notation in order to generalize the results that follow.

Remarks

Since $\hat{\mathbf{x}}$ computed with $\mathbf{y}$ is optimal, any preprocessing of $\mathbf{y}$ (via the transformation $\mathbf{T}$) may result in loss of information, that is, $\tilde{\mathbf{x}}$ may be less accurate than $\hat{\mathbf{x}}$, or $\tilde{\mathbf{P}}$ may be larger than $\mathbf{P}$. Two questions arise: (a) can one prove $\tilde{\mathbf{P}} \geq \mathbf{P}$ true in general? (b) If so, under what condition will the equality hold? The answers are provided in the following theorem.

7.A.2   A Fundamental Theorem

Theorem 7.A.1 (Theorem on Estimation with Transformed Measurements): Let

$\mathbf{H} \colon \mathbb{R}^n \to \mathbb{R}^m$, $\mathbf{T} \colon \mathbb{R}^m \to \mathbb{R}^p$

where $m \geq p \geq n$,

Rank $(\mathbf{T}) = p$, and

$\mathbf{R}$ is a positive definite and symmetric matrix in $\mathbb{R}^m$ with dimension $m \times m$.

Then for all $\mathbf{x} \in \mathbb{R}^n$,

(a) $\langle \mathbf{R}^{-1}\mathbf{H}\mathbf{x}, \mathbf{H}\mathbf{x} \rangle \geq \langle [\mathbf{T}\mathbf{R}\mathbf{T}^*]^{-1}\mathbf{T}\mathbf{H}\mathbf{x}, \mathbf{T}\mathbf{H}\mathbf{x} \rangle$,

(b) Equality of (a) holds for all $\mathbf{x}$ *iff* $\mathcal{N}(\mathbf{T}) \subset \mathcal{N}(\mathbf{H}^*\mathbf{R}^{-1})$,

(c) Condition (b) is equivalent to $\mathbf{H}^*\mathbf{R}^{-1}\mathbf{H} = \mathbf{H}^*\mathbf{T}^*[\mathbf{T}\mathbf{R}\mathbf{T}^*]^{-1}\mathbf{T}\mathbf{H}$,

where $\langle .,. \rangle$ denotes the inner product of the enclosed vectors and $\mathcal{N}(.)$ denotes the null space of the enclosed operator.

The relationships among the three vector spaces are illustrated in Figure 7.13. $\mathbf{T}^\dagger$ denotes the pseudo-inverse operator going from $\mathbb{R}^p$ to $\mathbb{R}^m$. The transformations shown from $\mathbb{R}^m$ to $\mathbb{R}^n$ and $\mathbb{R}^p$ to $\mathbb{R}^n$ are the WLS estimators without the a priori knowledge.

Proof: Let $\mathbf{w}$, $\mathbf{y}$ be two vectors in $\mathbb{R}^m$. Define a new inner product

$$[\mathbf{w}, \mathbf{y}] \triangleq \langle \mathbf{R}^{-1}\mathbf{w}, \mathbf{y} \rangle.$$

Let $\mathbf{T}^\dagger$ be a pseudo-inverse operator of $\mathbf{T}$ w.r.t. $[.,.]$. It can be shown that the pseudo-inverse given the minimum norm solution in $\mathbb{R}^m$ has the expression[7]

$$\mathbf{T}^\dagger = \mathbf{R}\mathbf{T}^* [\mathbf{T}\mathbf{R}\mathbf{T}^*]^{-1}$$

---

7   See page 165 of [18] for the theory of pseudo-inversion operator in ordinary inner products among vector spaces.

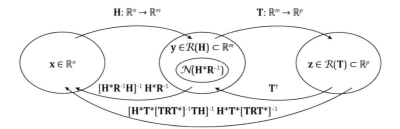

Figure 7.13   Illustration of transformations among the vector spaces of the theorem.

where $\mathbf{T}^\dagger$ satisfies $\mathbf{TT}^\dagger\mathbf{T} = \mathbf{T}$, $\mathbf{T}^\dagger\mathbf{TT}^\dagger = \mathbf{T}^\dagger$, $\mathbf{TT}^\dagger$ gives the projection on the range of $\mathbf{T}$ denoted as $\mathcal{R}(\mathbf{T}) \subset \mathbb{R}^p$, and $\mathbf{T}^\dagger\mathbf{T}$ given the projection on the subspace perpendicular to $\mathcal{N}(\mathbf{T}) \subset \mathbb{R}^m$ denoted as $\mathcal{N}^\perp(\mathbf{T})$.

To prove (a), for all $\mathbf{y} \in \mathcal{R}(\mathbf{H}) \subset \mathbb{R}^m$ and the fact that $[\mathbf{y}, \mathbf{y}] \geq [\mathbf{T}^\dagger\mathbf{Ty}, \mathbf{y}]$, $\langle \mathbf{R}^{-1}\mathbf{y}, \mathbf{y}\rangle \triangleq [\mathbf{y}, \mathbf{y}] \geq [\mathbf{T}^\dagger\mathbf{Ty}, \mathbf{y}] \triangleq \langle \mathbf{R}^{-1}\mathbf{T}^\dagger\mathbf{Ty}, \mathbf{y}\rangle$.
Substituting the expression for $\mathbf{T}^\dagger$ and applying $\mathbf{y} = \mathbf{Hx}$ on both sides of the inequality, we obtain

$$\langle \mathbf{R}^{-1}\mathbf{T}^\dagger\mathbf{Ty}, \mathbf{y}\rangle = \langle \mathbf{R}^{-1}\mathbf{RT}^*[\mathbf{TRT}^*]^{-1}\mathbf{THx}, \mathbf{Hx}\rangle = \langle \mathbf{T}^*[\mathbf{TRT}^*]^{-1}\mathbf{THx}, \mathbf{Hx}\rangle$$
$$= \langle [\mathbf{TRT}^*]^{-1}\mathbf{THx}, \mathbf{THx}\rangle$$

and

$$\langle \mathbf{R}^{-1}\mathbf{Hx}, \mathbf{Hx}\rangle \geq \langle [\mathbf{TRT}^*]^{-1}\mathbf{THx}, \mathbf{THx}\rangle,$$

which is statement (a).

To prove (b), one must find the necessary and sufficient condition for all $\mathbf{y} \in \mathcal{R}(\mathbf{H}) \subset \mathbb{R}^m$ w.r.t. $[.,.]$ in which $\mathbf{T}^\dagger\mathbf{T}$ returns all $\mathbf{y} \in \mathcal{R}(\mathbf{H})$. This is the same as requiring $\mathbf{y} \in \mathcal{R}(\mathbf{H}) \subset \mathcal{R}(\mathbf{T}^\dagger\mathbf{T})$. This is equivalent to the existence of a $\mathbf{B}: \mathbb{R}^n \to \mathbb{R}^m$ such that $\mathbf{H} = \mathbf{T}^\dagger\mathbf{TB}$, thus

$$\mathbf{H} = \mathbf{RT}^*[\mathbf{TRT}^*]^{-1}\mathbf{TB}$$

or

$$\mathbf{H}^* = \mathbf{B}^*\mathbf{T}^*[\mathbf{TRT}^*]^{-1}\mathbf{TR}.$$

Moving $\mathbf{R}$ to the left yields

$$\mathbf{H}^*\mathbf{R}^{-1} = \mathbf{B}^*\mathbf{T}^*[\mathbf{TRT}^*]^{-1}\mathbf{T}.$$

This proves that condition $\mathcal{N}(\mathbf{T}) \subset \mathcal{N}(\mathbf{H}^*\mathbf{R}^{-1})$ is necessary.

We now assume that this is true, then there exists a $\mathbf{C}: \mathbb{R}^p \to \mathbb{R}^n$ such that $\mathbf{H}^*\mathbf{R}^{-1} = \mathbf{CT}$. This gives $\mathbf{H}^* = \mathbf{CTR}$ or $\mathbf{H} = \mathbf{RT}^*\mathbf{C}^*$. Let $\mathbf{y} = \mathbf{y}_1 + \mathbf{y}_0$ where $\mathbf{y}_1 \in \mathcal{R}(\mathbf{H})$ and $\mathbf{y}_0 \in \mathcal{N}(\mathbf{T})$; thus, $\mathbf{y}_1 = \mathbf{Hx} = \mathbf{RT}^*\mathbf{C}^*\mathbf{x}$ and

$$[\mathbf{y}_1, \mathbf{y}_0] = \langle \mathbf{R}^{-1}\mathbf{y}_1, \mathbf{y}_0 \rangle = \langle \mathbf{R}^{-1}\mathbf{RT}*\mathbf{C}*\mathbf{x}, \mathbf{y}_0 \rangle = \langle \mathbf{T}*\mathbf{C}*\mathbf{x}, \mathbf{y}_0 \rangle = \langle \mathbf{C}*\mathbf{x}, \mathbf{Ty}_0 \rangle = \langle \mathbf{C}*\mathbf{x}, \mathbf{0} \rangle = 0$$

This proves that $\mathcal{N}(\mathbf{T}) \subset \mathcal{N}(\mathbf{H}*\mathbf{R}^{-1})$ is sufficient.

To prove (c), recall from Section 7.A.2 that the covariances of the optimal estimate and the estimate based upon the transformed measurements are: $\mathbf{P} = [\mathbf{H}^*\mathbf{R}^{-1}\mathbf{H}]^{-1}$ and $\tilde{\mathbf{P}} = \left[\mathbf{H}*\mathbf{T}*[\mathbf{TRT}*]^{-1}\mathbf{TH}\right]^{-1}$, this proves statement (c).   Q.E.D.

Remarks

1. The theorem above was described using operators because it admits more generality in an operator theory context. It is very useful for estimation problems in the matrix context.

2. Some example conditions for equality are

   a. $\mathbf{T}$ is an identity matrix,

   b. $\mathbf{T}^{-1}$ is unique and $\mathbf{T}^{-1}\mathbf{T} = \mathbf{TT}^{-1} = \mathbf{I}$,

   c. $\mathbf{T}$ is the optimal estimator, $[\mathbf{H}^*\mathbf{R}^{-1}\mathbf{H}]^{-1} [\mathbf{H}^*\mathbf{R}^{-1}]$.

   One can easily show using algebraic manipulation that any of the above conditions will allow one to obtain the optimal estimate $\hat{\mathbf{x}}$ using $\mathbf{z}$.

An Illustrative Example for Theorem 7.A.1

Let $\mathbf{y} = (y_1, y_2, y_3)$ denote three independent measurements of a scalar parameter x with measurement noise $N(0, 1)$. This means $x \in \mathbb{R}^1$ and $\mathbf{y} \in \mathbb{R}^3$. The weighted least squares estimate of x using $\mathbf{y}$ is $\hat{x} = \frac{1}{3}(y_1 + y_2 + y_3)$; with variance $\frac{1}{3}$. Let $\hat{x}^{1,3}$ and $\hat{x}^{2,3}$ denote the estimates of x using (y_1, y_3) and (y_2, y_3), respectively, where $\hat{x}^{1,3} = \frac{1}{2}(y_1 + y_3)$ and $\hat{x}^{2,3} = \frac{1}{2}(y_2 + y_3)$. $\hat{x}^{1,3}$ and $\hat{x}^{2,3}$ form a $\mathbf{z}$ vector in $\mathbb{R}^2$ space, that is, $\mathbf{z} = \left(\hat{x}^{1,3}, \hat{x}^{2,3}\right)^T$. The covariance of $\mathbf{z}$ is

$$Cov\{\mathbf{z}\} = \begin{bmatrix} \dfrac{1}{2} & \dfrac{1}{4} \\ \dfrac{1}{4} & \dfrac{1}{2} \end{bmatrix}$$

Let $\tilde{x}$ be the WLS estimate of x using $\hat{x}^{1,3}$ and $\hat{x}^{2,3}$. The variance of $\tilde{x}$ is $\frac{3}{8}$, bigger than the variance of $\hat{x}$, which is $\frac{1}{3}$. In fact, one cannot obtain $\hat{x}$ with any estimators that rely only on $\hat{x}^{1,3}$ and $\hat{x}^{2,3}$.

Now let us apply the above theorem to this example. This is done by selecting a vector in $\mathbb{R}^3$ such that it does not satisfy condition (b). The relationship among x, $\mathbf{y}$, and $\mathbf{z}$ defines the transformation $\mathbf{T}$

$$\mathbf{T} = \begin{bmatrix} \dfrac{1}{2} & 0 & \dfrac{1}{2} \\ 0 & \dfrac{1}{2} & \dfrac{1}{2} \end{bmatrix}$$

and $\mathbf{H}^*\mathbf{R}^{-1} = [1,1,1]$. Let $\mathbf{u} = [1,1,-1]^T$ be a vector in $\mathbb{R}^3$. Then $\mathbf{Tu} = [0,0]^T$ but $\mathbf{H}^*\mathbf{R}^{-1}\mathbf{u} = 1$. Clearly $\mathbf{u} \in \mathcal{N}(\mathbf{T})$ but $\notin \mathcal{N}(\mathbf{H}^*\mathbf{R}^{-1})$. This violates condition (b), and thus $Cov\{\tilde{x}\} > Cov\{\hat{x}\}$. In fact, one cannot obtain $\hat{x}$ with any estimators using $\hat{x}^{1,3}$ and $\hat{x}^{2,3}$ alone. If additional information is used, such as using some components of $\mathbf{y}$ in addition to the $\mathbf{z}$ vector, then $\hat{x}$ can be recovered. Under this condition, a new transformation $\mathbf{T}^N$ can be defined. As long as $\mathbf{T}^N$ is uniquely invertible, $\hat{x}$ can be recovered using the theorem discussed above. The details are left as a homework problem for the reader.

Remarks

The above theorem and example demonstrate that preprocessing data may result in the loss of information. For example, if the preprocessing involves preliminary estimates with overlapping time windows, the final estimate using preprocessed data may result in larger estimation errors.

7.A.3   Extension to Measurement Fusion versus State Fusion

Corollary 7.A.1 (Corollary of Theorem on Estimation with Transformed Measurements): Let $\mathbf{P}_{k|k} = Cov\{\hat{\mathbf{x}}_{k|k}\}$ and $\tilde{\mathbf{P}}_{k|k} = Cov\{\tilde{\mathbf{x}}_{k|k}\}$, then

$$\mathbf{P}_{k|k} \leq \tilde{\mathbf{P}}_{k|k}.$$

Equality holds if $\boldsymbol{\mu}_k = \mathbf{0}$ for all k and $\hat{\mathbf{x}}_{0|0}^i$ and $\hat{\mathbf{x}}_{0|0}^j$ are uncorrelated for all i and j.

Proof: Consider the two-sensor case with initial condition $\hat{\mathbf{x}}_{0|0}$, and two measurements at time $k = 1$, $\mathbf{y}_1^1$ and $\mathbf{y}_1^2$. It can be shown that the global estimate (via KF) with measurement fusion $\hat{\mathbf{x}}_{1|1}$ is

$$\hat{\mathbf{x}}_{1|1} = \left[\mathbf{I} - \mathbf{K}^1\mathbf{H}^1 - \mathbf{K}^2\mathbf{H}^2\right]\hat{\mathbf{x}}_{1|0} + \mathbf{K}^1\mathbf{y}_1^1 + \mathbf{K}^2\mathbf{y}_1^2,$$

and the local estimates $\hat{\mathbf{x}}_{1|1}^1$ and $\hat{\mathbf{x}}_{1|1}^2$ are

$$\hat{\mathbf{x}}_{1|1}^1 = \left[\mathbf{I} - \mathbf{K}^1\mathbf{H}^1\right]\hat{\mathbf{x}}_{1|0} + \mathbf{K}^1\mathbf{y}_1^1$$

$$\hat{\mathbf{x}}_{1|1}^2 = \left[\mathbf{I} - \mathbf{K}^2\mathbf{H}^2\right]\hat{\mathbf{x}}_{1|0} + \mathbf{K}^2\mathbf{y}_1^2.$$

Clearly

$$\mathbf{T} = \begin{bmatrix} \mathbf{I} - \mathbf{K}^1\mathbf{H}^1 & \mathbf{K}^1 & \mathbf{0} \\ \mathbf{I} - \mathbf{K}^2\mathbf{H}^2 & \mathbf{0} & \mathbf{K}^2 \end{bmatrix}.$$

Applying Theorem 7.A.1, it can be shown that $\tilde{\mathbf{P}}_{1|1} \geq \mathbf{P}_{1|1}$. It can be extended from k to $k + 1$ using the same approach. The corollary is thus proved by mathematical induction. A detailed exploration is left to the reader in Homework Problem 7.

Homework Problems

1. Following the example in Section 7.5, extend your simulation work to include two radars. Use a simulation trajectory of your choice, with sensor locations that can be moved around.

    a. See if two sensors achieve better accuracy.

    b. Move one of the two sensors to several different locations, and see if you can identify any difference in estimation covariance. Explain why.

    c. Add biases to one of the two sensors. Build the filter by either

        (i) increasing the measurement covariance commensurate with the known magnitude of the bias, or

        (ii) applying the SKF (Section 4.5.2).

    Compare the filter root mean square (RMS) error with covariance to check consistency.

2. Analytically show that the parallel filter, sequential filter, and data compression filters discussed in Section 7.2.1 are mathematically identical.

3. Derive Equations (7.3-1) and (7.3-2).

4. Build CRB for your two radar scenarios, and compare your simulation results with CRB.

5. Derive the expression for pseudo-inverse operator $\mathbf{T}^\dagger$ of $\mathbf{T}$, w.r.t. $[.,.]$ as $\mathbf{T}^\dagger = \mathbf{RT}^*[\mathbf{TRT}^*]^{-1}$.

6. Expand the illustrative example of Theorem 7.A.1 with more general matrix equations. Consider a three-sensor case and let $\hat{\mathbf{x}}$ be the WLS estimate using three vectors $(\mathbf{y}_1, \mathbf{y}_2, \mathbf{y}_3)$ and $\hat{\mathbf{x}}^{1,2}$ and $\hat{\mathbf{x}}^{2,3}$ the estimates of using measurement vectors $(\mathbf{y}_1, \mathbf{y}_2)$ and $(\mathbf{y}_2, \mathbf{y}_3)$, respectively.

   a. Develop matrix equations to show that the estimate $\tilde{\mathbf{x}}$ obtained using $\hat{\mathbf{x}}^{1,2}$ and $\hat{\mathbf{x}}^{2,3}$ has higher covariance than that of $\hat{\mathbf{x}}$.

   b. Assume that any one of the three $\mathbf{y}$ vectors are available to use with $\hat{\mathbf{x}}^{1,2}$ and $\hat{\mathbf{x}}^{2,3}$ for computing $\tilde{\mathbf{x}}$. Show that in this case $\hat{\mathbf{x}}$ and $\tilde{\mathbf{x}}$ are the same.

7. Following the Corollary 7.A.1 in Appendix 7.A.3,

   a. Identify the three vector spaces and $\mathbf{H}^*\mathbf{R}^{-1}$.

   b. Show that they do not satisfy condition (b) of Theorem 7.A.1.

   c. Derive the state fusion equation (joint estimate).

   d. Show that when the system is deterministic (process noise is zero), the equality holds.

References

1. S. Blackman and R. Popoli, *Design and Analysis of Modern Tracking Systems*. Norwood, MA: Artech House, 1999.

2. D. Willner, C.B. Chang, and K.P. Dunn, "Kalman Filter Algorithms for a Multi-Sensor System," in *Proceedings of 1976 IEEE Conference on Decision and Control*, pp. 570–574, Dec. 1976.

3. D. Willner, C.B. Chang, and K.P. Dunn, "Kalman Filter Configuration for a Multiple Radar System," MIT Lincoln Laboratory Technical Note TN-1976-21, April 14 1976.

4. Y. Bar-Shalom, "Update with Out-of-Sequence Measurements in Tracking, Exact Solution," *IEEE Transactions on Aerospace Electronic Systems*, vol. AES-38, pp. 769–778, July 2002.

5. Y. Bar-Shalom, P.K. Willett, and X. Tian, *Tracking and Data Fusion: A Hand Book of Algorithms*. Storrs, CT, YBS Publishing, 2011.

6. C.Y Chong, S. Mori, W. Parker, and K.C. Chang, "Architecture and Algorithm for Track Association and Fusion," *IEEE AES Systems Magazine*, Jan. 2000.

7. Y. Bar-Shalom, "On the Track-to-Track Correlation Problem," *IEEE Transactions on Automatic Control*, vol. AC-26, pp. 571–572, April 1981.

8. Y. Bar-Shalom and L. Campo, "The Effect of the Common Process Noise on the Two-Sensor Fused Track Covariance," *IEEE Transactions on Aerospace Electronic Systems,* vol. AES-22, pp. 803–805, Nov. 1986.

9. K.C. Chang, R.K. Saha, and Y. Bar-Shalom, "On Optimal Track-to-Track Fusion," *IEEE Transactions on Aerospace Electronic Systems*, vol. AES-33, pp. 1271–1276, Oct. 1997.

10. C.Y. Chong, "Hierarchical Estimation," in *Proceedings of MIT/ONR Workshop on C3*, 1979.

11. C.Y. Chong, S. Mori, and K.C. Chang, "Information Fusion in Distributed Sensor Networks," in *Proceedings of the 1985 American Control Conference*, pp. 830–835, June 1985.

12. C.Y. Chong, K.C. Chang, and S. Mori, "Fundamentals of distributed estimation," in *Distributed Data Fusion for Network-Centric Operations*, D.L. Hall, C.-Y. Chong, J. Llinas, and M. Liggins, II, Eds., Boca Raton: CRC Press, pp. 95–124, 2012.

13. S. Mori, K.C. Chang, and C.Y. Chong, "Essence of Distributed Target Tracking; Track Fusion and Track Association," in *Distributed Data Fusion for Network-Centric Operations*, D.L. Hall, C.-Y. Chong, J. Llinas, and M. Liggins, II, Eds., CRC Press, pp. 125–160, 2012.

14. S. Mori, W.H. Barker, C.-Y. Chong, K.C. and Chang, "Track Association and Track Fusion with Non-deterministic Target Dynamics," *IEEE Transactions on Aerospace and Electronic Systems*, vol. 38, pp. 659–668, 2002.

15. K.C. Chang, T. Zhi, S. Mori, and C.-Y. Chong, "Performance Evaluation for MAP State Estimate Fusion," *IEEE Transactions on Aerospace and Electronic Systems*, vol. AES-40, pp. 706–714, 2004.

16. K.C. Chang, Z. Tian, and R.K. Saha, "Performance Evaluation of Track Fusion with Information Matrix Filter," *IEEE Transactions on Aerospace and Electronic Systems,* vol. AES-38, pp. 455–466, 2002.

17. C.B. Chang and R.B. Holmes, "On Linear Estimation with Transformed Measurements," *IEEE Transactions on Automatic Control*, vol. AC-28, pp. 242–244, Feb. 1983.

18. D.G. Luenberger, *Optimization by Vector Space Methods*. New York: Wiley, 1966.

8

Estimation and Association with Uncertain Measurement Origin

8.1 Introduction

The focus of the book from this chapter forward will be on target tracking. Special terminology used in the tracking community is reviewed in Appendix C. Target tracking requires decision making in assigning measurements to tracks. A track consists of a collection of past measurements and the resulting state estimate and covariance using these measurements. Uncertainties on measurement-to-track assignment arise due to the presence of multiple targets generating multiple measurements resulting in multiple tracks. Techniques discussed here apply equally well to single- or multiple-sensor systems.

In previous chapters, fundamental theories and algorithms for the problem of estimation were developed. In all cases, an estimator was used to process a set of measurements with the assumption that all measurements came from the same object or target. There are cases, however, when multiple individual objects are so closely spaced that the assumption that the time-sequenced measurements come from the same object can no longer be determined with sufficient confidence. When objects are spaced closer than estimation and measurement uncertainties, the problem of track ambiguity arises. The process to determine whether a sequence of measurements has originated from a common object is referred to as association. The process for determining whether tracks or measurements from different sensors came from a common target is referred to as correlation.[1] The case where measurements may have originated from different objects is referred to as the problem of measurements with uncertain

---

[1] The meaning of association and correlation as described by these two sentences is commonly adopted in the tracking community. As will be shown in the ensuing development, the underlying mathematics used for addressing these two types of problems are the same.

origin. Estimators would produce erroneous results when the measurements processed did not come from the same target. Similarly, the ability to correctly perform association or correlation depends on the accuracy and consistency of the state estimates. The problems of estimation and association (or correlation) are fundamentally different. The problem of estimation is analytical and handled with conventional calculus, in which the traditional mathematical approaches in optimization apply. The problem of association or correlation, similar to the classical problem of signal detection, is sometimes handled as a binary problem, in which a decision is made on an either/or basis. This is known as making a hard decision. The method to blend all measurements for use to update track is referred to as a soft decision. Both methods will be covered later. The situation when measurement-to-track association, or track-to-track correlation, cannot be made uniquely with high confidence is referred to as track with ambiguity. Techniques to address track ambiguity are the subject of multiple target tracking (MTT), which is the subject of the next three chapters. There is a wealth of literature in this subject area [see, e.g., 1–21].

Due to differences in problem formulations, ambiguity conditions, and solution objectives, different MTT algorithms were developed. There are algorithms that only consider a single scan of measurements with a single track (independent of the existence of other tracks) such as nearest neighbor (NN) association [16, 20]; algorithms that allow for track splitting [16, 18–19]; and algorithms that blend all measurements such as probabilistic data association filter (PDAF) [3, 5, 17]. Some algorithms deal with multiple track jointly such as global nearest neighbor (GNN) [16, 20] and joint probabilistic data association filter (JPDAF) [3, 17]; and others consider multiple tracks with multiple scans of measurement jointly such as the multiple hypothesis tracker (MHT) [1, 7, 9, 10, 13, 16]. The taxonomy of MTT approaches will be given in Section 8.3 and a brief discussion of each approach will be provided.

As stated in the earlier part of this book, during the ensuing development, whenever possible the state and measurement system is assumed to be linear Gaussian and the system and measurement noise processes are mutually uncorrelated and also uncorrelated in time. Under this assumption, the Kalman filter (KF) is the fundamental filter for use. When the system is nonlinear, most of our analyses are extendable to using the extended Kalman filter (EKF) as approximations.

8.1.1 Track Ambiguity Illustration

Examples of track ambiguity are illustrated in Figure 8.1. When there is no confusion in measurement-to-track association, such as the case of a single target with only one

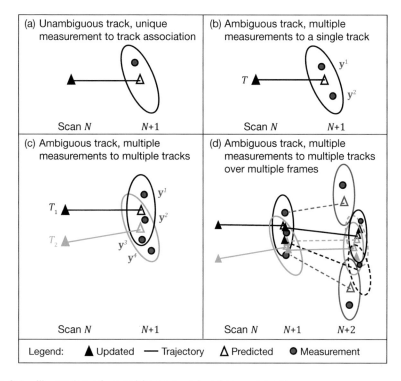

Figure 8.1   Illustration of unambiguous and ambiguous tracks.

measurement appearing in the predicted uncertainty region,[2] the track is unambiguous, as shown in Figure 8.1(a). Track ambiguity may be caused by multiple new measurements that fall within the predicted uncertainty region of an initially unambiguous track, as shown in Figure 8.1(b). Another case may have two initially unambiguous tracks with predicted uncertainty regions that overlap and multiple new measurements fall both inside and outside of the overlapped region, as shown in Figure 8.1(c). An overlapping situation could persist over multiple scans, as in Figure 8.1(d). When a track is unambiguous, the estimation algorithms of previous chapters apply. Techniques to address track ambiguity are the subject of MTT, which is the subject of the next three chapters. There is a large body of literature on this subject [see, e.g., 1–21].

---

2   Referred to as the acceptance gate (AG), which will be the topic of Section 8.1.2.

8.1.2   Acceptance Gate

Given the fact that sensors may have a very broad field of view and that targets may be widely separated over large geographical areas, not all measurements are candidates for association with all tracks. The ellipses drawn around a predicted measurement location in Figure 8.1 are meant to illustrate that the region within which measurements fall describe candidates for association with that track. This region is referred to as the acceptance gate (AG). The AG is therefore the first line of processing in MTT. AG is referred to as a validation region (VR) in [16]. All measurements that fall within the AG of a track are a candidate for association with that track. Let $\hat{\mathbf{y}}_{k|k-1} \triangleq \mathbf{H}_k \hat{\mathbf{x}}_{k|k-1}$ denote the predicted measurement of $\mathbf{y}_k$. The vector $\boldsymbol{\gamma}_k = \mathbf{y}_k - \hat{\mathbf{y}}_{k|k-1}$ is the filter residual process with covariance $\boldsymbol{\Gamma}_k = \left[\mathbf{H}_k \mathbf{P}_{k|k-1} \mathbf{H}_k^T + \mathbf{R}_k\right]$.[3] The random variable λ, defined as $\lambda = \frac{1}{m} \boldsymbol{\gamma}_k^T \boldsymbol{\Gamma}_k^{-1} \boldsymbol{\gamma}_k$, follows the normalized χ^2 distribution,[4] where m is the dimension of $\boldsymbol{\gamma}_k$. With the above equation, λ is a weighted distance square between $\mathbf{y}_k$ and $\hat{\mathbf{y}}_{k|k-1}$, and is known as the Mahalanobis distance. The χ^2 distribution property holds when $\mathbf{y}_k$ comes from the same object as all prior measurements used to compute $\hat{\mathbf{x}}_{k|k-1}$, that is, all $\{\mathbf{y}_1, \mathbf{y}_2, \dots, \mathbf{y}_k\}$ originate from the same object. This property is used to define candidate measurements that may be associated to a track. Assume that there are N measurements at time t_k and are denoted by $\mathbf{y}_k^i$ for $i = 1, \dots N$. A total of N Mahalanobis distances can be computed as

$$\lambda_i = \left(\mathbf{y}_k^i - \hat{\mathbf{y}}_{k|k-1}\right)^T \boldsymbol{\Gamma}_k^{-1} \left(\mathbf{y}_k^i - \hat{\mathbf{y}}_{k|k-1}\right). \tag{8.1-1}$$

Those measurements having λ_i less than a threshold $\mathcal{T}$ are declared as candidates for association with the track represented by $\hat{\mathbf{y}}_{k|k-1}$, that is, select the measurements with their $\lambda_{\hat{i}} \leq \mathcal{T}$ where $\hat{i}$ corresponds to the subset of $i = 1, \dots, N$ satisfying the threshold $\mathcal{T}$. The threshold $\mathcal{T}$ is selected such that the true target is in the AG with probability P_G, to be referred to as the gate probability.

8.2   Illustration of the Multiple Target Tracking Problem

Now consider the case that a sensor has collected k scans of measurements. On the kth scan, the sensor detects N_k distinct measurements, each given by a vector $\mathbf{y}_k^{n_k}$, where n_k is the index of measurements detected within the kth scan. The set of all measurements in scan k is denoted by

$$\mathbf{Y}_k = \left\{\mathbf{y}_k^1, \mathbf{y}_k^2, \mathbf{y}_k^3, \dots, \mathbf{y}_k^{N_k}\right\}. \tag{8.2-1}$$

---

3   Refer to Remark 3 in Section 2.6.
4   See any standard textbook on probability and statistics [e.g., 22].

Let $\mathbf{Y}_{1:k}$ denote the set of measurements from the 1st through the kth scan, that is,

$$\mathbf{Y}_{1:k} = \{\mathbf{Y}_1, \mathbf{Y}_2, \mathbf{Y}_3, \ldots, \mathbf{Y}_k\}. \tag{8.2-2}$$

Thus $\mathbf{Y}_{1:k}$ has a total of $N_1 + N_2 + N_3 + \ldots + N_k$ measurement vectors. The number of measurement vectors not being equal across the scan is because (a) there may be new targets detected, (b) some targets were not detected because they have moved out of the sensor detection range or the sensor field of view, (c) some missed detections may be due to low signal to noise ratio, (d) some targets may be unresolved by the measurement sensors, thus a measurement may have contained more than one target, and (e) some of the detections may be due to false alarms.

Regardless of the reason for an unequal number of detections, the problem is on determining which sets of measurements across scans constitute tracks (or from the same target). Figure 8.2 is used to illustrate the concept of MTT using a one-dimensional (1D) representation for measurements. The discrete time or scan number is plotted on the horizontal axis and the measurement is plotted on the vertical axis. Detections across scans connected by lines are potential tracks. Without constraints, all detections can be connected throughout all scans as tracks, as is represented by all blue lines in Figure 8.2. In an ideal situation, all targets are detected, resolved, and there is no false detection, that is, $P_d = 1$ and $P_{fa} = 0$; the maximum number of possible tracks is the product $N_1 N_2 N_3 \ldots N_k$, which can be a prohibitively large number when the number of detections per frame is large.[5] The detections connected with red lines in Figure 8.2 show likely tracks, which are selected with the constraint that targets are moving in approximately straight lines while allowing a certain curvature for measurement uncertainties and possible target maneuvers. This concept will be further illustrated in a tracking algorithm presented in Section 8.7.

Let each possible measurement combination be denoted by a track, $\mathbf{y}_{1:k}^{m_k}$. The measurement vectors contained in $\mathbf{y}_{1:k}^{m_k}$ are

$$\mathbf{y}_{1:k}^{m_k} \triangleq \left\{ \mathbf{y}_1^{n_1}, \mathbf{y}_2^{n_2}, \mathbf{y}_3^{n_3}, \ldots, \mathbf{y}_k^{n_k} \right\} \in \mathbf{Y}_{1:k}, \tag{8.2-3}$$

and

$$\boldsymbol{\theta}_{1:k}^{m_k} \triangleq \left\{ \mathbf{y}_{1:k}^{m_k} \text{ is a true track} \right\}, \tag{8.2-4}$$

---

5   Design factors that can be used to curtail the growth of a large number of tracks are (a) only consider those detections that fall within AG, (b) target motion must follow certain physics constraints, and (c) limit the number of scans for decision, that is, choose a small k using a sliding window approach, and so on.

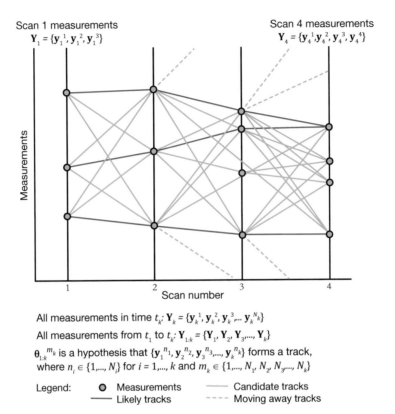

All measurements in time t_k: $\mathbf{Y}_k = \{\mathbf{y}_k^1, \mathbf{y}_k^2, \mathbf{y}_k^3 \dots \mathbf{y}_k^{N_k}\}$
All measurements from t_1 to t_k: $\mathbf{Y}_{1:k} = \{\mathbf{Y}_1, \mathbf{Y}_2, \mathbf{Y}_3, \dots, \mathbf{Y}_k\}$
$\theta_{1:k}^{m_k}$ is a hypothesis that $\{\mathbf{y}_1^{n_1}, \mathbf{y}_2^{n_2}, \mathbf{y}_3^{n_3}, \dots, \mathbf{y}_k^{n_k}\}$ forms a track,
where $n_i \in \{1, \dots, N_i\}$ for $i = 1, \dots, k$ and $m_k \in \{1, \dots, N_1, N_2, N_3, \dots, N_k\}$

Legend: ○ Measurements   —— Candidate tracks
          —— Likely tracks   - - - Moving away tracks

Figure 8.2   Illustration of multiple target tracks.

where $n_i \in \{1, \dots, N_i\}$ for $i=1, \dots, k$ and $m_k \in \{1, \dots, N_1 N_2 N_3 \dots N_k\}$.[6] When each possible sequence of measurements is processed by a filter, the probability of $\mathbf{y}_{1:k}^{m_k}$ given $\theta_{1:k}^{m_k}$ can be computed recursively using

$$\Lambda\left(\theta_{1:k}^{m_k}\right) \triangleq p\left(\mathbf{y}_{1:k}^{m_k} | \theta_{1:k}^{m_k}\right) = p\left(\mathbf{y}_k^{n_k} | \mathbf{y}_{1:k-1}^{m_{k-1}}, \theta_{1:k}^{m_k}\right) p\left(\mathbf{y}_{1:k-1}^{m_{k-1}} | \theta_{1:k}^{m_k}\right) = \prod_{j=1}^{k} p\left(\mathbf{y}_j^{n_j} | \mathbf{y}_{1:j-1}^{m_{j-1}}, \theta_{1:k}^{m_k}\right), \quad (8.2\text{-}5)$$

where $p\left(\mathbf{y}_k^{n_k} | \mathbf{y}_{1:k-1}^{m_{k-1}}, \theta_{1:k}^{m_k}\right)$ is the residual density of the measurement vector $\mathbf{y}_k^{n_k}$ with the filter matched to $\mathbf{y}_{1:k-1}^{m_{k-1}}$. To enumerate the increment from m_{k-1} to m_k, one needs to count all $n_k \in \{1, \dots, N_k\}$. The complexity of the above equation lies in the "book-keeping," or being able to correctly account for all possible combinations. Note the

---

6   For the purpose of illustration, when the number of measurement vectors in each scan is the same and equals
    N, then $m_k \in \{1, \dots, N^k\}$ where N^k denotes the kth power of N.

similarity of this formulation to the multiple model estimation algorithm (MMEA) with switching parameters discussed in Chapter 5.

Remarks

1. $p\left(\mathbf{y}_k^{n_k} | \mathbf{y}_{1:k-1}^{m_{k-1}}, \boldsymbol{\theta}_{1:k}^{m_k}\right)$ is also the likelihood that measurement $\mathbf{y}_k^{n_k}$ comes from the track with prior measurements represented by $\boldsymbol{\theta}_{1:k}^{m_k}$. It is the key function for track quality monitoring and track pruning.

2. The total number of potential tracks can be very large. Assume that the number of candidate association measurements in each scan is the same and each measurement in a scan can only be used once, then the total number of combination is $N!$. When the true number of targets is N, then there should only be N tracks. One can easily show that when all measurements were well-separated,[7] such that there is only one detection in each AG, then there will only be N tracks.[8]

3. The total number of $\boldsymbol{\theta}_{1:k}^{m_k}$ is $N_1 N_2 N_3 \ldots N_k$ or N^k for a constant N. Let $\mathbb{F}$ denote a set of tracks in which each measurement is only used once. Entries of $\mathbb{F}$ are referred to as feasible sets of assignments. The total number of tracks in $\mathbb{F}$ for a constant N is $N!$, a large number but much less than N^k. When the true number of targets is N, then there should only be N tracks. A possible solution for track selection is to find the subset of N tracks in $\mathbb{F}$ in which the sum of their hypothesis probability is maximum, that is,

$$\text{Max}_{\boldsymbol{\theta}_{1:k}^{\ell}} \left\{ \sum_{\boldsymbol{\theta}_{1:k}^{\ell} \subset \mathbb{F}} \text{Pr}\left\{\boldsymbol{\theta}_{1:k}^{\ell} | \mathbf{Y}_{1:k}\right\} \right\}.$$

This constitutes a k-dimensional assignment problem, which is sometimes referred to as a multidimensional assignment (MDA) approach to multiple target tracking [16]. The computational burden for large N and k is clearly nontrivial. When $k = 2$, that is, only two scans of measurements are considered, the solution is the same as the track to measurement assignment problem for a given scan, as will be shown in Section 8.4.

8.3 A Taxonomy of Multiple Target Tracking Approaches

The discussion in Section 8.2 illustrates the basic issues and the solution concept with MTT. In this section, some approaches in addressing the MTT problem are outlined and are summarized in Table 8.1, a taxonomy of MTT approaches. The first column

---

7 *Well-separated* means that the distance between measurements is bigger than a multiplication factor (a system design parameter) of the covariance of the measurement residual process.

8 In the true MHT formulation, this simple concept will no longer hold because there are missed detections and false alarms. It is used here to illustrate the complexity of MTT even with ideal sensors and cooperative environments.

Table 8.1 Multiple Target Tracking Taxonomy

	Single Scan	Multiple Scan
Multiple tracks processed independently	*Category I* 1. Nearest neighbor (NN) measurement-to-track assignment 2. Probabilistic data association filter (PDAF)	*Category III* Use *m/n* track initiation, with each initial detection processed independently. Using track split when using multiple measurements in AG; and delete track when the track likelihood is below threshold.
Multiple tracks processed jointly	*Category II* 1. Global nearest neighbor (GNN) multiple measurements to multiple track assignment 2. Joint probabilistic data association filter (JPDAF)	*Category IV* MHT 1. Measurement-oriented (Reid) 2. Track-oriented (Kurien)

of the table, Categories I and II, considers approaches when only a single scan of measurements is used at one time. The second column, Categories III and IV, considers multiple scans of measurement. The first row consists of the Categories I and III where tracks are processed independent of each other even when measurements fall within their overlapped acceptance regions. Categories II and IV are tracks processed jointly when measurements do fall within overlapped acceptance regions. Most of these MTT approaches will work only with a synchronous measurement case. Even for a single sensor, measurements of multiple targets do not happen at the same time. It is necessary to time-align these measurements in order to apply these MTT algorithms by means of extrapolation and/or interpolation. Elements of these four categories are briefly explained below with further exploration in later sections.

Nearest Neighbor Association

There may be multiple measurements falling within an AG when the target density is high. If only one measurement is to be chosen, then the $\mathbf{y}_k^i$ corresponding to the smallest λ_i (see Equation [8.1-1]) is selected, which is known as the nearest neighbor (NN) association. The red dot closer to the predicted measurement in Figure 8.1(b) is assigned to this track using the NN association rule.

Track Split

Due to the fact that the target may maneuver and measurements are noisy, the measurement that is closest (in the weighted sense) to the predicted measurement may not be the right choice. If the objective is to minimize leakage with allowance to a higher number of false tracks, then all measurements within the AG are accepted for track update. One option with this idea is to split the track. When multiple measurements fall within the acceptance gate of a track, this track is split into multiple tracks, each using a measurement for track update. In this case, the scenario shown in Figure 8.1(b) will result in two tracks each updated with a separate red dot. The track split method is used for both coordinated and uncoordinated track processing. The coordinated track split method is used in MHT. The uncoordinated track split method, that is, when each track is processed as if others do not exist, is the approach listed in Category III of Table 8.1. This method is simple and effective with a modest level of target density. The track split method will lead to an increase in the number of tracks with some of them being false tracks. After a few scans, a subset of the tracks will continue because one of them is the true track and the others are closely spaced with the true track, thus maintaining a small track error. A false track will eventually drop out, because it may no longer find new measurements in its AG. Erratic tracks with poor track accuracy will be deleted as measured by the track likelihood function.

Probabilistic Data Association Filter

An alternative to track split is to use all the measurements in the AG to update the given track. All measurements in the AG are summed up with the a posteriori probability as weighting factors and used for update. Probabilities of missed detection and false alarms are also considered in computing weighting factors. This is known as PDAF.

The Global Nearest Neighbor Assignment

It is possible that some measurements in an AG may also be in the AG of neighboring tracks. This constitutes a multiple-measurement to multiple-track assignment problem. If one is to conduct the assignment in series in which the measurement that is closest to a track is assigned to that track first, this is an extension of the NN assignment approach illustrated above and is known as the greedy algorithm. A better approach, in the sense of minimum assignment error probability, is to consider the assignment jointly such that the total sum of weighted distances of all assignments is minimum. This is known as the GNN assignment. Both NN and GNN association methods are hard decisions mentioned in Section 8.1.1.

Joint Probabilistic Data Association Filter

Similar to the fact that GNN is an extension of NN, The JPDAF is an extension of PDAF in which it considers all tracks with shared measurements jointly. Both PDAF and JPDAF are soft decisions mentioned in Section 8.1.1.

Multiple Hypothesis Tracker

A well-known technique for MTT is the MHT. The concept of the MHT has been in formation for some time before it was solidified by Reid in his 1977 IEEE paper [9]. Reid's method has been referred to as measurement-based MHT. Its complexity lies in the data file management. Later the track-oriented MHT due to Kurien [10] provided a way for simplification. In the traditional MHT sense, a hypothesis is defined as one possible combination of tracks, where each track consists of a unique set of measurements, and a measurement is only used once in a hypothesis. In the MHT formulation, each new measurement is considered as (a) a continuation of an existing track, (b) the start of a new track, or (c) a false alarm. An existing track may continue with or without a new measurement because the expected measurement of this track may be missing (not in the acceptance gate or not detected by the sensor). With all these considered, a multiple number of hypotheses and tracks will result even with the simplest and most unambiguous case as shown in Figure 8.1(a). (For more details see the example in Section 9.2.) Many publications dated before and after Reid's paper offer algorithm refinements with applications, while not all of them strictly follow the framework of MHT [see, e.g., 1–8, 10–17] and the references therein. The MHT concept considering multiple scans of data with multiple tracks jointly is illustrated in Figure 8.1(d).

There are many MTT algorithms designed and applied to real-world problems. A practical algorithm is often a combination of approaches shown above including variations using engineering judgments. The taxonomy shown in Table 8.1 represents a menu of choices for system designers. These techniques will be further explained. Topics in Categories I, II, and III are presented here in Chapter 8. A practical algorithm applying some of these techniques representing the culmination of the authors' experience in MTT algorithm design will be shown in Section 8.7 based on [18–21]. The subject of MHT, Category IV, will be discussed in Chapter 9.

8.4   Track Split

Consider $p\left(\mathbf{y}_k^{n_k}|\mathbf{y}_{1:k-1}^{m_{k-1}},\boldsymbol{\theta}_{1:k}^{m_k}\right)$ in Equation (8.2-5), which is the likelihood that measurement $\mathbf{y}_k^{n_k}$ comes from the same target that forms the track with prior measurements represented by $\mathbf{y}_{1:k-1}^{m_{k-1}}$. Using the notation $\Lambda\left(\boldsymbol{\theta}_{1:k}^{m_k}\right)$ for likelihood

$$\Lambda\left(\boldsymbol{\theta}_{1:k}^{m_k}\right)\triangleq p\left(\mathbf{y}_{1:k}^{m_k}|\boldsymbol{\theta}_{1:k}^{m_k}\right).$$

Using the chain rule of probability, from Equation (8.2-5), one can show that

$$\Lambda\left(\boldsymbol{\theta}_{1:k}^{m_k}\right)=\prod_{j=1}^{k}p\left(\mathbf{y}_j^{n_j}|\mathbf{y}_{1:j-1}^{m_{j-1}},\boldsymbol{\theta}_{1:k}^{m_k}\right),$$

where $n_i\in\{1,\dots,N_i\}$ for $i=1,\dots k$ and $m_k\in\{1,\dots,N_1N_2N_3\dots N_k\}$ for a given sequence of measurement vectors in $\mathbf{Y}_{1:k}$, such that

$$\mathbf{y}_{1:k}^{m_k}\triangleq\left\{\mathbf{y}_1^{n_1},\mathbf{y}_2^{n_2},\mathbf{y}_3^{n_3},\dots,\mathbf{y}_k^{n_k}\right\}.$$

Since $p\left(\mathbf{y}_k^{n_k}|\mathbf{y}_{1:k-1}^{m_{k-1}},\boldsymbol{\theta}_{1:k}^{m_k}\right)$ is the residual density of the measurement vector $\mathbf{y}_k^{n_k}$ with the filter matched to $\boldsymbol{\theta}_{1:k-1}^{m_{k-1}}$, we arrive at

$$p\left(\mathbf{y}_k^{n_k}|\mathbf{y}_{1:k-1}^{m_{k-1}},\boldsymbol{\theta}_{1:k}^{m_k}\right)=c_{m_k}\times\exp\left\{-\frac{1}{2}\left(\mathbf{y}_k^{n_k}-\hat{\mathbf{y}}_{k|k-1}^{m_{k-1}}\right)^T\Gamma_{\boldsymbol{\theta}_{1:k}^{m_k}}^{-1}\left(\mathbf{y}_k^{n_k}-\hat{\mathbf{y}}_{k|k-1}^{m_{k-1}}\right)\right\},\qquad(8.4\text{-}1)$$

where $\hat{\mathbf{y}}_{k|k-1}^{m_{k-1}}$ is the predicted measurement based on the tracking using measurements in $\mathbf{y}_{1:k-1}^{m_{k-1}}$ and c_{m_k} is a normalization constant related to $\boldsymbol{\theta}_{1:k}^{m_k}$. For the purpose of simplification, without losing generality, the following notation not explicitly showing indices of measurement in a scan will be used: $\boldsymbol{\gamma}_k=\mathbf{y}_k^{n_k}-\hat{\mathbf{y}}_{k|k-1}^{m_{k-1}}$ and $\Gamma_k=Cov\{\boldsymbol{\gamma}_k\}$ track hypothesis $\boldsymbol{\theta}_{1:k}^{m_k}$. With this simplification we obtain the following log likelihood function of Equation (8.2-5)

$$-\ln\Lambda\left(\boldsymbol{\theta}_{1:k}^{m_k}\right)=\sum_{i=1}^{k}\boldsymbol{\gamma}_i^T\Gamma_i^{-1}\boldsymbol{\gamma}_i+\text{const.}=\sum_{i=1}^{k}\lambda_i+\text{const.},\qquad(8.4\text{-}2)$$

where λ_i is the Mahalanobis distance defined in Equation (8.1-1). Equation (8.4-2) is the summation of residual errors of a track plus a constant. This is a very important equation because it provides the health check of a track. If all measurements of a track are from the same target, then Equation (8.4-2) follows a χ^2 distribution with kmth degrees of freedom where k is the time index and m is the dimension of the measurement vector $\mathbf{y}$. If some of the measurements are not from the same target, then its value computed with Equation (8.4-2) will not follow the χ^2 distribution, and thus will be a candidate for deletion or pruning. For simplicity, the constant term in Equation (8.4-2) can be ignored based upon the assumption that the constant term in Equation (8.4-2) does not depend on the measurement and predicted measurement

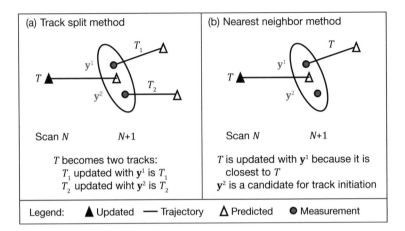

Figure 8.3   Illustration of track split and nearest neighbor methods.

vectors. For nonlinear systems, this is generally not true. The use of Mahalanobis distance alone will not be sufficient for track assignment, because the log likelihood function, Equation (8.4-2), will have different constant terms that will be a function of $\boldsymbol{\theta}_{1:k}^{m_k}$.

This equation is part of the track scoring method used in the MHT algorithm, which will be discussed in the next chapter. An illustration of track split and nearest neighbor concepts are depicted in Figure 8.3.

8.5   The Nearest Neighbor and Global Nearest Neighbor Assignment Algorithms

Consider the case in which N_1 measurements are to be assigned to N_2 tracks. The total number of possible pairings of measurement-to-track is $N_1 N_2$.

To uniquely assign measurement-to-track is analogous to an assignment problem in operations research, such as assigning n jobs to m individuals. Let the cost[9] of person i assigned to job j be denoted $\lambda_{i,j}$. The objective of the many-on-many assignments is to achieve the minimum total cost, which can be formulated as the solution of

$$\text{Min}_{i,j} \sum_{i,j} \lambda_{i,j} \tag{8.5-1}$$

---

9   Or inversely the performance.

where each $\lambda_{i,j}$ can only be used once, corresponding to the assumption that each person can only be assigned to one job; thus, it is a feasible set of assignments. Note that n and m do not need to be equal. This corresponds to the fact that there may be more jobs than people or vice versa. This constitutes an optimization problem in which the best set of the assignment is the one that gives the minimum total cost. This is analogous to the problem where multiple measurements are to be assigned to multiple tracks in which a track (or measurement) is constrained to accept only one measurement (or track).[10]

For the tracking problem, the assignment solution can be formulated as follows: Let $\hat{\mathbf{y}}_{k|k-1}^{j}$ denote the jth predicted measurements from a total of n tracks having covariance $\mathbf{P}_{\hat{\mathbf{y}},k|k-1}^{j} = \mathbf{H}_{k}^{j}\mathbf{P}_{k|k-1}^{j}\mathbf{H}_{k}^{jT}$ and $\mathbf{y}_{k}^{i}$ the ith measurement from a total of m measurements having covariance $\mathbf{R}_{k}^{i}$, all at time k. Following the notation used in Equation (8.1-1), the weighted distance may be used as a measure of cost for the assignment problem

$$\lambda_{i,j} = \left(\mathbf{y}_{k}^{i} - \hat{\mathbf{y}}_{k|k-1}^{j}\right)^{T}\left[\mathbf{H}_{k}^{j}\mathbf{P}_{k|k-1}^{j}\mathbf{H}_{k}^{jT} + \mathbf{R}_{k}^{i}\right]^{-1}\left(\mathbf{y}_{k}^{i} - \hat{\mathbf{y}}_{k|k-1}^{j}\right), \qquad (8.5\text{-}2)$$

where $\lambda_{i,j}$ is the Mahalanobis distance between $\hat{\mathbf{y}}_{k|k-1}^{j}$ and $\mathbf{y}_{k}^{i}$, and represents the cost of assigning measurement i to predicted measurement j (depicted in Figure 8.4).[11] The optimum assignment is selected as those entries giving $\text{Min}_{i,j} \sum_{i,j}\lambda_{i,j}$ under the constraint that each column/row can be assigned to a row/column once. Such a constraint is often referred to as the feasible set. It can be shown that the Mahalanobis distance between $\hat{\mathbf{y}}_{k|k-1}^{j}$ and $\mathbf{y}_{k}^{i}$ is a sufficient statistic representing the solution of a maximized likelihood function for the track to measurement assignment problem when the determinant of the covariance $\mathbf{P}_{k|k-1}^{j}$ does not depend on $\hat{\mathbf{y}}_{k|k-1}^{j}$ and $\mathbf{y}_{k}^{i}$. This is true only for linear systems.

Nearest Neighbor

For the case when multiple measurements do not fall in the overlapped AGs of multiple tracks, one is only going to select one measurement for a track (many-for-one, i.e., there is only one $\hat{\mathbf{y}}_{k|k-1}$ to be considered), selecting the $\mathbf{y}_{k}^{i}$ that is closest to $\hat{\mathbf{y}}_{k|k-1}$ as the NN solution. See Figure 8.3(b) for illustration. For the case in which multiple measurements are in the overlapped AGs of multiple tracks, as represented by Figure 8.4, the NN method can still be applied by going through one column at a time and

---

10   Because a measurement sensor is not perfect, for example detection probability P_d is less than one and false alarm probability P_{fa} is greater than zero and the sensor has limited resolution, this constraint may be relaxed in practical systems.

11   More precisely, the cost function should be the log likelihood of the assignment as shown in Equation (8.4-2).

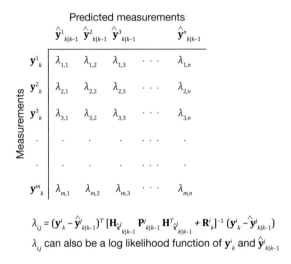

Predicted measurements

$$\lambda_{i,j} = (\mathbf{y}^i_k - \hat{\mathbf{y}}^j_{k|k-1})^T [\mathbf{H}_{\mathbf{x}^j_{k|k-1}} \mathbf{P}^j_{k|k-1} \mathbf{H}^T_{\mathbf{x}^j_{k|k-1}} + \mathbf{R}^i_k]^{-1} (\mathbf{y}^i_k - \hat{\mathbf{y}}^j_{k|k-1})$$

$\lambda_{i,j}$ can also be a log likelihood function of $\mathbf{y}^i_k$ and $\hat{\mathbf{y}}^j_{k|k-1}$

Figure 8.4   The assignment matrix: cost of assigning measurement i to predicted measurement j.

selecting the $\mathbf{y}^i_k$ that is closest to $\hat{\mathbf{y}}^j_{k|k-1}$ in column j. One could choose to limit a measurement to be assigned to only one track.[12] As such, a selected $\mathbf{y}^i_k$ must be deleted from the assignment matrix before continuing on to the next column. This method is known as a greedy algorithm. The solution of a greedy algorithm is made unique by choosing the row and column that correspond to the minimum entry of the assignment, and what remains of the assignment matrix.

Global Nearest Neighbor

When the many-on-many problems and measurements to track assignments are considered together, as shown in Figure 8.4, the $\mathrm{Min}_{i,j} \sum_{i,j} \lambda_{i,j}$ solution is the solution in the sense of the generalized likelihood ratio test, and is known as the GNN solution. A brute force approach to solve this optimization problem is to enumerate all possible combinations and choose the one with the minimum sum. This corresponds to a total of $n!/(n - m)!$ operations for $n \geq m$ or $m!/(m - n)!$ for $m \geq n$. An efficient algorithm to solve this problem was first published by Kuhn [23–24], which was named the Hungarian algorithm because this algorithm was largely based on the work of two

---

12   If this constraint is not used, then there may be measurements selected by multiple tracks.

Hungarian mathematicians, Konig and Egervary. In 1957, Munkres published a refinement of this algorithm [25] showing that the computational counts will at most be equal to $(11n^3 + 12n^2 + 31n)/6$ (when $n = m$) thus representing a substantial savings in computation for large n. This algorithm has been known as the Kuhn–Munkres algorithm or the Munkres assignment algorithm. Interested readers can consult the cited references for a description of this algorithm.

The optimal solution for the one-scan assignment problem defined above is referred to as immediate resolution, or decision with a single scan of measurement. In the MHT formulation to be introduced in the next chapter, each possible selection from the assignment matrix represents a track. The total number of possible combinations in a single scan grows larger than $n!/(n − m)!$ for $n \geq m$ or $m!/(m − n)!$ for $m \geq n$, as will be shown in the next chapter. When such a large number of tracks are propagated forward to associate with the measurements of the next scan, the number of tracks grows combinatorially. The reason for maintaining a large number of tracks is that the best assignment in any single scan is not necessarily the correct choice. The immediate resolution method and the multiple-scan resolution method are analogous to filtering versus smoothing in state estimation.

8.6   The Probabilistic Data Association Filter and the Joint Probabilistic Data Association Filter

The PDAF and JPDAF approaches are due to Bar-Shalom et al. [see 3, 5, 16–17]. Reference [17] is a more recent review and tutorial article. The concepts of PDAF and JPDAF are illustrated in Figure 8.5.

Probabilistic Data Association Filter

The assumptions of PDAF are listed below.

Assumptions:

1. Each track is processed independently (ignoring the existence of other tracks even if some measurements may be shared).

2. Only one of the detections in the AG is from the true target associated with $\hat{\mathbf{y}}_{k|k-1}$. All others come from receiver noise or clutter.[13] Let N_g denote the number of

---

[13]   Clutter is referred to as extra detections from environmental effects such as waves, trees, buildings, and so on. It can also be detections from the same targets. Athans et al. [6] considered the case where the trailing wake of a target may generate multiple detections, all of which are associated with the target of interest. They applied the PDAF for tracking. This phenomenology is similar to the case when a target's physical size is bigger than sensor resolution, and then multiple detections can be originated from the same target.

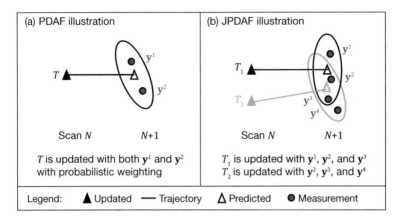

Figure 8.5   Illustration of PDAF and JPDAF.

detections in AG. All detections within AG are to be used for filter update because one cannot discern which one is true due to the fact that $P_d < 1$ and $P_{fa} > 0$.

3.   Detections from clutter, to be referred to as false alarms (FA), are assumed to be independent identically distributed (i.i.d.) with uniform spatial distribution. The number of FA detections could be described by (a) a Poisson distribution with known density λ; or (b) a diffuse prior that is equal to the number of detections minus one in AG divided by the volume of the AG. In either case, it will be denoted as δ_{FT}.

4.   The past of the tracks is summarized by an approximate sufficient statistics to the current scan. This is referred to as the first order generalized pseudo-Bayesian (GPB1), that is, only dependent on one scan of measurement. As stated in Section 8.1.2, the residual process is Gaussian, $\gamma_k \sim N(\mathbf{0}, \boldsymbol{\Gamma}_k)$, where $\gamma_k = \mathbf{y}_k - \hat{\mathbf{y}}_{k|k-1}$ with covariance $\boldsymbol{\Gamma}_k = \left[\mathbf{H}_k \mathbf{P}_{k|k-1} \mathbf{H}_k^T + \mathbf{R}_k \right]$. This condition applies to all measurements within the measurement AG, defined by Equation (8.1-1).

Some terms and relations are first defined before describing the PDAF algorithm. Let $\hat{\mathbf{x}}_{k|k-1}$ and $\hat{\mathbf{y}}_{k|k-1}$ denote predicted state and measurement, respectively, for scan k and $\mathbf{Y}_k$, the collection of all measurements in scan k contained in the AG. Using the same notation as in Equation (8.2-1), the measurements in $\mathbf{Y}_k$ are denoted as $\left\{ \mathbf{y}_k^1, \mathbf{y}_k^2, \ldots, \mathbf{y}_k^{N_g} \right\}$. Each $\mathbf{y}_k^i$ has a corresponding residual process $\gamma_k^i = \mathbf{y}_k^i - \hat{\mathbf{y}}_{k|k-1}$ with covariance $\boldsymbol{\Gamma}_k$. Then the state estimate updated with γ_k^i is denoted as $\hat{\mathbf{x}}_{k|k}^i$ with covariance $\mathbf{P}_{k|k}^c = \mathbf{P}_{k|k-1} - \mathbf{P}_{k|k-1} \mathbf{H}_k^T \boldsymbol{\Gamma}_k^{-1} \mathbf{H}_k \mathbf{P}_{k|k-1}$. Let A_i denote the event that measurement $\mathbf{y}_k^i$ is associated with $\hat{\mathbf{y}}_{k|k-1}$, A_0 denote the event that none of these measurements in AG is from the target, and A the collection of all association events related to $\hat{\mathbf{y}}_{k|k-1}$. Let β_k^i denote the probability of A_i given all measurements at the kth scan, that is, $\beta_k^i \triangleq \Pr(A_i|\mathbf{Y}_k)$. Then the conditional mean estimate of $\mathbf{x}_k$ using all measurements in AG is

$$\hat{\mathbf{x}}_{k|k} = E\{\mathbf{x}_k|\mathbf{Y}_k\} = E\{E\{\mathbf{x}_k|\mathbf{Y}_k, A\}|\mathbf{Y}_k\} = \sum_{A_i \in A} E\{\mathbf{x}_k|\mathbf{Y}_k, A_i\}\Pr\{A_i|\mathbf{Y}_k\} = \sum_{i=0}^{N_g} \beta_k^i \hat{\mathbf{x}}_{k|k}^i,$$
(8.6-1)

with $\hat{\mathbf{x}}_{k|k}^0 = \hat{\mathbf{x}}_{k|k-1}$. The covariance of $\hat{\mathbf{x}}_{k|k}$ is

$$\mathbf{P}_{k|k} = \sum_{i=0}^{N_g} \beta_k^i \left[\mathbf{P}_{k|k}^i + \left(\hat{\mathbf{x}}_{k|k}^i - \hat{\mathbf{x}}_{k|k}\right)\left(\hat{\mathbf{x}}_{k|k}^i - \hat{\mathbf{x}}_{k|k}\right)^T \right].$$
(8.6-2)

Note the similarity of the above equations to that of the MMEA with the constant model case.

Data Association Probability

The association probability is used as weights for filter update. In addition to using the residual process, or the likelihood function of a measurement, the probability of detection and of false alarms are also used for the association probability computation. Assume that there are N_g measurements in the AG as defined in Section 8.1.2. The AG threshold $\mathcal{T}$ is selected such that the probability that the true target will be contained in AG is P_G, the gate probability. The association probabilities β_k^i for all measurements are

$$\beta_k^i = \frac{\Lambda^i}{1 - P_d P_G + \sum_{j=1}^n \Lambda^j}, \text{ for } i = 1, \dots, N_g$$
(8.6-3)

and

$$\beta_k^0 = \frac{1 - P_d P_G}{1 - P_d P_G + \sum_{j=1}^n \Lambda^j}$$

where β_k^0 is the probability that the true target is not contained in the AG. Note that when $P_d = P_G = 1$, then $\beta_k^0 = 0$, which means that it is impossible for the true target not to be included in the measurement set. The likelihood ratio Λ^i of the hypothesis that the ith measurement being true to the corresponding hypothesis of being false is

$$\Lambda^i = \frac{f(\gamma_k^i) P_d}{\delta_{\text{FT}}},$$
(8.6-4)

where $f\left(\mathbf{\gamma}_k^i\right) = \frac{1}{c|\Gamma_k|}\exp\left\{-\frac{1}{2}\mathbf{\gamma}_k^{iT}\mathbf{\Gamma}^{i^{-1}}\mathbf{\gamma}_k^i\right\}$ is the density function of the residual process $\mathbf{\gamma}_k^i$ and δ_{FT} is the density of false track. This is either the density of the assumed Poisson process or the total number of detections minus one in the AG divided by the volume of the AG. Their impact to the change in β_k^i value is small when both P_d and P_G are close to one. For derivations see [16, 17].

Probabilistic Data Association Filter Algorithm

There is no difference on the prediction equations for PDAF from the regular KF/EKF algorithms. The updated equations can be derived after some algebraic manipulations starting with Equations (8.6-1) and (8.6-2). The final algorithms presented in [16, 17] are shown below.

The state update equation is

$$\hat{\mathbf{x}}_{k|k} = \hat{\mathbf{x}}_{k|k-1} + \mathbf{K}_k\mathbf{\gamma}_k. \tag{8.6-5}$$

Using the weighted sum of residuals

$$\mathbf{\gamma}_k = \sum_{i=1}^{N_g}\beta_k^i\left(\mathbf{y}_k^i - \hat{\mathbf{y}}_{k|k-1}\right). \tag{8.6-6}$$

Filter gain $\mathbf{K}_k$ is independent of which measurement is used

$$\mathbf{K}_k = \mathbf{P}_{k|k-1}\mathbf{H}_k^T\mathbf{\Gamma}_k^{-1}, \tag{8.6-7}$$

while the covariance equation is

$$\mathbf{P}_{k|k} = \beta_k^0\mathbf{P}_{k|k-1} + \left(1 - \beta_k^0\right)\mathbf{P}_{k|k}^c + \tilde{\mathbf{P}}_{k|k}, \tag{8.6-8}$$

where

$$\mathbf{P}_{k|k}^c = \mathbf{P}_{k|k-1} - \mathbf{P}_{k|k-1}\mathbf{H}_k^T\mathbf{\Gamma}_k^{-1}\mathbf{H}_k\mathbf{P}_{k|k-1}, \tag{8.6-9}$$

and

$$\tilde{\mathbf{P}}_{k|k} = \mathbf{K}_k\left[\left[\sum_{i=1}^{N_g}\beta_k^i\mathbf{\gamma}_k^i\mathbf{\gamma}_k^{iT}\right] - \mathbf{\gamma}_k\mathbf{\gamma}_k^T\right]\mathbf{K}_k^T. \tag{8.6-10}$$

The covariance for the PDAF state estimate is Equation (8.6-8). Note that the probability of the result that none of the measurements is correct is represented by β_k^0. When this is true, the predicted covariance comes through in Equation (8.6-8) weighted by the corresponding probability. When the corresponding measurements are used, the normal KF update equation comes through and is jointly weighted by $1 - \beta_k^0$. The sample data covariance is represented by $\tilde{\mathbf{P}}_{k|k}$. Because no measurement in the AG is discarded, the state is likely being updated with the true target measurement. The inclusion of all measurements increases the covariance of the updated state estimate, Equation (8.6-8), making the covariance more likely to be consistent with the true error.

Joint Probabilistic Data Association Filter

JPDAF is an extension of the PDAF in which multiple measurements appear in the overlapped AGs. Tracks with shared measurements are to be considered jointly; see illustration in Figure 8.5(b). For notational simplicity, the time index k will be omitted in the following discussion.

Assumptions Pertaining to JPDAF

In addition to assumptions to PDAF, the following apply to JPDAF.

1. Measurements from a target can fall in the AG of a neighboring target.

2. The measurement-to-target association probabilities are computed jointly across all targets in the AG.

3. The state estimate is updated either separately for each target as in PDAF, that is, in a decoupled manner, resulting in a JPDAF, or in a coupled manner using a stacked state vector, resulting in a joint probabilistic data association coupled filter (JPDACF). In this discussion, only JPDAF will be shown.

Joint Association Event

We define a joint association event, A, using the intersection operator $\cap$,

$$A = \bigcap_{j=1}^{m} A^{j,t_j}, \tag{8.6-11}$$

where A^{j,t_j} is the event that measurement j is originated from target t; where $j = 1, \ldots, m$ and $t = 0, 1, \ldots, n$; where $t=0$ means for none of the targets; and t_j is the index of that target to which measurement j is associated in the event A^{j,t_j}; and the total number of targets is n. On the other hand, $A^{j,t_j} = \bigcup_{A^{j,t_j} \in A} A$.

We define an event matrix, $\Omega = [\omega_{j,t}]$ for $j = 1, \ldots, m$ and $t = 0, 1, \ldots, n$ where $\omega_{j,t} = 1$ indicates $A^{j,t}$ is a valid event and $\omega_{j,t} = 0$ otherwise. Not every event matrix is validated within a given scenario. $\Omega = [\omega_{j,t}]$ is referred to as the validation matrix. Consider the case illustrated in Figure 8.5(b) with two tracks and four measurements in which two measurements are in the overlapped AG of these two tracks. In this case $m = 4$ and $n = 2$. The event matrix that is validated for the case shown in Figure 8.5(b) is

$$\Omega = \begin{bmatrix} 1 & 1 & 0 \\ 1 & 1 & 1 \\ 1 & 1 & 1 \\ 1 & 0 & 1 \end{bmatrix}.$$

Targets are listed as columns with first column corresponding to $t = 0$, which means none of the targets is assigned to this measurement, thus receiving all 1's for the column.

Furthermore, let us define a feasible event matrix, A, such that

$$\hat{\Omega}(A) = [\hat{\omega}_{j,t}(A)],$$

where $\hat{\omega}_{j,t}(A) = 1$, if $A^{j,t} \in A$ and $\hat{\omega}_{j,t}(A) = 0$ otherwise, and such that it satisfies the following constraints

(a) a measurement can only have one source, that is, for each measurement the sum of $\hat{\omega}_{j,t}(A)$ over all targets (including $t = 0$) is one, and

(b) at most one measurement can originate from a target, that is, for each track, the sum of $\hat{\omega}_{j,t}(A)$ over all measurement is less than or equal to one.

The summation of $\hat{\omega}_{j,t}(A)$ across all measurements is a target detection indicator,

$$\delta_t(A) \triangleq \sum_{j=1}^{m} \hat{\omega}_{j,t}(A)$$

with its counterpart as the measurement association indicator

$$\tau_j(A) \triangleq \sum_{t=1}^{n} \hat{\omega}_{j,t}(A).$$

With this definition, the number of false (i.e., those that are unassociated) measurements in event A is

$$\phi(A) = \sum_{j=1}^{m} (1 - \tau_j(A)).$$

A feasible event matrix $\hat{\Omega}(A) = [\hat{\omega}_{j,t}(A)]$ has the same meaning as a feasible assignment defined in Section 8.2 and used in Section 8.5. The entries of $\hat{\Omega}(A) = [\hat{\omega}_{j,t}(A)]$ are zeros and ones with only one "1" for each row and column, which means each measurement is assigned to only one track in a feasible assignment.

The Association Probability

The association probability for a given target t, $\beta^{j,t}$ with all its measurements in the AG is

$$\beta^{j,t} \triangleq \Pr\{A^{j,t}|\mathbf{Y}_k\} = \sum_A \Pr\{A|\mathbf{Y}_k\}\hat{\omega}_{j,t}. \qquad (8.6\text{-}12)$$

The computation of $\Pr\{A^{j,t}|\mathbf{Y}_k\}$ involves using residual densities of all measurements in target t's AG, the sensor's detection probability P_d and miss detection probability $(1 - P_d)$ raised to the power of number of detections, and the combinatorics of false detections. Similar to the PDAF, there are two models for the false measurements. The number of the false alarms, ϕ, is assumed to follow a Poisson distribution

$$\mu_F(\phi) = e^{-\lambda V}\frac{(\lambda V)^{\phi}}{\phi!},$$

where λ is the spatial density of false alarm and V is the volume of AG. Following the derivation of [16] we obtain the following expression

$$\Pr\{A|\mathbf{Y}_k\} = \frac{1}{c_p}\prod_j\left\{\lambda^{-1}f(\boldsymbol{\gamma}^{j,t_j})\right\}^{\tau_j}\prod_t P_d^{\delta_t}(1 - P_d)^{1-\delta_t}, \qquad (8.6\text{-}13)$$

where $f(\boldsymbol{\gamma}^{j,t_j}) \propto \exp\left\{-\frac{1}{2}(\mathbf{y}_k^j - \mathbf{y}_{k|k-1}^t)^T\boldsymbol{\Gamma}_k^{j,t^{-1}}(\mathbf{y}_k^j - \mathbf{y}_{k|k-1}^t)\right\}$ is a Gaussian density and c_p is the normalization constant for the Poisson model. When the false alarms are represented as a diffuse prior for the number of false alarms, ϕ over the volume of AG, the above equation is modified to

$$\Pr\{A|\mathbf{Y}_k\} = \frac{1}{c_d}\phi!\prod_j\left\{Vf(\boldsymbol{\gamma}^{j,t_j})\right\}^{\tau_j}\prod_t P_d^{\delta_t}(1 - P_d)^{1-\delta_t}, \qquad (8.6\text{-}14)$$

where c_d is the normalization constant for the diffuse prior mode. The $\Pr\{A|\mathbf{Y}_k\}$ of either Equation (8.6-13) or Equation (8.6-14) is used to compute $\beta^{j,t}$ in Equation (8.6-12).

For a given t, $\beta^{j,t}$ is then used to compute updated state estimate and covariance, in the same way as in Equations (8.6-5) through (8.6-10). It will therefore not be repeated here.

The JPDAF computation procedure is outlined below:

1. For a given scenario, form its validation matrix $\Omega = [\omega_{j,t}]$.

2. Use Ω to enumerate all feasible association events. For each feasible event, A, there is an event matrix, $\hat{\Omega}(A) = [\hat{\omega}_{j,t}(A)]$.

3. Compute $\Pr\{A|\mathbf{Y}_k\}$ using either Equation (8.6-13) or (8.6-14) for each feasible association event.

4. For each association pair (j, t), sum up those $\Pr\{A|\mathbf{Y}_k\}$ associated with $\hat{\omega}_{j,t}(A)=1$, and this is $\beta^{j,t}$.

5. For a given t, $\beta^{j,t}$ is used to compute updated state estimate and covariance, in the same way as in Equations (8.6-5) through (8.6-10).

Illustration for $\beta^{j,t}$ computation:

1. The validation matrix for the case shown in Figure 8.5(b) is $\Omega = \begin{bmatrix} 1 & 1 & 0 \\ 1 & 1 & 1 \\ 1 & 1 & 1 \\ 1 & 0 & 1 \end{bmatrix}$.

2. The feasible association events for measurement 1 with target 1 are:

$$\begin{bmatrix} 0 & 1 & 0 \\ 0 & 0 & 1 \\ 1 & 0 & 0 \\ 1 & 0 & 0 \end{bmatrix}, \begin{bmatrix} 0 & 1 & 0 \\ 1 & 0 & 0 \\ 0 & 0 & 1 \\ 1 & 0 & 0 \end{bmatrix}, \begin{bmatrix} 0 & 1 & 0 \\ 1 & 0 & 0 \\ 1 & 0 & 0 \\ 0 & 0 & 1 \end{bmatrix}, \begin{bmatrix} 0 & 1 & 0 \\ 1 & 0 & 0 \\ 1 & 0 & 0 \\ 1 & 0 & 0 \end{bmatrix}.$$

3. Compute $\Pr\{A|\mathbf{Y}_k\}$ for each association event listed above; their sum is $\beta^{1,1}$.

4. Follow the same procedure for the rest of $\beta^{j,t}$.

Remarks

1. PDAF and JPDAF treat extra detections as "clutters," which is not the same assumption as the case of a multiple, dense target environment.

2. PDAF and JPDAF use all detections in AG for updates, thus preventing track initiation using detections in the AG.

3. The feasible association event is the same as the feasible solution for the assignment problem defined in Section 8.5.

   o In an assignment problem, only one feasible solution is chosen, and thus it is a "hard" decision.

   o In PDAF and JPDAF, all feasible association events are generated with their association probabilities used to update the state and covariance, and thus it is a "soft" decision.

4. The MHT algorithms, which will be explored in the next chapter, will also enumerate all possible solutions. The JPDAF assumes that extra measurements are clutters (or false detections) while MHT treats all measurements as track continuation, new detection, or false alarm.

5. Weighted combination of all detection in the AG is not used in MHT.

8.7   A Practical Set of Algorithms

In this section, some engineering solutions to the MTT problem are summarized. They are presented in terms of a set of approaches or algorithms representing a compilation of the author's experience in solving practical MTT problems [18–21]. Two cascaded processes, namely, track initiation and track continuation, are used. As noted by their names, the first one is a process to get tracks started while the second one is to maintain tracks. The division between these two is somewhat arbitrary. In this text, the following definitions for initiation process and continuation process are used.

Initiation Process: A quick process using simple dynamic models to connect measurements from successive scans in order to start up tracks. Whenever possible, a linear target motion model in measurement space is used. The goal is to minimize missed tracks at the expense of a higher false track rate.

Continuation Process: Extending the tracks formed in the previous step using more precise dynamic models and more sophisticated filter algorithms. The objective for track continuation is to obtain a more accurate state estimate while maintaining low leakage and false track rates. More accurate dynamic models will result in better accuracy (smaller covariance, or AG), and reduce the probability of false association.

8.7.1   Initiation Process

Consider an initiation process, that is, a start-up process, in which the sensor is first turned on and/or when new targets move into the sensor field of view or begin to become detectable by a sensor. This process employs a simple target motion model, such as a polynomial fit to the observations (first or second order). Particular applications, in terms of sensor measurement accuracy and target density, will determine the time to convergence for this process before switching to a continuation (track maintenance mode) process. Consider, for example, an M out of N approach, where if a track is persistent for greater than or equal to M out of N scans, the initiation is complete and is ready to be moved to the continuation stage. What is suggested here is more of an engineering approach than rigorous mathematics. It represents the author's experience in solving similar problems in practical applications. The process is illustrated with the following example.

Process Illustration

The initiation process example used here is for $M = 6$ and $N = 7$. It is illustrated via a sequence of figures. Figure 8.6 presents a 1D (position) measurement space (vertical

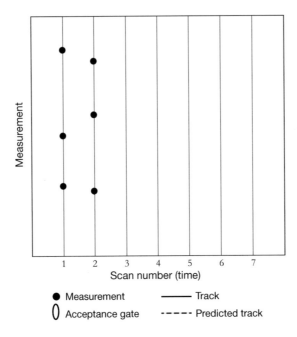

Figure 8.6   First two scans of measurement.

axis) versus time (horizontal axis). The black dots represent measurements, and the parallel vertical lines denote measurement scan times. Two scans of measurements are shown.

The first step is to connect all measurements of the first scan with those of the second scan subject to a maximum velocity constraint, as shown in Figure 8.7. Solid lines, from left to right as time advances, connect measurements that associate with a potentially viable track. Note that each measurement of the first scan is connected to the closest two measurements of the second scan, with the assumption that the targets can only travel so far during the scan time.

Dashed lines denote potential tracks, as can be seen in Figure 8.8. At this stage of the initiation process in which tracks are formed for all possible measurement-to-measurement connections (subject to constraints) they are referred to as tentative tracks. A model such as constant velocity (CV) can be used to compute the predicted measurement location. The covariance of the predicted measurement is computed using the same model (see Homework Problem 15 in Chapter 1, derivation of estimator and covariance for polynomial models). Specific equations for a dish-radar coordinate and a radar centered Cartesian state coordinate are shown in Appendix 8.A.

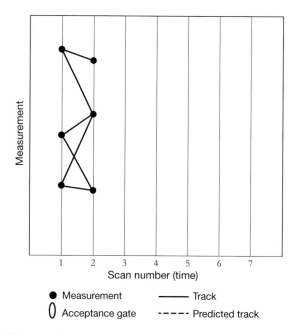

Figure 8.7   Possible associations of the first two scans.

Narrow ellipses are used to denote the new measurement acceptance gates (or regions) around the predicted measurement location. The use of the predicted covariance to form acceptance gates is discussed in Section 8.1.2. Figure 8.8 shows a total of six partially overlapping acceptance gates. It is clear that this process has enumerated more possibilities than number of targets so that some of the gates may not contain any new measurements, as shown in Figure 8.9. Three new measurements are found in the third scan. There is only one new measurement that falls unambiguously in one acceptance gate. The other measurements will have to be shared by two tracks, as illustrated in Figure 8.10.

The same process is repeated when moving to Scan 4 as shown in Figure 8.11. Again there are three measurements in Scan 4. One measurement falls completely outside of any acceptance region, and becomes a candidate for a new track. One measurement is again shared by two tracks, thus two tracks will continue with the same measurement (track split). The track that is moving upward has no measurement in its acceptance region. A track like this will be kept and continued to the next scan to account for the possibility of a missed detection. Notice that the track that is moving downward is moving out of the measurement field-of-view, and in this case is not

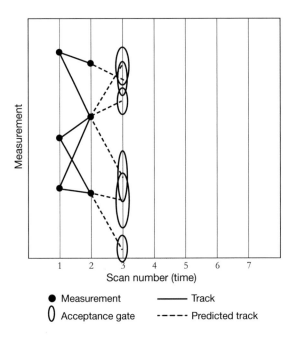

Figure 8.8 Prediction to the third scan.

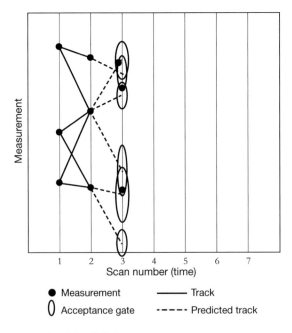

Figure 8.9 Measurements of the third scan.

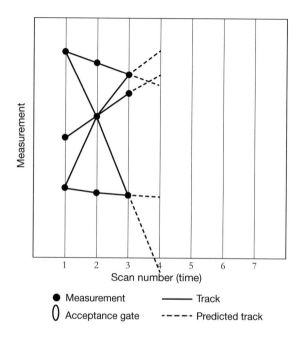

Figure 8.10   Accepting measurements, allowing for shared measurements.

continued as the responsibility of this sensor. The resulted tracks with predictions to the fifth scan are illustrated in Figure 8.12. Also shown are the new measurements of this scan.

As shown in Figure 8.12, the acceptance region for the track with one missing measurement is bigger due to longer prediction time. In Scan 5, two acceptance gates do not contain a target. The track with two missing measurements in a row (the one that is going upward) will be dropped. The measurement that does not fall into any acceptance gate is connected with the new detection found in Scan 4.

This process continues and moves on to the sixth scan, as shown in Figure 8.13. Note that the track with a missing measurement in a scan has a larger acceptance gate. In this scan all new measurements fall unambiguously in an acceptance region. The tracks are thus continuing into the seventh scan as shown in Figure 8.14.

The initiation process described in Figures 8.6 through 8.14 is summarized in the flow diagram in Figure 8.15. The first two blocks represent the tentative track formation with two scans of data. The third block with the decision block on its right-hand side illustrates the decision on whether any measurements fall within the acceptance region. If yes, a track is updated. If no and the miss is consecutive, a track is dropped.

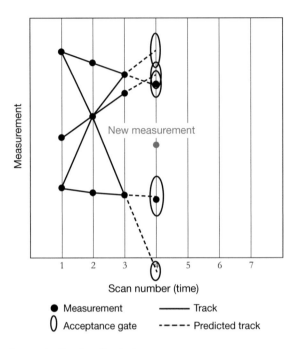

Figure 8.11   Moving on to the fourth scan.

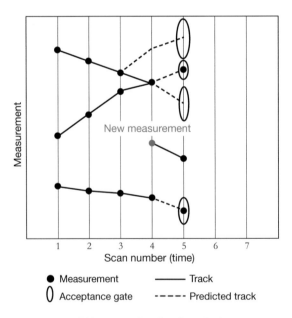

Figure 8.12   Continuing to the fifth scan, allowing for missing measurements and new tracks in the previous scan to form a new track.

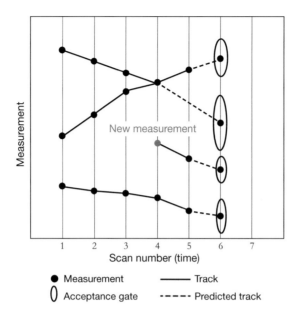

Figure 8.13   Continuing to the sixth scan.

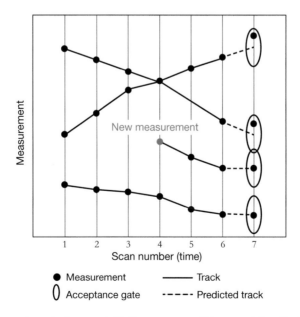

Figure 8.14   For a six out of seven initiation process, this completes the track initiation for some targets.

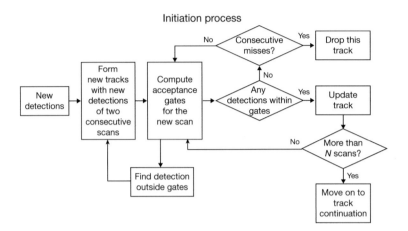

Figure 8.15   Initiation process flow diagram.

For nonconsecutive misses, the track goes back with another cycle of prediction and checking. If a track has been persistent for M through N scans, it is ready to be moved on to the next stage: track continuation. When new measurements fall outside of all acceptance gates, they become candidates for new tracks.

An explicit set of equations of estimation with polynomial dynamics are shown in Appendix 8.A. These equations can be used for filter initial condition computation. When the initiation process is completed, the states in the measurement space are transformed to radar-centered Cartesian coordinates. The transformation equations are also given in Appendix 8.A. The estimate covariance is used for the acceptance region calculation, as was shown in Section 8.1.2.

8.7.2   Continuation Process

Tracks that have met the M out of N criteria in the initiation process are ready to be moved on to the continuation process. Since the choice of M and N are problem dependent, there is obviously no clear line between initiation and continuation. Tracks moved into continuation use more precise dynamic models and more sophisticated filter algorithms. State estimates generated by the continuation process are more accurate with smaller covariance, and consequently have a lower probability of associating measurements from other targets.

The progression of an unambiguous track is illustrated in Figure 8.16. The solid triangle represents an updated track state in the measurement space. It is propagated to the next scan ($N + 1$) to provide a predicted measurement location, represented by an open triangle. The acceptance gate (or region) is illustrated by an ellipse around the

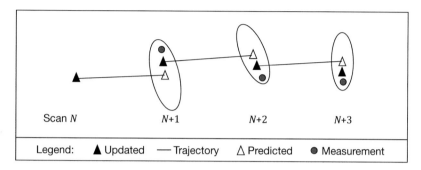

Figure 8.16   An unambiguous track.

predicted measurement (open triangle). A new measurement, represented by the red dot, falls within the acceptance gate. The mathematics developed in Section 8.1.2, the acceptance gate, is used to test whether a measurement is accepted for track continuation processing. The acceptance gate as described in Equation (8.1-1) is determined by both the estimation and measurement covariances. It is not explicitly shown in the figure for the purpose of making the illustration easier to read. Since in this case there is only one new measurement within the gate, the association is unambiguous and the new measurement is used to update the filter. The process continues as shown. Because there are no competing tracks or measurements in this case, it represents the simplest track continuation case, referred to as an unambiguous track. In the following sections, a set of figures is used to illustrate the logic for measurement association and ambiguity resolution when observations from multiple targets fall within the same acceptance gate.

Track Split

Next, consider the case when multiple measurements fall within the acceptance region. This is illustrated in Figure 8.17. Two measurements are contained within the acceptance gate in scan $N + 2$. One choice is to associate each measurement with the track and propagate separate track files for each of the new measurements. This is known as a track split. This may be a reasonable choice if the separation of the new measurements is small relative to sensor resolution. In this case they may have come from two unresolved objects in the previous scans. For example, both objects may be traveling in a certain formation with their separation smaller than the sensor resolution. During this period, a sensor can only detect one object. When these two objects begin to separate, their trajectories move away from each other, and they become resolved by the sensor and multiple detections result. Track split logic fits well with such a scenario.

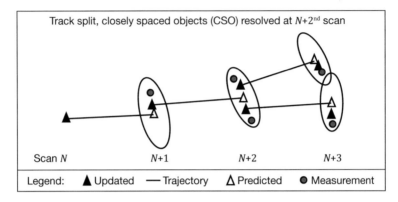

Figure 8.17   Track split.

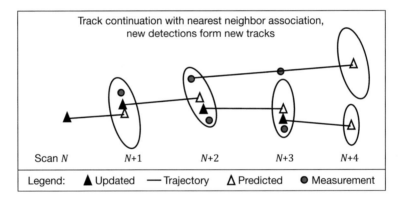

Figure 8.18   Association using nearest neighbor rule and forming new tracks for un-selected measurement(s) within the acceptance gate.

An alternative approach to this scenario is illustrated in Figure 8.18. If one chooses to accept only one measurement for a track, then the NN algorithm described in Section 8.5 can be applied. With one measurement selected for the update, the extra measurement becomes a candidate for track initiation (see Section 8.3.1) that is then linked with an unassociated measurement of the next scan to form a new track.

Ambiguity Resolution

Now consider the case where multiple tracks compete for multiple measurements. Figure 8.19 shows two tracks that have overlapping predicted acceptance gates at Scan

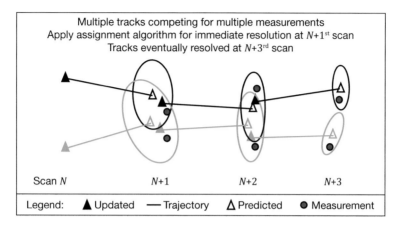

Figure showing: Multiple tracks competing for multiple measurements. Apply assignment algorithm for immediate resolution at $N+1^{st}$ scan. Tracks eventually resolved at $N+3^{rd}$ scan. With axis labels Scan N, $N+1$, $N+2$, $N+3$. Legend: ▲ Updated — Trajectory △ Predicted ● Measurement

Figure 8.19   Multiple tracks with multiple measurements, immediate resolution.

$N + 1$ and two new measurements that fall within the intersection of the two acceptance gates, resulting in an ambiguous situation. If the ambiguity must be resolved at the current time scan, it is termed an immediate resolution. If the resolution decision can be deferred, resulting in track splits (i.e., multiple tentative assignments) at this scan, it is referred to as delayed resolution. Delayed resolution allows for temporary false tracks, and uses target dynamics along with future measurements to jointly resolve the ambiguity that occurred in the previous scan. This is illustrated in Figure 8.20. The delayed resolution is similar to the multiple hypothesis decision case for three-scan resolution of measurements. The immediate resolution is the same as two-scan resolution. The delayed resolution has a higher chance of containing the correct answer because it enumerates more possibilities. The disadvantage of delayed resolution is slower reaction time for a practical system (delayed decision) and a higher computational burden for carrying multiple track hypotheses.

The association logic discussed above is summarized in Figure 8.21. Note that (a) the unambiguous case is the simplest case where only one measurement is contained within the acceptance region of a single track, (b) employs a track split for multiple measurements associating with a single track, (c) shows immediate resolution for the single track case, forcing the track to select one measurement from multiple associated measurements using the NN rule, (d) shows immediate resolution for multiple tracks with multiple measurements, using the global nearest neighbor rule (i.e., assignment algorithm to uniquely assign measurement-to-track file), and (e) depicts delayed resolution for multiple tracks and multiple measurements.

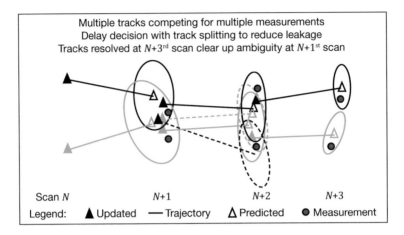

Figure 8.20   Multiple tracks with multiple measurements, delayed resolution.

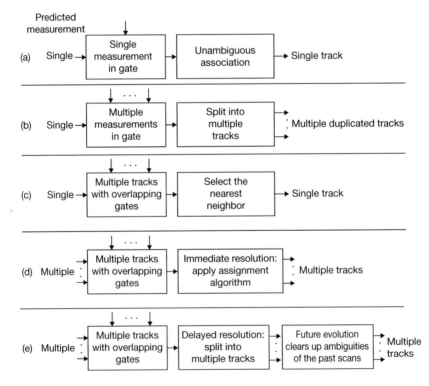

Figure 8.21   A menu of association logics.

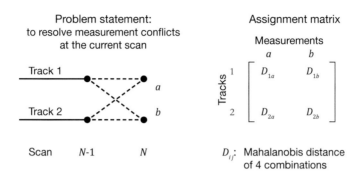

Figure 8.22   Illustration of immediate resolution method.

8.7.3   Illustration of Immediate and Delayed Resolution

In this section, the ambiguity resolution methods are further illustrated, Ref. [18]. In Figure 8.22, two track files with measurements up to the $N - 1$st scan are propagated to the Nth scan and discover that both measurements a and b fall within the joint acceptance region. An assignment matrix is formed to resolve this ambiguity immediately, as shown in Figure 8.22. For example, $D_{1,a}$ is the Mahalanobis distance between the predicted measurement of the first track and measurement a. The optimum assignment consists of those pairs of D_{ij}'s that give the minimum sum, Equation (8.5-1), providing the immediate resolution.

Figure 8.23 illustrates the situation if one simply splits both tracks and propagates to the $N + 1$ scan. If the two new measurements c and d fall within the intersection of all acceptance regions, there is a total of eight possible tracks but only two will be selected. The optimum assignment with two new scans of measurement is to form a two-layered matrix of a posteriori hypothesis probabilities with dimension $2 \times 2 \times 2$, which is an expansion of Equation (8.5-1) into three dimensions. The resulting optimization problem is now a 3D version of Equation (8.5-1). The optimum solution consists of the admissible pairs (a measurement at a given time can only be assigned to one track) such that the sum of the entries is minimized. Due to the fact that the computation of finding the minimum of the expanded Equation (8.5-1) with a large number of scans is tedious when the number of tracks and measurements are large, an alternative procedure for a three-scan problem, referred to as the one-scan delayed resolution technique, is suggested, and is illustrated in Figure 8.23.

There are two possible tracks that include track 1 and measurement a of the Nth scan, namely (*1,a,c*) and (*1,a,d*). The track with smaller residual (or larger a posteriori

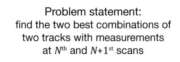

Problem statement:
find the two best combinations of
two tracks with measurements
at N^{th} and $N+1^{st}$ scans

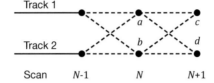

Two-layered assignment matrix

For tracks with measurement c For tracks with measurement d

Measurements   Measurements

$$\text{Tracks}\begin{array}{c}1\\2\end{array}\begin{array}{cc}a & b\\ \left[\begin{array}{cc}Pr_{1ac} & Pr_{1bc}\\ Pr_{2ac} & Pr_{2bc}\end{array}\right]\end{array}\qquad \text{Tracks}\begin{array}{c}1\\2\end{array}\begin{array}{cc}a & b\\ \left[\begin{array}{cc}Pr_{1ad} & Pr_{1bd}\\ Pr_{2ad} & Pr_{2bd}\end{array}\right]\end{array}$$

Pr_{ijk}: the a posteriori hypothesis probability of i going through
measurement j of Nth scan and end at measurement k of
the $N+1^{st}$ scan

Figure 8.23   Illustration of ambiguity resolution with two scans of measurements.

probability) is chosen. This procedure is repeated for all tracks and all measurements of Nth scan. Note that for illustration purposes, the chosen tracks in Figure 8.24 are $(1,a,c)$, $(1,b,d)$, $(2,a,c)$, and $(2,b,d)$. The residuals of these four tracks are entries of the assignment matrix on the right. The final set of tracks selected is those feasible pairs that achieve the minimum sum.

The one-scan delayed resolution is an attempt to reduce a three-scan assignment problem to a 2D assignment problem. The suggested procedure is suboptimal but is computationally straightforward and gives improved performance relative to the immediate resolution method, as demonstrated by the numerical example in the next section. The one-scan deferred resolution case is similar in concept to the one-step lagged fixed-lag smoothing algorithm used in estimation.

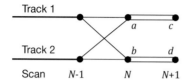
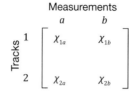

Problem statement:
to resolve measurements of
N^{th} scan with the help of
measurements at $N+1^{st}$ scan

Assignment matrix

χ_{ij}: the smallest residual of track i
going through measurement j
of Nth scan and end at $N+1^{st}$
scan

Figure 8.24   Illustration of a one-scan delayed ambiguity resolution.

8.7.4   A Joint Multiscan Estimation and Decision Process

If past measurement scans are available for reassignment and the application problem of interest can afford the time-delay incurred with using past data, then one may consider reprocessing and reselecting past measurements to improve upon the accuracy of a track file (i.e., to increase the percentage of the measurements of a track originated from the same object). This approach is analogous to a fixed-interval smoother (FIS) for estimation. An iterative process for implementing this concept, the joint multiscan estimation and decision algorithm (JMSEDA), is suggested in Figure 8.25. Measurements in initial track files are first obtained using the algorithm suggested in Sections 8.7.1 through 8.7.3. They are further processed by a batch filter (either weighted least squares, WLS, or an FIS, see Chapter 2) to obtain smoothed state estimates having improved and smaller covariances. Some of the previously accepted measurements may now lay outside of the (smaller) n-covariance criterion where n is a choice of the algorithm designer for a particular application. Those measurements can be rejected and new measurements can be reselected from the stored measurement data until no more measurements are found outside of the n-covariance region. This algorithm is used in a numerical example later to show that it further reduces correlation error. The shortcoming of this algorithm is the time-delayed decision, as is the case in any multiple hypothesis algorithms employing past data to achieve a final decision. Exactly which choice to make is clearly application dependent, and the decision of the system designers.

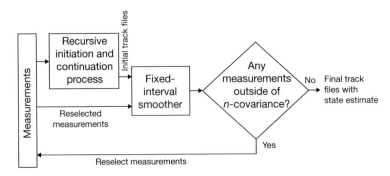

Figure 8.25 An iterative algorithm for joint estimation and correlation with batch processing.

Remarks

The suggested JMSEDA is an approximation of the MDA approach depicted in Section 8.2 to improve both the state estimates and the correct correlation rate. If time delay is not a concern, this is a recommended approach.

8.8   Numerical Examples

In this section, the performance of the algorithm described above is illustrated with two examples:

Example 1

The purpose of this example [18] is to illustrate the progressive improvement of the three ambiguity resolution methods described in Section 8.7, namely immediate resolution, one-scan deferred resolution, and JMSEDA (referred to as batch processing for short). A multiple object trajectory simulation and tracking program is used for this comparison. All three ambiguity resolution methods described above were implemented and run on the same set of data. In a dense target environment, targets in sensor field of view are often unresolved, and are referred to as closely spaced objects (CSOs). A functional model characterizing and simulating the CSO problem [shown in 26] was used to model the unresolved target situations. Trajectories of a large number of objects traveling in space (using an exoatmospheric motion model) are simulated and observed by a sensor over multiple scans. The track initiation and continuation processes described in Sections 8.7.1 and 8.7.2 were applied to produce track files.

The scoring method to be described is for simulation studies where the ground truth is known. This method is often used for algorithm comparison during a design process and is not suitable for real-time application where the ground truth is not known.

A subset of track files after exercising the multiple object simulation and tracking program is shown in Figure 8.26. The entries in this table are the true target identification number (ID). The top row gives the measurement scan number. The left column gives the track file (TF) identification number. Only 12 scans of measurements are shown for this example. TF 100 contains Target 20 for the 12 scans as shown. This track scores with a 100% consistency because all entries in the track are from the same target. TF 101 contains Target 22. It begins at Scan 4 and ends at Scan 12 with a missed measurement at Scan 7. This track file is also 100% consistent, although it has missed measurements. TF 102 and 103 should be examined together. Targets 31 and 32 form an unresolved closely spaced object cluster at Scans 1, 2, 3, 4, and 7. When these two targets begin to get resolved at Scans 5 and 6, and beyond Scan 7, these two track files have successfully tracked them. TF 104 and 105 show a similar situation except that the resulting track files are much less consistent after the two targets in the CSO became resolved (see Scan 8 and beyond). Measurements containing more than two unresolved targets can also be found in the simulated sensor data (not shown in the figure).

Based upon the above observations, a target oriented scoring (performance evaluation) scheme can be defined. All track files containing the same target are identified. Using Figure 8.26, target number 31 is contained in TFs 102 and 103. A track file containing this target most often is assigned to represent this target, in this case, TF 102 is chosen to represent Target 31. Applying this logic, TF 100, 101, 102, 103, 104, and 105 are assigned to represent Targets 20, 22, 31, 32, 27, and 26, respectively. A performance for tracking each target can be evaluated using the assigned track files. Consequently, Targets 20, 22, 31, 32, 26, and 27 each meet 100%, 100%, 100%, 100%, 83.33%, and 83.33%, performance, respectively. Using Target 26 as an example, it appeared for 12 scans. The track file assigned to represent Target 26 is TF 105 which contains Target 26 for 10 scans. This means that the performance on tracking Target 26 is $10/12 = 83.33\%$.

The above scoring scheme is applied to the entire data set and the results are shown in Figure 8.27. The horizontal axis gives performance criterion described in the above

TF	Scan number											
#	1	2	3	4	5	6	7	8	9	10	11	12
100	20	20	20	20	20	20	20	20	20	20	20	20*
101	0	0	0	22	22	22	0	22	22	22	22	22
102	31 32	31 32	31 32	31 32	31	31	31 32	31	31	31	31	31
103	31 32	31 32	31 32	31 32	32	32	31 32	32	32	32	32	32
104	26 27	26 27	26 27	26 27	26	26	26 27	27	27	27	27	27
105	26 27	26 27	26 27	26 27	27	27	26 27	26	26	26	26	26

* Entries are true target IDs, multiple IDs indicate unresolved
measurements from the target pairs

Figure 8.26   Illustration of typical track files.

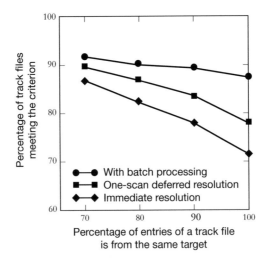

Figure 8.27   Performance comparison of three ambiguity resolution methods.

paragraph and the vertical axis gives the percentage of targets satisfying a given criterion. The three-track ambiguity resolution methods described in Section 8.7, namely, immediate resolution, one-scan deferred resolution, and batch processing, are compared. For illustration, examining the data of Figure 8.26, one can see that four out of six targets meet the 100% performance criterion and all targets meet the 80% criterion.

In Figure 8.27, the immediate resolution method represents a quick and crude process and achieves a certain level of performance. The one-scan deferred resolution has some improvement over the immediate resolution method. Clearly, the algorithm that makes use of all the data (the batch processing) achieves the best performance, which comes with a cost of more processing and delay in decision. Also note that the biggest gain in using a sophisticated algorithm is at the 100% performance criterion level. This is because the precision trajectory estimation begins by using track files produced by the previous algorithm. In order to have the precision estimation succeed, the initial track files need to be reasonably consistent. The batch processing method is still limited in performance due to the limit of sensor resolution. When some of the targets are not resolved, no processing techniques can completely successfully sort them out.

Example 2

The purpose of this example [19] is to illustrate tracking performance as a function of sensor accuracy and resolution. Similar to Example 1, trajectories of a large number of space objects were simulated and their measurements processed by the three algorithms previously described.

Due to the presence of closely spaced unresolved objects, scoring of tracking performance requires careful thinking. Two scoring criteria were used in [18, 19]. A target oriented scoring scheme was used in [18]. This scheme, used in the previous example, assigns each target a track file that contains the most measurements from this target. The track file is then evaluated based upon the total number of occurrences of that target in the entire track interval. This scheme is very useful when tracking performance associated with a specific type of target is important. A less sophisticated scheme, the so-called file oriented scoring scheme [19], is used to present results of this example. The file oriented scoring scheme assigns each track file with a performance number

$$x = M_T/M_{\text{TF}},$$

Normalized sensor parameters	Test case				
	A	B	C	D	E
σ_θ	1	1	1	2	3
Δ_θ	1	1.75	2.5	1.75	1.75

Figure 8.28   Test cases as a function of sensor accuracy and resolution.

where M_T is the maximum number of measurements in the track file that come from the same target; and M_{TF} is the total number of measurements in the track file.

The results of applying the tracking algorithm are represented by two statistics, namely, the percentage of correct correlations $P_c(x)$, and the percentage of false correlations $P_f(x)$ for track files with a performance number greater than or equal to x, that is,

$$P_c(x) = N_c(x)/N_T$$

and

$$P_f(x) = 1 - N_c(x)/N_{TF},$$

where $N_c(x)$ is the number of track files with performance number greater than or equal to x; x is the performance number defined as M_T/M_{TF}; N_T is the total number of targets in the field of view; and N_{TF} is the total number of track files.

The set of sensor parameters used in this example is shown in Figure 8.28 where σ_θ is sensor accuracy and Δ_θ is sensor bias.

Entries in Figure 8.28 are test cases with various sensor accuracy and resolution, in which Case A is the baseline (the best) case while others are various levels of degradation from the baseline. Sensor measurements were simulated and processed by the tracking algorithm.

Some intermediate results such as target scenes and a target statistics summary as a function of scan number are shown in Figures 8.29 and 8.30, respectively. The results of applying the tracking algorithm are shown in Figure 8.31. For a fixed sensor, the angular accuracy performance degrades as the sensor angular resolution degrades. The same trend is true for a fixed sensor angular resolution. When the performance number (the horizontal axis) increases, a smaller and smaller number of tracks can meet this criterion, and the performance in terms of correct and false correlation thus decreases.

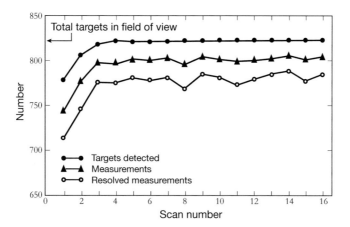

Figure 8.29   Target/measurement statistics.

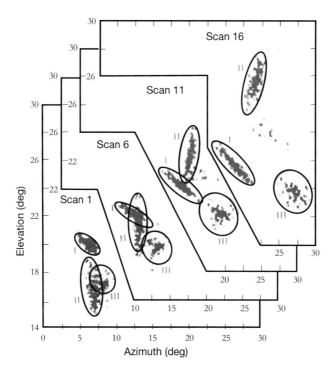

Figure 8.30   Scan to scan target scenes.

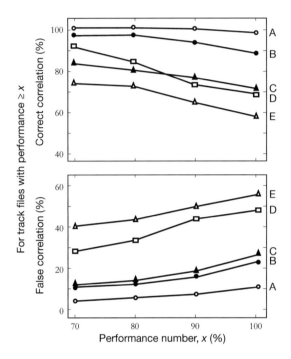

Figure 8.31   Correct and false correlations for a given track file performance number for various sensor accuracies and resolutions.

Appendix 8.A   Example Track Initiation Equations

8.A.1   Applying the Fixed Interval Smoother to Compute Initial Conditions

Consider a discrete linear system

$$\mathbf{x}_k = \mathbf{\Phi}_{k,k-1}\mathbf{x}_{k-1} + \mathbf{\mu}_{k-1}$$

$$\mathbf{y}_k = \mathbf{H}_k\mathbf{x}_k + \mathbf{v}_k$$

where $\mathbf{x}, \mathbf{\mu} \in \mathbb{R}^n$, $\mathbf{y}, \mathbf{v} \in \mathbb{R}^m$, $\mathbf{\Phi}_{k,k-1}$ and $\mathbf{H}_k$ are known matrices, and $\mathbf{x}_0: \sim N(\hat{\mathbf{x}}_0, \mathbf{P}_0)$, $\mathbf{\mu}_k : \sim N(\mathbf{0}, \mathbf{Q}_k)$, and $\mathbf{v}_k : \sim N(\mathbf{0}, \mathbf{R}_k)$ are mutually independent.

Consider the case where there is no prior knowledge of the initial state, thus $\hat{\mathbf{x}}_0$ and $\mathbf{P}_0$ are not given. One must compute them using the measurements $\mathbf{y}_k$, $k=1 \ldots K$. Consider the algorithm in Section 2.9.5, the FIS for a deterministic system. It described a WLS estimator for a noise-free system. The deterministic system assumption is valid when the length of the data window used for the WLS is short. Applying Equations (2.9-13) and (2.9-14) by setting $\mathbf{P}_0^{-1} = \mathbf{0}$ (no a priori information) we obtain

$$\hat{\mathbf{x}}_{k\text{II}:K} = \left[\sum\nolimits_{i=1}^{K} \mathbf{\Phi}_{i,k}^{T} \mathbf{H}^{T} \mathbf{R}_{i}^{-1} \mathbf{H} \mathbf{\Phi}_{i,k} \right]^{-1} \left(\sum\nolimits_{i=1}^{K} \mathbf{\Phi}_{i,k}^{T} \mathbf{H}^{T} \mathbf{R}_{i}^{-1} \mathbf{y}_{i} \right), \qquad (8.A-1)$$

with covariance

$$\mathbf{P}_{k\text{II}:K} = \left[\sum\nolimits_{i=1}^{K} \mathbf{\Phi}_{i,k}^{T} \mathbf{H}^{T} \mathbf{R}_{i}^{-1} \mathbf{H} \mathbf{\Phi}_{i,k} \right]^{-1}. \qquad (8.A-2)$$

The data window length K should be taken to be as short as possible in order to make the deterministic system assumption valid. For instance, for $\mathbf{x} \in \mathbb{R}^n$, and $\mathbf{y} \in \mathbb{R}^m$, K can be $\text{int}\left(\frac{n}{m}\right) + 1$, where $\text{int}\left(\frac{n}{m}\right)$ denotes the integer of the fraction $\frac{n}{m}$. The resulting $\hat{\mathbf{x}}_{k\text{II}:K}$ and $\mathbf{P}_{k|1:K}$ can be used to initiate a KF with initial time K.

The FIS with a deterministic system introduced here is for linear systems. It will also work for nonlinear systems. For nonlinear systems, the above system and measurement matrices are replaced with Jacobian matrices (the first derivatives). To obtain a more accurate estimate, an iterative solution for nonlinear systems introduced in Chapter 3 on nonlinear estimation can be used.

8.A.2   Applying First Order Polynomial Smoothing[14] to Radar Measurements to Obtain Initial State Estimate and Covariance for a Tracking Filter in Cartesian Coordinates

Consider the case where a radar measures range (r), azimuth (a), and elevation (e). For the initiation process, CV motion that is independent in each of the three axes (r, a, e) is assumed. This is the same as using a first order polynomial to represent the target's equation of motion. Given a batch of measurements, the target's position and velocity estimates can be computed independently in each of the three axes (r, a, e). Let

$$r_i, i = 1, \dots N$$

$$a_i, i = 1, \dots N$$

$$e_i, i = 1, \dots N \qquad (8.A-3)$$

denote a set of N radar measurements with time between each measurement as T, then the total time interval of data is $(N-1)T$. The estimate of $r, \dot{r}, a, \dot{a}, e, \dot{e}$ corresponding

---

14   Polynomial modeling is the subject of Homework Problem 15 in Chapter 1.

to the center of the data interval can be obtained using a first order polynomial (straight line) model.[15] Since the measurements on r, a, e are independent, they can be fitted separately. For range and range rate, $r, \dot{r}$ we obtain

$$
\begin{bmatrix} \hat{r} \\ \hat{\dot{r}} \end{bmatrix} = \begin{bmatrix} \sum_{i=1}^{N} r_i \\ \sum_{i=1}^{N} T\left(i - 1 - \frac{(N-1)}{2}\right) r_i \end{bmatrix}
\tag{8.A-4}
$$

with covariance

$$
\mathbf{P}_r = \sigma_r^2 \begin{bmatrix} \frac{1}{N} & 0 \\ 0 & \frac{12}{T^2(N-1)N(N+1)} \end{bmatrix}.
\tag{8.A-5}
$$

The corresponding $\hat{a}, \hat{\dot{a}}, \hat{e}, \hat{\dot{e}}$ and covariance can be obtained the same way. Let $\hat{\mathbf{z}}_0$ denote the polynomial smoothed radar measurements with the subscript corresponding to the center of the data window

$$
\hat{\mathbf{z}}_0 = \begin{bmatrix} \hat{r} \\ \hat{a} \\ \hat{e} \\ \hat{\dot{r}} \\ \hat{\dot{a}} \\ \hat{\dot{e}} \end{bmatrix} = \begin{bmatrix} \sum_{i=1}^{N} r_i \\ \sum_{i=1}^{N} a_i \\ \sum_{i=1}^{N} e_i \\ \sum_{i=1}^{N} T\left(i - 1 - \frac{(N-1)}{2}\right) r_i \\ \sum_{i=1}^{N} T\left(i - 1 - \frac{(N-1)}{2}\right) a_i \\ \sum_{i=1}^{N} T\left(i - 1 - \frac{(N-1)}{2}\right) e_i \end{bmatrix}
\tag{8.A-6}
$$

with covariance $\mathbf{P}_{\mathbf{z}_0}$

$$
\mathbf{P}_{\mathbf{z}_0} = \begin{bmatrix} c_1\sigma_r^2 & 0 & 0 & 0 & 0 & 0 \\ 0 & c_1\sigma_a^2 & 0 & 0 & 0 & 0 \\ 0 & 0 & c_1\sigma_e^2 & 0 & 0 & 0 \\ 0 & 0 & 0 & c_2\sigma_r^2 & 0 & 0 \\ 0 & 0 & 0 & 0 & c_2\sigma_a^2 & 0 \\ 0 & 0 & 0 & 0 & 0 & c_2\sigma_e^2 \end{bmatrix},
\tag{8.A-7}
$$

---

15   Homework Problem 15 in Chapter 1 builds a general set of relationships for polynomial modeling with specific equations for first and second order polynomials.

where $c_1 = 1/N$ and $c_2 = 12/T^2(N-1)N(N+1)$. Before transforming to the state coordinate system, first propagate $\hat{\mathbf{z}}_0$ and $\mathbf{P}_{\mathbf{z}_0}$ to the desired time (denoted as t) to match the new measurement time for updating the filter. Using the transition matrix to accomplish this task,

$$\hat{\mathbf{z}}_t = \mathbf{\Phi}_{t,0}\hat{\mathbf{z}}_0 \tag{8.A-8}$$

where

$$\mathbf{\Phi}_{t,0} = \begin{bmatrix} 1 & 0 & 0 & t-(N-1)T/2 & 0 & 0 \\ 0 & 1 & 0 & 0 & t-(N-1)T/2 & 0 \\ 0 & 0 & 1 & 0 & 0 & t-(N-1)T/2 \\ 0 & 0 & 0 & 1 & 0 & 0 \\ 0 & 0 & 0 & 0 & 1 & 0 \\ 0 & 0 & 0 & 0 & 0 & 1 \end{bmatrix} \tag{8.A-9}$$

Remarks: Two-Pulse Initiation

If only two measurements are used for track initiation, that is, $N=2$, then for the simplest case

$$\hat{r} = (r_2 + r_1)/2, \sigma_{\hat{r}}^2 = \sigma_r^2/2$$

$$\hat{\dot{r}} = (r_2 - r_1)/T, \sigma_{\hat{\dot{r}}}^2 = 2\sigma_r^2/T$$

$$\hat{a} = (a_2 + a_1)/2, \sigma_{\hat{a}}^2 = \sigma_a^2/2$$

$$\hat{\dot{a}} = (a_2 - a_1)/T, \sigma_{\hat{\dot{a}}}^2 = 2\sigma_a^2/T$$

$$\hat{e} = (e_2 + e_1)/2, \sigma_{\hat{a}}^2 = \sigma_a^2/2$$

$$\hat{\dot{e}} = (e_2 - e_1)/T, \sigma_{\hat{\dot{e}}}^2 = 2\sigma_e^2/T$$

This is sometimes referred to as two-pulse initiation. Although straightforward, it is also the noisiest case. Exactly what N to pick is a trade-off between accuracy and time to convergence for the initiation process.

Once the initial state estimate and covariance are obtained in (r, a, e), they can be transformed to Cartesian coordinates. In the case where the tracking filter is constructed in radar centered Cartesian coordinates, that is, East (x), North (y), Up (z), then the state vector $\mathbf{x} = (x, y, z, \dot{x}, \dot{y}, \dot{z})^T$ and radar measurement variables are related by

$$x = r\cos(e)\sin(a)$$

$$y = r\cos(e)\cos(a)$$

$$z = r\sin(e) \tag{8.A-10}$$

and their first derivatives with respect to time yield

$$\dot{x} = \dot{r}\cos(e)\sin(a) - r\sin(e)\sin(a)\dot{e} + r\cos(e)\cos(a)\dot{a}$$

$$\dot{y} = \dot{r}\cos(e)\cos(a) - r\sin(e)\cos(a)\dot{e} - r\cos(e)\sin(a)\dot{a}$$

$$\dot{z} = \dot{r}\sin(e) + r\cos(e)\dot{e}. \tag{8.A-11}$$

Let $\mathbf{x}$ denote the corresponding state in Cartesian coordinates and let

$$\mathbf{x} = \mathbf{g}(\mathbf{z}),$$

Where $\mathbf{z} = (r, a, e, \dot{r}, \dot{a}, \dot{e})^T$.

The specific nonlinear functions $\mathbf{g}(.)$ are defined in Equations (8.A-10) and (8.A-11). As shown in Chapter 3, when the nonlinear transformation is defined, the Jacobian matrices can be used to compute the first order approximation of the covariance matrix of the transformed vector. Denote the Jacobian of $\mathbf{g}(.)$ as $\mathbf{G}$, then

$$\mathbf{G} = \left[\frac{\partial \mathbf{g}(\mathbf{z})}{\partial \mathbf{z}} \right]_{\mathbf{z}}$$

and in this case $\mathbf{G}$ is a 6×6 matrix. The covariance of $\mathbf{x}$, $\mathbf{P}_x$, is

$$\mathbf{P}_x = \mathbf{G}\mathbf{P}_z\mathbf{G}^T. \tag{8.A-12}$$

The initial estimate for the extended Kalman filter (EKF), $\hat{\mathbf{x}}_{t_0}$ and $\mathbf{P}_{\hat{\mathbf{x}}_{t_0}}$, where time t_0 is the end of the data interval, $(N - 1)T/2$.

Note the details of the $\mathbf{g}(.)$ and $\mathbf{G}$ matrices

$$\mathbf{x} = \mathbf{g}(\mathbf{z})$$

$$\mathbf{z} = \begin{bmatrix} r \\ a \\ e \\ \dot{r} \\ \dot{a} \\ \dot{e} \end{bmatrix}$$

$$\mathbf{x} = \begin{bmatrix} x \\ y \\ z \\ \dot{x} \\ \dot{y} \\ z \end{bmatrix} = \mathbf{g}(\mathbf{z})$$

$$x = r\cos(e)\sin(a)$$

$$y = r\cos(e)\cos(a)$$

$$z = r\sin(e)$$

$$\dot{x} = \dot{r}\cos(e)\sin(a) - r\sin(e)\sin(a)\dot{e} + r\cos(e)\cos(a)\dot{a}$$

$$\dot{y} = \dot{r}\cos(e)\cos(a) - r\sin(e)\cos(a)\dot{e} - r\cos(e)\sin(a)\dot{a}$$

$$\dot{z} = \dot{r}\sin(e) + r\cos(e)\dot{e}$$

$$\mathbf{G} = \frac{\partial \mathbf{g}(\mathbf{z})}{\partial \mathbf{z}} = \begin{bmatrix} \dfrac{\partial x}{\partial r} & \dfrac{\partial x}{\partial a} & \dfrac{\partial x}{\partial e} & \dfrac{\partial x}{\partial \dot{r}} & \dfrac{\partial x}{\partial \dot{a}} & \dfrac{\partial x}{\partial \dot{e}} \\[2mm] \dfrac{\partial y}{\partial r} & \dfrac{\partial y}{\partial a} & \dfrac{\partial y}{\partial e} & \dfrac{\partial y}{\partial \dot{r}} & \dfrac{\partial y}{\partial \dot{a}} & \dfrac{\partial y}{\partial \dot{e}} \\[2mm] \dfrac{\partial z}{\partial r} & \dfrac{\partial z}{\partial a} & \dfrac{\partial z}{\partial e} & \dfrac{\partial z}{\partial \dot{r}} & \dfrac{\partial z}{\partial \dot{a}} & \dfrac{\partial z}{\partial \dot{e}} \\[2mm] \dfrac{\partial \dot{x}}{\partial r} & \dfrac{\partial \dot{x}}{\partial a} & \dfrac{\partial \dot{x}}{\partial e} & \dfrac{\partial \dot{x}}{\partial \dot{r}} & \dfrac{\partial \dot{x}}{\partial \dot{a}} & \dfrac{\partial \dot{x}}{\partial \dot{e}} \\[2mm] \dfrac{\partial \dot{y}}{\partial r} & \dfrac{\partial \dot{y}}{\partial a} & \dfrac{\partial \dot{y}}{\partial e} & \dfrac{\partial \dot{y}}{\partial \dot{r}} & \dfrac{\partial \dot{y}}{\partial \dot{a}} & \dfrac{\partial \dot{y}}{\partial \dot{e}} \\[2mm] \dfrac{\partial \dot{z}}{\partial r} & \dfrac{\partial \dot{z}}{\partial a} & \dfrac{\partial \dot{z}}{\partial e} & \dfrac{\partial \dot{z}}{\partial \dot{r}} & \dfrac{\partial \dot{z}}{\partial \dot{a}} & \dfrac{\partial \dot{z}}{\partial \dot{e}} \end{bmatrix}$$

$$\frac{\partial x}{\partial r} = \cos(e)\sin(a)$$

$$\frac{\partial x}{\partial a} = r\cos(e)\cos(a)$$

$$\frac{\partial x}{\partial e} = -r\sin(e)\cos(a)$$

$$\frac{\partial x}{\partial \dot{r}} = \frac{\partial x}{\partial \dot{a}} = \frac{\partial x}{\partial \dot{e}} = 0$$

$$\frac{\partial y}{\partial r} = \cos(e)\cos(a)$$

$$\frac{\partial y}{\partial a} = -r\cos(e)\sin(a)$$

$$\frac{\partial y}{\partial e} = -r\sin(e)\cos(a)$$

$$\frac{\partial y}{\partial \dot{r}} = \frac{\partial y}{\partial \dot{a}} = \frac{\partial y}{\partial \dot{e}} = 0$$

$$\frac{\partial z}{\partial r} = \sin(e)$$

$$\frac{\partial z}{\partial a} = 0$$

$$\frac{\partial z}{\partial e} = r\cos(e)$$

$$\frac{\partial z}{\partial \dot{r}} = \frac{\partial z}{\partial \dot{a}} = \frac{\partial z}{\partial \dot{e}} = 0$$

$$\frac{\partial \dot{x}}{\partial r} = -\sin(e)\sin(a)\dot{e} + \cos(e)\cos(a)\dot{a}$$

$$\frac{\partial \dot{x}}{\partial a} = \dot{r}\cos(e)\cos(a) - r\sin(e)\cos(a)\dot{e} - r\cos(e)\sin(a)\dot{a}$$

$$\frac{\partial \dot{x}}{\partial e} = -\dot{r}\sin(e)\sin(a) - r\cos(e)\sin(a)\dot{e} - r\sin(e)\cos(a)\dot{a}$$

$$\frac{\partial \dot{x}}{\partial \dot{r}} = \cos(e)\sin(a)$$

$$\frac{\partial \dot{x}}{\partial \dot{a}} = r\cos(e)\cos(a)$$

$$\frac{\partial \dot{x}}{\partial \dot{e}} = -r\sin(e)\sin(a)$$

$$\frac{\partial \dot{y}}{\partial r} = -\sin(e)\cos(a)\dot{e} - \cos(e)\sin(a)\dot{a}$$

$$\frac{\partial \dot{y}}{\partial a} = -\dot{r}\cos(e)\sin(a) + r\sin(e)\sin(a)\dot{e} - r\cos(e)\cos(a)\dot{a}$$

$$\frac{\partial \dot{y}}{\partial e} = -\dot{r}\sin(e)\cos(a) - r\cos(e)\cos(a)\dot{e} + r\sin(e)\sin(a)\dot{a}$$

$$\frac{\partial \dot{y}}{\partial \dot{r}} = \cos(e)\cos(a)$$

$$\frac{\partial \dot{y}}{\partial \dot{a}} = -r\cos(e)\sin(a)$$

$$\frac{\partial \dot{y}}{\partial \dot{e}} = -r\sin(e)\cos(a)$$

$$\frac{\partial \dot{z}}{\partial r} = \cos(e)\dot{e}$$

$$\frac{\partial \dot{z}}{\partial a} = 0$$

$$\frac{\partial \dot{z}}{\partial e} = \dot{r}\cos(e) - r\sin(e)\dot{e}$$

$$\frac{\partial \dot{z}}{\partial \dot{r}} = \sin(e)$$

$$\frac{\partial \dot{z}}{\partial \dot{a}} = 0$$

$$\frac{\partial \dot{z}}{\partial \dot{e}} = r\cos(e).$$

Homework Problems

1. Extend Homework Problem 1 in Chapter 3 to a multiple-target case. Build an MTT based on Section 8.4. Be creative, fine tune the algorithm, and make it robust against various target densities (a parameter that you should vary). Target sets to be considered should include:

    a. All straight and parallel

    b. All straight and crossing with varying crossing angles

    c. Mixed straight lines and maneuvering targets. You can create maneuvering targets that are either turning on a circle, or weaving, or both.

2. Apply your algorithms to space trajectories. Build the algorithm illustrated in Figure 8.25.

    a. When limited to space trajectories the system is deterministic, a condition where FIS will work well.

    b. Make the time between measurements long, 5 to 10 s. This will reduce the total number of measurements per trajectory for a long time window of observation. This condition will enable you to obtain good estimates while achieving reasonable computer runtime.

References

1. R.W. Sittler, "An Optimal Data Association Problem in Surveillance Theory," *IEEE Transactions on Military Electronics*, vol. MIL-8, pp. 125–139, Apr. 1964.

2. R.A. Singer, R.G. Sea, and K.B. Housewright, "Derivation and Evaluation of Improved Tracking Filters for Use in Dense Multi-Target Environments," *IEEE Transactions on Information Theory*, vol. IT-20, pp. 423–432, July 1974.

3. Y. Bar-Shalom, "Extension of the Probabilistic Data Association Filter in Multiple-Target Tracking," in *Proceedings of the 5th Symposium on Nonlinear Estimation*, pp. 16–21, Sept. 1974.

4. D.L. Alspach, "A Gaussian Sum Approach to the Multi-Target Identification-Tracking Problem," *Automatica*, vol. 11, pp. 285–296, 1975.

5. Y. Bar-Shalom and E. Tse, "Tracking in a Cluttered Environment with Probabilistic Data Association," *Automatica*, vol. 11, pp. 451–460, 1975.

6. M. Athans, R.H. Whiting, and M. Gruber, "A Suboptimal Estimation Algorithm with Probabilistic Editing for False Measurements with Applications to

Target Tracking with Wake Phenomena," *IEEE Transactions on Automatic Control*, vol. AC-22, 372–385, June 1977.

7. C.L. Morefield, "Application of 0-1 Integer Programming to Multi-Target Tracking Problem," *IEEE Transactions on Automatic Control*, vol. AC-22, 302–312, June 1977.

8. Y. Bar-Shalom, "Tracking Methods in a Multi-Target Environment," *IEEE Transactions on Automatic Control*, vol. AC-23, pp. 618–626, Aug. 1978.

9. D.B. Reid, "An Algorithm for Tracking Multiple Targets," *IEEE Transactions on Automatic Control*, vol. AC-24, pp. 843–854, Dec. 1979.

10. T. Kurien, "Issues in the Design of Practical Multi-Target Tracking Algorithms," in *Multitarget Multisensor Tracking: Advanced Applications*, Y. Bar-Shalom, Ed., pp. 43–83, Norwood, MA: Artech House, 1990.

11. M.K. Chu, "Target Breakup Detection in the Multiple Hypothesis Tracking Formulation," M.E. thesis, MIT, 1996.

12. S.S. Blackman, "Multiple Hypothesis Tracking for Multiple Target Tracking," *IEEE Transactions on Aerospace and Electronic Systems Magazine*, vol. 19, pp. 5–18, Jan. 2004.

13. S.S. Blackman, *Multiple-Target Tracking with Radar Applications*. Norwood, MA: Artech House, 1986.

14. S.S. Blackman and R. Popoli, *Design and Analysis of Modern Tracking Systems*. Norwood, MA: Artech House, 1999.

15. Y. Bar-Shalom, X. Rong Li, and T. Kirubarajan, *Estimation with Applications to Tracking and Navigation*. New York: Wiley, 2001.

16. Y. Bar-Shalom, P. Willett, and X. Tian, *Tracking and Data Fusion: A Handbook of Algorithms*. Storrs, CT: YBS Publishing, 2011.

17. Y. Bar-Shalom, F. Daum, and J. Huang, "The Probabilistic Data Association Filter," *IEEE Control System Magazine*, Dec. 2009.

18. C.B. Chang and L.C. Youens, "An Algorithm for Multiple Target Tracking and Data Correlation," MIT Lincoln Laboratory Technical Report TR-643, June 1983.

19. C.B. Chang, K.P. Dunn, and L.C. Youens, "A Tracking Algorithm for Dense Target Environment," in *Proceedings of the 1984 American Control Conference*, pp. 613–618, June 1984.

20. C.B. Chang and L.C. Youens, "Measurement Correlation for Multiple Sensor Tracking in a Dense Target Environment," *IEEE Transactions on Automatic Control*, vol. AC-27, pp. 1250–1252, Dec. 1982.

21. M.J. Tsai, L.C. Youens, and K.P. Dunn, "Track Initiation in a Dense Target Environment Using Multiple Sensors," in *Proceedings of the 1989 SPIE: Signal and Data Processing of Small Targets*, pp. 144–151, Mar. 1989.

22. J.A. Gubner, *Probability and Random Processes for Electrical and Computer Engineers*. Cambridge, UK: Cambridge University Press, 2006.

23. H.W. Kuhn, "The Hungarian Method for the Assignment Problem," *Naval Research Logistics Quarterly*, vol. 2, pp. 83–97, 1955.

24. H.W. Kuhn, "Variants of the Hungarian Method for Assignment Problems," *Naval Research Logistics Quarterly*, vol. 3, pp. 253–258, 1956.

25. J. Munkres, "Algorithms for the Assignment and Transportation Problems," *Journal of the Society for Industrial and Applied Mathematics*, vol. 5, pp. 32–38, Mar. 1957.

26. C.B. Chang and K.P. Dunn, "A Functional Model for the Closely Spaced Object Resolution Process," MIT Lincoln Laboratory Technical Report TR-611, May 1982.

9

Multiple Hypothesis Tracking Algorithm

9.1   Introduction

Following the multiple target tracking (MTT) taxonomy shown in Table 8.1, five approaches for addressing the MTT problem were introduced in Chapter 8. The nearest neighbor (NN) and global nearest neighbor (GNN) methods are in the category of assignment approach, which assigns measurements to tracks in a single scan. It was shown that a single scan assignment approach is a simplification of the multiscan assignment approach introduced in Section 8.2. The single scan assignment approach is significant because it is used in many practical MTT algorithms. The probabilistic data association filter (PDAF) and joint probabilistic data association filter (JPDAF) are in the category of the target-oriented approach in which all the measurements that fall within a target's acceptance gate (AG) are used to update the track of a target. This method has the feature that the uncertainties of the measurement origin are accounted for with enlarged covariance. When multiple detections in an AG persist over several scans, filter performance may deteriorate rapidly because a larger covariance exacerbates an association problem. Because all measurements in an AG are used for track updates, new tracks can be initiated only for those measurements that fall outside of all AGs. The track split method is suboptimal but simple to implement and works well for modest levels of target density. A practical algorithm was suggested in the second half of this chapter, which uses a combination of track split and assignment methods over several scans. Equations for filter initial condition computation for dish radar coordinates with state vectors in East-North-Up Cartesian coordinates are given in Appendix 8.A.

A well-known technique for MTT is the multiple hypothesis tracker (MHT), which is the subject of this chapter. The MHT approach considers the association of sequences of measurements and scores the track/hypotheses using their probabilities.

MHT keeps track of multiple possible combinations of measurements across time. A hypothesis in MHT is one possible data association of all the measurements[1] encountered up to that scan. Hypotheses are formed using tracks with the constraint that each measurement can only be used once, thus no two tracks in the same hypothesis can share any measurements. Three possibilities are always considered for each new measurement: (a) it is the continuation of an existing track; (b) it is the starting point of a new track; or (c) it is a false alarm. With these possibilities, there will always be a hypothesis that a track continuing without being updated with a new measurement. MHT is ideally considered as optimum for MTT because it enumerates all possibilities, and therefore the set of true tracks must be contained within one of the hypotheses. As will be illustrated later, exhaustive enumeration has its practical limits,[2] thus implementable MHT algorithms are nevertheless suboptimum. Because MHT considers measurements across time and allows for association revocation, it is equivalent to smoothing in estimation.

The concept of MHT was in development for some time before it was solidified by Reid in his 1977 IEEE paper [1]. Many publications dated before and after Reid's paper offer different approaches to algorithms and applications. Not all of them follow the framework of MHT [see, e.g., 2–16 and references therein]. Reid's method has been referred to as a measurement-oriented MHT. Its implementation is a data file management problem and its complexity is in the hypothesis enumeration where the increase of the total number of hypotheses is very rapid (see examples later). One of the alternatives to measurement-oriented MHT is track-oriented MHT, see Kurien [8]. Both methods are illustrated in Section 9.2 using examples. Multiple hypotheses and tracks will result even when the problem is in the simplest and most unambiguous case, as shown in Figure 8.1(a). For more details see the examples in Section 9.2.

This chapter is organized as follows. The concept of MHT will be introduced in Section 9.2. Both measurement-oriented and track-oriented approaches will be illustrated using examples. Their corresponding hypothesis probabilities will be derived. Two implementation methods, with the goal of reducing computational burden found in [9–12], will be briefly discussed. Basic equations for track and hypothesis scoring and pruning based upon the description in [14] are explained in Section 9.3 with a numerical example. The realization of MHT requires exhaustive enumeration of all possibilities. A technique known as the Nassi–Shneiderman chart [17, 18] may be applicable for MHT implementation and is briefly mentioned in Section 9.4.

---

1   To simplify the problem, only tracks with shared AG are processed jointly with the MHT technique.
2   The computational burden for this class of discrete programming problems is referred to as NP-hard.

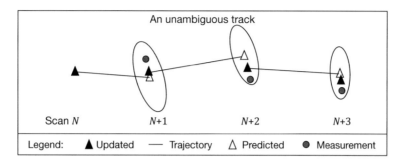

Figure 9.1   An unambiguous track, with one measurement per scan, and with the assumption that the detection probability is one and the false alarm probability is zero.

9.2   Multiple Hypothesis Tracking Illustrations

For comparison purposes, a simple case is shown in Figure 9.1, which is reproduced from Figure 8.16 in Chapter 8 as an unambiguous track case. In this case, in each of the four scans, only one measurement appears in the AG. Assuming that the sensor detection probability is one ($P_d = 1$) and the false alarm and the new target probabilities are both zero ($P_{fa} = P_{NT} = 0$), then as shown in Figure 9.1, there is only one track. However, when the ideal detection situation (in terms of P_d, P_{fa}, and P_{NT}) is no longer true, the track continuation with MHT becomes much more complex and is illustrated below.[3]

9.2.1   Measurement-Oriented MHT

In a measurement-oriented approach, every possible track is listed for each measurement. Each measurement is treated as (a) a new detection; (b) a continuation of a track; or (c) a false alarm. New hypotheses are generated encompassing all of these possibilities.

A very high level flow diagram of an MHT is shown Figure 9.2. The predicted measurement locations and uncertainties based on previously updated tracks are used to set up the AG, and the measurements that fall within overlapped AGs are used to form assignment matrices. New hypotheses are formed with their probabilities computed. Tracks are the constituents of these hypotheses. Hypotheses with low

---

3   It is a system design engineer's trade-off in comparing the complexity of MHT and when P_d is nearly one and P_{fa} and P_{NT} are nearly zero.

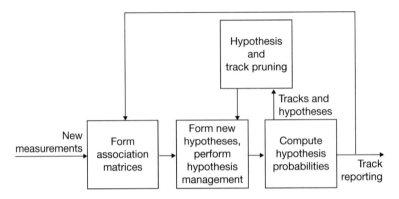

Figure 9.2   A flow diagram for the measurement-oriented MHT algorithm.

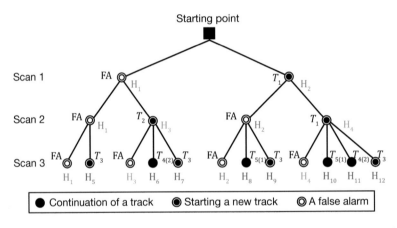

Figure 9.3   Hypothesis evolution for measurement-oriented MHT with one detection per scan for three scans.

probabilities are pruned and the surviving hypotheses are used for track reporting and processing at the next measurement scan.

The evolution of a hypothesis tree for a three-scan case, with a single measurement of each scan, is illustrated in Figure 9.3. This case shares the same scenario as that of Figure 9.1. The black square at the top of the tree indicates the starting point. On the first scan, a measurement is obtained. This measurement is treated as a false alarm[4] or

---

4   When a measurement is designated as a false alarm, it will not be used as part of any track.

the starting point of a new track (see labels FA and T_1 and their corresponding hypotheses, H_1 and H_2). In the hypothesis enumeration method illustrated in [3] and [14], it takes two scans to start a new track therefore the detection of the second scan cannot be counted as a continuation of the previous detection (e.g., T_1, T_2, and T_3). With this track initiation rule, the tree grows to four nodes. The branch with the FA assumption continues with the previous hypothesis designation as shown. Two new hypotheses, H_3 and H_4 are added. Moving on to the third scan, growing from each of the four tree nodes, the new measurement is treated as FA, a continuation of previous track(s), or a new track. This results in a total of 12 hypotheses.

The MHT evolution is further explained as follows. Consider the notation T_i for a track file: $T_i = (i_1, i_2, ..., i_K)$ where i_k indicates the i_kth detection at the kth scan and $i_k = 0$ indicates no detection for track T_i at the kth scan. For example, $T_i = (0, 1)$ means track T_i has a measurement at the second scan and does not have any measurement in the first scan. $T_0 = (0, 0, ..., 0)$ is a track that contains no measurement. It is included for the sake of completeness.

With this notation, the hypotheses and tracks for all three scans are listed below for the purpose of a more detailed illustration on the hypothesis evolution of measurement-oriented MHT.

After the second scan, one has

H_1: (0, 0) T_0

H_2: (1, 0) T_1

H_3: (0, 1) T_2

H_4: (1, 0) and (0, 1) T_1 and T_2

Moving on to the third scan, the tree branches begin to grow differently. Going from left to right of Figure 9.3,

1. Two branches grow from the (FA, H_1) root

    (1) FA

    (2) New track (T_3)

2. Three branches grow from the (T_2, H_3) root

    (1) FA

    (2) T_2 is updated with this measurement, becoming T_4

    (3) New track (T_3)

3. Three branches grow from the (FA, H_2) root

    (1) FA

    (2) T_1 is updated with this measurement, becoming T_5

    (3) New track (T_3)

4. Four branches grow from T_2 and H_4 root

    (1) FA

    (2) T_1 is updated with this measurement, becoming T_5

    (3) T_2 is updated with this measurement, becoming T_4

    (4) New track (T_3)

The end of each branch is a hypothesis. Note that some tracks are listed at multiple branches. That is because they are included in different hypotheses. The contents of each hypothesis are shown below, going from left to right using Figure 9.3.

H_1: (0, 0, 0) T_0

H_5: (0, 0, 1) T_3

H_3: (0, 1, 0) T_2

H_6: (0, 1, 1) T_4

H_7: (0, 1, 0); (0, 0, 1) T_2, T_3

H_2: (1, 0, 0) T_1

H_8: (1, 0, 1) T_5

H_9: (1, 0, 0); (0, 0, 1) T_1, T_3

H_4: (1, 0, 0); (0, 1, 0) T_1, T_2

H_{10}: (1, 0, 1); (0, 1, 0) T_5, T_2

H_{11}: (1, 0, 0); (0, 1, 1) T_1, T_4

H_{12}: (1, 0, 0); (0, 1, 0); (0, 0, 1) T_1, T_2, T_3

   This is an example of a very simple case with a single detection in each of the three scans. The number of hypotheses grows to 12 after three scans. The number of tracks grows to six including the all false alarm case (0, 0, 0). In the case when the sensor performs perfectly, there should only be one track: (1, 1, 1) as illustrated in Figure 9.1. This example is chosen to illustrate the complexity of MHT when the sensor imperfection in detection and false alarm is considered. With multiple detections per scan, as this is normally the case in a real problem, the number of hypotheses can quickly grow out of hand. Our example above is for one measurement per scan, while the example shown in [14] was for two measurements per scan. After two scans, the number of hypotheses grows to 34. After three scans, the number of hypothesis grows to be more than 500 [14, p. 290]. As elegant as the MHT formulation is, its application to real problems is severely limited unless some simplification measure can be put in place.

Derivation of Measurement-Oriented MHT Algorithm   The derivation in this section follows the development of Reid [1].

Notation

Consider the case where a sensor has collected k scan of measurements. On the kth scan, the sensor detects N_k distinct measurements, each given by a vector $\mathbf{y}_k^{n_k}$ where n_k is the index of measurements detected within the kth scan. The set of all measurement in scan k is denoted by

$$\mathbf{Y}_k = \left\{ \mathbf{y}_k^1, \mathbf{y}_k^2, \mathbf{y}_k^3, \dots, \mathbf{y}_k^{N_k} \right\} \cup \varnothing.^5$$

Let $\mathbf{Y}_{1:k}$ denote the set of measurements from the first to the kth scan, that is,

$$\mathbf{Y}_{1:k} = \{ \mathbf{Y}_1, \mathbf{Y}_2, \mathbf{Y}_3, \dots, \mathbf{Y}_k \}.$$

We define the following hypothesis notation.

$\boldsymbol{\theta}_k \triangleq \left\{ \theta_k^1, \theta_k^2, \theta_k^3, \dots, \theta_k^{M_k} \right\}$, the set of all hypotheses at the kth scan, where each hypothesis is a collection of tracks found in $\mathbf{Y}_{1:k}$, and M_k is the total number of hypotheses generated by the evolution of measurement-oriented MHT illustrated above. For example, M_k is 12 for the illustrative example in Figure 9.1 for $k=3$.

Remarks

1. Although $\boldsymbol{\theta}_k$ appears to be similar to the global hypothesis defined in Chapter 5 for MMEA, it is actually very different. For example, there is no corresponding local hypothesis here as was the case for MMEA.

2. $\boldsymbol{\theta}_k$ is a collection of hypotheses, $\left\{ \theta_k^1, \theta_k^2, \theta_k^3, \dots, \theta_k^{M_k} \right\}$, where θ_k^i contains a collection of tracks and is a hypothesis in $\boldsymbol{\theta}_k$.

3. Both $\boldsymbol{\theta}_k$ and its elements θ_k^i are different from $\theta_{1:k}^{m_k}$, which is defined in Section 8.2. $\theta_{1:k}^{m_k}$ is a hypothesis containing only one track, $\mathbf{y}_{1:k}^{m_k}$, where measurements in $\mathbf{y}_{1:k}^{m_k}$ are $\left\{ \mathbf{y}_1^{n_1}, \mathbf{y}_2^{n_2}, \mathbf{y}_3^{n_3}, \dots, \mathbf{y}_k^{n_k} \right\}$.

---

5   $\varnothing$ denotes an empty set that represents the case where no measurement in $\left\{ \mathbf{y}_k^1, \mathbf{y}_k^2, \mathbf{y}_k^3, \dots, \mathbf{y}_k^{N_k} \right\}$ is used at the kth scan.

Hypothesis Evolution

When a new set of measurements $\mathbf{Y}_{k+1} = \left\{ \mathbf{y}_{k+1}^1, \mathbf{y}_{k+1}^2, \mathbf{y}_{k+1}^3, ..., \mathbf{y}_{k+1}^{N_{k+1}} \right\}$ is received at the $k + 1$ scan, a new set of hypotheses $\boldsymbol{\theta}_{k+1}$ is formed with the following procedure.

1. Let $\overline{\boldsymbol{\theta}}_{k+1}^0 = \boldsymbol{\theta}_k$.

2. A new set of hypotheses $\overline{\boldsymbol{\theta}}_{k+1}^n$ is repetitively formed for each prior hypothesis $\overline{\boldsymbol{\theta}}_{k+1}^{n-1}$ with each measurement vector in $\mathbf{Y}_{k+1}$, $\mathbf{y}_{k+1}^n$.

3. Once all measurements in $\mathbf{Y}_{k+1}$ have been processed, then $\boldsymbol{\theta}_{k+1} = \overline{\boldsymbol{\theta}}_{k+1}^{N_{k+1}}$.

Hypothesis Probability Calculation

In the following, the hypothesis probability equations developed by Reid [1] are derived. This result is for a Type 1 sensor [1], in which the number of measurement types (continuation of a track, a false alarm, or a new track) is provided and all the measurement data are available and processed together.

Let $\Pr\left\{ \theta_k^i | \mathbf{Y}_{1:k} \right\}$ denote the probability that θ_k^i is true given all data up to scan k. Let A_g^i denote the association event of all measurements (i.e., each measurement associated with a measurement type designation) for θ_k^i in an AG at scan k that is related to the jth hypothesis θ_{k-1}^j at scan $k - 1$. The hypothesis θ_k^i is equal to $\theta_{k-1}^j \cap A_g^i$. Applying Bayes' rule yields the following probability relationship

$$\Pr\left\{ \theta_k^i | \mathbf{Y}_k, \mathbf{Y}_{1:k-1} \right\} = \Pr\left\{ \theta_{k-1}^j \cap A_g^i | \mathbf{Y}_k, \mathbf{Y}_{1:k-1} \right\}$$
$$= \frac{1}{c} p\left(\mathbf{Y}_k | \theta_{k-1}^j, A_g^i, \mathbf{Y}_{1:k-1} \right) \Pr\left\{ A_g^i | \theta_{k-1}^j, \mathbf{Y}_{1:k-1} \right\} \Pr\left\{ \theta_{k-1}^j | \mathbf{Y}_{1:k-1} \right\}, \quad (9.2\text{-}1)$$

where c is a normalization constant. The right-hand side of the above equation is the product of three terms: the first term consists of densities for those measurements that are designated with A_g^i, the second term consists of the probabilities of measurement types designated by A_g^i over the surveillance volume V, and the third term is the prior probability.

The First Term on the Right-Hand Side of Equation (9.2-1), $p\left(\mathbf{Y}_k | \theta_{k-1}^j, A_g^i, \mathbf{Y}_{1:k-1} \right)$: The first term is equal to the product of residual densities for those measurements designated as a continuation of a track, and the probabilities of measurements designated as a false alarm or a new track over the surveillance volume V, that is,

$$p\left(\mathbf{Y}_k | \theta_{k-1}^j, A_g^i, \mathbf{Y}_{1:k-1} \right) = \prod_{n=1}^{N_k} f\left(\mathbf{y}_k^n \right), \quad (9.2\text{-}2)$$

where

$f\left(\mathbf{y}_k^n\right) = 1/V$ if the nth measurement is a false alarm or a new target,

$f\left(\mathbf{y}_k^n\right) = N\left(\mathbf{y}_k^n - \hat{\mathbf{y}}_{k|k-1}, \mathbf{\Gamma}_{k|k-1}\right)$ if $\mathbf{y}_k^n$ is from a confirmed track represented by $\hat{\mathbf{y}}_{k|k-1}$ with covariance $\mathbf{\Gamma}_{k|k-1}$.

The Second Term on the Right-Hand Side of Equation (9.2-1), $\Pr\left\{A_g^i | \theta_{k-1}^j, \mathbf{Y}_{1:k-1}\right\}$:   This term $\Pr\left\{A_g^i | \theta_{k-1}^j, \mathbf{Y}_{1:k-1}\right\}$ is to count the probability of occurrence of false alarms and new tracks for a given A_g^i,

$$\Pr\left\{A_g^i | \theta_{k-1}^j, \mathbf{Y}_{1:k-1}\right\} = \Pr\left\{N_{\mathrm{DT}}, N_{\mathrm{FT}}, N_{\mathrm{NT}} | \theta_{k-1}^j, \mathbf{Y}_{1:k-1}\right\} \times \Pr\left\{\text{configuration} | N_{\mathrm{DT}}, N_{\mathrm{FT}}, N_{\mathrm{NT}}\right\}$$

$$\times \Pr\left\{\text{Assignment} | \text{configuration}\right\}. \tag{9.2-3}$$

The following is an expanded list of variables used to define the probabilities on the right-hand side of Equation (9.2-3).

P_d sensor detection probability

N_{TGT} number of targets

N_{k} number of measurements (or sensor reports)

N_{DT} number of detected targets

N_{FT} number of false targets

N_{NT} number of new targets

β_{FT} density of false targets

β_{NT} density of new targets

$F_N(\lambda)$ Poisson probability for N events with an average rate of λ, that is,

$$F_N(\lambda) = \lambda^N / N!$$

Given the definition of the number of detected, false, and new targets, and the total number of measurements we have

$$N_{\mathrm{k}} = N_{\mathrm{DT}} + N_{\mathrm{FT}} + N_{\mathrm{NT}}$$

The probability of a given set of N_{DT}, N_{FT}, and N_{NT} values given θ_{k-1}^j, $\mathbf{Y}_{1:k-1}$ is

$$\Pr\left\{N_{\mathrm{DT}}, N_{\mathrm{FT}}, N_{\mathrm{NT}} | \theta_{k-1}^j, \mathbf{Y}_{1:k-1}\right\} = \binom{N_{\mathrm{TGT}}}{N_{\mathrm{DT}}} P_d^{N_{\mathrm{DT}}} \left(1 - P_d^{N_{\mathrm{DT}}}\right)^{N_{\mathrm{TGT}} - N_{\mathrm{DT}}} F_{N_{\mathrm{FT}}}\left(\beta_{\mathrm{FT}} V\right) F_{N_{\mathrm{NT}}}\left(\beta_{\mathrm{NT}} V\right).$$

$$\tag{9.2-4}$$

where $\begin{pmatrix} N_{\text{TGT}} \\ N_{\text{DT}} \end{pmatrix} = \dfrac{N_{\text{TGT}}!}{N_{\text{DT}}!(N_{\text{TGT}} - N_{\text{DT}})!}$ denotes the number of combinations of N_{DT} given

the total number of targets N_{TGT}. Multiplying $\begin{pmatrix} N_{\text{TGT}} \\ N_{\text{DT}} \end{pmatrix}$ with $P_d^{N_{\text{DT}}} \left(1 - P_d^{N_{\text{DT}}}\right)^{N_{\text{TGT}} - N_{\text{DT}}}$ gives

the probability of N_{DT} detected targets. The last two terms give the probability of false target and new target.

The total number of combinations for N_k measurements given N_{DT}, N_{FT}, and N_{NT} is,

$$\begin{pmatrix} N_k \\ N_{\text{DT}} \end{pmatrix} \begin{pmatrix} N_k - N_{\text{DT}} \\ N_{\text{FT}} \end{pmatrix} \begin{pmatrix} N_k - N_{\text{DT}} - N_{\text{FT}} \\ N_{\text{NT}} \end{pmatrix} = \frac{N_k!}{N_{\text{DT}}! N_{\text{FT}}! N_{\text{NT}}!}.$$

The probability of a specific configuration is the ratio of one to the total number of configuration, for example,

$$\Pr\{\text{configuration}|N_{\text{DT}}, N_{\text{FT}}, N_{\text{NT}}\} = \frac{N_{\text{DT}}! N_{\text{FT}}! N_{\text{NT}}!}{N_k!}. \tag{9.2-5}$$

The total number of assignment of N_{DT} to N_{TGT} is $\dfrac{N_{\text{TGT}}!}{(N_{\text{TGT}} - N_{\text{DT}})!}$, thus the probability of an assignment for a given configuration is

$$\Pr\{\text{Assignment}|\text{configuration}\} = \frac{(N_{\text{TGT}} - N_{\text{DT}})!}{N_{\text{TGT}}!}. \tag{9.2-6}$$

Substituting Equations (9.2-4) through (9.2-6) into Equation (9.2-3) yields

$$\Pr\left\{A_g^i | \theta_{k-1}^j, \mathbf{Y}_{1:k-1}\right\} = \frac{N_{\text{FT}}! N_{\text{NT}}!}{N_k!} P_d^{N_{\text{DT}}} \left(1 - P_d^{N_{\text{DT}}}\right)^{N_{\text{TGT}} - N_{\text{DT}}} F_{N_{\text{FT}}}\left(\beta_{\text{FT}} V\right) F_{N_{\text{NT}}}\left(\beta_{\text{NT}} V\right). \tag{9.2-7}$$

The Updated Hypothesis Probability Equation for Association Event A_g^i Substituting Equations (9.2-2) and (9.2-7) into Equation (9.2-1) yields the final result

$$\Pr\left\{\theta_{k-1}^j \cap A_g^i | \mathbf{Y}_k, \mathbf{Y}_{1:k-1}\right\} =$$

$$\frac{1}{c} P_d^{N_{\text{DT}}} \left(1 - P_d^{N_{\text{DT}}}\right)^{N_{\text{TGT}} - N_{\text{DT}}} \beta_{\text{FT}}^{N_{\text{FT}}} \beta_{\text{NT}}^{N_{\text{NT}}} \prod_{n=1}^{N_{\text{DT}}} N\left(\mathbf{y}_k^n - \hat{\mathbf{y}}_{k|k-1}, \mathbf{\Gamma}_{k|k-1}\right) \Pr\left\{\theta_{k-1}^j | \mathbf{Y}_{1:k-1}\right\}, \tag{9.2-8}$$

where c is the normalization factor. Note we have taken the liberty to include the notation for Gaussian density, $N\left(\mathbf{y}_k^n - \hat{\mathbf{y}}_{k|k-1}, \mathbf{\Gamma}_{k|k-1}\right)$ in the equation. The Poisson probability, $F_{N_{\text{type}}}\left(\beta_{\text{type}} V\right)$, combined with $f\left(\mathbf{y}_k^n\right)$ for false and new targets becomes $\beta_{\text{type}}^{N_{\text{type}}}$, thus this equation no longer depends on the surveillance volume V.

Remarks

1. Equation (9.2-8) is Reid's main contribution to MHT for the Type 1 sensor [1].

2. The Type 2 sensor as defined in [1] does not send the number of target type information. The new measurements are processed one at a time and the new target density is not changed after each measurement report. The results are similar and the likelihood of a measurement given the number of target types depends on the average (expected) number of detections, false alarms, and new targets. The results will not be discussed any further here.

3. The MHT defined above depends on a specific model of the sensor detection capability (number of detections, number of false detections), clutter field characteristics (number of false alarms), and actual target scenarios (number of new targets). The MTT based on MHT will work well when the assumptions for a parametric model using P_d, β_{FT}, β_{NT}, N_{TGT}, and Poisson distributions are true. In reality, the selection of those values is an art, not a science.

4. The most important part of the performance measure is $\prod_{n=1}^{N_{DT}} N(\mathbf{y}_k^n - \hat{\mathbf{y}}_{k|k-1}, \Gamma_{k|k-1})$, the metric accuracies that contributed to the total hypothesis performance. This measure is the same as Equation (8.4-2), the summation of normalized residual errors of a track. The importance of this performance measure is such that it does not depend on the parametric model mentioned in point 3 above and is a more realistic characterization of track/hypothesis scores.

5. The way that the measurement-oriented MHT is formulated is complex. A possible simplification is the track-oriented MHT due to Kurien [8], which is developed next.

9.2.2   Track-Oriented MHT

In the track-oriented method, every possible measurement is listed for each track. The basic move of the track-oriented MHT of [8] is track split. The same assumption is made, that a measurement can be a continuation of this track, a false alarm, or a new target. Following the discussion in [8], the track-oriented approach constructs a target tree for each target. The root of each target tree represents the birth of the target, and the branches represent the different dynamic models that may be used by the target and the various reports with which the target may be associated in subsequent scans. A trace of successive branches from the root to a leaf of the tree represents a potential track. Typically, fewer hypotheses are generated in the track-oriented MHT than in the measurement-oriented MHT.

The track-oriented MHT allows for tracking of the tree growth while keeping the number of hypotheses to a much more manageable level. From the remark stated in

[2], the computational burden for a track-oriented approach is more manageable. The decision on which approach to use is application dependent and will be left to the application engineer.

Another important feature of the track-oriented MHT is such that different target dynamics models (such as maneuvering or not maneuvering) are part of the multiple track formation. This is a very useful feature. The means to incorporate this feature, while maintaining the computational burden under control, is a current development issue [2, 8].

The track-oriented MHT concept is illustrated in Figure 9.4. The state of an existing track is split into three different predictions: the target is nonmaneuvering, maneuvering, or terminating. Using measurements contained in the AG, tracks are further split using measurements for track continuation, missed detection, and initiation of new tracks.

Each target tree is a track hypothesis. A global hypothesis is formed by combining tracks from different target trees by picking at most one track from each target tree. This is illustrated in Figure 9.5.

The track-oriented MHT example is presented in Figure 9.6, using the same scenarios as the measurement-oriented MHT shown in Figure 9.1. For the purpose of simplicity in illustration, this model does not include target termination or the use of multiple motion models.

Contents listed under Scan 1 through Scan 4 in Figure 9.6 are the track indicator files. For example, (m, n, o, p) indicates measurements absent or present using either 0 or 1 for its entries, respectively. All possible tracks are listed as a function of scan numbers. The entry of the first column represents an existing track containing one measurement, denoted by (1). On the second scan, a new measurement is obtained. The new measurement has three possibilities; it (a) may not be assigned to the existing

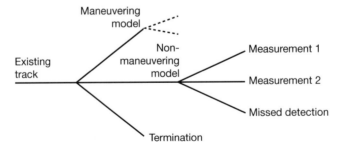

Figure 9.4   Illustration of the track-oriented MHT concept.

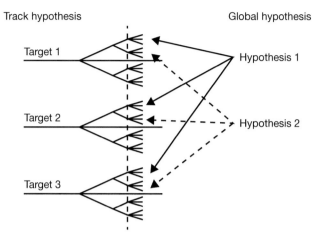

Figure 9.5   Illustration of track and global hypotheses.

Scan number	1	2	3	4
Number of measurements	One existing track	1	1	1
Contents in track indicating files: $(m, n, o, p, ...)$ m: measurement indication of the first scan, n: measurement indication of the second scan, ... 0: no measurement is included 1: there is a measurement at the corressponding scan	(1)	(1, 0) (1, 1) (0, 1)	(1, 0, 0) (1, 1, 0) (0, 1, 0) (1, 0, 1) (1, 1, 1) (0, 1, 1) (0, 0, 1)	(1, 0, 0, 0) (1, 1, 0, 0) (0, 1, 0, 0) (1, 0, 1, 0) (1, 1, 1, 0) (0, 1, 1, 0) (0, 0, 1, 0) (1, 0, 0, 1) (1, 1, 0, 1) (0, 1, 0, 1) (1, 0, 1, 1) (1, 1, 1, 1) (0, 1, 1, 1) (0, 0, 1, 1) (0, 0, 0, 1)
Number of tracks	1	3	7	15

Figure 9.6   Illustration of track-oriented MHT for four scans with one track starting and one measurement for each scan.

track, (1, 0); (b) may be the continuation of the existing track, (1, 1); or (c) may be the start of a new track, (0, 1). With this logic, the first three tracks in Scan 3 of Figure 9.6 are tracks from Scan 2 that do not have the new measurement assigned. Tracks 4 to 6 are tracks with the new measurement assigned. Track 7 represents the new measurement starting a new track, (0, 0, 1). The reader can continue the track forming process to generate all tracks appearing in Scan 4. Using the track-oriented approach, the number of possible tracks generated from a single target with a single measurement at each scan grows exponentially as $2^K - 1$, where K is the current scan number.

We now relate the scan number definitions of Figure 9.6 with those of Figure 9.1; the measurement at the first scan in Figure 9.6 is an existing track at Scan N in Figure 9.1. The measurements of Scans 2 through 4 in Figure 9.6 correspond to the measurements of scans $N + 1$, $N + 2$, and $N + 3$ in Figure 9.1. The number of tracks at the fourth scan is $2^4 - 1$ or 15. The -1 in the formula $2^k - 1$ is due to the fact that this process starts with one existing track in Scan 1 (or Scan N) where k is the scan index. The number of tracks in Figure 9.1 remains one, which is track (1, 1, 1, 1) in Figure 9.6. This is the case when the track using the new measurement is the only continuing track.

Next, consider the hypothesis forming methodology using the example of Figure 9.6 in which a single starting track continues with a single measurement at each scan. This is illustrated in Figure 9.7. The same track numbers will continue only when they are not assigned with a new measurement such as T_1, as shown in Figure 9.7(b). A new track number is given when a new measurement is assigned, such as T_2 in Figure 9.7 (b). Consider the case where there is only one track at the first scan, denoted by T_1.

1. At the second scan, a new measurement $\mathbf{y}_2^1$ is obtained, which is the first detection in the second scan, as shown in Figure 9.7(a). Three possible tracks result, as shown in Figure 9.7(b).

   a. T_1: (1, 0) continuation of T_1 without assigning new measurements,

   b. T_2: (1, 1) continuation of T_1 assigned with $\mathbf{y}_2^1$,

   c. T_3: (0, 1) a new track starting with $\mathbf{y}_2^1$.

   These result in two hypotheses, as shown in Figure 9.7(c):

   a. H_1: T_1, T_3,

   b. H_2: T_2.

A hypothesis may contain multiple tracks and must account for all past and current measurements. A measurement can only be used once in a hypothesis. In this case, there is an initial track, T_1, and one measurement at the current scan, $\mathbf{y}_2^1$. The first hypothesis, H_1, consists of T_1 with no measurement assigned at the current scan and T_3

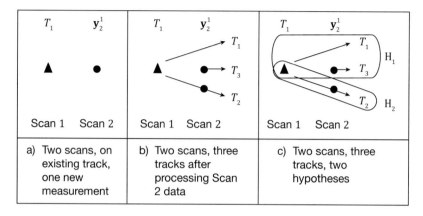

Figure 9.7   Illustration of evolution of tracks and hypotheses, including two scans, one starting track, and one measurement.

where T_3 is a new track with only the new measurement $\mathbf{y}_2^1$. The second hypothesis, H_2, consists of T_2, which is the continuation of T_1 assigned with $\mathbf{y}_2^1$. Both hypotheses thus contain the past and new measurement. Note that the boxes identifying H_1 and H_2 in Figure 9.7(c) are used for illustration purposes.

2. At the third scan, a new measurement $\mathbf{y}_3^1$ is obtained. The number of tracks is increased to seven. The track evolution is shown as follows.

   a. T_1: $(1, 0, 0)$ continuation of T_1 without assigning a new measurement,

   b. T_2: $(1, 1, 0)$ continuation of T_2 without assigning a new measurement,

   c. T_3: $(0, 1, 0)$ continuation of T_3 without assigning a new measurement,

   d. T_4: $(1, 0, 1)$ T_1 assigned with $\mathbf{y}_3^1$,

   e. T_5: $(1, 1, 1)$ T_2 assigned with $\mathbf{y}_3^1$,

   f. T_6: $(0, 1, 1)$ T_3 assigned with $\mathbf{y}_3^1$,

   g. T_7: $(0, 0, 1)$ $\mathbf{y}_3^1$ starts as a new track.

   These result in five hypotheses:

   a. H_1: T_1, T_3, T_7,

   b. H_2: T_2, T_7,

   c. H_3: T_3, T_4,

   d. H_4: T_1, T_6,

   e. H_5: T_5.

372 Chapter 9: Multiple Hypothesis Tracking Algorithm

Similar to the previous step, a hypothesis may contain multiple tracks. These tracks must account for all past and current measurements. A measurement can only be used once in a hypothesis. T_1 through T_3 are continuations of the same tracks without assigning a new measurement. T_4, T_5, and T_6 are tracks T_1, T_2, and T_3 from a previous scan assigned with $\mathbf{y}_3^1$, respectively. T_7 is the new track using the new measurement $\mathbf{y}_3^1$. Each hypothesis contains a combination of tracks, which in total, account for all the measurements for the three scans where a measurement can only be used once.

Based upon these observations, the logic on track and hypothesis generation can be summarized as:

1. Track number generation

   a. Tracks keep the same number when they have no new measurements assigned to them

   b. The next set of new track numbers is generated for tracks assigned with new measurements

   c. A new set of tracks is generated with each new measurement of the current scan

2. Hypothesis number generation

   a. The lowest hypothesis number contains the greatest number of tracks

   b. The hypotheses containing tracks with various combinations follow with decreasing number of tracks

   c. The last hypothesis contains tracks that consist of all measurements of all scans (for the example above, there was only one measurement per scan resulting in only one such track in this hypothesis)

   Generation of the fourth scan is left as an exercise for the reader.

Remarks

For sensors with low false alarm and high detection probabilities, the obvious answer of the correct track is T_2 for the second scan and T_5 for the third scan, because they contain all measurements up to that point. Scoring of a track will thus depend on at least four factors (a) accuracy of the state estimate; (b) track status, for example, new tracks, tracks without an assigned measurement for certain number of scans, and so on; (c) the sensor's detection probabilities; and (d) density of new targets. These factors and how they are used for track scoring will be discussed in a later section.

Track-Oriented Multiple Hypothesis Tracker Probability Calculation

Following [8], a global hypothesis $\theta_k^i \in \boldsymbol{\theta}_k$ given all data $\mathbf{Y}_{1:k}$ can be computed recursively using Bayes' rule

$$\Pr\left\{\theta_k^i | \mathbf{Y}_k, \mathbf{Y}_{1:k-1}\right\} = \frac{1}{c} p\left\{\mathbf{Y}_k | A_g^i, \theta_{k-1}^j, \mathbf{Y}_{1:k-1}\right\} \Pr\left\{A_g^i | \theta_{k-1}^j, \mathbf{Y}_{1:k-1}\right\} \Pr\left\{\theta_{k-1}^j | \mathbf{Y}_{1:k-1}\right\},$$

Similar to the development in Section 9.2.1, the right-hand side of the above equation is the product of three terms: the first term consists of the residual densities and sensor detection, missed detection, and false alarm probabilities, the second term consists of combinations of target and measurement types, and the third term is the prior probability. The derivation for the final equation is similar to that described in Section 9.2.2. The details will not be repeated here. The key variables are defined below before showing the final hypothesis probability equation.

Track-oriented MHT considers additional target characteristics, that is, target termination and multiple target motion models, and so on. They are included in the hypothesis probability equation. The following is an expanded list of variables.

P_d sensor detection probability

P_T target termination probability

P_M target maneuvering probability

N_k number of measurements (or sensor reports)

N_{TGT} number of targets

N_{DT} number of detected targets

N_E number of existing targets (continuing from the previous scan)

N_T number of terminated targets

N_M number of maneuvering targets

N_{FT} number of false targets

N_{NT} number of new targets

β_{FT} density of false targets

β_{NT} density of new targets

$F_N(\lambda)$ Poisson probability for N events with an average rate of λ (i.e., $F_N(\lambda) = \lambda^N/N!$).

λ_{FA} false alarm rate of the Poisson probability

λ_{NT} new target rate of the Poisson probability

The second term in the above equation considers the combinatorics of various conditions, which include the following six variables.

N_1 number of combinations in which N_T out of N_{TGT} is designated as terminated

N_2 number of combinations in which N_D out of N_{TGT} is designated as detected

N_3 number of combinations in which N_M out of N_{DT} is designated as maneuvering

N_4 number of combinations in which N_E out of N_{DT} is designated as continuing from previous scan

N_5 number of combinations in which N_{NT} out of $N_k - N_E$ is designated as a new target

N_6 number of combinations in which N_{DT} out of N_k is designated as a detected target

The final probability equation is written as the product of five terms, with the fifth term being the prior probability

$$\Pr\left\{\theta_k^i | \mathbf{Y}_k, \mathbf{Y}_{1:k-1}\right\} = \frac{1}{c} P_1 P_2 P_3 P_4 \Pr\left\{\theta_{k-1}^j | \mathbf{Y}_{1:k-1}\right\},$$

P_1 through P_4 represent combinatorics due to the number of target types (P_1) and the residual density of maneuvering targets (P_2), nonmaneuvering targets (P_3), and new targets (P_4). These are expressed as follows [8]

$$P_1 = P_T^{N_T}\left((1-P_T)(1-P_d)(1-P_M)\right)^{N_{TGT}-N_{DT}-N_M},$$

$$P_2 = \prod_{i \in J_M} \frac{(1-P_T)P_d P_M p_T(\mathbf{y}_k^i | \theta_{k-1}^j, \mathbf{Y}_{1:k-1})}{\lambda_{FA} p_{FT}(\mathbf{y}_k^i | \theta_{k-1}^j, \mathbf{Y}_{1:k-1})},$$

$$P_3 = \prod_{i \in J_{NonM}} \frac{(1-P_T)P_d(1-P_M)p_T(\mathbf{y}_k^i | \theta_{k-1}^j, \mathbf{Y}_{1:k-1})}{\lambda_{FA} p_{FT}(\mathbf{y}_k^i | \theta_{k-1}^j, \mathbf{Y}_{1:k-1})},$$

$$P_4 = \prod_{i \in J_{NT}} \frac{\lambda_{NT} p_{NT}(\mathbf{y}_k^i | \theta_{k-1}^j, \mathbf{Y}_{1:k-1})}{\lambda_{FA} p_{FT}(\mathbf{y}_k^i | \theta_{k-1}^j, \mathbf{Y}_{1:k-1})}.$$

where J_M, J_{NonM}, and J_{NT} are the indices for maneuvering, nonmaneuvering, and new target sets, respectively. The density functions $p_T(\mathbf{y}_k^i | \theta_{k-1}^j, \mathbf{Y}_{1:k-1})$, $p_{FT}(\mathbf{y}_k^i | \theta_{k-1}^j, \mathbf{Y}_{1:k-1})$, and $p_{NT}(\mathbf{y}_k^i | \theta_{k-1}^j, \mathbf{Y}_{1:k-1})$ are residual densities for targets, false targets, and new targets, respectively.

9.2.3   Track and Hypothesis Generation Example; Multiple Target Case

Consider the case where two tracks share an overlapped acceptance gate, as shown in Figure 9.8(a). As before, target termination and multiple target motion models are not considered in this example. Three new measurements are detected in this scan with two contained in the overlapped gate region and one in the gate of T_1. An assignment matrix is illustrated in Figure 9.8(b). This matrix is similar to the one introduced in Figure 8.4 but with some extensions, as explained below. Four regions labeled A, B, C, and D are shown in this assignment matrix. The $\lambda_{i,j}$'s in region A are the same as defined in Section 8.5 with the exception of $(T_2, \mathbf{y}_{N+1}^1)$, which is denoted by ∞ because $\mathbf{y}_{N+1}^1$ does not fall within T_2's acceptance gate, as shown in Figure 9.8(a). Region B denotes missed detections. Region C denotes new tracks. Region D denotes the area in which any pairing should not impact any decision and carries the entry 0. Entries with ∞ indicate impossible pairings.

We define β_{NT} as the density of new targets; and β_{FT} as the density of false targets. Then the two constants η and κ represent missed detections and new targets, respectively. They are defined by the following equations [14].

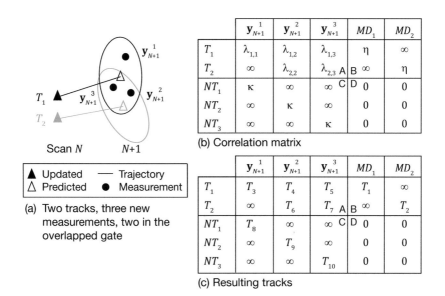

Figure 9.8   Illustration of multiple tracks with multiple measurements.

$\eta = -\ln(1 - P_d)$, for missed detection (MD)

$\kappa = -\ln(\beta_{NT}/\beta_{FT})$, for new target (NT),

where P_d is a sensor detection probability. For a radar with high detection probability, for example, $P_d = 0.99$, then $\eta = 4.6$. For a more noisy radar with $P_d = 0.5$, then $\eta = 0.69$. These values are to be evaluated against $\lambda_{i,j}$, when the assigned target-measurement pairs are from the same target, then $\lambda_{i,j}$ follows a normalized χ^2 density. We can apply similar observations to target density and false alarm density represented by κ.

Using the same convention as in Section 9.2.2, we obtain the following tracks after considering new measurements,

T_1 $(T_1, 0)$ continuation of T_1 without assigning a new measurement[6]

T_2 $(T_2, 0)$ continuation of T_2 without assigning a new measurement,

T_3 $(T_1, 1)$ T_1 assigned with $\mathbf{y}_{N+1}^1$,

T_4 $(T_1, 2)$ T_1 assigned with $\mathbf{y}_{N+1}^2$,

T_5 $(T_1, 3)$ T_1 assigned with $\mathbf{y}_{N+1}^3$,

T_6 $(T_2, 2)$ T_2 assigned with $\mathbf{y}_{N+1}^2$,

T_7 $(T_2, 3)$ T_2 assigned with $\mathbf{y}_{N+1}^3$,

T_8 $(0, 1)$ new track starts with $\mathbf{y}_{N+1}^1$,

T_9 $(0, 2)$ new track starts with $\mathbf{y}_{N+1}^2$,

T_{10} $(0, 3)$ new track starts with $\mathbf{y}_{N+1}^3$.

Note that T_2 cannot be assigned with $\mathbf{y}_{N+1}^1$ because $\mathbf{y}_{N+1}^1$ is not contained within the acceptance gate. Based upon these tracks, 10 hypotheses are formed and are the unique combinations of tracks in Figure 9.8(c) under the constraint that each measurement can only be used once in a hypothesis.

1. H_1: T_1, T_2, T_8, T_9, T_{10},

2. H_2: T_1, T_6, T_8, T_{10},

3. H_3: T_1, T_7, T_8, T_9,

4. H_4: T_2, T_3, T_9, T_{10},

5. H_5: T_2, T_4, T_8, T_{10},

6. H_6: T_2, T_5, T_8, T_9,

---

6   As defined before, T_1 is used to represent $(T_1, 0)$ because it is the continuation of T_1 with no measurement added.

7. H_7: T_3, T_6, T_{10},

8. H_8: T_3, T_7, T_9,

9. H_9: T_4, T_7, T_8,

10. H_{10}: T_5, T_6, T_8.

Each hypothesis contains the complete set of existing tracks and new measurements. The number of tracks for an N track by M measurement association problem is $(N * M + N + M)$ for the case that all M measurements fall within the intersection of all acceptance gates of N tracks. The reader can prove that the number of the hypothesis is equal to the number of unique combinations of all measurements and tracks. The hypotheses listed above are the same as the feasible association events shown in Section 8.6.

It is clear that the above example gives a track extension and hypothesis formulations for a single scan. The track history offers the connection through time. Hypotheses are extracted from tracks such that each hypothesis contains a unique combination of measurements in time history. It offers multiple solutions and is therefore different from choosing a single set of assignments in a given scan as shown in Section 8.5.

9.2.4 Additional Implementation Methods

The full realization of MHT demands a high computational burden. Ways to implement it within the capability of computing machines is of interest. Here, two implementation concepts that have drawn considerable interest are briefly described below, namely, the m-best [9, 10] and the Lagrangian relaxation method [11, 12].

m-best Implementation of Reid's Algorithm   As illustrated in Section 9.2.2, Reid's algorithm forms a large number of hypotheses quickly. The quality of each hypothesis is characterized by the corresponding hypothesis probability. As the hypothesis number grows larger, with the exception of some dominating hypotheses, the rest of the hypotheses have much smaller values in hypothesis probability. A method to generate the hypotheses in a ranking order is desirable. Murty developed an algorithm to rank all assignments in the order of increasing cost to address problems in operations research [9], known as Murty's algorithm. Cox and Hingorani [10] applied the Murty's algorithm to generate the m-best hypotheses for MHT. The choice of m value is application dependent and the decision of the system designer.

The detail of Murty's algorithm is beyond the scope of this book. Similar to Munkre's algorithm, Murty's algorithm is an efficient method in generating multiple hypotheses for solving assignment problems.

Lagrangian Relaxation for Solving the Multidimensional Assignment Problem   It was shown in Section 8.2 that the MTT problem can be formulated as a multidimensional assignment (MDA) problem. The optimum set of tracks is the solution of those, which achieve the best sum of a posteriori probabilities. The computational burden using an exhaustive search is prohibitively large. Deb [11] and Poor [12] and their associates independently recognized that the Lagrangian relaxation method can be applied to find the most likely hypothesis or a set of the *m*-best hypotheses.

The algorithm takes in a set of tracks with their scores. These tracks must be compatible. Track compatibility is defined as a constraint such that no tracks in a hypothesis can share the same measurements (the feasible set). The basic principle of the Lagrangian relaxation method is to replace the constraints by Lagrange multipliers in the objective function, which is the sum of track scores. The actual implementation of this method depends on the proper choice of Lagrange multipliers in such a way that the final solution is approaching the optimum solution.

The exact details of this method are very complex and beyond the scope of this book. Interested readers should consult [11, 12] and the cited references. For some applications using this method, see the discussion in [2].

Remarks

1. The idea behind MHT is to find all possible combinations of measurements, given time history and type, such that the true combination must be contained in one of the hypotheses. This is thought possible because it is a discrete enumeration problem.

2. The difference between the measurement-oriented approach and the track-oriented approach is in the method of building up hypotheses and tracks, and the way to enumerate all possibilities. The measurement-oriented approach starts by building hypotheses. The track-oriented approach starts by building tracks. The track-oriented approach is more intuitively appealing, with less computational burden [2].

3. The MHT method not only enumerates all measurement combinations in time history, it also accounts for sensor imperfections (missed detection) in a noncooperative environment (false detection due to clutter and emergence of new targets). When sensor false alarm and new target statistics are assumed with parametric models, the fidelity of hypothesis/track probability is dependent on the validity of such assumptions.

4. The most important part of the performance measure is the metric accuracy, which appears as the residual density function in the hypothesis probability equation. This measure does not depend on the assumptions on the parametric model mentioned in Remark 3.

9.3   Track and Hypothesis Scoring and Pruning

In practical implementation of MHT, a score is assigned to each track. The score of a hypothesis is the sum of scores of all tracks contained within the hypothesis. The scoring equations to be introduced here are due to [14]. The method is similar to the probability equations developed in Section 9.2 but with a very significant departure; while the method of [14] makes an assumption on the density of false targets and new targets, but it does not depend on the assumptions of target types (see Section 9.2.1 and 9.2.2 for definitions of target types). Additional terms such as track status and track length are included to account for practical aspects of multitarget tracking. Track and hypothesis scores are used for track and hypothesis pruning in order to keep the total number of tracks and hypotheses under control.

Track Pruning

Scores for tracks can differ by orders of magnitude. For example, the score of a track with many missing measurements will decrease exponentially. Tracks that included measurements from different targets will potentially result in poor metric performance (large $\lambda_{i,j}$ values), and thus lower scores. Tracks with extremely low scores can be deleted. A deletion threshold is based on a comparison with the highest scored track. The threshold setting is an engineering decision based on the system application.

Hypothesis Pruning

The hypothesis score is also used for pruning. Two methods have been considered.

1.   Apply a preselected threshold to the hypothesis score

2.   Keep the K_H best hypotheses (the hypotheses with the K_H highest scores)

The first method may result in a very small or a very large number of hypotheses retained. It is not ideal in either way. A reasonable number of hypotheses must be kept in order to maintain the essential benefit of MHT while not creating an excessive computational burden. The purpose of the second method is to maintain a certain fixed number of hypotheses. In some cases, several hypotheses maintain high scores and the rest of them drop to a low plateau with about equal values. The selection of K_H value in this case becomes important. Pruned hypotheses lead to track deletion; tracks that appear only in pruned hypotheses are deleted.

Hypothesis Merging

Hypothesis merging is based upon an N-scan criterion. Any two or more hypotheses containing the same tracks, after track pruning for the most recent N scans, are merged.

9.3.1   Definition of Track Status

Blackman gave a definition for track life stage that is useful for track scoring that is explored below with some additions and modifications [14, p. 262].

Potential Track

A potential track is a single point track. In the MHT formulation, all detections, whether in an acceptance gate or not, can form new tracks. Also, as part of the MHT formulation, all tracks may continue without being assigned with new measurements. The initial score of a new potential track is determined by new target and false target densities, β_{NT} and β_{FT}, and sensor detection probability. When a potential track is not assigned with a new measurement, its score is decreased exponentially as the number of missed measurements increases.

Tentative Track

A tentative track is an initial group of two or more measurements for which a state vector can be computed but the estimation accuracy is still poor and the number of measurements in the track is still small.

Confirmed Track

A confirmed track is a track that contains a sufficiently large number of measurements and the estimation accuracy (or track score) has met a certain threshold. Both track length (e.g., number of measurements) and estimation accuracy (track score) thresholds are set by system engineers for their applications.

Deleted Track

A track is deleted when it has not received any measurements for a specified number of scans and/or the score has fallen below a threshold. The number of scans allowed for missing measurements is lower for tentative tracks than confirmed tracks. When a potential or tentative track is deleted, it is considered to be a false track. When a confirmed track is deleted, it is considered to be a lost track or a track that moved out of the field of view.

9.3.2   Track and Hypothesis Scoring

Assume that a tracking sensor has collected K scans of measurements labeled 1 to K, $\mathbf{Y}_{1:K}$, such that

$$\mathbf{Y}_{1:K} = \{\mathbf{Y}_1, \mathbf{Y}_2, \mathbf{Y}_3, .., \mathbf{Y}_K\},$$

where $\mathbf{Y}_k = \left\{ \mathbf{y}_k^1, \mathbf{y}_k^2, \mathbf{y}_k^3, .., \mathbf{y}_k^{N_k} \right\} \cup \varnothing$ and $\mathbf{y}_k^{i_k}$ is the i_kth measurement of the kth scan in $\mathbf{Y}_k$.
Let T_i denote the track data indicator of the ith track of a hypothesis θ_K^j

$$T_i = (i_1, i_{1, ..., } i_K),$$

where $i_k = 1$ indicates at the kth scan, $\mathbf{y}_k^{i_k}$ is the measurement vector in T_i, and $i_k = 0$ indicates there is no measurement at the kth scan in T_i. Let ℓ_i be the length of T_i, that is, the total number of scans represented by T_i, in this case, $\ell_i = K$. Let the number of detected measurements in T_i (i.e., the total number of nonzero entries in T_i) be N_d, generally $N_{d_i} \leq \ell_i$ and $N_{d_i} = \ell_i$ when there is no missed detection.

The track scoring method developed by Blackman [14] is stated below.

Potential Track

A potential track consists of a single measurement. The probability of a potential track (P_{PT}) is the sum of the probability of a false track (P_{FT}) and the probability of a new track (P_{NT}), which is

$$P_{PT} = P_{FT} + P_{NT} = \beta_{FT} + \beta_{NT}\left(1 - P_d\right)^{N_m} = \beta_{FT}\left(1 + \frac{\beta_{NT}}{\beta_{FT}}\left(1 - P_d\right)^{N_m}\right),$$

where β_{NT} and β_{FT} are new and false target densities, P_d is the sensor detection probability, and N_m is the number of subsequent missed detections.

Given the above expression, Blackman defined the score of a potential track as

$$L_{PT} = \ln\left[\frac{\beta_{NT}}{\beta_{FT}}\left(1 - P_d\right)^{N_m}\right] = \ln\left(\frac{\beta_{NT}}{\beta_{FT}}\right) + N_m \ln\left(1 - P_d\right).$$

Tentative and Confirmed Track

Both tentative and confirmed tracks are those which have received more than a single measurement so that the state estimate can be computed. The difference is that a confirmed track has passed a certain track length (number of measurements in the track) threshold. The concept of tentative and confirmed tracks mirrors the concept of track initiation and continuation from Section 8.7. The choice of track length threshold is somewhat arbitrary but meaningful. A confirmed track has received many scans of measurements and it is less likely to be the result of false detections. The score for the ith track $T_i = (i_1, i_2, ..., i_K)$ of the hypothesis θ_k^j, $L_{T_i}^j$, is

$$L_{T_i}^j = \ln\left(\frac{\beta_{NT}}{\beta_{FT}}\right) + N_m \ln(1 - P_d) + \sum_{k=1}^{K}\left[\ln\left(\frac{P_d}{\beta_{FT}(2\pi)^{\frac{m}{2}}|\mathbf{S}_{i,k}|^{\frac{1}{2}}}\right) - \frac{\lambda_{i,k}}{2}\right],$$

where $\lambda_{i,k}$ is a normalized filter residual $\mathbf{y}_k^{i_k} - \hat{\mathbf{y}}_{k|k-1}^i$,

$$\lambda_{i,k} = \left(\mathbf{y}_k^{i_k} - \hat{\mathbf{y}}_{k|k-1}^i\right)^T [\mathbf{S}_{i,k}]^{-1}\left(\mathbf{y}_k^{i_k} - \hat{\mathbf{y}}_{k|k-1}^i\right),$$

and $\mathbf{S}_{i,k}$ is the covariance of $\mathbf{y}_k^{i_k} - \hat{\mathbf{y}}_{k|k-1}^i$,

$$\mathbf{S}_{i,k} = \mathbf{H}_{\hat{\mathbf{x}}_{k|k-1}^i}\mathbf{P}_{k|k-1}^i\mathbf{H}_{\hat{\mathbf{x}}_{k|k-1}^i}^T + \mathbf{R}_k^{i_k},$$

with $\hat{\mathbf{x}}_{k|k-1}^i$, $\mathbf{P}_{k|k-1}^i$, $\hat{\mathbf{y}}_{k|k-1}^i$, and $\mathbf{S}_{i,k}$ as the prediction statistics based on measurements in T_i up to scan $k-1$ and $\mathbf{y}_k^{i_k}$ as the measurement vectors contained in track T_i at scan k.

The score for the hypothesis θ_k^j is therefore the summation of scores of all tracks in that hypothesis

$$L^j = \sum_{T_i \in \theta_k^j} L_{T_i}^j.$$

The above equations are summarized in Table 9.1 with some observations given.

In some applications where a track may evolve into multiple tracks, called track breakup in [3], then the MHT scoring method will need to be modified to accommodate the track complexity issue that is equivalent to the model complexity problem for model identification. A track score consists of the metric quality

$$\sum_{k=1}^{K}\left[\ln\left(\frac{P_d}{\beta_{FT}(2\pi)^{\frac{m}{2}}|\mathbf{S}_{i,k}|^{\frac{1}{2}}}\right) - \frac{\lambda_{i,k}}{2}\right],$$

Table 9.1 Track Scoring

Definition	Scoring equation	Observations
Potential track	$\ln\left(\dfrac{\beta_{NT}}{\beta_{FT}}\right)$	Start of new track depends on target and false target densities
Missed detections in a track	$(\ell_i - N_{d_i})\ln(1 - P_d)$	A negative number where the magnitude is increasing as the number of detections in a file N_{d_i} is decreasing, results in the decrement of track score
Track (metric) quality of T_i	$\sum_{k=1}^{K}\left[\ln\left(\dfrac{P_d}{\beta_{FT}(2\pi)^{\frac{m}{2}}\lvert \mathbf{S}_{i,k}\rvert^{\frac{1}{2}}}\right) - \dfrac{\lambda_{i,k}}{2}\right]$ $\lambda_{i,k} = \left(\mathbf{y}_k^{i_k} - \hat{\mathbf{y}}_{k\lvert k-1}^i\right)^T [\mathbf{S}_{i,k}]^{-1}\left(\mathbf{y}_k^{i_k} - \hat{\mathbf{y}}_{k\lvert k-1}^i\right)$ $\mathbf{S}_{i,k} = \mathbf{H}_{\hat{\mathbf{x}}_{k\lvert k-1}^{i_k}}\mathbf{P}_{k\lvert k-1}^i\mathbf{H}_{\hat{\mathbf{x}}_{k\lvert k-1}^{i_k}}^T + \mathbf{R}_k^{i_k}$ with $\hat{\mathbf{x}}_{k\lvert k-1}^i$, $\mathbf{P}_{k\lvert k-1}^i$, $\hat{\mathbf{y}}_{k\lvert k-1}^i$, and $\mathbf{S}_{i,k}$ as the prediction statistics based on the measurements in T_i up to scan $k-1$ and $\mathbf{y}_1^{i_k}$ as the measurement vectors contained in track T_i at scan k.	$\lambda_{i,k}$ is the normalized filter residual that is also the Mahalanobis distance between the measurement and the predicted measurement. Poorer track quality has larger $\lambda_{i,k}$ resulting in decrement of track score.

and track complexity, which includes models for track breakup, lost track, false track, track with missing measurements, and so on. The Akaike criterion is applied in [3] to balance the tendency to over-fit the data with the spawning of more new tracks (higher order and more complex model) versus track breakup (low order and less complex model). The interested reader should consult this reference [3] for details.

9.3.3 Track Scoring Example

An example of track scoring is presented in this section. For the sake of simplicity, measurement noise is assumed to be zero, and thus the track metric quality part of the score is ignored. In this case, the results emphasize the impact of missing measurements. An extension to include measurement noise is straightforward but complicated because a target motion model must be incorporated. The simplest track evolution example, one initial track and one measurement per scan for five scans (one more than shown in Figure 9.6), is used. The extension to multiple target cases is tedious but

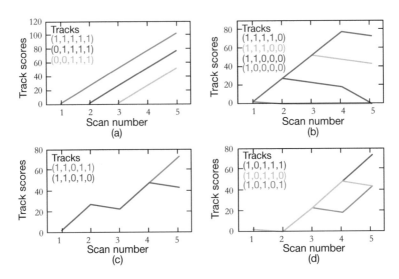

Figure 9.9   Track file and track score examples.

straightforward. In this example, a track is promoted from tentative to a confirmed track after receiving four consecutive scans of data. A tentative track becomes a false track when it has three scans of missing measurements (the three scans do not have to be consecutive).

The results are shown in Figure 9.9. The corresponding track files and track scores are color coded and should be easily identifiable. The track file convention is the same as in Figure 9.6. For example, in Figure 9.9(a), (1, 1, 1, 1, 1) indicates all five scans contain measurements, (0, 1, 1, 1, 1) with measurements missing at the first scan, and (0, 0, 1, 1, 1) with measurements missing at the first and the second scan, and so on. The corresponding scores show three parallel lines each reflecting one scan delayed track start. Figure 9.9(b) shows that scores decrease linearly as measurements are missing at the end of each track. After missing three scans of measurements, the track score reaches zero. Figure 9.9(c) shows that scores go down with missing measurements but rebound with new measurements. Figure 9.9(d) shows the same track scores for those with an equal number of measurements and missed measurements regardless of their timeline distribution; see the blue and green tracks.

9.4   Multiple Hypothesis Tracker Implementation Using Nassi–Shneiderman Chart

On a functional level, the core MHT is depicted in the flow diagram shown in Figure 9.2. At each new scan time existing tracks are extrapolated to the time where new

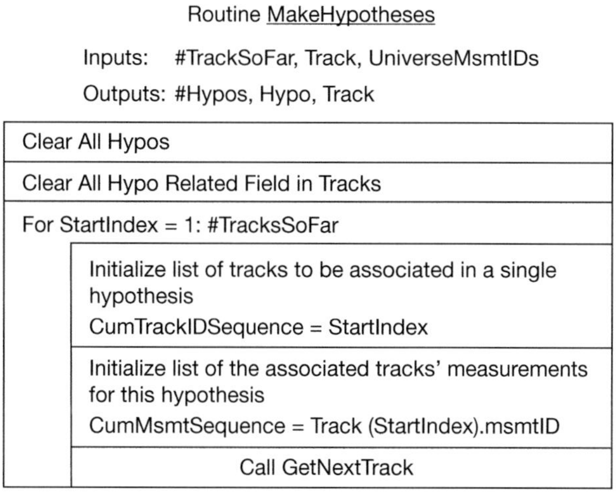

Routine <u>MakeHypotheses</u>

Inputs: #TrackSoFar, Track, UniverseMsmtIDs
Outputs: #Hypos, Hypo, Track

Clear All Hypos
Clear All Hypo Related Field in Tracks
For StartIndex = 1: #TracksSoFar

	Initialize list of tracks to be associated in a single hypothesis CumTrackIDSequence = StartIndex
	Initialize list of the associated tracks' measurements for this hypothesis CumMsmtSequence = Track (StartIndex).msmtID
	Call GetNextTrack

Figure 9.10   Nassi–Shneiderman chart: process driver for making hypotheses.

measurements are expected, with their association gates computed. Together with the new measurements at the new scan time, associations between existing tracks and the new data are determined, and tracks and hypotheses are formed. The scan is completed with track and hypothesis pruning. The process is repeated for each new scan time.

A convenient tool for structured design involving recursion is the Nassi–Shneiderman chart [17, 18]. It's application to MHT implementation is illustrated here.[7]

In hypothesis formation, tracks are associated with hypotheses, and hypotheses are associated with tracks. Inputs to the hypothesis formation subroutine, MakeHypotheses, consist of a list of track structures, their number, and the number of measurements obtained to date. The Nassi–Shneiderman chart for MakeHypotheses is shown in Figure 9.10.

Starting with the first candidate track in the list, subroutine GetNextTrack, shown in Figure 9.11, is called, where the next track on the list is checked for duplication of measurements. If there is no duplication, GetNextTrack is called again, recursively (from within itself), and again and again as necessary until all measurements to date are accounted for. If not all measurements can be accounted for without duplication,

---

7   The Nassi–Shneiderman tracker implementation method was brought to the authors' attention by Fannie Rogal of MIT Lincoln Laboratory. The associated materials in this section and the track-scoring example of Section 9.3.3 were provided by her.

Routine <u>GetNextTrack</u>

Inputs: #TrackSoFar, Track, StartIndex, #Hypos, Hypo, UniverseMsmtIDs,
CumTrackIDSequence, CumMsmtSequence

Outputs: #Hypos, Hypo, Track, CumTrackIDSequence, CumMsmtSequence

Does Input CumMsmtSequence Contain Duplicates?			
Y			N
Input Tracks are not disjoint, Not a Valid Set of Tracks	Are All Measurements to Date (UniversMsmtIDs) Accounted for in the set of input tracks?		
	Y		N
	Make a new Hypo: Assign Hypo the CumTrackID Sequence and the score that is the sum of its constituent tracks' scores	Save Copies of CumTrackIDSequence and CumMsmtSequence	
	Update Track to show this Hypo Uses It and Increment Hypo Counter in Track	For I = 1: #TracksSoFar	
		CumTrackIDSequence and CumMsmtSequence = Saved Copies	
		Increment StartIndex	
		StartIndex ≤ #TracksSoFar?	
		Y	N
		Add StartIndex to CumTrackIDSequence Add Track (StartIndex) Msmt IDs to CumMsmt Seqence Call GetNextTrack	Reached end of adding possible tracks but couldn't get all msmts represented Clear CumTrackID and CumMsmtSequences

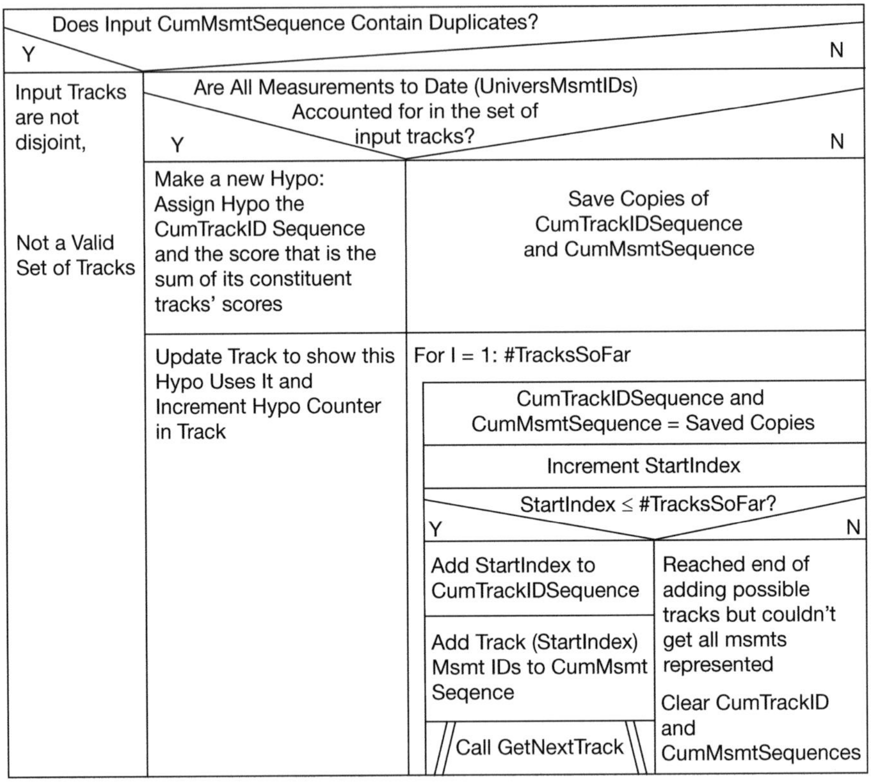

Figure 9.11   Nassi–Shneiderman chart: process to associate input tracks with hypotheses.

the potential hypothesis is invalid and exits the thread of recursion. If they can, then a legitimate hypothesis is formed, and its score is the sum of its constituent tracks. In both cases, the return is to MakeHypotheses.

The second candidate track in the list calls GetNextTrack, and the process is repeated.

In MakeHypotheses, likewise each remaining candidate track results in a call to GetNextTrack.

Outputs from MakeHypotheses consist of a list of hypothesis structures, their number, and updated track structure fields as to the number of hypotheses to which they are constituent.

The design shown here is rudimentary, and does not leverage the construction of the prior scan's hypotheses. Further, it does not consider possible measurement-sharing as may be the case in crossing track or in closely-spaced object (CSO) conditions, or hypothesis merging. It is shown here to stress the fact that software engineering tools may be available to handle a complicated enumeration problem such as MHT.

9.5   Extending It to Multiple Sensors with Measurement Fusion

Following the development of Chapter 7, State Estimation for Multiple Sensors, the extension of MHT to multiple sensors is straightforward. When measurements of different sensors are used directly to update the state estimate, as in Section 7.2.1, the only thing that must be made clear is the difference in transformation between the state coordinate and the measurement coordinate, Equations (7.2-1) through (7.2-5). The block diagram in Figure 9.2 still applies. When computing the predicted measurement and acceptance gate, it must be checked that they are in the same coordinates as the respective sensor.

9.6   Concluding Remarks

The essence of MHT was introduced in this chapter. The purpose is to help readers grasp the principles of MHT. MHT exhaustively enumerates all possible measurement combinations and has the capability to use future measurements to correct past measurement assignments. It is therefore expected that MHT will make fewer mistakes than other traditional methods. This is, however, achieved with a heavy computational burden. The exact choice of approach will be dependent on particular applications.

It is evident that MHT can quickly expand into a large number of tracks and hypotheses even with a small total number of targets. Targets spread over large geographical areas and traveling to different directions need not be jointly processed by MHT. For example, an air traffic control system may apply MHT to the group of objects approaching the airport. This group does not have to be processed together with objects that are only in transit through the same air traffic control area. In this case objects may be preclustered so that MHT is applied only to objects in the same cluster; this is necessary for the purpose of saving computation time and is suggested in the MHT literature [e.g., 1, 2].

MHT is considered to be an optimum MTT technique that was developed with more mathematical rigor. Its implementation for application problems still appears in new publications. Interested readers are encouraged to explore them for further research.

Homework Problems

1. In Section 9.2, the track and hypothesis generation in Figure 9.3 up to Scan 3 were illustrated. Use the same approach to complete the track and hypothesis generation to Scan 4.

2. There are $N * M$ entries of an N by M matrix. The number of choices for the set of all unambiguous pairing is $N!/(N - M + 1)!$ for $N \geq M$. Using Figure 9.9 as a guideline, derive the number of tracks and number of hypotheses for N existing tracks associated with M new measurements.

3. Apply the correlation matrix method illustrated in Figure 9.9 to the track and hypothesis evolution examples of Section 9.2. Show that the results obtained either way are the same. Write a flow diagram to show how one can evolve new tracks using the entries of the assignment matrix and new hypotheses as the combinations of the entries of the matrix.

4. Repeat Problem 1 of Chapter 8 with MHT. Compare results with

   a. Track purity

   b. False track rate

   c. Track consistent rate

   d. Computer runtime

References

1. D.B. Reid, "An Algorithm for Tracking Multiple Targets," *IEEE Transactions on Automatic Control*, vol. AC-24, pp. 843–854, Dec. 1979.

2. S.S. Blackman, "Multiple Hypothesis Tracking for Multiple Target Tracking," *IEEE Transactions on Aerospace and Electronic Systems Magazine*, vol. 19, pp. 5–18, Jan. 2004.

3. M.K. Chu, "Target Breakup Detection in the Multiple Hypothesis Tracking Formulation," M.E. thesis, MIT, 1996.

4. R.W. Sittler, "An Optimal Data Association Problem in Surveillance Theory," *IEEE Transactions on Military Electronics*, vol. MIL-8, pp. 125–139, Apr. 1964.

5. R.A. Singer, R.G. Sea, and K.B. Housewright, "Derivation and Evaluation of Improved Tracking Filters for Use in Dense Multi-Target Environments," *IEEE Transactions on Information Theory*, vol. IT-20, pp. 423–432, July 1974.

6. C.L. Morefield, "Application of 0-1 Integer Programming to Multi-Target Tracking Problem," *IEEE Transactions on Automatic Control*, vol. AC-22, vol. 302–312, June 1977.

7. Y. Bar-Shalom, "Tracking Methods in a Multi-Target Environment," *IEEE Transactions on Automatic Control*, vol. AC-23, pp. 618–626, Aug. 1978.

8. T. Kurien, "Issues in the Design of Practical Multi-Target Tracking Algorithms," in *Multitarget Multisensor Tracking: Advanced Applications*, Y. Bar-Shalom, Ed., pp. 43–83, Norwood, MA: Artech House, 1990.

9. K.G. Murty, "An Algorithm for Ranking All the Assignments in Order of Increasing Cost," *Operations Research*, 16, 682–687, 1968.

10. I.J. Cox and S.L. Hingorani, "An Efficient Implementation of Reid's Multiple Hypotheses Tracking Algorithm and Its Evaluation for the Purposes of Visual Tracking," *IEEE Transactions on Pattern Analysis and Machine Intelligence*, vol. 18, pp. 138–150. Feb. 1996.

11. S. Deb, M. Yeddanapudi, K. Pattipati, and Y. Bar-Shalom, "A Generalized S-D Assignment Algorithm for Multi-Sensor-Multi State Estimation," *IEEE Transactions on Aerospace and Electronic Systems*, vol. AES-33, 523–537, April 1997.

12. A.B. Poore and A. J. Robertson, "A New Lagrangian Relaxation Based Algorithm for a Class of Multidimensional Assignment Problems," *Computational Optimization and Applications*, vol. 8, pp. 129–150, Sept. 1997.

13. S.S. Blackman and R. Popoli, *Design and Analysis of Modern Tracking Systems*. Norwood, MA: Artech House, 1999.

14. S.S. Blackman, *Multiple-Target Tracking with Radar Applications*. Norwood, MA: Artech House, 1986.

15. Y. Bar-Shalom, X. Rong Li, and T. Kirubarajan, *Estimation with Applications to Tracking and Navigation*. New York: Wiley, 2001.

16. Y. Bar-Shalom, P. Willett, and X. Tian, *Tracking and Data Fusion: A Handbook of Algorithms*. Storrs, CT: YBS Publishing, 2011.

17. I. Nassi and B. Shneiderman, "Flowchart Techniques for Structured Programming," *ACM SIGPLAN Notices*, vol. 8, pp. 12–26, Aug. 1973.

18. C.M. Yoder and M.L. Schrag, "Nassi–Shneiderman Charts; An Alternative to Flowcharts for Design," in *Proceedings of ACM SIGSOFT/BIGMETRICS Software and Assurance Workshop*, pp. 386–393, Nov. 1978.

10

Multiple Sensor Correlation and Fusion with Biased Measurements

10.1 Introduction

Estimation algorithms for multiple sensor systems were introduced in Chapter 7. Two fusion architectures were discussed: measurement fusion and state fusion. Measurement fusion achieves optimal state estimation and is more straightforward algorithmically. State fusion on the other hand is suboptimal but would require less communication bandwidth if the local estimates are being distributed less frequently than the sensor measurements. Several alternate approaches for reducing data rate were discussed for both fusion architectures.

There are many advantages to using a multiple sensor system. Some of them were mentioned in Chapter 7: improved track accuracy through geometric diversity, improved coverage of shared surveillance areas resulting in higher probability of detection, continuous coverage across contiguous surveillance regions resulting in better track continuity and object identification, the ability to view the same object complex at different aspect angles with possibly different frequencies (different sensors operating at different wave bands) resulting in a rich set of observables to allow for more robust exploitation of phenomenology differences, and enhanced object identification and possible resolution to track ambiguities. Specifically within the scope of this book, improvement in estimation accuracy is of interest. The ability to achieve these advantages depends on at least two factors:

1. Ability to handle track ambiguities; a problem of association and correlation.

2. Ability to estimate and mitigate sensor bias; the subject of this chapter.

Association/correlation is the subject of Chapters 8 and 9, in which several approaches were discussed in detail, see Table 8.1, Multiple Target Tracking Taxonomy. In this chapter, multiple sensor estimation and correlation in the presence of sensor bias[1] will be discussed. The goal is to estimate both state and bias at the same time.

This chapter is organized as follows. Section 10.2 compares two approaches for jointly estimating state and bias using sensor measurements. Both joint state and bias estimation approaches are in the category of measurement fusion as defined in Chapter 7. It is assumed that the measurement to state association has already been accomplished. Methods for measurement to state association for this problem are the same as those discussed in Chapters 8 and 9 and will therefore not be repeated here. The focus is on joint bias and state estimation. Several approaches addressing this type of problem are addressed in the literature [1–5]. In this section, joint bias and state estimation for a space trajectory using the state augmentation method is illustrated with the Cramer–Rao bound (CRB).

Section 10.3 discusses the joint bias estimation and state-to-state correlation problem. An individual sensor can estimate the target states of multiple targets while ignoring the presence of bias in their sensors, resulting in biased state estimates. When one is to estimate the biases using state estimates from multiple sensors, correlation and bias estimation becomes a joint problem. The approach in this category is state fusion, covered in Chapter 7. State fusion with biased measurement has drawn some attention in the literature [6–20].

10.2   Bias Estimation Directly with Sensor Measurements

Two approaches on joint bias and state estimation are compared by means of numerical examples. The mathematics of both approaches is straightforward, but the general conclusion is very different and illuminating. The content of this section is a revision of the paper shown in Corwin et al. [1]. Readers are encouraged to build computer models to understand the results and develop their own interpretations.

10.2.1   Problem Statement

Consider a space trajectory described by the following deterministic difference equation of motion

$$\mathbf{x}_k = \mathbf{f}(\mathbf{x}_{k-1}, t_k). \tag{10.2-1}$$

---

1   The Schmidt–Kalman filter (SKF) introduced in Chapter 4 was designed to account for the residual bias, which is what is left after bias estimation and correction. The covariance of estimated bias is used as input to SKF. The purpose of SKF is to maintain state covariance consistency with the actual error.

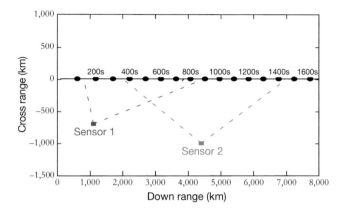

Figure 10.1   Geometry of target trajectory and two radars.

A simple space trajectory with a spherical earth model was suggested in Homework Problem 2 in Chapter 4. Readers interested in more refined space trajectory dynamics can consult Bate et al. [21]. We emphasize that a more detailed trajectory model will not change the conclusion on joint bias and state estimation, that is, the simple model gives the same qualitative conclusion as a higher fidelity model.

Two radars providing measurements on $\mathbf{x}_k$ in discrete time t_k are represented by the following relationship

$$\mathbf{y}_k^1 = \mathbf{h}^1\left(\mathbf{x}_k\right) + \mathbf{b}_k^1 + \mathbf{v}_k^1, \tag{10.2-2}$$

$$\mathbf{y}_k^2 = \mathbf{h}^2\left(\mathbf{x}_k\right) + \mathbf{b}_k^2 + \mathbf{v}_k^2, \tag{10.2-3}$$

where $\mathbf{v}_k^1$ and $\mathbf{v}_k^2$ are measurement noise with known covariance, and $\mathbf{b}_k^1$ and $\mathbf{b}_k^2 \in \mathbb{R}^m$ are the measurement biases of the two respective sensors. Both measurement biases are assumed as unknown constants. Simplified radar measurement equations can be found in Homework Problem 1 in Chapter 3.

The measurement geometry of this example is shown in Figure 10.1. Two radars take measurements on the trajectory independently with the following specifications.[2]

---

2   These choices are purely arbitrary. Readers are welcome to make choices of their own. The goal here is to use an example to draw a qualitative conclusion.

Sensor 1

Position	1100 km down range (DR), −700 km cross range (CR)
Update rate	10 Hz
Measurement noise standard deviation	10 m range, 0.5 mrad Az/El
Bias	5 mrad Az, 0.2 mrad El

Sensor 2

Position	4400 km DR, −1000 km CR
Update rate	10 Hz
Measurement noise standard deviation	10 m range, 0.5 mrad Az/El
Bias	−0.3 mrad Az, −0.8 mrad El

10.2.2   Comparison of Two Bias Estimation Approaches

In all cases presented in this section, it is assumed that the measurement to state associations have already been accomplished, for example, using the methods in Chapters 8 and 9. Readers who would like to dig deeper in the area of joint association and bias estimation can consult [3–5].

Two approaches on state and bias estimations are now compared.

1.   Bias and state estimated jointly but by individual sensors,

2.   Bias and state estimated jointly with data from both sensors.

The two approaches are compared using the space trajectory and radar specifications described in Section 10.2.1.

State and Bias Estimation in Local Sensors

In this approach, each sensor works independently. Bias estimation is obtained by augmenting biases as part of the state vector (see Chapter 4). Let $\mathbf{x}_k^1$ and $\mathbf{x}_k^2$ denote the augmented state vectors (ASV) for the first and second radar, that is, $\mathbf{x}_k^\ell = \left[\mathbf{x}_k^T, \mathbf{b}_k^{\ell^T} \right]^T$ for $\ell = 1,2$. Equation (10.2-1) is modified to become

$$\mathbf{x}_k^\ell = \begin{bmatrix} \mathbf{f}\left(\mathbf{x}_{k-1}, \ t_k\right) \\ \mathbf{b}_{k-1}^\ell \end{bmatrix} \qquad (10.2\text{-}4)$$

$$\mathbf{y}_k^\ell = \mathbf{h}^\ell\left(\mathbf{x}_k\right) + \mathbf{b}_k^\ell + \mathbf{v}_k^\ell \qquad (10.2\text{-}5)$$

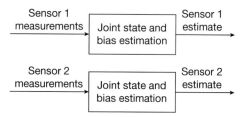

Radars estimate their bias and state independently

Figure 10.2   Each sensor estimating its own bias and state.

for both sensors, $\ell = 1,2$. With this model, state and bias estimates are obtained for each sensor individually as illustrated in Figure 10.2. Note that in this method no attempts are made to fuse the estimates from individual sensors.

Multisensor Joint State and Bias Estimation

In the second approach, an ASV with biases from both sensors is used to jointly estimate state and biases using measurements from both sensors. Let $\mathbf{x}_k^a$ denote the ASV with both biases, that is, $\mathbf{x}_k^a = \left[\mathbf{x}_k^T, \mathbf{b}_k^{1^T}, \mathbf{b}_k^{2^T}\right]^T$, resulting in the following state and measurement equations

$$\mathbf{x}_k^a = \begin{bmatrix} \mathbf{x}_k \\ \mathbf{b}_k^1 \\ \mathbf{b}_k^2 \end{bmatrix} = \begin{bmatrix} \mathbf{f}\left(\mathbf{x}_{k-1}, t_k\right) \\ \mathbf{b}_{k-1}^1 \\ \mathbf{b}_{k-1}^2 \end{bmatrix} \tag{10.2-6}$$

$$\mathbf{y}_k^a = \begin{bmatrix} \mathbf{y}_k^1 \\ \mathbf{y}_k^2 \end{bmatrix} = \begin{bmatrix} \mathbf{h}^1\left(\mathbf{x}_k\right) + \mathbf{b}_k^1 \\ \mathbf{h}^2\left(\mathbf{x}_k\right) + \mathbf{b}_k^2 \end{bmatrix} + \begin{bmatrix} \mathbf{v}_k^1 \\ \mathbf{v}_k^2 \end{bmatrix}. \tag{10.2-7}$$

Note that the joint ASV has a larger dimension than the ASV of the individual sensors. The state estimate obtained with joint ASV is the fused state estimate. The block diagram for this approach is shown in Figure 10.3.

Comparison of Results

Figure 10.4 compares the CRB on the covariance of bias estimates. The blue and green curves are the results of using only Sensors 1 or 2, respectively. Note that the vertical drops correspond to the starting time of these two sensors (Section 10.2.1).

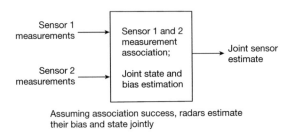

Figure 10.3   Multisensor joint bias and state estimation (measurement fusion).

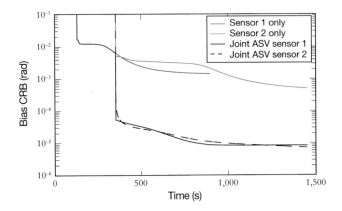

Figure 10.4   Cramer–Rao bound on bias estimation error.

The initial conditions corresponding to the bias components in the CRB were chosen to be very large, representing nearly no a priori information. The estimation error (represented by the CRB) of the individual sensors hit a plateau at about 2 mrad then reduced very slowly. At the end of Sensor 1 coverage, the estimation error is about 1.5 mrad and that of Sensor 2 is about 0.5 mrad. The values are comparable to the magnitudes of the bias. The bias estimation error when the data from both sensors are used jointly drops nearly vertically when the second sensor begins to contribute at about 370 s. The estimation error for the biases of Sensor 1 and Sensor 2 are shown with black solid and black dashed curves, respectively. Note that before the start of the second sensor, the estimation error for Sensor 1 bias follows that when Sensor 1 is operating alone—this is expected. It is interesting to note that when the object moves out of Sensor 1 coverage (at about 900 s), the estimation error for Sensor 1 bias stays constant while that of Sensor 2 still improves (although only slightly). At this point, the errors are about 0.008 to 0.009 mrad. Based upon these observations, one concludes that angular bias is not observable with individual radars but angular biases of both (not co-located) radars are observable when data from both sensors are used jointly for estimation.

The CRB on the position estimation error is shown in Figure 10.5. As shown in the examples in previous chapters, the position error is computed as the scalar value of the CRB given by the square root of the sum of the diagonal terms that correspond to the components of the vector. The observations and conclusions made with Figure 10.4 on bias estimation error can be qualitatively made the same way using the CRB for position estimation error: the estimation error for the individual sensor case plateaus on the order of kilometers, while in the joint estimation case it is reduced to meters.

Remarks

Sensor bias estimation is an important subject in practice. A sensor must be able to estimate its bias and characterize the bias estimation error when sharing its measurement with other sensors. Given the example explored above, the following conclusions can be made.

1. Bias cannot be estimated successfully with state augmentation when sensors operate individually because the ASV with individual bias using individual sensor measurements is unobservable.

2. When multiple uncolocated sensors are used jointly on the same object, bias can be estimated successfully with state augmentation because the joint ASV with both biases is observable.

3. This conclusion is true when the object in track follows a deterministic and completely known trajectory, such as an object traveling in space. Trajectory uncertainties due to target maneuvering or even just air turbulence will reduce the accuracy in bias estimation. In this case, one may consider estimating relative basis, that is, the bias between a pair of sensors.

4. Readers are encouraged to extend this example to two targets or more in order to examine whether the same conclusion still holds.

5. Readers are encouraged to build a trajectory simulation followed with estimators and compare this with the result of the CRB.

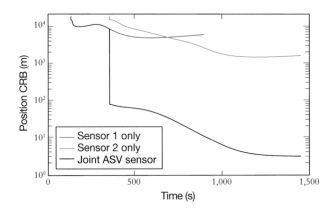

Figure 10.5   Cramer–Rao bound on position estimation error.

10.3   State-to-State Correlation and Bias Estimation

Consider two sensor systems each tracking N and M objects, respectively, where

$\hat{\mathbf{x}}_{k|k}^{1,i}$ and $\mathbf{P}_{k|k}^{1,i}$ are the state estimate and covariance of the first sensor with $i = 1,.., N$, and

$\hat{\mathbf{x}}_{k|k}^{2,j}$ and $\mathbf{P}_{k|k}^{2,j}$ are the state estimate and covariance of the second sensor with $j = 1,.., M$,

and only a subset of the objects is observed by both sensors. The objective is to find and correlate the common objects. This can be formulated as a problem in state-to-state correlation. When there is no measurement bias, the state estimates $\hat{\mathbf{x}}_{k|k}^{1,i}$ and $\hat{\mathbf{x}}_{k|k}^{2,j}$ are also unbiased, and the information needed for decision is in the state estimate and covariance. The solution for this problem is mathematically identical to that of the problem of associating measurements to predicted measurements as discussed in Section 8.5. When measurements are biased and the estimator does not include bias estimation, the resulting state estimates are biased. With biased estimates, the optimization problem formulated in Section 8.5 must be reformulated as a joint bias estimation and assignment problem. The extension of Section 8.5 to state-to-state correlation without bias is addressed first in Section 10.3.1. Bias estimation with a given correlation is presented as a parameter estimation problem in Section 10.3.2. The joint correlation and bias estimation problem is a much more difficult problem. It will be formulated in Section 10.3.3 with two solutions suggested.

Joint correlation and bias estimation has drawn considerable attention in the literature [6–20]. Similar to approaches to multiple target tracking (MTT) problems, the problem of multiple-target multiple-sensor correlation with sensor bias can be formulated as a single scan decision problem or a multiple scan decision problem such as multidimensional assignment and/or multiple hypothesis correlation. In this section, only the single scan correlation approach is presented. Interested readers can consult [16–20] for further studies on using multiple scans.

10.3.1   Review of Fundamental Approaches to Correlation Without Bias

The state-to-state correlation problem with two sensors is analogous to a two-scan association problem that was formulated in Section 8.5 as an assignment problem. In such problems, if $\lambda_{i,j}$ is used to denote the cost of assigning person i to job j, then the objective of the many-on-many assignment is to achieve the minimum total cost (or the inverse, the maximum total performance) that can be formulated as the solution of

$$\text{Min}_{i,j}\Sigma_{i,j}\lambda_{i,j} \tag{10.3-1}$$

where each $\lambda_{i,j}$ can only be used once corresponding to the assumption that each person can only be assigned one job. Note that n and m do not need to be equal. This corresponds to the fact that there may be more jobs than people or vice versa. This constitutes an optimization problem where the best set of assignments is the one that gives the minimum total cost.

Let $\boldsymbol{\theta}$ be the hypothesis that the ith track of the first sensor $(\hat{\mathbf{x}}_{k|k}^{1,i}, \mathbf{P}_{k|k}^{1,i})$ and the jth track of the second sensor $(\hat{\mathbf{x}}_{k|k}^{2,j}, \mathbf{P}_{k|k}^{2,j})$ belong to the same object. Similar to Equation (8.4-1) for the measurement-to-track association problem, the likelihood function of $\hat{\mathbf{x}}_{k|k}^{1,i}$ and $\hat{\mathbf{x}}_{k|k}^{2,j}$ given $\boldsymbol{\theta}$ is

$$\Lambda_{i,j} \triangleq p\left(\hat{\mathbf{x}}_{k|k}^{1,i}, \hat{\mathbf{x}}_{k|k}^{2,j} | \boldsymbol{\theta}\right) = \frac{1}{\left((2\pi)^{2n}\left|\mathbf{P}_{k|k}^{1,i}\right|\left|\mathbf{P}_{k|k}^{2,j}\right|\right)^{1/2}} \exp\left\{-\frac{1}{2}\lambda_{i,j}\right\}, \tag{10.3-2}$$

and

$$\ln\Lambda_{i,j} = -\frac{1}{2}\lambda_{i,j} + \text{constant}, \tag{10.3-3}$$

where

$$\lambda_{i,j} = \left(\hat{\mathbf{x}}_{k|k}^{1,i} - \hat{\mathbf{x}}_{k|k}^{2,j}\right)^T \left[\mathbf{P}_{k|k}^{1,i} + \mathbf{P}_{k|k}^{2,j}\right]^{-1} \left(\hat{\mathbf{x}}_{k|k}^{1,i} - \hat{\mathbf{x}}_{k|k}^{2,j}\right) \tag{10.3-4}$$

and $|\mathbf{A}|$ denotes the determinant of matrix $\mathbf{A}$.

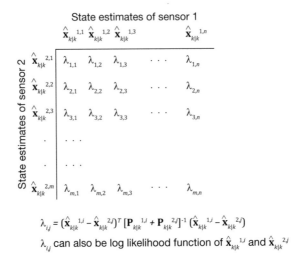

State estimates of sensor 1

$$\lambda_{i,j} = (\hat{\mathbf{x}}_{k|k}^{1,i} - \hat{\mathbf{x}}_{k|k}^{2,j})^T \, [\mathbf{P}_{k|k}^{1,i} + \mathbf{P}_{k|k}^{2,j}]^{-1} \, (\hat{\mathbf{x}}_{k|k}^{1,i} - \hat{\mathbf{x}}_{k|k}^{2,j})$$

$\lambda_{i,j}$ can also be log likelihood function of $\hat{\mathbf{x}}_{k|k}^{1,i}$ and $\hat{\mathbf{x}}_{k|k}^{2,j}$

Figure 10.6   Correlation matrix: the cost of assigning track i to track j.

For the tracking problem, $\lambda_{i,j}$ is formulated as the cost[3] of correlating track i of the first sensor to track j of the second sensor. Following the development in Section 8.5, it can be shown that $\lambda_{i,j}$ is the Mahalanobis distance between the state estimate and covariance pairs: $\{\hat{\mathbf{x}}_{k|k}^{1,i}, \mathbf{P}_{k|k}^{1,i}\}$ and $\{\hat{\mathbf{x}}_{k|k}^{2,j}, \mathbf{P}_{k|k}^{2,j}\}$, and $\lambda_{i,j}$ forms the correlation matrix as depicted in Figure 10.6. The optimum correlation is selected as those entries giving $\text{Min}_{i,j} \, \Sigma_{i,j} \, \lambda_{i,j}$ under the constraint that each column/row can be assigned to a row/column once—such a constraint is sometimes referred to as a feasible set. The use of the Mahalanobis distance for the correlation solution is appropriate when the determinants in Equation (10.3-2) do not depend on the state estimates. When this is not true, the Mahalanobis distance cannot be used and one must revert to the likelihood function, Equation (10.3-2), to seek the optimal solution. For the following development, it is assumed that the use of $\lambda_{i,j}$ is sufficient.

Similar to the discussion in Section 8.5, the optimum solution obtained as the set of (i, j) giving $\text{Min}_{i,j} \Sigma_{i,j} \lambda_{i,j}$ with (i, j) in the feasible set is the global nearest neighbors (GNN) solution. A row and column elimination solution (greedy) method is known as nearest neighbor (NN) and is suboptimal (i.e., it does not always find the minimum solution). The GNN solution can be obtained via Munkres algorithm. The total

---

3   More precisely, the cost function should be the log likelihood of the assignment in Equation (10.3-3).

number of possible correlation solutions is $N!/(N - M)!$ for $N \geq M$ or $M!/(M - N)!$ for $M \geq N$. Similar to MHT, multiple hypothesis correlation (MHC) is when multiple possible selections are maintained pending resolution with future data [7, 16].

10.3.2 Bias Estimation

Consider the case of two sensor systems with each sensor tracking N and M objects, respectively, and where $\hat{\mathbf{x}}_{k|k}^{1,i}$ and $\mathbf{P}_{k|k}^{1,i}$ are the state estimate and covariance of the first sensor and $\hat{\mathbf{x}}_{k|k}^{2,j}$ and $\mathbf{P}_{k|k}^{2,j}$ are the state estimate and covariance of the second sensor. The difference between this case and Section 10.3.1 is that $\hat{\mathbf{x}}_{k|k}^{1,i}$ and $\hat{\mathbf{x}}_{k|k}^{2,j}$ are biased estimates of $\mathbf{x}_k$. The definitions of absolute bias and relative bias are stated below with $\mathbf{b}^\ell \in \mathbb{R}^n$ for $\ell = 1,2$. Note that the definition of sensor bias is in the state space, $\mathbb{R}^n$, which is different from the definition in the measurement space, $\mathbb{R}^m$, in Section 10.2.1.

Absolute Bias

Absolute bias means the bias of each sensor measured in inertial space. As such each bias vector is independent of the other. Biases of the first and second sensor will be denoted as $\mathbf{b}^1$ and $\mathbf{b}^2$.

Relative Bias

For relative bias, the estimation is for the sensor bias relative to the other sensor bias, that is, estimating $\mathbf{b} = \mathbf{b}^1 - \mathbf{b}^2$.

Next consider the cases in estimating either type of bias with a given pair of assignment.

Absolute Bias Estimation

With biases in the individual state estimates, the Mahalanobis distance of Equation (10.3-4) is modified to become

$$\lambda_{i,j} = \left(\hat{\mathbf{x}}_{k|k}^{1,i} - \hat{\mathbf{x}}_{k|k}^{2,j} + \mathbf{b}^1 - \mathbf{b}^2\right)^T \left[\mathbf{P}_{k|k}^{1,i} + \mathbf{P}_{k|k}^{2,j}\right]^{-1} \left(\hat{\mathbf{x}}_{k|k}^{1,i} - \hat{\mathbf{x}}_{k|k}^{2,j} + \mathbf{b}^1 - \mathbf{b}^2\right). \quad (10.3\text{-}5)$$

In some cases one may have a priori distribution of the unknown bias. The a priori distribution of $\mathbf{b}^\ell$ is assumed to be Gaussian with zero mean and known covariance $\mathbf{S}^\ell$, that is, $\mathbf{b}^\ell: \sim N(\mathbf{0}, \mathbf{S}^\ell)$, where $\ell = 1,2$. Applying the weighted least squares (WLS) parameter estimator formulation of Section 1.7.2 results in the following minimization problem for a given pair of (i, j).

$$J_{(i,j)}\left(\mathbf{b}^1, \mathbf{b}^2\right) = \left(\hat{\mathbf{x}}_{k|k}^{1,i} - \hat{\mathbf{x}}_{k|k}^{2,j} + \mathbf{b}^1 - \mathbf{b}^2\right)^T \left[\mathbf{P}_{k|k}^{1,i} + \mathbf{P}_{k|k}^{2,j}\right]^{-1} \left(\hat{\mathbf{x}}_{k|k}^{1,i} - \hat{\mathbf{x}}_{k|k}^{2,j} + \mathbf{b}^1 - \mathbf{b}^2\right)$$
$$+ \mathbf{b}^{1^T} \mathbf{S}^{1^{-1}} \mathbf{b}^1 + \mathbf{b}^{2^T} \mathbf{S}^{2^{-1}} \mathbf{b}^2. \tag{10.3-6}$$

The absolute bias estimates $\hat{\mathbf{b}}^1$ and $\hat{\mathbf{b}}^2$ are the values of $\mathbf{b}^1$ and $\mathbf{b}^2$ in Equation (10.3-6) that minimize $J_{(i,j)}\left(\mathbf{b}^1, \mathbf{b}^2\right)$. Taking the derivative of $J_{(i,j)}\left(\mathbf{b}^1, \mathbf{b}^2\right)$ with respect to $\mathbf{b}^1$ and $\mathbf{b}^2$, and after some manipulation we obtain the solutions for $\hat{\mathbf{b}}^1$ and $\hat{\mathbf{b}}^2$ as

$$\hat{\mathbf{b}}^1 = \left[\left[\mathbf{P}_{k|k}^{1,i} + \mathbf{P}_{k|k}^{2,j}\right]^{-1} + \mathbf{S}^{1^{-1}}\right]^{-1} \left(\left[\mathbf{P}_{k|k}^{1,i} + \mathbf{P}_{k|k}^{2,j}\right]^{-1} \left(\hat{\mathbf{x}}_{k|k}^{2,i} - \hat{\mathbf{x}}_{k|k}^{1,j}\right)\right) \tag{10.3-7}$$

$$\hat{\mathbf{b}}^2 = \left[\left[\mathbf{P}_{k|k}^{1,i} + \mathbf{P}_{k|k}^{2,j}\right]^{-1} + \mathbf{S}^{2^{-1}}\right]^{-1} \left(\left[\mathbf{P}_{k|k}^{1,i} + \mathbf{P}_{k|k}^{2,j}\right]^{-1} \left(\hat{\mathbf{x}}_{k|k}^{1,j} - \hat{\mathbf{x}}_{k|k}^{2,i}\right)\right) \tag{10.3-8}$$

with covariance

$$Cov\left\{\hat{\mathbf{b}}^{\ell}\right\} = \left[\left[\mathbf{P}_{k|k}^{1,i} + \mathbf{P}_{k|k}^{2,j}\right]^{-1} + \mathbf{S}^{\ell^{-1}}\right]^{-1} \text{ for } \ell = 1, 2. \tag{10.3-9}$$

Remarks

1. The above results indicate that the two bias estimates $\hat{\mathbf{b}}^1$ and $\hat{\mathbf{b}}^2$ differ by their a priori distribution and, in this case, their respective covariance $\mathbf{S}^1$ and $\mathbf{S}^2$. $\hat{\mathbf{b}}^1$ and $\hat{\mathbf{b}}^2$ are opposite in sign with magnitudes weighted by track covariance and $\mathbf{S}^1$ and $\mathbf{S}^2$.

2. When there is no a priori knowledge, that is, $\mathbf{S}^{1^{-1}} = \mathbf{S}^{2^{-1}} = \mathbf{0}$ (no information), then $\hat{\mathbf{b}}^1$ is given as $\hat{\mathbf{b}}^1 = -\hat{\mathbf{b}}^2 = \hat{\mathbf{x}}_{k|k}^{2,i} - \hat{\mathbf{x}}_{k|k}^{1,j}$, a simple difference of the two state estimates and has covariance $\mathbf{P}_{k|k}^{1,i} + \mathbf{P}_{k|k}^{2,j}$. This indicates that the absolute biases of these sensors are not independent and thus not individually observable, similar to the conclusion in Section 10.2.2. The remaining discussion on state-to-state correlation will therefore only be concerned with relative bias estimation, that is, estimating $\mathbf{b} = \mathbf{b}^1 - \mathbf{b}^2$.

Relative Bias Estimation

For relative bias estimation, given the a priori distribution $\mathbf{b}: \sim N(\mathbf{0}, \mathbf{S})$, the performance index of Equation (10.3-6) is updated as

$$J_{(i,j)}\left(\mathbf{b}\right) = \left(\hat{\mathbf{x}}_{k|k}^{1,i} - \hat{\mathbf{x}}_{k|k}^{2,j} + \mathbf{b}\right)^T \left[\mathbf{P}_{k|k}^{1,i} + \mathbf{P}_{k|k}^{2,j}\right]^{-1} \left(\hat{\mathbf{x}}_{k|k}^{1,i} - \hat{\mathbf{x}}_{k|k}^{2,j} + \mathbf{b}\right) + \mathbf{b}^T \mathbf{S}^{-1} \mathbf{b}. \tag{10.3-10}$$

Note that the problem definition given in Equation (10.3-10) assumed a bias invariant property, that is, the bias for all state estimates is the same. The solution to Equation (10.3-10) is

$$\hat{\mathbf{b}} = \left[\left[\mathbf{P}_{k|k}^{1,i} + \mathbf{P}_{k|k}^{2,j} \right]^{-1} + \mathbf{S}^{-1} \right]^{-1} \left[\mathbf{P}_{k|k}^{1,i} + \mathbf{P}_{k|k}^{2,j} \right]^{-1} (\hat{\mathbf{x}}_{k|k}^{2,i} - \hat{\mathbf{x}}_{k|k}^{1,j}) \qquad (10.3\text{-}11)$$

with covariance

$$Cov\{\hat{\mathbf{b}}\} = \left[\left[\mathbf{P}_{k|k}^{1,i} + \mathbf{P}_{k|k}^{2,j} \right]^{-1} + \mathbf{S}^{-1} \right]^{-1}. \qquad (10.3\text{-}12)$$

Similarly, when $\mathbf{S}^{-1} = \mathbf{0}$, then $\hat{\mathbf{b}}$ is given as $\hat{\mathbf{b}} = (\hat{\mathbf{x}}_{k|k}^{2,i} - \hat{\mathbf{x}}_{k|k}^{1,j})$ having covariance $\mathbf{P}_{k|k}^{1,i} + \mathbf{P}_{k|k}^{2,j}$. Next, we consider the estimation of relative bias with a given set of correlations. Similar to the observation stated in Section 10.2.2, absolute biases are not observable, only estimation of relative bias is possible. Let $\mathcal{C}$ denote the set of all correlated pairs of tracks (i, j) of Sensor 1 and 2, respectively. $\mathcal{C}$ obtained without considering bias estimation jointly would not be optimal, but for the purpose of current development, it is assumed to have been provided. Mori and Chong [6] gives a survey on bias removal techniques with track-to-track correlation. Given $\mathcal{C}$, the bias estimate is the $\mathbf{b}$ that minimizes

$$J_{\mathcal{C}}(\mathbf{b}) = \sum_{(i,j)\in\mathcal{C}} \left(\hat{\mathbf{x}}_{k|k}^{1,i} - \hat{\mathbf{x}}_{k|k}^{2,j} + \mathbf{b} \right)^{T} \left[\mathbf{P}_{k|k}^{1,i} + \mathbf{P}_{k|k}^{2,j} \right]^{-1} \left(\hat{\mathbf{x}}_{k|k}^{1,i} - \hat{\mathbf{x}}_{k|k}^{2,j} + \mathbf{b} \right) + \mathbf{b}^{T}\mathbf{S}^{-1}\mathbf{b}. \qquad (10.3\text{-}13)$$

The solution is

$$\hat{\mathbf{b}}_{\mathcal{C}} = \left[\sum_{(i,j)\in\mathcal{C}} \left[\mathbf{P}_{k|k}^{1,i} + \mathbf{P}_{k|k}^{2,j} \right]^{-1} + \mathbf{S}^{-1} \right]^{-1} \left(\sum_{(i,j)\in\mathcal{C}} \left[\mathbf{P}_{k|k}^{1,i} + \mathbf{P}_{k|k}^{2,j} \right]^{-1} \left(\hat{\mathbf{x}}_{k|k}^{2,i} - \hat{\mathbf{x}}_{k|k}^{1,j} \right) \right) \qquad (10.3\text{-}14)$$

with covariance

$$Cov\{\hat{\mathbf{b}}_{\mathcal{C}}\} = \left[\sum_{(i,j)\in\mathcal{C}} \left[\mathbf{P}_{k|k}^{1,i} + \mathbf{P}_{k|k}^{2,j} \right]^{-1} + \mathbf{S}^{-1} \right]^{-1}. \qquad (10.3\text{-}15)$$

When there is no a priori knowledge, we can set $\mathbf{S}^{-1} = \mathbf{0}$ from Equations (10.3-14) and (10.3-15). Note that in this case the bias estimate is the weighted average of pairwise differences of all correlated states.

10.3.3   Joint Correlation and Bias Estimation

Joint correlation and bias estimation has attracted considerable interest in the technical community. A broad range of approaches to this problem area is available in the literature [6–16]. In this section, this problem is formulated mathematically and two solution algorithms are suggested.

The joint bias estimation and correlation problem is to find the feasible set of (i, j)'s and the vector $\mathbf{b}$ that minimizes the following cost function

$$J = \mathrm{Min}_{(i,j),\,\mathbf{b}} \left\{ \sum_{(i,j)} \left(\hat{\mathbf{x}}_{k|k}^{1,i} - \hat{\mathbf{x}}_{k|k}^{2,j} + \mathbf{b} \right)^T \left[\mathbf{P}_{k|k}^{1,i} + \mathbf{P}_{k|k}^{2,j} \right]^{-1} \left(\hat{\mathbf{x}}_{k|k}^{1,i} - \hat{\mathbf{x}}_{k|k}^{2,j} + \mathbf{b} \right) + \mathbf{b}^T \mathbf{S}^{-1} \mathbf{b} \right\}.$$

(10.3-16)

This is a difficult joint optimization problem because it is a combined analytical and integer programming problem. Two solution methods, namely an exhaustive search algorithm and an iterative optimization algorithm, are suggested below.

Exhaustive Search

Assuming the bias invariant property that biases between all pairs of correlated states are identical, an exhaustive search procedure is described below.

1. Select $\hat{\mathbf{x}}_{k|k}^{1,i}$ and $\hat{\mathbf{x}}_{k|k}^{2,j}$ as the correlated pair, then an initial bias estimate is $\tilde{\mathbf{b}}^{(i,j)} = \hat{\mathbf{x}}_{k|k}^{1,i} - \hat{\mathbf{x}}_{k|k}^{2,j}$.

2. Apply $\tilde{\mathbf{b}}^{(i,j)}$ to all state estimates to form the bias corrected state estimates.

3. Find the best set of assignments (e.g., the GNN solution) using the bias corrected state estimates.

4. Repeat Steps 1 through 3 for a different pair of (i, j), for $N \times M$ times, which is the total number of possible pairings. For each pairing, there is a total assignment cost. The best correlation solution is the one with the smallest total assignment cost.

5. Using the "best" correlation solution, compute the final bias estimate using Equation (10.3-14).

Remarks

1. Since this is an exhaustive search method, the GNN algorithm is expected to obtain the solution that is closest to the truth. Exceptions are when target spacing is comparable or smaller than the estimation error.

2. One concern for this method is that when N and M are large, the computational burden could be very high.

An Iterative Method

In this section, an alternative algorithm is suggested with the intent of reducing the computational burden of the exhaustive search algorithm.

Let $\mathcal{C}$ denote the set of all correlated pairs of tracks (i, j) of Sensor 1 and Sensor 2, respectively. The iterative method separates the minimization problem posed in Equation (10.3-12) into two steps.

Step 1: Obtain $\mathcal{C}$ for a given $\hat{\mathbf{b}}$ (set to some engineering guess initially),

Step 2: Solve for $\hat{\mathbf{b}}_{\mathcal{C}}$ for the set $\mathcal{C}$ obtained in Step 1.

Steps (1) and (2) are repeated until converged (for discussion on convergence, see Remarks at the end of this section). This algorithm is described below.

Initiation:

Step 1: Set $\mathbf{b} = \mathbf{0}^4$ in Equation (10.3-16), and find the GNN solution for

$$J = \underset{(i,j)}{\text{Min}} \sum_{(i,j)} \left(\hat{\mathbf{x}}_{k|k}^{1,i} - \hat{\mathbf{x}}_{k|k}^{2,j} \right)^T \left[\mathbf{P}_{k|k}^{1,i} + \mathbf{P}_{k|k}^{2,j} + \mathbf{S}^0 \right]^{-1} \left(\hat{\mathbf{x}}_{k|k}^{1,i} - \hat{\mathbf{x}}_{k|k}^{2,j} \right), \qquad (10.3\text{-}17)$$

where $\mathbf{S}^0$ denotes the covariance of the residual bias. Note that it is inserted into Equation (10.3-17) with the covariance of the state estimates as opposed to an additive term for minimization as in Equation (10.3-16). This method represents a covariance inflation technique for correlation. Let the correlation solution be denoted as $\mathcal{C}_0$, then the remaining minimization is

$$\text{Min}_\mathbf{b} \left\{ \hat{J} = \sum_{(i,j) \in \mathcal{C}_0} \left(\hat{\mathbf{x}}_{k|k}^{1,i} - \hat{\mathbf{x}}_{k|k}^{2,j} + \mathbf{b} \right)^T \left[\mathbf{P}_{k|k}^{1,i} + \mathbf{P}_{k|k}^{2,j} + \mathbf{S}^0 \right]^{-1} \left(\hat{\mathbf{x}}_{k|k}^{1,i} - \hat{\mathbf{x}}_{k|k}^{2,j} + \mathbf{b} \right) \right\}. \qquad (10.3\text{-}18)$$

Step 2: Take the derivative of $\hat{J}$ with respect to $\mathbf{b}$ and set it to zero to find $\hat{\mathbf{b}}$:

$$\sum_{(i,j) \in \mathcal{C}_0} \left[\mathbf{P}_{k|k}^{1,i} + \mathbf{P}_{k|k}^{2,j} + \mathbf{S}^0 \right]^{-1} \left(\hat{\mathbf{x}}_{k|k}^{1,i} - \hat{\mathbf{x}}_{k|k}^{2,j} + \mathbf{b} \right) - \left[\sum_{(i,j) \in \mathcal{C}_0} \left[\mathbf{P}_{k|k}^{1,i} + \mathbf{P}_{k|k}^{2,j} + \mathbf{S}^0 \right]^{-1} \right] \mathbf{b} = \mathbf{0}.$$

The solution is

$$\hat{\mathbf{b}} = \mathbf{S}^1 \left[\sum_{(i,j) \in \mathcal{C}_0} \left[\mathbf{P}_{k|k}^{1,i} + \mathbf{P}_{k|k}^{2,j} + \mathbf{S}^0 \right]^{-1} \left(\hat{\mathbf{x}}_{k|k}^{1,i} - \hat{\mathbf{x}}_{k|k}^{2,j} \right) \right]. \qquad (10.3\text{-}19)$$

---

4   Or set to be some initial engineering guess. For the purpose of convenience, $\mathbf{0}$ will be used for algorithm description.

Set $\hat{\mathbf{b}}^1 = \hat{\mathbf{b}}$ as the estimate of $\mathbf{b}$ with first iteration and

$$\mathbf{S}^1 = \left[\sum\nolimits_{(i,j)\in \mathcal{C}_0} \left[\mathbf{P}_{k|k}^{1,i} + \mathbf{P}_{k|k}^{2,j} + \mathbf{S}^0 \right]^{-1} \right]^{-1}. \tag{10.3-20}$$

is the covariance of $\hat{\mathbf{b}}^1$. Note that the superscript of $\hat{\mathbf{b}}$ is used to denote the first estimate of $\mathbf{b}$ as opposite to the bias estimate of the first sensor used in Section 10.3.2.

Update: Obtain a New Correlation and Bias Solution Using the Previous Bias Estimate

Step 1: With given $\hat{\mathbf{b}}^1$ and $\mathbf{S}^1$, obtain the next correlation solution by applying GNN to

$$J = \min_{(i,j)} \sum\nolimits_{(i,j)} \left(\hat{\mathbf{x}}_{k|k}^{1,i} - \hat{\mathbf{x}}_{k|k}^{2,j} + \hat{\mathbf{b}}^1 \right)^T \left[\mathbf{P}_{k|k}^{1,i} + \mathbf{P}_{k|k}^{2,j} + \mathbf{S}^1 \right]^{-1} \left(\hat{\mathbf{x}}_{k|k}^{1,i} - \hat{\mathbf{x}}_{k|k}^{2,j} + \hat{\mathbf{b}}^1 \right). \tag{10.3-21}$$

Let the solution be denoted by $\mathcal{C}_1$, resulting in the minimized J as

$$\hat{J}^1 = \sum\nolimits_{(i,j)\in \mathcal{C}_1} \left(\hat{\mathbf{x}}_{k|k}^{1,i} - \hat{\mathbf{x}}_{k|k}^{2,j} + \hat{\mathbf{b}}^1 \right)^T \left[\mathbf{P}_{k|k}^{1,i} + \mathbf{P}_{k|k}^{2,j} + \mathbf{S}^1 \right]^{-1} \left(\hat{\mathbf{x}}_{k|k}^{1,i} - \hat{\mathbf{x}}_{k|k}^{2,j} + \hat{\mathbf{b}}^1 \right). \tag{10.3-22}$$

The next step is to obtain an updated estimate of bias. The least squares estimator is applied again to solve for the updated $\mathbf{b}$ that minimizes

$$\hat{J}^1 = \sum\nolimits_{(i,j)\in \mathcal{C}_1} \left(\hat{\mathbf{x}}_{k|k}^{1,i} - \hat{\mathbf{x}}_{k|k}^{2,j} + \mathbf{b} \right)^T \left[\mathbf{P}_{k|k}^{1,i} + \mathbf{P}_{k|k}^{2,j} + \mathbf{S}^1 \right]^{-1} \left(\hat{\mathbf{x}}_{k|k}^{1,i} - \hat{\mathbf{x}}_{k|k}^{2,j} + \mathbf{b} \right). \tag{10.3-23}$$

Step 2: Take the derivative of $\hat{J}^1$ with respect to $\mathbf{b}$ and set it to zero to obtain $\hat{\mathbf{b}}$.

Skipping the detailed matrix algebra, the updated bias estimate and covariance $\hat{\mathbf{b}}^2$ and $\mathbf{S}^2$ are

$$\hat{\mathbf{b}}^2 = \mathbf{S}^2 \left[\sum\nolimits_{(i,j)\in \mathcal{C}_1} \left[\mathbf{P}_{k|k}^{1,i} + \mathbf{P}_{k|k}^{2,j} + \mathbf{S}^1 \right]^{-1} \left(\hat{\mathbf{x}}_{k|k}^{1,i} - \hat{\mathbf{x}}_{k|k}^{2,j} \right) \right]. \tag{10.3-24}$$

and

$$\mathbf{S}^2 = \left[\sum\nolimits_{(i,j)\in \mathcal{C}_1} \left[\mathbf{P}_{k|k}^{1,i} + \mathbf{P}_{k|k}^{2,j} + \mathbf{S}^1 \right]^{-1} \right]^{-1}. \tag{10.3-25}$$

Continue the Iterative Solution Process

Step n: Use $\hat{\mathbf{b}}^n$ and $\mathbf{S}^n$ computed in the previous step to obtain an updated set of $\mathcal{C}_n$ by applying GNN to

$$J^n = \min_{(i,j)} \sum_{(i,j)} \left(\hat{\mathbf{x}}_{k|k}^{1,i} - \hat{\mathbf{x}}_{k|k}^{2,j} + \hat{\mathbf{b}}^n\right)^T \left[\mathbf{P}_{k|k}^{1,i} + \mathbf{P}_{k|k}^{2,j} + \mathbf{S}^n\right]^{-1} \left(\hat{\mathbf{x}}_{k|k}^{1,i} - \hat{\mathbf{x}}_{k|k}^{2,j} + \hat{\mathbf{b}}^n\right). \quad (10.3\text{-}26)$$

The corresponding bias estimate is the $\mathbf{b}$ minimizing $\hat{J}^n$

$$\hat{J}^n = \min_{\mathbf{b}} \left\{\sum_{(i,j)\in\mathcal{C}_n} \left(\hat{\mathbf{x}}_{k|k}^{1,i} - \hat{\mathbf{x}}_{k|k}^{2,j} + \mathbf{b}\right)^T \left[\mathbf{P}_{k|k}^{1,i} + \mathbf{P}_{k|k}^{2,j} + \mathbf{S}^n\right]^{-1} \left(\hat{\mathbf{x}}_{k|k}^{1,i} - \hat{\mathbf{x}}_{k|k}^{2,j} + \mathbf{b}\right)\right\}. \quad (10.3\text{-}27)$$

Take the derivative of $\hat{J}^n$ with respect to $\mathbf{b}$ and set it to zero to find $\hat{\mathbf{b}}^{n+1}$

$$\hat{\mathbf{b}}^{n+1} = \mathbf{S}^{n+1} \left[\sum_{(i,j)\in\mathcal{C}_n} \left[\mathbf{P}_{k|k}^{1,i} + \mathbf{P}_{k|k}^{2,j} + \mathbf{S}^n\right]^{-1} \left(\hat{\mathbf{x}}_{k|k}^{1,i} - \hat{\mathbf{x}}_{k|k}^{2,j}\right)\right] \quad (10.3\text{-}28)$$

with covariance

$$\mathbf{S}^{n+1} = \left[\sum_{(i,j)\in\mathcal{C}_n} \left[\mathbf{P}_{k|k}^{1,i} + \mathbf{P}_{k|k}^{2,j} + \mathbf{S}^n\right]^{-1}\right]^{-1}. \quad (10.3\text{-}29)$$

The initiation step is when $n = 0$. The iteration is terminated when

$$\left|\hat{J}^{n+1} - \hat{J}^n\right| \leq \varepsilon_1,$$

or a weighted L^2 norm

$$\left\|\left(\hat{\mathbf{b}}^{n+1} - \hat{\mathbf{b}}^n\right)^T \left[\mathbf{S}^{n+1} + \mathbf{S}^{\mathbf{n}}\right]^{-1} \left(\hat{\mathbf{b}}^{n+1} - \hat{\mathbf{b}}^n\right)\right\| \leq \varepsilon_2,$$

or simply an L^2 norm

$$\left\|\left(\hat{\mathbf{b}}^{n+1} - \hat{\mathbf{b}}^n\right)^T \left(\hat{\mathbf{b}}^{n+1} - \hat{\mathbf{b}}^n\right)\right\| \leq \varepsilon_3.$$

is satisfied, where ε_l for $l = 1, 2, 3$ are selected small numbers.

Remarks

1. The above algorithm represents a partitioning approach to the joint optimization problem of Equation (10.3-16). It is separated into two consecutive steps: Step 1: obtaining $\mathcal{C}$ for a $\hat{\mathbf{b}}$ (set to zero initially); and Step 2: solving for $\hat{\mathbf{b}}$ for the set of $\mathcal{C}$ obtained in Step 1. Step 1 and Step 2 are repeated until convergence.

2. There is no mathematical proof of convergence for this procedure. An alternative procedure that may help the convergence has been suggested. This involves changing Equations (10.3-28) and (10.3-29) to

$$\hat{\mathbf{b}}^{n+1} = \mathbf{S}^{n+1} \left[\sum\nolimits_{(i,j)\in\mathcal{C}_n} \left[\mathbf{P}_{k|k}^{1,\,i} + \mathbf{P}_{k|k}^{2,\,j} + \mathbf{S}^0 \right]^{-1} \left(\hat{\mathbf{x}}_{k|k}^{1,i} - \hat{\mathbf{x}}_{k|k}^{2,j} \right) \right] \qquad (10.3\text{-}30)$$

$$\mathbf{S}^{n+1} = \left[\sum\nolimits_{(i,\,j)\in\mathcal{C}_n} \left[\mathbf{P}_{k|k}^{1,\,i} + \mathbf{P}_{k|k}^{2,\,j} + \mathbf{S}^0 \right]^{-1} \right]^{-1} \qquad (10.3\text{-}31)$$

Note that the above algorithm keeps $\mathbf{S}$ on the right-hand side of the equation as $\mathbf{S}^0$, not as part of the iteration.

3. While difficult to prove convergence for either case, a steady state analysis is offered for the algorithm of Equations (10.3-28) and (10.3-29). The steady state (converged) solution of $\mathbf{S}$ must satisfy $\mathbf{S}^{n+1} = \mathbf{S}^n = \mathbf{S}$ in Equation (10.3-29)

$$\mathbf{S} = \left[\sum\nolimits_{(i,\,j)\in\mathcal{C}_n} \left[\mathbf{P}_{k|k}^{1,i} + \mathbf{P}_{k|k}^{2,\,j} + \mathbf{S} \right]^{-1} \right]^{-1}. \qquad (10.3\text{-}32)$$

Further assume that there are a total of N tracks and all tracks have the same covariance, that is, $\mathbf{P}_{k|k}^{1,i} = \mathbf{P}_{k|k}^{2,j} = \mathbf{P}$ for all $(i,j) \in \mathcal{C}_n$. Using these simplifications in Equation (10.3-32) results in $\mathbf{S} = [N(2\mathbf{P}+\mathbf{S})^{-1}]^{-1} = \frac{1}{N}(2\mathbf{P}+\mathbf{S})$. The solution for $\mathbf{S}$ is therefore

$$\mathbf{S} = \frac{2}{N-1}\mathbf{P}.$$

In this case, $2\mathbf{P}$ is the covariance of the difference of two correlated state vectors, thus the covariance of a single sample bias estimate. The covariance of the bias estimate from all the correlated tracks is reduced by $N-1$ as expected in the theory of sample statistics. Substituting $\mathbf{S}$ into Equation (10.3-28) gives the steady state solution for $\hat{\mathbf{b}}$

$$\hat{\mathbf{b}} = \frac{1}{N}\sum\nolimits_{i=1}^{N}\mathbf{b}_i, \qquad (10.3\text{-}33)$$

where $\mathbf{b}_i$ is the difference of two correlated tracks from two sensors, thus the overall bias estimate is the average of all individual bias estimates. This analysis shows that the suggested algorithm in Equations (10.3-28) and (10.3-29) make intuitive and technical sense.

4. Further understanding of performance and convergence of this procedure is left for the reader by conducting numerical studies, Homework Problem 2.

Homework Problems

1. Build computer models to show the results in Section 10.2.

2. Extend the results of Section 10.2 with two or more targets as suggested in the remarks of Section 10.2.

3. Conduct a computer simulation study to characterize both the performance of the exhaustive search algorithm and the iterative bias estimation and correlation algorithm in Section 10.3.3.

Setup:

Step 1. Simulate the space trajectory used in Section 10.2.

Step 2. Assume two radar locations, also as used in Section 10.2, generate track covariance using CRB. Radar sensitivity and measurement covariance can be chosen with some nominal values. Select several relative sensor bias errors in range and angles. Bias values should span the range relative to CRB from much smaller to much larger than CRB.

Step 3. Select a modest number of targets (between 10 to 20) going in parallel with spatial distribution spanning from small to large relative to the CRB.

Step 4. The number of targets tracked by the first and the second sensor should not be equal; select a subset that they track in common.

Step 5. Generate simulated state estimates with statistical errors represented by CRB. The covariance of the state estimates is the CRB.

Step 6. Generate Monte Carlo samples of state estimates by repeating the process in Step 5.

Study:

Apply the exhaustive search method and the iterative method in Section 10.3.3 to data generated in Step 6 above. Obtain sample statistics with Monte Carlo runs and compare results using the following metrics:

1. Percent correct correlation

2. Percent incorrect correlation

3. Rate of convergence of the iterative method

Repeat by varying (a) target spatial distribution, (b) size of relative biases, and (c) the number of common targets to both sensors.

Observation:
Let the number of targets tracked by the first and second sensors be N and M, respectively. The total number of trials of the exhaustive method is $N \times M$. For example, if $N = M = 10$, then the total trial is 100. If the iterative method converges to the same solution (statistically) and the total number of iteration for convergence is much smaller than $N \times M$, then the iterative method is shown to be useful.

References

1. B. Corwin, D. Choi, K.P. Dunn, and C.B. Chang, "Sensor to Sensor Correlation and Fusion with Biased Measurements," in *Proceedings of MSS National Symposium on Sensor and Data Fusion*, 2005.

2. X. Lin and Y. Bar-Shalom, "Multisensor Bias Estimation with Local Tracks without a priori Association," in *Proceedings of the 2003 SPIE: Signal and Data Processing of Small Targets,* vol. 5204, 2003.

3. D.F. Crouse, Y. Bar-Shalom, and P. Willett, "Sensor Bias Estimation in the Presence of Data Association Uncertainties," in *Proceedings of the 2009 SPIE: Signal and Data Processing of Small Targets*, vol. 7445 74450P-1, 2009.

4. S. Danford, B.D. Kragel, and A.B. Poore, "Joint MAP Bias Estimation and Data Association: Algorithms," in *Proceedings of the SPIE Conference on Signal and Data Processing of Small Targets*, Vol. 6699, 2007.

5. S. Danford, B.D. Kragel, and A.B. Poore, "Joint MAP Bias Estimation and Data Association: Simulations," in *Proceedings of the SPIE Conference on Signal and Data Processing of Small Targets*, Vol. 6699, 2007.

6. S. Mori and C. Chong, "Comparison of Bias Removal Algorithms in Track-to-Track Associations," in *Proceedings of the 2007 SPIE: Signal and Data Processing of Small Targets,* vol. 6699, 2007.

7. S. Mori, K.C. Chang, C.Y. Chong, and K.P. Dunn, "Tracking Performance Evaluation," in *Proceedings of the 1990 SPIE: Signal and Data Processing of Small Targets*, vol. 1305, 1990.

8. C.J. Humke, "Bias Removal Techniques for the Target-Object Mapping Problem," MIT Lincoln Laboratory Technical Report 1060, July 2002.

9. S. Mori and C. Chong, "Effects of Unpaired Objects and Sensor Biases on Track-to-Track Association: Problems and Solutions," in *Proceedings of MSS National Symposium on Sensor and Data Fusion*, vol. 1, pp. 137–151, June 2000.

10. L. D. Stone, M. L. Williams, and T.M. Tran, "Track-To-Track Associations and Bias Removal," in *Proceedings of the 2002 SPIE: Signal and Data Processing of Small Targets*, vol. 4728, pp. 315–329, Apr. 2002.

11. J.P. Ferry, "Exact Bias Removal for the Track-To-Track Association Problem," in *Proceedings of the 12th International Conference on Information Fusion*, pp. 1642–1649, July 2009.

12. M. Levedahl, "An Explicit Pattern Matching Assignment Algorithm," in *Proceedings of the 2002 SPIE: Signal and Data Processing of Small Targets*, vol. 4728, Apr. 2002.

13. S.S. Blackman and N.D. Banh, "Track Association Using Correction for Bias and Missing Data," in *Proceedings of 1994 SPIE: Signal and Data Processing of Small Targets*, vol. 2235, pp. 529–539, Apr. 1994.

14. S.M. Herman and A.B. Poore, "Nonlinear Least-Squares Estimation for Sensor and Navigation Biases," in *Proceedings of the 2006 SPIE: Signal and Data Processing of Small Targets*, vol. 6236, Apr. 2006.

15. S. Mori and C. Chong, "BMD Mid-Course Object Tracking: Track Fusion under Asymmetric Conditions," in *Proceedings of MSS National Symposium on Sensor and Data Fusion*, June 2001.

16. S. Herman, J. Johnson, A. Shaver, B. Kragel, S. Miller, G. Norgard, K. Obermeyer, and E. Schmidt, "Multiple Hypothesis Correlation (MHC) Algorithm Description," unpublished report of Numerica Corporation, Ft. Collins, CO, 2014.

17. B. Kragel, S. Herman, and N. Roseveare, "A Comparison of Methods for Estimating Track-to-Track Assignment Probabilities," *IEEE Transactions on Aerospace and Electronic Systems*, vol. 48, pp. 1870–1888, 2012.

18. A.B. Poore and N. Rijavec, "A Lagrangian Relaxation Algorithm for Multidimensional Assignment Problems Arising from Multitarget Tracking," *SIAM Journal on Optimization*, Vol. 3, pp. 545–563, 1993.

19. A.B. Poore and A.J. Robertson, III, "A New Class of Lagrangian Relaxation Based Algorithms for a Class of Multidimensional Assignment Problems," *Journal of Computational Optimization and Applications*, Vol. 8, pp. 129–150, 1997.

20. A.B. Poore, "Multidimensional Assignment Formulation of Data Association Problems Arising from Multi-Target and Multi-Sensor Tracking," *Computational Optimization and Applications*, vol. 3, pp. 27–57, 1994.

21. R.R. Bate, D.D. Mueller, J.E. White, and W.W. Saylor, *Fundamentals of Astrodynamics*, 2nd ed. New York: Dover Books on Physics, 2015.

Concluding Remarks

Estimation and association has been a rich field of research and development for decades. The theory for linear state estimation was developed in the 1950s with landmark papers by Kalman, and Kalman and Bucy. Much effort was devoted to extend the Kalman–Bucy theory to nonlinear estimation. Publications during the 1960s and 1970s encompass both theoretical and practical results. Although much of the theory for estimation has been developed, applying it to solve practical problems still continues.

In looking back on the content of this book, note that it was arranged progressively starting with the theory, and then progressing to practical applications in the overall subject of state estimation and association. Chapters 1–3 were designed to establish the fundamental theory for estimation as it applies to either parameter or state estimation, and were treated in the context of both linear and nonlinear systems. Chapters 4–7 focused on applying the basic theory of estimation to practical problems such as systems having model uncertainties including the case of maneuvering targets, approximating the a posteriori density function evolution as an approach to nonlinear estimation, and developing estimation algorithms for multiple sensor systems. Chapters 8–10 dealt with multiple objects and address measurements with uncertain origin, track ambiguity, and concluded with the correlation and fusion of estimates of multiple objects in a multiple sensor system containing sensor biases.

The areas for estimation and for association are briefly summarized below.

Estimation

Notable application areas for estimation include maneuvering target tracking, sensor bias estimation and mitigation, implementation of nonlinear estimators, and architecture for multiple sensor systems.

Solutions for estimating the state in the presence of model uncertainty can still be improved. Methods for filter tuning, adaptation, responding to rapid changes in system uncertainties, unexpected system internal failures, or external disturbances still challenge system design engineers (Chapter 4). In the 1970s, a class of decision directed algorithms that incorporated switching among system models when the filter sensed changes was applied to many practical problems with various degrees of success. Because this approach implemented a hard switch among systems, it sometimes suffered from detection delay and filter transient error. The multiple model (MM) approach allows for a soft switch using a weighted (by the a posteriori probabilities) sum of the output from a bank of filters built assuming different models, and appears able to circumvent the effects of hard switching (Chapter 5). The more rigorous MM algorithm is, however, complicated or impossible to implement. Future work providing simplified algorithms with minimum degradation in performance would be a significant contribution.

In a multiple sensor environment, estimation in the presence of intersensor bias is still a challenging problem. In some practical systems, the use of auxiliary means to calibrate the sensors and to estimate their biases are applied. As shown in Chapter 10, state and bias estimations are sometimes not separable, and must be approached simultaneously. Once biases have been estimated and removed from the measurement systems, the residual effects of the bias (the estimated bias has nonzero covariance) need to be properly accounted for. Currently, the Schmidt–Kalman filter (SKF) has been shown to be an effective approach to account for such residual bias in such a way that the filter covariance remains consistent (Chapter 4).

Nonlinear estimation can best be approached by approximating the a posteriori density function (Chapters 3 and 6). The conventional approach is approximating the conditional probabilities as Gaussian densities with mean and covariance derived from the Taylor series expansion given the prior statistics such as the extended Kalman filter (EKF). Two deterministic sampling techniques were introduced in Chapter 6. One of them is a popular technique using the unscented transformation to obtain the mean and covariance of the a posteriori density function, the unscented KF. A random sample technique to realize such an estimator, known as a particle filter, is also discussed in Chapter 6. Particle filter implementation requires large computational and memory resources. Certain properties and limitations of the approach also prevent it from becoming a widely useful technique. It is, however, found to be useful in areas where extremely fine details of system characteristics need to be estimated, and when the conventional approach such as the EKF fails.

Two architectures for estimation with multiple sensor systems were introduced in Chapter 7. Trade-offs on architecture choice, involving both system complexity and performance evaluation, challenge system design engineers in a variety of application areas.

It is the authors' goal that after students have learned the materials taught in these chapters, they will be equipped to solve problems in the general area of estimation.

Association/Correlation

An estimator is used to process a set of measurements with the assumption that all measurements came from the same object. In the case when multiple individual objects are closely spaced, the assumption that the time-sequenced measurements originate from the same object may no longer be held in high confidence. Objects closely spaced over successive observation frames constitute the main cause for track ambiguity. The process for determining whether a measurement or a sequence of measurements originated from the same object is referred to as association. The process to determine whether state estimates from multiple sensors are from the same object is called correlation. It was show that the basic mathematics for association and correlation are the same. Estimation would produce erroneous results if the measurements processed were not from the same object. Similarly, the ability to correctly perform association and correlation depends on the accuracy and consistency of the state estimates. The problems of estimation and association or correlation are fundamentally different. The problem of estimation is analytical and handled with conventional calculus, in which the traditional mathematical approaches of optimization apply. The problem of association or correlation, similar to the classical problem of signal detection, is usually handled as a binary problem, in which an either/or decision is made.[1] In the situation where the problem of association/correlation and estimation must be solved together, the solution process becomes complicated. To date, the combined optimal solution for estimation and correlation/association is still open.

The multiple hypothesis tracker (MHT) has drawn a lot of attention from practitioners in multitarget tracking. In this approach a track is a collection of measurements over the entire time window of tracking. The total number of tracks is the combination of all possible tracks in a time history, which allows a measurement in a scan to be used by multiple tracks. A hypothesis is a collection of tracks such that a

---

1   The exception is the probabilistic data association filter, which uses a weighted average.

measurement can only appear in one of the tracks in this hypothesis. When all possible tracks have been obtained, then the true tracks must be contained in the answers. As the number of measurement scans increases, the number of combinations becomes enormously large (Chapter 9), which is one of the limitations of this approach. MHT computes hypothesis probabilities (used as scores for hypotheses and tracks) for pruning and track reporting purposes. The hypothesis probability computation uses a parametric model on sensor characteristics, environmental effects, target motions, object counts, and so on, and as such, the validity of the computed hypothesis probability depends on the validity of the underlying parametric model. The selection of the parameter values used in the probability calculation is not exactly a settled science (Chapter 9). A corresponding approach for a multiple sensor system using state fusion is known as the multiple hypothesis correlator (MHC). MHC and MHT are mathematically equivalent.

A number of possible approaches for multiple target tracking were introduced in Chapter 8. Many of them are not tightly coupled with parametric models for performance calculation. Culminating several approaches, a simple multiple target tracker was suggested. In this approach, the decision on measurement to state association is typically done on the same scan time. Some extensions to this method in terms of delayed decision and track file smoothing were also suggested in Chapter 8. It is felt that this approach is simple, intuitive, and does not depend on parametric performance modeling. Examples for measurement fusion and state fusion are discussed in Chapter 10. The goal here is to motivate thinking on possible approaches since there is no theoretically based right answer.

In all the approaches above, the solution is not obtained as a joint optimization of estimation and correlation. This is still an open issue.

If this book helps students to master the basics, both in estimation and in association/correlation, such that they are able to make choices on their own and investigate proper solutions, then the authors have achieved their goal.

Appendix A

Matrix Inversion Lemma

Several matrix identities are very useful for deriving estimation algorithms known as the matrix inversion lemma. They are stated and proved here as reference for derivations used in the main book.

Matrix Inversion Lemma

Let $\mathbf{A}$ be a square and nonsingular matrix with dimension (n, n). $\mathbf{B}$ and $\mathbf{C}$ are rectangular matrices with dimension (n, m) and (m, n), respectively, such that $[\mathbf{A} + \mathbf{BC}^T]$ is nonsingular. Then the following identity is true

$$[\mathbf{A} + \mathbf{BC}^T]^{-1} = \mathbf{A}^{-1} - \mathbf{A}^{-1}\mathbf{B}[\mathbf{I} + \mathbf{C}^T\mathbf{A}^{-1}\mathbf{B}]^{-1}\mathbf{C}^T\mathbf{A}^{-1}. \tag{A.1}$$

This can be shown to be true by showing that the multiplication of a matrix with its inverse results in an identity matrix $\mathbf{I}$. Let us postmultiply $[\mathbf{A} + \mathbf{BC}^T]$ with the right-hand side of Equation (A.1).

$$[\mathbf{A} + \mathbf{BC}^T][\mathbf{A}^{-1} - \mathbf{A}^{-1}\mathbf{B}[\mathbf{I} + \mathbf{C}^T\mathbf{A}^{-1}\mathbf{B}]^{-1}\mathbf{C}^T\mathbf{A}^{-1}] = \mathbf{I} - \mathbf{B}[\mathbf{I} + \mathbf{C}^T\mathbf{A}^{-1}\mathbf{B}]^{-1}\mathbf{C}^T\mathbf{A}^{-1} + \mathbf{BC}^T\mathbf{A}^{-1}$$
$$- \mathbf{BC}^T\mathbf{A}^{-1}\mathbf{B}[\mathbf{I} + \mathbf{C}^T\mathbf{A}^{-1}\mathbf{B}]^{-1}\mathbf{C}^T\mathbf{A}^{-1} = \mathbf{I} - \mathbf{B}[\mathbf{I} + \mathbf{C}^T\mathbf{A}^{-1}\mathbf{B}]^{-1}\mathbf{C}^T\mathbf{A}^{-1} + \mathbf{B}[\mathbf{I} + \mathbf{C}^T\mathbf{A}^{-1}\mathbf{B}][\mathbf{I} + \mathbf{C}^T\mathbf{A}^{-1}\mathbf{B}]^{-1}\mathbf{C}^T\mathbf{A}^{-1} - \mathbf{BC}^T\mathbf{A}^{-1}\mathbf{B}[\mathbf{I} + \mathbf{C}^T\mathbf{A}^{-1}\mathbf{B}]^{-1}\mathbf{C}^T\mathbf{A}^{-1}$$
$$= \mathbf{I} + [-\mathbf{B} + \mathbf{B}[\mathbf{I} + \mathbf{C}^T\mathbf{A}^{-1}\mathbf{B}] - \mathbf{BC}^T\mathbf{A}^{-1}\mathbf{B}][\mathbf{I} + \mathbf{C}^T\mathbf{A}^{-1}\mathbf{B}]^{-1}\mathbf{C}^T\mathbf{A}^{-1} = \mathbf{I}.$$

This completes the derivation. One can also show this by premultiplying $[\mathbf{A} + \mathbf{BC}^T]$ with the right-hand side of Equation (A.1), the details are omitted here.

The above identity is a general form of the matrix inversion lemma. Two immediate extensions that are more obviously related to filter derivations are stated below.

Extension 1

In addition to the definition of matrix $\mathbf{A}$ and $\mathbf{B}$, let $\mathbf{D}$ be a square and nonsingular matrix with dimension (m, m). Then the following identity is true.

$$[\mathbf{A}^{-1} + \mathbf{B}^T\mathbf{D}^{-1}\mathbf{B}]^{-1} = \mathbf{A} - \mathbf{A}\mathbf{B}^T[\mathbf{B}\mathbf{A}\mathbf{B}^T + \mathbf{D}]^{-1}\mathbf{B}\mathbf{A}. \tag{A.2}$$

Note the similarity of Equation (A.2) to Equation (A.1). It can be shown to be true in the same way as above and the detail will be omitted.

Extension 2

With the same definition of matrices $\mathbf{A}$, $\mathbf{B}$, and $\mathbf{D}$, the following identity is true.

$$[\mathbf{A}^{-1} + \mathbf{B}^T\mathbf{D}^{-1}\mathbf{B}]^{-1}\mathbf{B}^T\mathbf{D}^{-1} = \mathbf{A}\mathbf{B}^T[\mathbf{B}\mathbf{A}\mathbf{B}^T + \mathbf{D}]^{-1}. \tag{A.3}$$

The above equation can be derived using extension 1 of the matrix inversion lemma. Substituting Equation (A.2) into the left-hand side of Equation (A.3) yields

$$
\begin{aligned}
[\mathbf{A}^{-1} + \mathbf{B}^T\mathbf{D}^{-1}\mathbf{B}]^{-1}\mathbf{B}^T\mathbf{D}^{-1} &= \mathbf{A}\mathbf{B}^T\mathbf{D}^{-1} - \mathbf{A}\mathbf{B}^T[\mathbf{B}\mathbf{A}\mathbf{B}^T + \mathbf{D}]^{-1}\mathbf{B}\mathbf{A}\mathbf{B}^T\mathbf{D}^{-1} \\
&= \mathbf{A}\mathbf{B}^T[\mathbf{D}^{-1} - [\mathbf{B}\mathbf{A}\mathbf{B}^T + \mathbf{D}]^{-1}\mathbf{B}\mathbf{A}\mathbf{B}^T\mathbf{D}^{-1}] \\
&= \mathbf{A}\mathbf{B}^T[\mathbf{B}\mathbf{A}\mathbf{B}^T + \mathbf{D}]^{-1}[[\mathbf{B}\mathbf{A}\mathbf{B}^T + \mathbf{D}]\mathbf{D}^{-1} - \mathbf{B}\mathbf{A}\mathbf{B}^T\mathbf{D}^{-1}] \\
&= \mathbf{A}\mathbf{B}^T[\mathbf{B}\mathbf{A}\mathbf{B}^T + \mathbf{D}]^{-1}.
\end{aligned}
$$

This completes the proof.

As will be shown during filter derivations, Equation (A.2) gives two identical methods to compute the updated covariance and Equation (A.3) shows two ways to compute the Kalman filter gain. In addition, these identities are very useful for filter equation manipulations.

Appendix B

Notation and Variables

x,y: bold and lower case variables are column vectors.

$\mathbf{A}_l$, $\mathbf{B}_l$, $\mathbf{H}_k$, $\mathbf{\Phi}_{k,k-1}$: bold upper case variables are matrices.

x,y: italic and nonbold variables are used to denote scalars, elements of a column vector, elements of a matrix, and so on.

(...): variables and terms enclosed by parentheses are vectors.

[...]: variables and terms enclosed by brackets are matrices.

$[.]_{i,j}$: (i,j)th element of the enclosed matrix.

$|a|$: the absolute value of variable a.

$|\mathbf{A}|$: the determinant of the matrix $\mathbf{A}$.

$a \triangleq b$: a is defined to be b.

$\ni$: such that, a logical statement connecting two arguments.

$\forall \mathbf{a}$: for all $\mathbf{a}$.

$\mathbb{R}^n$: a real-valued vector space with dimension n.

$\mathbf{a} \in \mathbf{S}$: $\mathbf{a}$ is an element of a set or vector space $\mathbf{S}$.

$\mathbf{A} \subset \mathbf{B}$: $\mathbf{A}$ is a subset or subspace of a set or space $\mathbf{B}$.

$\langle \mathbf{a},\mathbf{b} \rangle$: the inner product of the enclosed vectors $\mathbf{a}$ and $\mathbf{b} \in \mathbb{R}^n$.

$\|\mathbf{a}\|$: the norm of the enclosed vector $\mathbf{a} \in \mathbb{R}^n$, for example, $\|\mathbf{a}\|^2 = \langle \mathbf{a},\mathbf{a} \rangle$.

$\mathbf{T}$: a transformation from $\mathbb{R}^n$ to $\mathbb{R}^m$, written as: $\mathbf{T}:\mathbb{R}^n \to \mathbb{R}^m$.

$\mathbf{T}^*$: the adjoint operator of $\mathbf{T}$, which is a transformation $\mathbf{T}^*:\mathbb{R}^m \to \mathbb{R}^n \ni \langle \mathbf{Ta},\mathbf{b} \rangle = \langle \mathbf{a},\mathbf{T}^*\mathbf{b} \rangle$, $\forall \mathbf{a} \in \mathbb{R}^n, \forall \mathbf{b} \in \mathbb{R}^m$. For finite dimensional vector spaces, $\mathbf{T}$ is an $m \times n$ matrix and $\mathbf{T}^*$ is the transpose of $\mathbf{T}$, $\mathbf{T}^T$, such that $[\mathbf{T}^T]_{j,i} = [\mathbf{T}]_{i,j}$.

$\mathcal{R}(\mathbf{T}) \subset \mathbb{R}^m$: $\mathcal{R}(\mathbf{T})$ is the range space of $\mathbf{T}$, which is a subspace of $\mathbb{R}^m$.

$\mathcal{N}(\mathbf{T}) \subset \mathbb{R}^n$: $\mathcal{N}(\mathbf{T})$ is the null space of $\mathbf{T}$, which is a subspace of $\mathbb{R}^n$.

$\mathcal{N}^{\perp}(\mathbf{T})$: the subspace of $\mathbb{R}^n$ that is perpendicular to $\mathcal{N}(\mathbf{T})$.

$\mathbf{T}^{-1}$: the inverse of $\mathbf{T}$. If it exists then it is uniquely defined and $\mathbf{T}\mathbf{T}^{-1} = \mathbf{T}^{-1}\mathbf{T} = \mathbf{I}$.

$\mathbf{T}^{\dagger}$: the pseudo-inverse of $\mathbf{T}$. The Moore–Penrose pseudo-inverse satisfies $\mathbf{T}\mathbf{T}^{\dagger}\mathbf{T} = \mathbf{T}$ and $\mathbf{T}^{\dagger}\mathbf{T}\mathbf{T}^{\dagger} = \mathbf{T}^{\dagger}$.

$\mathbf{x}$: a parameter vector, when time-invariant, or a state vector when varying according to a state model.

$\mathbf{y}$: a measurement vector of $\mathbf{x}$. $\mathbf{y}$ and $\mathbf{x}$ can be related linearly or nonlinearly.

$p(a)$: probability density function of random variable a.

σ_a^2: variance of a.

$p(\mathbf{x})$: probability density function of random vector $\mathbf{x}$. This is referred to as the a priori density function.

$\Pr\{A\}$: probability of an event A or a set $\mathbf{A}$.

$E_u\{.\}$: expected value of the enclosed variable or function with respect to a random variable or vector u.

$Cov\{.\}$: covariance of the enclosed variable or function.

$N(\bar{\mathbf{a}}, \mathbf{P})$: a Gaussian density function with mean $\bar{\mathbf{a}}$ and covariance $\mathbf{P}$.

$\mathbf{a}$: $\sim N(\bar{\mathbf{a}}, \mathbf{P})$: a random vector $\mathbf{a}$ has a Gaussian density $N(\bar{\mathbf{a}}, \mathbf{P})$.

$p(\mathbf{y}; \mathbf{x})$: probability density function of random vector $\mathbf{y}$ given parameter $\mathbf{x}$. This is referred to as the likelihood function.

$p(\mathbf{x}|\mathbf{y})$: conditional probability density function of random vector $\mathbf{x}$ given random vector $\mathbf{y}$. In this book, $\mathbf{y}$ is the measurement of $\mathbf{x}$. It is referred to as the a posteriori density function.

$\ln p(\mathbf{y}; \mathbf{x})$ or $\ln p(\mathbf{y}|\mathbf{x})$: natural log of the likelihood function, referred to as log likelihood.

J: performance index or cost function.

$\hat{\mathbf{x}}$: estimate of $\mathbf{x}$.

$\mathbf{x}_t$, $\mathbf{x}_k$: time-varying state vectors in either continuous-time t or discrete-time k.

$\mathbf{y}_k$: measurement vector of $\mathbf{x}_k$ taken in discrete-time k. In this book, measurements are always taken in discrete time.

$\mathbf{f}(.), \mathbf{h}(.)$: column vectors, often used to denote nonlinear system and measurement relations.

$\mathbf{A}_t$: system matrix for a continuous linear system.

$\mathbf{B}_t$: input distribution matrix for a continuous linear system.

$\mathbf{H}_k$: measurement matrix for a discrete linear system.

$\boldsymbol{\Phi}_{k,k-1}$: state transition matrix, the discrete representation of continuous system $\mathbf{A}_t$.

$\boldsymbol{\xi}_t$, $\boldsymbol{\mu}_k$: system input vectors corresponding to continuous-time and discrete-time systems, respectively, often used to denote unknown disturbances or uncharacterized system models, referred to as system noise or process noise with corresponding covariance $\boldsymbol{\Sigma}_t$ and $\mathbf{Q}_k$, respectively.

$\mathbf{v}_k$: measurement noise vector with covariance $\mathbf{R}_k$.

$\mathbf{y}_{1:k}$: collection of a random sequence or measurements from time 1 through k, $\mathbf{y}_{1:k} = \{\mathbf{y}_1, \mathbf{y}_2, \dots, \mathbf{y}_k\}$.

$\hat{\mathbf{x}}_{i|j}$: estimate of $\mathbf{x}_i$ given $\mathbf{y}_{1:j}$.

$\mathbf{P}_{i|j}$: covariance of $\hat{\mathbf{x}}_{i|j}$.

$\mathbf{P_{xx}}$: covariance of vector $\mathbf{x}$.

$\mathbf{P_{xy}}$: cross covariance of vector $\mathbf{x}$ and vector $\mathbf{y}$.

$\left[\dfrac{\partial \mathbf{f}(\mathbf{x})}{\partial \mathbf{x}}\right]_{\hat{\mathbf{x}}} \triangleq \mathbf{F}_{\hat{\mathbf{x}}}$: Jacobian of vector-valued system function $\mathbf{f}(\mathbf{x})$ evaluated at $\mathbf{x} = \hat{\mathbf{x}}$.

$\left[\dfrac{\partial \mathbf{h}(\mathbf{x})}{\partial \mathbf{x}}\right]_{\hat{\mathbf{x}}} \triangleq \mathbf{H}_{\hat{\mathbf{x}}}$: Jacobian of vector-valued measurement function $\mathbf{h}(\mathbf{x})$ evaluated at $\mathbf{x} = \hat{\mathbf{x}}$.

$\chi_{k|k}^{i\,2}$: χ^2 variable of the ith trial of estimation error at time k: $\tilde{\mathbf{x}}_{k|k}^i$.

$\overline{\chi_{k|k}^2} \triangleq \dfrac{1}{M}\sum_{i=1}^{M} \chi_{k|k}^{i\,2}$: the averaged χ^2 of M Monte Carlo trials.

$\delta(.)$: Kronecker delta function (with finite value).

$\delta_D(.)$: Dirac delta function (with infinite value at the origin).

$\Pr\{\theta^i\}$: probability of the hypothesis θ^i is true.

θ_k^i: the hypothesis that a parameter vector $\mathbf{p}=\mathbf{p}_i$ is true, for $i \in I$ at time k. This is referred to as a local hypothesis (in terms of time).

$\theta_{1:k}^j$: a hypothesis that represents a sequence of parameter vectors from time 1 through time k, thus a global hypothesis. A global hypothesis contains a set of local hypotheses such that $\theta_{1:k}^j = \{\theta_1^{j_1}, \theta_2^{j_2}, \dots, \theta_k^{j_k}\} \forall j_1, j_2, \dots, j_k \in I$.

$\Pr\{\theta_k^i | \mathbf{y}_{1:k}\}$: the probability of θ_k^i being true given $\mathbf{y}_{1:k}$ at k.

$\Pr\{\theta_{1:k}^j | \mathbf{y}_{1:k}\}$: the probability of $\theta_{1:k}^j = \{\theta_1^{j_1}, \theta_2^{j_2}, \dots, \theta_k^{j_k}\}$ being true given $\mathbf{y}_{1:k}$ at k.

$\hat{\mathbf{x}}_{k|k}^{\ell}$: updated state estimate for model ℓ at time k given $\mathbf{y}_{1:k}$.

$\mathbf{P}_{k|k}^{\ell}$: covariance of $\hat{\mathbf{x}}_{k|k}^{\ell}$.

$\hat{\mathbf{x}}_{k|k}^{\ell m}$: updated state estimate for model m at time k given $\hat{\mathbf{x}}_{k-1|k-1}^{\ell}$ at $k-1$.

$\mathbf{P}_{k|k}^{\ell m}$: covariance of $\hat{\mathbf{x}}_{k|k}^{\ell m}$.

$\mathbf{x}^i \sim q(\mathbf{x})$: i.i.d. sample points drawn from a density function, $q(\mathbf{x})$.

$\left\{ \mathbf{x}_{0:k}^i : i = 1,\ldots,N_S \right\}$: a set of N_S sample points of random process $\mathbf{x}_{0:k} = \{\mathbf{x}_j : j = 0,\ldots, k\}$.

$\mathbf{Y}_k$: the set of all measurements in scan k, $\left\{ \mathbf{y}_k^1, \mathbf{y}_k^2, \mathbf{y}_k^3,\ldots, \mathbf{y}_k^{N_k} \right\} \cup \varnothing$. $\varnothing$ denotes an empty set, that represents the case where no measurement in $\left\{ \mathbf{y}_k^1, \mathbf{y}_k^2, \mathbf{y}_k^3,\ldots, \mathbf{y}_k^{N_k} \right\}$ is used at the kth scan.

$\mathbf{Y}_{1:k}$: the set of measurement sets from the first to the kth scan, $\{\mathbf{Y}_1, \mathbf{Y}_2, \mathbf{Y}_3,\ldots, \mathbf{Y}_k\}$.

$\mathbf{y}_{1:k}^{m_k}$: a track, $\left\{ \mathbf{y}_1^{n_1}, \mathbf{y}_2^{n_2}, \mathbf{y}_3^{n_3},\ldots, \mathbf{y}_k^{n_k} \right\} \in \mathbf{Y}_{1:k}$.

$\theta_{1:k}^{m_k}$: a hypothesis that $\mathbf{y}_{1:k}^{m_k}$ is a true track.

$\Lambda\left(\theta_{1:k}^{m_k} \right)$: the likelihood of $\mathbf{y}_{1:k}^{m_k}$ given $\theta_{1:k}^{m_k}$.

$\lambda_{i,j}$: Mahalanobis distance of two random vectors $\mathbf{x}_i$ and $\mathbf{x}_j$.

$A^{j,t}$: a hypothesis (an event) that measurement j is originated from target t.

A: a joint association event, $\bigcap_{j=1}^{m} A^{j,t} \forall t$.

$\hat{\Omega}(A)$: the event matrix of A where the (j,t)th element of $\hat{\Omega}(A)$, $\hat{\omega}_{j,t}(A) = 1$, if $A^{j,t} \in A$ and $\hat{\omega}_{j,t}(A) = 0$ otherwise. The summation of $\hat{\omega}_{j,t}(A)$ across all measurements is a target detection indicator, $\delta_t(A) \triangleq \sum_{j=1}^{m} \hat{\omega}_{j,t}(A)$ with its counterpart as measurement association indicator $\tau_j(A) \triangleq \sum_{t=1}^{n} \hat{\omega}_{j,t}(A)$.

$\phi(A)$: the number of false (i.e., unassociated) measurements in event A, $\sum_{j=1}^{m} (1 - \tau_j(A))$.

T_i: a track file $(i_1, i_2,\ldots, i_k)$ where i_k indicates the i_kth detection at the kth scan and $i_k = 0$ indicates no detection for the track T_i at the kth scan.

θ_k: a collection of hypotheses, $\left\{ \theta_k^1, \theta_k^2, \theta_k^3,\ldots, \theta_k^{M_k} \right\}$, where θ_k^i contains a collection of tracks and is a hypothesis in θ_k.

A_g^i: the association event of all measurement types for θ_k^i in an acceptance gate (AG) at scan k.

P_d: sensor detection probability.

P_{fa}: sensor false alarm probability.

P_{T}: target termination probability.

P_{M}: target maneuvering probability.

N_k: number of measurements (or sensor reports).

N_{TGT}: number of targets.

N_{DT}: number of detected targets.

N_E: number of existing targets (continuing from the previous scan).

N_T: number of terminated targets.

N_M: number of maneuvering targets.

N_{FT}: number of false targets.

N_{NT}: number of new targets.

β_{FT}: density of false targets.

β_{NT}: density of new targets.

$F_N(\lambda)$: the Poisson probability for N events with an average rate of λ (i.e., $F_N(\lambda) = \lambda^N/N!$).

λ_{FA}: the false alarm rate of the Poisson probability.

λ_{NT}: the new target rate of the Poisson probability.

$\mathcal{C}$: a set of all correlated pairs of tracks (i,j) of Sensor 1 and Sensor 2, respectively.

Appendix C

Definition of Terminology Used in Tracking

The subject of Chapters 1–7 is estimation with the assumption that all measurements come from the same object. In this case there is no issue on measurements not belonging to the same object, because there is only one object producing measurements. In practice, this is not always true; a sensor often detects more than one object and sometimes with false alarms. Chapters 8–10 discuss state estimation with measurements coming from multiple objects. In this case, it becomes a joint decision and estimation problem. To obtain a sequence of measurements that belong to the same object is a decision problem. Assigning measurements/detections to an object is necessary before state estimates can be conducted. In this Appendix, terminology often used in the tracking community is defined as used in Chapters 8–10.

Object or Target

An object or a target is the entity of interest. This interest could be in estimating its current state, finding out where it came from, and predicting where it is going. In a multiple object environment, measurements from the same object must be identified before use by an estimator. Sometimes estimation and measurement assignment are performed interactively. An object could be a person, a vehicle, an airplane, and so on. The target is generally referred to as the object of interest. Objects and targets are used interchangeably in this book.

Sensor Measurement of Detected Target

When the sensor signal processor determines that a signal has passed the detection threshold, a set of parameter values characterizing this detection is obtained, which is the measurement vector of the detected target. For radar systems, the measurement vector contains the signal time of arrival (range to object) and angles of arrival (angles to object). Some radar will have signal Doppler (range rate of object) as an additional component of the measurement vector. If the detected target does not belong to a real object, it is called a false alarm.

Scan and Frame

Typically a sensor does not take measurements of multiple objects in the same instance of time. Exceptions could be a scene picture (a snapshot) of an optical sensor. In most cases, whether it is for a radar or an optical sensor, a sensor system may complete the scan of a given volume within a given time window. The time it takes to complete a scan is called the frame time. In a given scan, or frame, there are multiple time-tagged measurements. In the multiple object tracking case, it is sometimes necessary to have measurement time aligned. This can be done using extrapolation and interpolation algorithms (numerical integration) to preprocess measurements. In Chapters 8–10, a set of time-aligned measurements (or states) for assignment processing is referred to as a scan of measurements, with the understanding that scan and frame may be used interchangeably.

Track and State

A track is a collection of a sequence of time-tagged measurements (it is not necessary for all measurements to belong to the same object). A state estimate can be obtained with these measurements with a specific algorithm. Track and state are used interchangeably in this book.

Acceptance Gate

A track can establish a boundary to anticipate the next measurement. The covariance of the predicted measurement and the covariance of the measurement are typically used to set up an acceptance gate (AG) around the predicted measurement vector. The gate size is proportional to the sum of the predicted measurement covariance and the measurement covariance of the sensor. It is chosen to satisfy certain probability that the anticipated object measurement will be contained within the gate.

Single Target Track

When there is only one target present, a measurement can be uniquely assigned to a track. This is referred to as the single target track. Even in a multiple target environment, when the spacing among targets is larger than the track and measurement uncertainties, individual targets can be tracked independently, then they are all single target tracks. All state estimation problems considered in Chapters 1–7 fall under this case.

Multiple Target Track

When there are multiple targets present and the spacing among targets is closer than track uncertainties and sensor measurement error, measurement-to-track assignment may become ambiguous; this gives rise to the problem of multiple target tracking (MTT).

Hypothesis

A hypothesis is a proposed explanation of a given data set that sets up a Bayesian decision problem.

Association and Correlation

The broadly accepted definition of association and correlation is as follows:

Association: assigning measurements to a track.

Correlation: assigning tracks from across sensors.

The mathematics for association and correlation are identical. Chapters 8 and 9 are concerned with association while Chapter 9 is also concerned with correlation.

Track Ambiguity

- A track is ambiguous when measurements cannot be uniquely assigned to this track with high confidence.

- A track is ambiguous when it cannot be uniquely assigned to tracks from another sensor with high confidence.

- Track ambiguity arises when the predicted track uncertainty of a track contains more than one measurement.

- Track ambiguity arises when multiple predicted track uncertainties overlap and contain at least one common measurement.

Measurement Classifications

A measurement may be considered as: (a) the continuation of an existing track, (b) the starting point of a new track for a new object, and (c) a false alarm. In cases (b) and (c), even when a track is associated with this measurement it will move on without updating. With this classification rule, an existing track with a new measurement will become three possible tracks and two hypotheses. This classification rule is used in multiple hypothesis tracking.

Measurement-to-Track Assignment

In a multiple target tracking problem with ambiguities, multiple tracks may be competing with multiple measurements. Methods to assign measurements to tracks in a given scan include (a) choosing the measurement which is closest to a track (nearest neighbor); and (b) jointly selecting the set of measurement-to-track which maximizes the likelihood function (global nearest neighbor). The joint assignment approach can be extended to multiple scans which is the multiscan assignment approach.

Probabilistic Data Association Filter and Joint Probabilistic Data Association Filter

When multiple measurements appear in the AG of a given track, the probabilistic data association filter (PDAF) combines all measurements weighted by the residual densities into a single measurement. When multiple measurements appear in the joint AG of more than one track, the joint probabilistic data association filter (JPDAF) combines measurements for update with each track in the joint AG region.

Multiple Hypothesis Tracker

A multiple hypothesis tracker (MHT) is a collection of all possible hypotheses, each containing a set of tracks that explains all collected measurements up to the current scan. The same track can appear in different hypotheses while a measurement can only appear in a hypothesis once. A complete set of hypotheses is an exhaustive list of all possible combinations of measurements over the entire track time history. MHT uses the aforementioned measurement classification rule.

There are two hypothesis formulations methods,

- Measurement-oriented MHT: every possible track is listed for each measurement.

- Track-oriented MHT: every possible measurement is listed for each track.

Index

Printed in the United States
by Baker & Taylor Publisher Services